Advances in Intelligent Systems and Computing

Volume 1280

The series "Advances in Intelligent Systems and Computing" contains publications on theory, applications, and design methods of Intelligent Systems and Intelligent Computing. Virtually all disciplines such as engineering, natural sciences, computer and information science, ICT, economics, business, e-commerce, environment, healthcare, life science are covered. The list of topics spans all the areas of modern intelligent systems and computing such as: computational intelligence, soft computing including neural networks, fuzzy systems, evolutionary computing and the fusion of these paradigms, social intelligence, ambient intelligence, computational neuroscience, artificial life, virtual worlds and society, cognitive science and systems, Perception and Vision, DNA and immune based systems, self-organizing and adaptive systems, e-Learning and teaching, human-centered and human-centric computing, recommender systems, intelligent control, robotics and mechatronics including human-machine teaming, knowledge-based paradigms, learning paradigms, machine ethics, intelligent data analysis, knowledge management, intelligent agents, intelligent decision making and support, intelligent network security, trust management, interactive entertainment, Web intelligence and multimedia.

The publications within "Advances in Intelligent Systems and Computing" are primarily proceedings of important conferences, symposia and congresses. They cover significant recent developments in the field, both of a foundational and applicable character. An important characteristic feature of the series is the short publication time and world-wide distribution. This permits a rapid and broad dissemination of research results.

Indexed by SCOPUS, DBLP, EI Compendex, INSPEC, WTI Frankfurt eG, zbMATH, Japanese Science and Technology Agency (JST), SCImago.

All books published in the series are submitted for consideration in Web of Science.

More information about this series at http://www.springer.com/series/11156

Debnath Bhattacharyya · N. Thirupathi Rao
Editors

Machine Intelligence and Soft Computing

Proceedings of ICMISC 2020

Springer

Editors
Debnath Bhattacharyya
Department of Computer Science
and Engineering
K. L. University
Guntur, Andhra Pradesh, India

N. Thirupathi Rao
Department of Computer Science
and Engineering
Vignan's Institute of Information
Technology
Visakhapatnam, Andhra Pradesh, India

ISSN 2194-5357 ISSN 2194-5365 (electronic)
Advances in Intelligent Systems and Computing
ISBN 978-981-15-9515-8 ISBN 978-981-15-9516-5 (eBook)
https://doi.org/10.1007/978-981-15-9516-5

This Springer imprint is published by the registered company Springer Nature Singapore Pte Ltd.
The registered company address is: 152 Beach Road, #21-01/04 Gateway East, Singapore 189721, Singapore

Conference Committee Members

General Chair

Osvaldo Gervasi, University of Perugia, Italy
Debnath Bhattacharyya, K. L. University, India

Advisory Board

Paul S. Pang, Unitec Institute of Technology, New Zealand
Andrzej Goscinski, Deakin University, Australia
Osvaldo Gervasi, Perugia University, Perugia, Italy
Jason Levy, University of Hawaii, Hawaii, USA
Tai-hoon Kim, Jiao Tong University, Shanghai, China
Sabah Mohammed, Lakehead University, Ontario, Canada
Jinan Fiaidhi, Lakehead University, Ontario, Canada
Y. Byun, Jeju National University, South Korea
Amiya Bhaumick, LUC KL, Malaysia
Divya, LUC KL, Malaysia
Dipti Prasad Mukherjee, ISI Kolkata, India
Sanjoy Kumar Saha, Jadavpur University, Kolkata, India
Sekhar Verma, IIIT Allahabad, India
C. V. Jawhar, IIIT Hyderabad, India
Pabitra Mitra, IIT Kharagpur, India
Joydeep Chandra, IIT Patna, India
L. Rathaiah, Vignan Group, Guntur, India
Krishna D. Lavu, Vignan Group, Guntur, India
Pavan Krishna, Vignan Group, Visakhapatnam, India
V. M. Rao, VFSTR University, Guntur, India

Editorial Board

N. Thirupathi Rao, Vignan's Institute of Information Technology, India

Program Chair

Tai-hoon Kim, BJTU, China

Finance Committee

Naga Mallik Raj, Vignan's Institute of Information Technology, India

Technical Programme Committee

Sanjoy Kumar Saha, Professor, Jadavpur University, Kolkata
Hans Werner, Associate Professor, University of Munich, Munich, Germany
Goutam Saha, Scientist, CDAC, Kolkata, India
Samir Kumar Bandyopadhyay, Professor, University of Calcutta, India
Ronnie D. Caytiles, Associate Professor, Hannam University, Republic of Korea
Y. Byun, Professor, Jeju National University, Jeju Island, Republic of Korea
Alhad Kuwadekar, Professor, University of South Walse, UK
Bapi Gorain, Professor, LUC, KL, Malaysia
Poulami Das, Assistant Professor, Heritage Institute of Technology, Kolkata, India
Indra Kanta Maitra, Associate Professor, BPPIMT, Kolkata, India
Divya Midhun Chakravarty, Professor, LUC, KL, Malaysia
F. C. Morabito, Professor, Mediterranea University of Reggio Calabria, Reggio Calabria RC, Italy
Bidyut Gupta, Professor, Southern Illinois University Carbondale, Carbondale, IL 62901, USA
Nancy A. Bonner, Professor, University of Mary Hardin-Baylor, Belton, TX 76513, USA
Alfonsas Misevicius, Professor, Kaunas University of Technology, Lithuania
Ratul Bhattacharjee, AVP, AxiomSL, Singapore
Lunjin Lu, Professor and Chair, Computer Science and Engineering, Oakland University, Rochester, MI 48309-4401, USA
Ajay Deshpande, CTO, Rakya Technologies, Pune, India
Debasri Chakraborty, BIET, Suri, West Bengal, India
Bob Fisher, Professor, The University of Edinburgh, Scotland

Alexandra Branzan Albu, University of Victoria, Victoria, Canada
Maia Hoeberechts, Associate Director, Ocean Networks Canada, University of Victoria, Victoria, Canada
MHM Krishna Prasad, Professor, UCEK, JNTUK Kakinada, India
Edward Ciaccio, Professor, Columbia University, New York, USA
Yang-sun Lee, Professor, Seokyeong University, South Korea
Yun-sik Son, Professor, Dongguk University, South Korea
Jae-geol Yim, Professor, Dongguk University, South Korea
Jung-yun Kim, Professor, Gachon University, South Korea
Mohammed Usman, King Khalid University, Abha, Saudi Arabia
Xiao-Zhi Gao, University of Eastern Finland, Finland
Tseren-Onolt Ishdorj, Mongolian University of Science and Technology, Mongolia
Khuder Altangerel, Mongolian University of Science and Technology, Mongolia
Jong-shin Lee, Professor, Chungnam National University, South Korea
Jun-kyu Park, Professor, Seoul University, South Korea
Wang Jin, Professor, Changsha University of Science and Technology, China
Goreti Marreiros, IPP/ISEP, Portugal
Mohamed Hamdi, Professor, Supcom, Tunisia

Keynote Speakers

Dr. Saptarshi Das, The Pennsylvania State University, USA
Dr. Lalit Garg, University of Malta, Malta
Dr. Ozen Ozer, Kırklareli University, Turkey

Preface

Knowledge in engineering sciences is about sharing our ideas of research to others. In engineering, it has many ways to exhibit, and in that, conference is the best way to propose your idea of research and its future scope; it is the best way to add energy to build strong and innovative future. So, here, we are to give a small support from our side to confer your ideas by an "International Conference on Machine Intelligence and Soft Computing (ICMISC-2020)", related to Electrical, Electronics, Information Technology and Computer Science. It is not confined to a specific topic or region, and you can exhibit your ideas in similar or mixed or related technologies bloomed from anywhere around the world because "An idea can change the future and its implementation can build it". VIIT College is a great platform to make your idea(s) penetrated into world. We give as best as we can in every aspect related. Our environment leads you to a path on your idea, our people will lead your confidence, and finally, we give our best to make yours. Our intention is to make intelligence in engineering to fly higher and higher. That is why we are dropping our completeness into event. You can trust us on your confidentiality. Our review process is double-blinded through Easy Chair.

At last, we pay the highest regard to the Vignan's Institute of Information Technology(A), Visakhapatnam, and Vignan's Foundation for Science, Technology & Research (Deemed to be University) Guntur, Andhra Pradesh, a "not-for-profit" society from Guntur and Visakhapatnam for extending support for financial management of ICMISC-2020.

Best wishes from

Guntur, India
Visakhapatnam, India

Debnath Bhattacharyya
N. Thirupathi Rao

Contents

About the Editors

Debnath Bhattacharyya is associated as a Professor with the Computer Science and Engineering Department, K.L. University, KLEF, Guntur 522502, India. Dr. Bhattacharyya is presently an Invited International Professor of Lincoln University College, KL, Malaysia. He is also working as a Visiting Professor, MPCTM, Gwalior, India, in the Department of Computer Science and Engineering. Dr. Bhattacharyya received his Ph.D. (Tech., Computer Science and Engineering) from the University of Calcutta, Kolkata. He is the Editor of many international journals (indexed by Scopus, SCI, and Web of Science). He published 163 Scopus Indexed Papers and 128 Web of Science Papers. His Research interests include security engineering, pattern recognition, biometric authentication, multimodal biometric authentication, data mining, and image processing. In addition, he is serving as a Reviewer of various international journals of Springer, Elsevier, IEEE, etc., and international conferences.

N. Thirupathi Rao is currently associated as Associate Professor in the Department of CSE, Vignan's Institute of Information Technology (Autonomous), Duvvada, Visakhapatnam, India. He did his PhD from Andhra University, India. His research interests include networking and security. He published 35 indexed research papers in Scopus and Web of Science. He is working as Editors in many international Journals as well.

A Comparative Study on Automated Detection of Malaria by Using Blood Smear Images

D. Sushma, N. Thirupathi Rao, and Debnath Bhattacharyya

Abstract Malaria is a parasitic disease or mosquito-borne blood disease. When the mosquito bites a human being, that particular parasite is freed into the human being bloodstream and infects the red blood cells which cause the malaria. We need to understand if the blood-related illness is malaria or not before we provide the right therapy. For this purpose, we must diagnose red blood cells by recognizing or counting red blood cells (erythrocytes). It is very difficult to manually count and recognize infected red blood cells while testing under a microscope by pathologists because maybe it leads to different variations. The current paper gives an overview of the comparison of three different papers with three different techniques used to identify that the red blood cells are infected or not with great accuracy and also to identify which methods are giving best result while performing the diagnosis automatically. With different techniques and methods like Otsu threshold method, global threshold method and classifiers like artificial neural network and support vector machines. All these techniques and methods are related to the diagnosis of the malaria automatically which will reduce the time taken for performing the diagnosis and also it improves the consistency and gives the accurate, rapid result in diagnosis. From the above three methods used, an attempt has been made to finalize the best method from the above three methods.

D. Sushma (✉) · N. Thirupathi Rao
Department of Computer Science and Engineering, Vignan's Institute of Information Technology (A), Visakhapatnam, AP 530049, India
e-mail: sushmadasari25@gmail.com

N. Thirupathi Rao
e-mail: nakkathiru@gmail.com

D. Bhattacharyya
Department of Computer Science and Engineering, K L Deemed to be University, KLEF, Guntur 522502, India
e-mail: debnathb@gmail.com

D. Bhattacharyya and N. Thirupathi Rao (eds.), *Machine Intelligence and Soft Computing*, Advances in Intelligent Systems and Computing 1280,
https://doi.org/10.1007/978-981-15-9516-5_1

Keywords Otsu threshold method · Blood smear cells · Malaria · Global threshold · Water threshold transform · Artificial neural networks · Support vector machines · Automation · Parasites · Plasmodium malaria · Feature extraction

1 Introduction

Malaria is a blood disorder or a parasitic moustache that transmits a parasite, known as plasmodium, through the bite of an infected Anopheles mosquito. When the mosquito attacks a human being, this parasite is released into the bloodstream of the human being and infects the red blood cells that cause malaria. Before we take the correct treatment, it must be understood if the blood-related disease is malaria. We need to detect or count red blood cells (erythrocytes) as a diagnosis of red cells. Due to distinct variants, it is very hard to manually count and acknowledge infected red blood cells by pathologists while testing them under a microscope. The most dangerous malaria parasite is plasmodium falciparum (PF) [1].

Based on the WHO analysis [2], up to the year 2017, two hundred nineteen million cases are related to this malaria. Among those million cases, mostly children are affecting with the disease. Millions of blood smears are examined every 12 months for malaria which involves the manual identification of parasites and abnormal red blood cells by a trained pathologist. This paper is about the detection of infected RBCs automatically by using the different methods. This study explains the image processing techniques for detecting malaria, and also it introduces necessary methods based on the features. The remaining paper is about the section literature review (Sect. 2) which we discuss about the different detection methods and classifiers for detecting malaria are caused or not. And Sect. 3 describes about the comparison between different methods and techniques of Sect. 2. After that Sect. 4 is about result and conclusion.

2 Literature Review and Analysis of the Three Papers Considered

In [1], Ahmedel Mubarak Bashir, Zeinab A. Mustafa, Islah Abdelhamid and Rimaz Ibrahem are the authors who introduced a method for identifying the malaria using digital image processing methods. This paper states the image classification to detect malaria parasites are infected or not infected based on the features of erythrocytes samples.

2.1 System Architecture Proposed

The authors implemented the system model by using the main six processes, i.e. acquisition of images, pre-processing of images, segmentation of images, extraction of features, comparison and classification (Fig. 1).

The methods that the authors had considered in the current article are as follows.

All the images which are purchased from CDC have a resolution of 300 * 300 pixels. Both CDC and captured images are converted from RGB to greyscale in order to reduce the processing time (Fig. 2).

The conversion from RGB to greyscale is done by the MATLAB™ function rgb2grey, which converts the images from one to another, i.e. RGB to greyscale by eliminating the hue and saturation information while retaining the luminance. Now for reducing the noise present in the images are filtered by using the square median filter which is a filtering operation. The length used by the square median filter is 7 by 7. In this for calculating some features, the pixel values from the image can be useful directly. There are two intensity features which are used, and in this they are variance and skewness (Fig. 3).

First, let $h(\vartheta) \in R^{L*1}$, $h(v) = \{h(v, i)\}_{i=1}^{L}$ denote the frequency of pixel intensity value (histogram), where $\vartheta = 0, \ldots, N,$ and $p(v, i)$ is the probability distribution

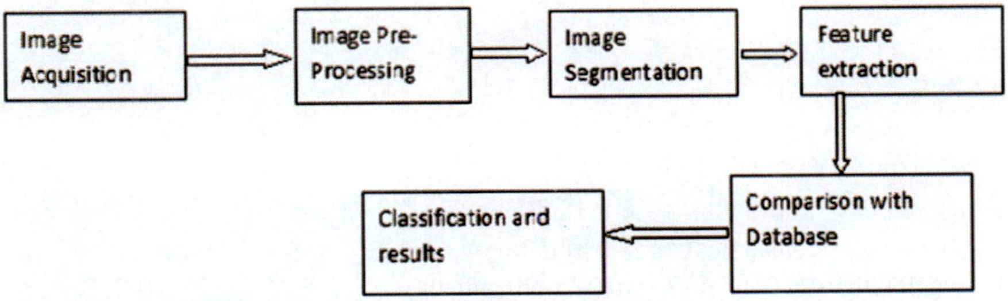

Fig. 1 Block diagram of the malaria diagnosis system [1]

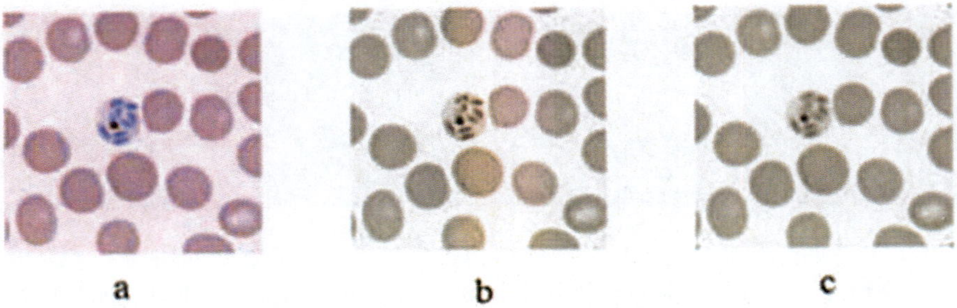

Fig. 2 Effect of image pre-processing [1]. **a** Rescaled images, **b** greyscale image, **c** filtered images

function, which is computed from the histogram by dividing it by the object's area "A" which is given by

$$A = \sum_{v} h(v) \tag{1}$$

The corresponding probability is calculated as

$$P(v, i) = h(v, i)/A \tag{2}$$

The mean μ_i and the σ_i^2 are computed as

$$\mu_i = \sum_{v=1}^{N} \vartheta P(v, i) \tag{3}$$

$$\text{and } \sigma_i^2 = \sum_{v=1}^{n} (v - \mu_i)^2 P(v, i) \tag{4}$$

Finally, the skewness is given by

$$\mu_3 = 1/\sigma_3^i \sum_{v=1}^{N} (v - \mu_i)^3 p(v, i) \tag{5}$$

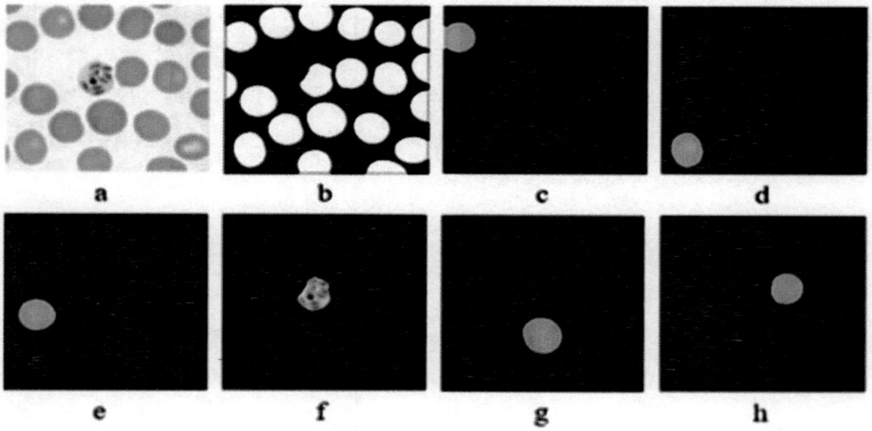

Fig. 3 RBCs extraction [1]. **a** The greyscale image, **b** the labelled image, **c–h** extracted objects (RBCs)

2.1.1 Result Analysis

There are some statistical measures, i.e. (1) sensitivity (2) specificity (3) accuracy.

By using above three measures, the performance was evaluated of the employed classification method. Here the sensitivity denoted by S_s, of a test used to define the probability of the positive test result given the presence of the disease

$$S_s = \frac{100\text{TP}}{\text{TP} + \text{FP}} \tag{6}$$

The specificity of a test is denoted by S_p, used to define the probability of a negative test result given the absence of the disease.

$$S_P = \frac{100\text{TN}}{\text{TN} + \text{FN}} \tag{7}$$

Finally, the accuracy is about how the measured value is close to the actual or true value.

$$A = \frac{100(\text{TP} + \text{TN})}{\text{TP} + \text{FP} + \text{TN} + \text{FN}} \tag{8}$$

where

TP True positive,
TN True negative,
FP False positive,
FN False negative.

The Calculation Process Table 1 from [1]

$$S_s = \frac{46}{46 + 0} \times 100 = 100\%$$

$$S_p = \frac{576}{576 + 2} \times 100 = 99.65\%$$

$$A = 46 + \frac{576}{46 + 576 + 0 + 2} \times 100 = 99.68\%$$

It is clear that from the performance evaluation table, the accuracy is 99.68 that is the ANN is giving more accurate result for the data which is used in this paper.

Table 1 Test results at various patients samples

	Abnormal	Normal	Total
Positive	46	2	48
Negative	0	576	576
Total	46	578	624

2.2 *About the paper Computer Aided System for Red Blood Cell Classification in Blood Smear Image*

Another relative paper [3], "Computer Aided System for Red Blood Cell Classification in Blood Smear Image" which is proposed by the authors, namely Razali Tomari et al. In this different methods like global threshold method and Otsu threshold method are used to extract the background of the erythrocyte images, and also filter is used to reduce the noise and unnecessary holes and also a classifier to classify whether the images of red blood cells are normal or abnormal [4].

2.2.1 System Overview

The method proposed to detect and classify RBC is included in the architecture of the system. It is made by first acquiring or buying the picture from an illuminated microscope with a static camera of the eye. The captured RGB picture is initially transformed into a single colour component to make next processing simple. Secondly, with a technique called an adaptive worldwide threshold and the low-level picture handling technique, the foreground is distinguished from the background, in order to decrease an acoustic preliminary pixel map. Later on, we group the linked parts of the foreground map before extracting individual object characteristics such as region, parameter and moment values in order to recognize overlapping and non-overlapping parts [5]. In this we have established an ANN classifier and then the algorithm of the object classification, using the extracted information before we use it for training and testing. Finally, we have evaluated the accuracy and performance of this classification procedure.

The methods that the authors had considered in the current article are as follows.

Image Acquisition

The images are acquired from light microscope that endowed with Dino Eye eyepiece camera. The image capture method involves blood film to the sample that has been prepared. Smear from the blood is a blood preparedness method that has been observed under the microscope on the slide. Digitization of the optical image with a 40-fold (40×) goal, equivalent to about 400 magnifications, is involved in the display of the RBC picture.

Colour Space Reduction

The shape of objects plays a significant part for the red blood cell classification, and colour data is low. As a result, the RGB image is changed into a single-channel colour representation in order to calculate a search table efficiently. In this project, an ideal colour canal is examined for the component red, green and blue to differentiate between red and background cells.

Illustration. Figure 5 shows the results of (a) presentation of the RGB image and (b), (c) and (d) presentation of the corresponding RGB image element. But the highest quality contrast (which is apparent darker) between RBC and the background is the green element of the green and blue channels.

RBC Segmentation and Post-processing

The segmentation of the images is mainly carried out to divide an image into a homogenous region which is the object of interest in the image. A RBC automated classification system's overall efficiency is significantly affected by its capacity to segment the region of RBC correctly in a given picture. For subsequent action, like analysis or the identification of objects, an exact extraction of the first-face images is required, making an important aspect of the scheme of image segmentation. This paper is used to distinguish between two classes of region in an adaptive threshold strategy [6] from Otsu in the green channel of a RBC picture.

In this paper, three methods are applied to eliminate unnecessary items, such as morphological, LCO and bounding filter. Three methods are used in this article. Morphological processes are used to create binary images to change size, form, structure and connectivity using an erosion/dilation structure and set operator. Erosion performs the function of 'diminishing' and 'thinning' picture objects while dilating items used in the photo 'growing' and 'thinning.' Both operators can be combined in order to remove, break, clear boundary and fill hole. In this project, the tiny noise and troughs in the cell are reduced by a double erosion, a two-times dilatation and contour-filling algorithm. The applicants are marked with linked element (CCL) labelling once such cells are in hand. The boundaries of the cell show the rectangular position minimum and maximum in the picture. As the cell object at the border is not valued for information, the box positions that affect the border of the image have been detected at the minimum and maximum x and y bindings. Upon computing the big cell area relative to normal cell, the overlap cell is ultimately acknowledged. Figure 6c shows outcomes after the above technique. The overlapping cells are labelled, and valuable classification information is obtained from the remainder.

Feature Extraction

The recognition of imagery objects can be done through the identification of an unidentified object as part of a collection of well-known objects for many apps. Typically, these characterizations are described by object measuring functions obtained from different kinds of images. The capacity of an object to represent its object exclusively from the data accessible determines the efficacy of the identification assignment. The object data in this project is obtained from a normal/abnormal sample as shown in Fig. 7. In this paper, approach to the anemias [4] relying on geometrical characteristics of the object, which are compact and invariant in time [7].

This kind of property has the benefit of actually differentiating normal cells from abnormal cell types, as our condition is very complex in comparison with normal cells. The fact that the compactness value is greater if its cell shape is more oval alone is not sufficient, though it is not complicated, to depend on the compactness value alone. A second function is therefore acquired, which presently contains invariant values. In the past, the technique used was used to analyse and identify the object type. One of their benefits is that the 2D transformation, such as translation, rotating, reflection and scaling, is easily available and invariantly. This property is very useful because such changes in blood smeared RBC images are very prevalent. Furthermore, the classification characteristics requirements are reduced as the invariant provides no data other than the initial moment values and reduces the complexity of the learning issue. For this project, we use 7 HU time features [7] to represent the RBC shape.

RBC Classification

A robust classifier should be used to distinguish between ordinary and abnormal RBC in the picture using the chosen characteristics. The classification module is done by using the classification of the artificial neural network (ANN). The ANNs are a biological brain mathematical approximation and have been recognized as a helpful framework for accurate nonlinear reaction modelling [8, 9]. It consists of several neurons that are attached to a network. Weight between neurons, i.e. weight the network's functionality lies in W_{ij} and W_{jk}. It needs education in order to be useful to the network. In essence, the training course will alter the weight to minimize the error between inputs and objectives. One of the fastest learning techniques is the Levenberg–Marquardt algorithm with a medium square error (MSE). The RBC feature, i.e. (compactness and invariant seven HU moments) data is presented here as normal/abnormal to the in feed neurons and the type of RBC to the target neurons in the training stage. Network set-up is regarded optimal for the highest detection velocity in both training and validation.

Table 2 Neural network performance [3]

	Sigmoid activation function								Tangent sigmoid activation function							
HN	2	3	4	5	6	7	8	9	2	3	4	5	6	7	8	9
IDs	96	100	100	100	100	100	100	100	96	96	100	100	100	100	100	100
VDs	100	90	100	90	100	90	81	81	81	100	63	54	54	36	36	36

Results

The effectiveness of the suggested RBC grading system for the type of RBC purchased from the blood flow is assessed in this chapter. It is screened in four blood cell samples in the light microscope. We first assessed ANN's effectiveness. Hundred samples are displayed in the figure. An optimal ANN system has been configured by the Razali Tomari and others/Procedia Computer Societies 42 (2014), 206–213 211(VD) set (VD) for training data collection (TDs) and by the new 50. We did this in an amount of hidden nodes (HNs) that are chosen with two distinct functions: sigmoid (S) and sigmoid (T). Table 2 summarizes the results of our job. TDs and VDs have a maximum effectiveness at 100 per cent and the sigmoid activation function in ANN, which was configured to four (2L-4S-S) and six (2L-6S-S). The performance of VDs deteriorates markedly when the tangent sigmoid is used with the enhanced amount of concealed nodes. Because reduced concealed nodes mean reduced calculation complexity, 2L-4S-S setting was chosen as the perfect networks for ordinary or abnormal RBC classifications.

We will then present an experiment carried out for evaluation of the results of the RBC classification scheme. Microsoft Visual Studio was developed with Open CV 2.4.7 and runs on a processor running 2.4 GHz i5-450M. The efficiency is evaluated on four distinct blood cell pictures, labelled in Figs. 4, 5, 6 and 7. Five based on RBC ordinary cells and exceptional cells can be identified and counted. The True Positive (TP), False Positive (FP) or False Negative parameters are used for the measured values for each class. Using the equation, precision, recurrence and precision are then assessed [8, 9]. Accuracy provides data on the number of fractional cells identified and reminds of the correct identification of numbers of cells from the whole picture in each class. On the other side, precision assesses the system's efficiency in relation to basic realities.

$$\text{Precision} = \frac{\text{TP}}{\text{TP} + \text{FP}}; \quad \text{Recall} = \frac{\text{TP}}{\text{TP} + \text{FN}};$$
$$\text{Accuracy} = \frac{\text{TP} + \text{TN}}{\text{TP} + \text{FP} + \text{TN} + \text{FN}} \tag{9}$$

Table 3 summarizes our results. Overall, the proposed technology is very good, averaging 83%, 82% of average normal precision and 76% of average recall. This means that most classes of objects have a correctly acknowledged acceptable error rate. The system also delivers a good detection result for abnormal RBC detection

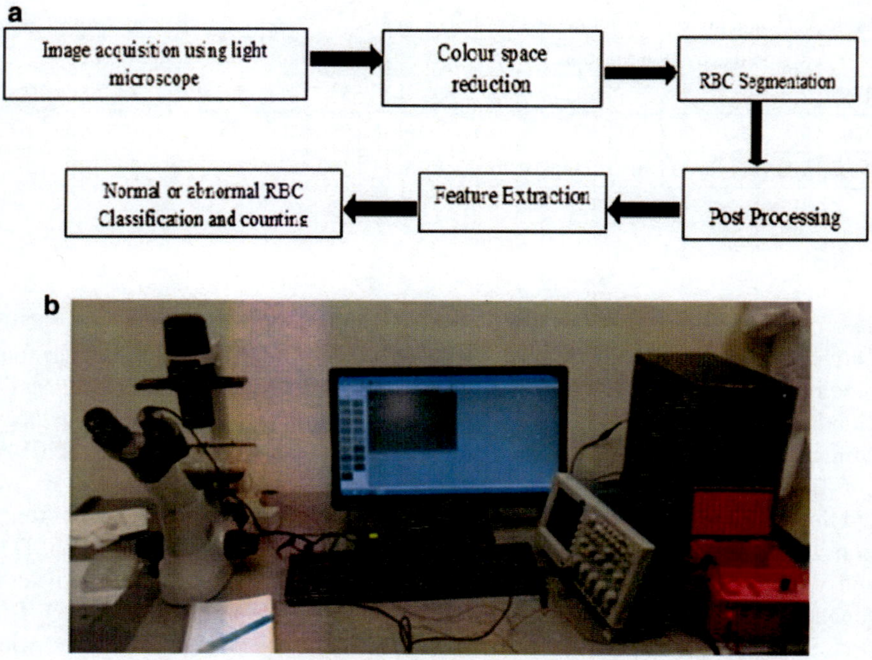

Fig. 4 **a** System architecture [3]. **b** image acquisition equipment [3]

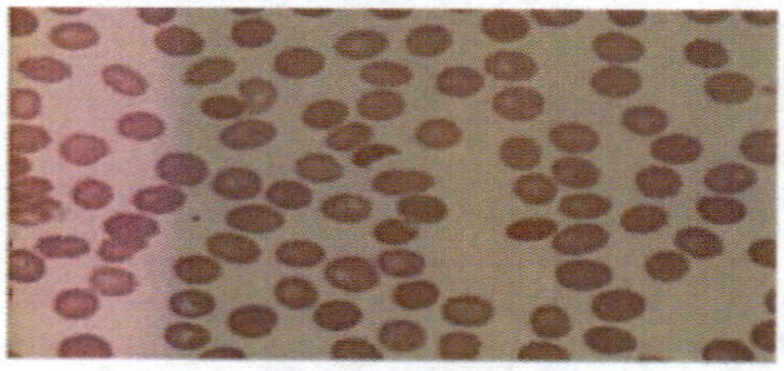

(a)RGB image

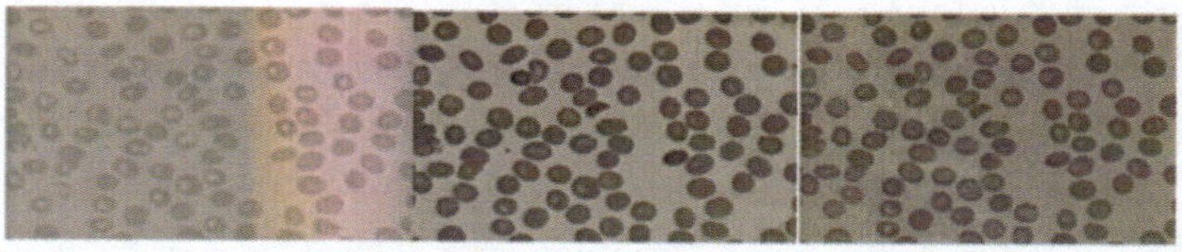

(b) Red Colour Component(c) Green ColourComponent (d) Blue Colour Component

Fig. 5 Colour component selection [3]

compared to the usual one. The truth is that the complexity of the abnormal RBC boundary needs to be distinguished. The normal RBC is usually incorrect because it is not complete with its segmentation and after processing. In the meantime, images Image 1 and Image 3 were the lowest in recall rates, 50% and 23%, respectively.

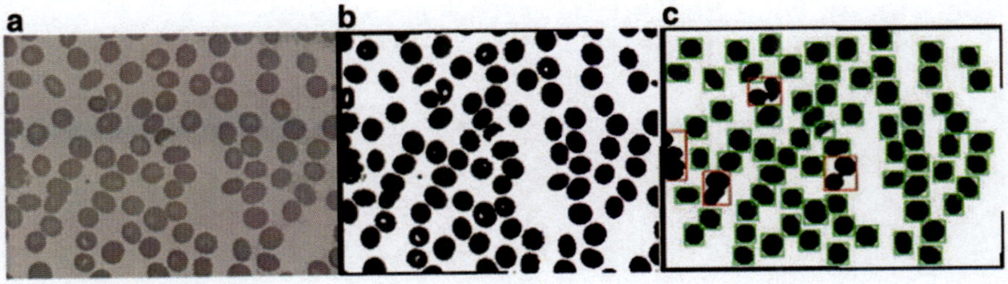

Fig. 6 **a** Green channel image **b** segmentation using Otsu method **c** result after series of post-processing operation [3]

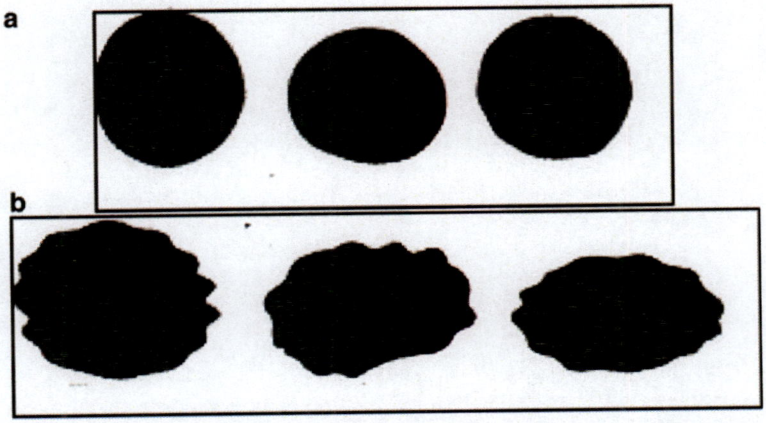

Fig. 7 **a** Normal RBC image **b** abnormal RBC image [3]

Table 3 Identification of normal and abnormal RBC [3]

Image	RBC type	TP	FP	TN	FN	Precision (%)	Recall (%)	Accuracy (%)
im_1	Normal	2	1	15	2	67	50	85
	Abnormal	15	2	2	1	88	94	
im_2	Normal	10	1	19	5	91	67	83
	Abnormal	19	5	10	1	79	95	
im_3	Normal	3	1	26	10	75	23	73
	Abnormal	26	10	3	1	72	96	
im_4	Normal	25	3	39	2	89	93	93
	Abnormal	39	2	25	3	95	93	

The main reason for the lower reminder speed is that the images are quite boring, and it is thus quite difficult to distinguish the cells. The reading of the microscope can be adjusted carefully to improve this problem later. The results of the method suggested are displayed in Fig. 8 during the test. The acknowledged RBC clusters

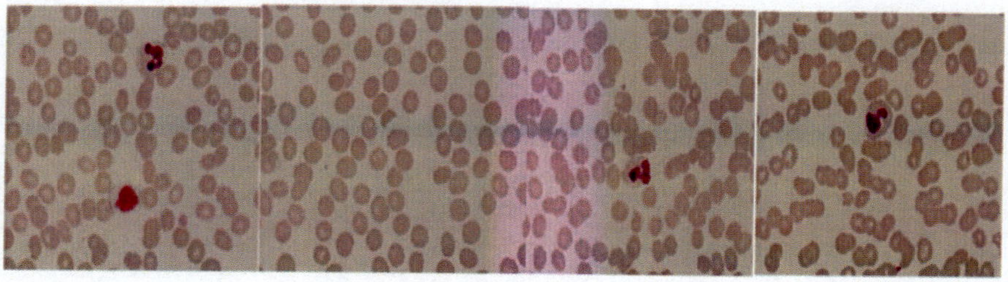

Fig. 8 System evaluation performance (im_1, im_2, im_3, im_4) [3]

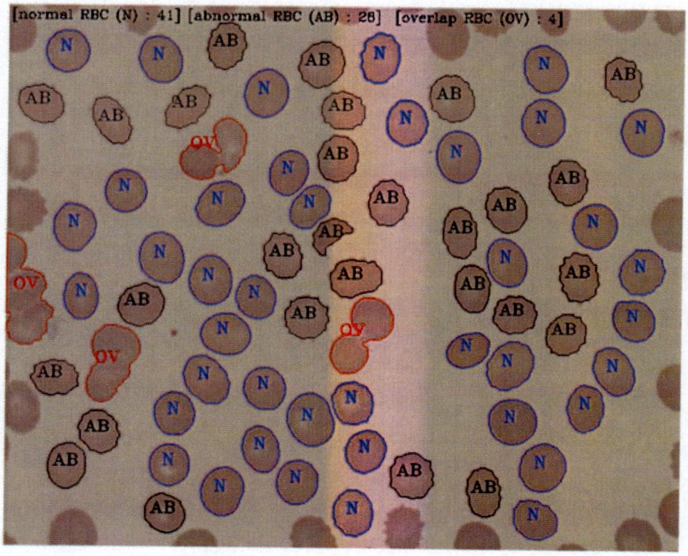

Fig. 9 Examples system for normal RBC and abnormal RBC [3]

were superimposed on the corresponding original image. The scheme allows to mark the cell, normal and abnormal RBC overlapping position and indicates an overall number of cells in the image (Fig. 9).

2.3 About the Paper Automatic System for Classification of Erythrocytes Infected with Malaria and Identification of Parasites Life Stage

Furthermore, in [3], "Automatic System for Classification of Erythrocytes Infected with Malaria and Identification of Parasites Life Stage" by the authors S. S. Savkarea*, S. P. Naroteb had discussed in detail.

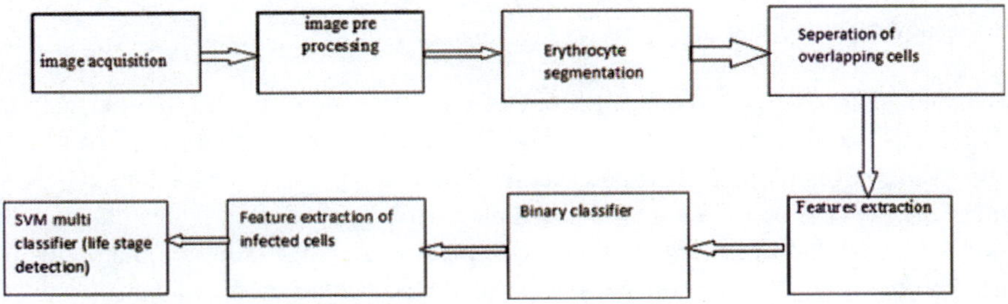

Fig. 10 Block diagram of malaria-infected cells identification model [2]

This paper uses the stained images for processing and also the RBC segmentation techniques and also to separate the overlapped cells, and the watershed algorithm is used and also the classification techniques or method to detect the infected erythrocytes and also for detection of parasite life stages.

2.3.1 System Architecture

Block diagram of a malaria-infected automated blood cell classification and life phase detection scheme of the parasite are presented in Fig. 10; this includes acquisition of a picture, pre-treatment and segmentation of erythrocytes, extraction and classification of functions.

The methods that the authors had considered in the current article are as follows.

Image Acquisition

The diaphragms are made with Giemsa operating systems. The image was captured by the digital microscope connection. Images used in the production of microscopes, light sources and staining impacts are distinct in magnification.

Image Pre-processing

The stage of pre-production includes noise reduction and image smoothing. In this document, we have used medium filters to flatten the colour image and for the cutting corners the Laplacian filter is used. This result is removed to enhance the image from the original. The median filter is a nonlinear process used to remove noise from photos using digital filters. The median of its neighbouring pixel values will be replaced by a median pixel filter. The Laplacian filter is required for the second-order pixel derivatives. After the image is pre-processed, send to the erythrocyte segmentation block.

Erythrocyte Segmentation

The first phase is to segment erythrocytes from the blood picture as the malaria virus lies in erythrocytes. The global and Otsu thresholds are used in the greyscale improved images in the foreground section of the background. Segmentation of low contrast images is on the improved green picture channel. Threshold results on both pictures are added to provide a binary cell picture. Average cell area that helps to remove tiny artefacts from the picture is calculated. Using median filter, unwanted pixels are removed. The range transform is implemented individually on each cell cluster for the separation of the overlapping cells. The watershed transform was then applied to separate cells. The next block is provided in this last binary image of the cells.

Feature Extraction

Given that the characteristics selected influence the efficiency of the classifier, it is just as essential to select characteristics to use for a particular information classification issue as the classifier itself. Features that prevail among ordinary and infected cells are recognized and used for practice. Geometric, colour and statistical characteristics are chosen.

$$\text{Skewness} = \frac{1}{\sigma^2}\sum_{b=0}^{L-1}(b-\bar{b})^3 \tag{10}$$

$$\text{Standard Deviation} = \left[\sum_{b=0}^{l-1}(b-\bar{b})\right]^{1/2} \tag{11}$$

The pixel amplitude value parameter b is the estimate for the first-order histogram. L is the top boundary of the amplitude rate quantified. For the extraction of functions, the above parameters are used. The statistical characteristics are based on grey pitch histogram and a pixel saddling histogram and, by means of such an assessment, the average value; corner 3, skewing, standard deviations are regarded as the characteristics.

SVM Classifier

The SVM is an effective solution to classification problems. An extremely great generalization and a powerful teaching method lead to a minimum of the defined error function and the main advantage of the SVM classification network. N-dimensional vector function x is defined by using the function $\varnothing(x)$ as $D(x) = w^t\varnothing(x) + b)$ for K-dimensional function space ($K > N$). The classification mode of teaching of the SVM network is designed to maximize the split between two classes. It is suggested

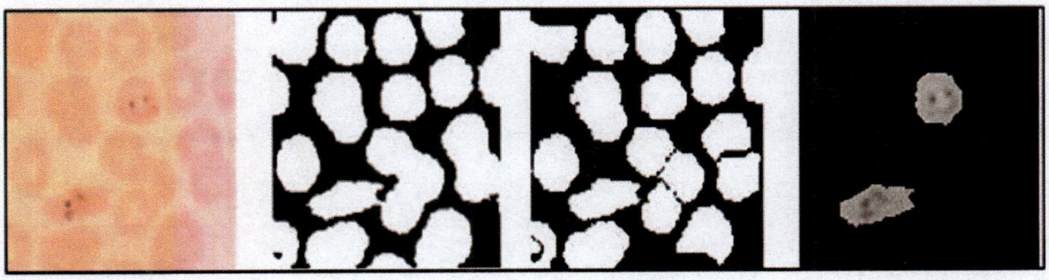

Fig. 11 **a** Original image, **b** segmentation of erythrocytes from image, **c** separation of overlapping cells, **d** recognition of malaria parasite infected cells [2]

that points be classified using the easier classification algorithm, by allocating them to the closer of two parallel levels (input or feature region). The SVMs that assign points for one half of two chambers are standard.

Extraction of Infected Cells

The malaria parasite displays three bloodstream life phases: the ring stage, the schizont stage and the gametocyte stage. Geometric and statistical characteristics are derived from infected erythrocytes for identification of the life phase of parasites. The bacteria are segmented from the picture with the use of a saturation plane, and the parasite overlaps with the infected erythrocyte overall region of the parasite. The amount of ##d pixels is divided as dark blue or red parasites.

Results

SVM binary classification is used to detect the infection or non-infection of erythrocytes. The total processing of 71 pictures is done via an automatic scheme. The infected erythrocyte region is bigger than the standard erythrocyte; the standard difference between the erythrocytes is very high, healthy cells are skewed to a maximum of 2, and infected cells are higher than 2. The described function extraction techniques generate a very wealthy set of parameters for SVM multi-classifier. The ring phase of erythrocyte is 20–50%, and the schizontal phase is 40–80%, with the whole erythrocyte being occupied by parasite in the gametocyte phase. Figure 11 indicates erythrocyte segmentation production, cell overlap separation and infected cell detection. Total erythrocyte count, complete cell count and life phase of a parasite appear in the control window. The ring phase of the parasite appears for the specified picture scheme.

Table 4 shows the SVM binary classifier sensitivity and specificity to the detection of infected erythrocytes; it also provides a summarized list of the correct phase life

Table 4 Results summary for 71 images [2]

Kernel	SVM binary classifier		SVM multi classifier correct detection rate (%)
	Sensitivity (%)	Specificity (%)	
Linear	94.85	89.96	93.87
Polynomial	96.26	99.09	90
RBF	96.26	99.09	96.42

identification of the linear, polynomial and RBF kernel parasite. Table 4 shows the results.

3 Comparison Between the Three Model Papers

This section is the main purpose of this paper because it explains which methods are giving best result while detecting the malaria automatically instead of detecting malaria manually. In this particular paper, we compared three papers which are used to detect the malaria automatically by using the blood film images and each paper used different methods and techniques. Table 5 will explain which paper is giving the best accurate result for detecting the malaria.

Table 5 states that the comparison between the three different papers for detecting the malaria, namely [1] by using the methods called the square median filter, greyscale image method and artificial neural network method for classification and by using these methods the malaria can be detected very accurately, i.e. with 99.68% of accuracy, [3] by using the methods called the global threshold method, Otsu threshold method, morphological filter and artificial neural network classifier and by using these methods the malaria can be detected with 83% of accuracy, [2] by using the methods called global threshold method, Otsu threshold method, watershed transform and support vector machine classifier and by using these methods the malaria can be detected with 96.42% of accuracy, respectively.

4 Conclusions

The malaria can be detected with great accuracy by using the three papers described above among those papers (1) is giving the best result with 99.68% accuracy by using methods known as the square median filter, the greyscale image process and the artificial neural network using the back-propagation method for classification. So, this (1) document is highly effective and decreases diagnosis time and results precisely and quickly, without manually diagnosing by the pathogens the RBC cells are tested automatically that the RBC cells are normal cells or abnormal cells.

Table 5 Comparison between above three models [2]

S.No.	Techniques	Methods and techniques used	Sensitivity/Recall (%)	Specificity/Precision (%)	Accuracy (%)	Overall accuracy
1	Technique 1	In this the square median filter, greyscale image method and artificial neural network method for classification are used	100	99.65	99.68	99.68% of accuracy
2	Technique 2	In this the global threshold method, Otsu threshold method, morphological filter and artificial neural network classifier are used	83	82	76	83% of accuracy
3	Technique 3	In this the global threshold method, Otsu threshold method, watershed transform and support vector machine classifier are used	96.26	99.09	96.42	96.42% of accuracy

References

1. A.M. Bashir, Z.A. Mustafa, I.A. Hameid, R. Ibrahem, Detection of malaria parasites using digital image processing, in *International Conference on Communication, Control, Computing and Electronics Engineering (ICCCCEE)*, Khartoum (IEEE, Sudan, 2017)
2. S.S. Savkarea*, S.P. Naroteb, Automatic system for classification of erythrocytes infected with malaria and identification of parasites life stage, in *2nd International Conference on Communication, Computing & Security (ICCCS)* (Elsevier, Canada, 2012)
3. R. Tomari, W.N.W. Zakaria, M.M.A. Jamil, F.M. Nor, N.F.N. Fuad, Computer aided system for red blood cell classification in blood smear image, in *International Conference on Robot Pride 2013–2014—Medical and Rehabilitation Robotics and Instrumentation, confPRIDE* (Elsevier, Canada, 2013–2014)
4. H.F. Bunn, Approach to the anemias, in *Cecil Medicine*, 24th edn., Chap. 161, ed. by L. Goldman, A.I. Schafer (Saunders Elsevier, Philadelphia, PA, 2011)
5. M. Habibzadeh, A. Krzyżak, T. Fevens, Comparative study of shape, intensity and texture features and support vector machine for white blood cell classification. J. Theor. Appl. Comput. Sci. **7**(1), 20–35 (2013)
6. E.A. Mohammed, M.M. Mohamed, B.H. Far, C. Naugler, Peripheral blood smear image analysis: a comprehensive review. J. Pathol. Inf. **5** (2014)
7. G. Lavanya, N. Thirupathi Rao, D. Bhattacharyya, Automatic identification of colloid cyst in brain through MRI/CT scan images, in *Third International Conference on SMARTDSC-2019, LNNS*, Visakhapatnam, vol. 105 (2019), pp. 45–52
8. K. Kim et al., Automatic cell classification in human's peripheral blood images based on morphological image processing, in *AI 2001: Advances in Artificial Intelligence*, ed. by M. Stumptner, D. Corbett, M. Brooks (Springer, Berlin, Heidelberg, 2001), pp. 225–236
9. B. Venkatalakshmi, K. Thilagavathi, Automatic red blood cell counting using Hough transforms, in *2013 IEEE Conference on Information & Communication Technologies (ICT)*, India (2013)

A Survey on Techniques for Android Malware Detection

Karampuri Navya, Karanam Madhavi, and Krishna Chythanya Nagaraju

Abstract Android is an open-source platform for numerous applications. It is playing a major role in the current world and used for handling users personal and confidential data. Android mobile users can download various applications and certainly upload them easily without any cost and authorization through Google play store. Due to this, Android threats may occur, and these are spreading easily leading to various types of Android malwares which are growing around the planet and effecting user's personal data, their systems, and reputed organizations harming the sensitive information. There are two mechanisms which propagate for Android malware detection. They are signature-based techniques and permission-based techniques. Signature-based techniques were used for detection of unknown malwares based on signature samples, but they cannot detect newly discovered threats like zero-day attacks, which are not known to the world before they are seen in the wild. This paper deals with a survey on malware detection techniques. In this survey, it is observed that Android malware detection techniques of permission-based systems perform more efficiently. There are discussions on various detection mechanisms and machine learning algorithms used in Android malware detection, and this paper highlights their advantages as well as disadvantages.

Keywords Android Malware · Machine learning algorithm · Malware detection techniques

K. Navya (✉) · K. Madhavi · K. C. Nagaraju
CSE Department, Gokaraju Rangaraju Institute of Engineering and Technology, Hyderabad, India
e-mail: karampurinavya@gmail.com

K. Madhavi
e-mail: bmadhaviranjan@yahoo.com

K. C. Nagaraju
e-mail: kcn_be@rediffmail.com

D. Bhattacharyya and N. Thirupathi Rao (eds.), *Machine Intelligence and Soft Computing*, Advances in Intelligent Systems and Computing 1280,
https://doi.org/10.1007/978-981-15-9516-5_2

1 Introduction

Android is an open source and is an easy platform for attackers to harm confidential data. It offers a wide range of possibilities at low cost and provides an easy interface to upload new applications without authorization. Google Play contains about 2.87 million applications, and there are about 50+ billion downloads till date [1]. Such kind of popularity attracts malware developers and challenges users not to lose their personal data. There are different types of Android malwares like Trojans, Spywares, Risk wares, and few more [1]. There are two main approaches for finding malware, one is Static and another is dynamic approach. Through them, application is categorized as either benign or malicious applications.

Without executing the code, static analysis extracts features, and dynamic analysis feature extraction is completely based on execution of code. Static analysis is more informative, particularly in cases of code when it is unclear [2]. Since, Android is the most utilized brilliant cell phone stage, it involves 82.8% of the overall industry. 'Lookout' is a recent report that shows the Android market is growing more than Apple's app store [3]. Table 1 shows 2018 statistics of app downloads using Google play store versus Apple's app store.

To detect Android malware, there are two techniques available one is signature-based and another is permission-based method. Signature-based method uses signatures which are compared with existing malware family signatures and categorizes code as malware. Due to the very limited size of the signature database, it fails in detection of unknown malwares [4]. Permission based is a fast-performing method, and it analyzes the manifest files.

Permission based detection is sub-categorized into four more types, and they are permissions which are normally asked, permissions which are dangerous, permissions with signatures, and system-based permissions [5]. Permission-based detection helps to find out malicious permissions which collect information like device ID, personal SMS, location details from mobile and send data to a remote server using malicious application (or) obtain financial gain over SMS and HTTP requests [5]. Some applications use certain actions including advantages and modify configurations and also monitor user activities like SD card storage content, network activity, etc. [5], So, Android security markets are highly based on permission-based detection mechanisms.

Table 1 App downloads on Google play store versus Apple's app store

Year	App download's using Google play store	App downloads using Apple's app store
2017	27.7 B	29.4 B
2018	8.3 B	8.9 B

2 Comparison Analysis

Malware applications in general use few penetration techniques to infect the devices that they are repackaging, updating, and downloading. Installing of malicious applications using Android as a platform is done with the help of repacking. Works through downloading apps which have more popularity and disassembles them. After disassembling, it follows with the addition of malicious code, code repacking, and uploading it again. In order to surpass the detection, malware authors change the signatures of these apps [6]. Updating, in general, builds a component containing wrapped inflicted code. It downloads the malware during the run time and does not directly include it in the application. Through visiting these kinds of online sites that contain malware content, users are tricked to download malicious applications in their mobile phones unknowingly [6]. Downloading deals with the installation of an encrypted application within the APK files, and after installation, malicious contents inside the app are decrypted on the user's device.

2.1 Different Detection Mechanisms

SigPID [7] is significant permission identification for Android malware detection. It is a static malware detection system used to manage rapid increase in Android malware, based on analysis of permissions that app requests. It performs multi-level data pruning technique and applies mining on the permission data to identify the most common permissions that can be effective in distinguishing the application as either benign or malicious. Then, it uses machine learning algorithm as classifier and performs few classification methods to classify different malware and benign families [7]. SigPID can be applied with various existing ML algorithms such as SVM, random forest, and decision trees. It reduces permissions from 135 to 22 after performing multi-level data pruning. When support vector machines are used as classifiers, it maintains over 90% of malware detection accuracy [7]. It is observed that, in a dataset of 91.4% unknown malware, SigPID showcases its efficiency by detecting 93.62% of unknown malicious apps. [7]. Machine learning algorithms can find, summarize, and extract the relevant data and also make predictions based on analysis data [7].Comparison among various surveyed detection mechanisms is portrayed below in Table 2.

AndroSimilar [8]. It extracts features that are statistically improbable, normalizes features in bloom filters, and generates signatures to detect malware Android applications. The method used in androsimilar brings out unseen variants from known malware [8]. It is a mechanism that finds areas of statistical similarity with the help of known malware to detect those unknown samples. It is observed as a faster detection approach than a fuzzy hashing approach [8]. It gives similarity to unknown applications with existing malware applications. But, signature-based mechanisms

Table 2 Comparison among various mechanisms

S. No.	Title of work	Aim	Type	Technique used	Dataset	Algorithm used	Best result
1.	SigPID	Controlling rapid increase of malware	Signature based	Multi-level data pruning technique and applies mining on the permission data	310,92 apps	SVM	93.62
2.	AndroSimilar	Detect Android malware applications	Signature based	Syntactic similarity with known samples	6779	Bloom filter	99.4
3.	DroidMat	Android malware detection	Signature based	API call tracing, SVD	1738	KNN	97.87
4.	DroidAnalytics	Find Android malware	Signature based	API call tracing	150,368	Multi-level signature algorithm	327 zero-day malwares
5.	DREBIN	Malware identification in Android	Permission based	Static investigation of applications	123,453	SVM	94
6.	RISK RANKER	Spot zero-day Android malwares	Permission based	Automated framework, fuzzing	118,318	CNN	322 zero day
7.	APK Auditor	Classify Android applications as either benign or malicious	Permission based	Signature data base, Android client and Central server is responsible for analyzing whole process	8792 applications	Logistic regression	88

(continued)

Table 2 (continued)

S. No.	Title of work	Aim	Type	Technique used	Dataset	Algorithm used	Best result
8.	Permission-based android malware detection system using feature selection with genetic algorithm	Android malware detection	Permission based	Dynamic analysis	1740	SVM and GA	98.45

are limited in the dataset and detect only known malwares. Sometimes fails to protect users' apps from unknown malwares.

DroidMat [9]. It is a static element-based malware detection mechanism performed through API call tracing and manifest in Android mobiles. To show Android behavior from each application's manifest file, parameters like permission data, action of components, messages passing with intent, and API calls are extracted. API calls related to permissions are traced using components like activity, service, receiver which are considered as entry points. Then applies *K*-means algorithm to improvise malware modeling capability and on a low rank approximation, the number of clusters using a method called singular value decomposition (SVD) are taken into account. Then, at last, it makes use of KNN algorithm and classifies application as either benign or malware with a good recall rate. But, it was observed that it is not so effective in identifying advertisement product tests [9].

DroidAnalytics [10]. It is a signature-based system which can thoroughly gather, cope, process, and find out Android malware. It allows analysts to associate and retrieve malicious actions at the op-code level. Initially at method level, it generates a signature and signatures of methods, application signatures along with API call traces. At a class level, these are used to generate signatures. DroidAnalytics is efficient using 150,368 applications, and from 102 different Android families, it successfully determines 2494 malware. The results show that DroidAnalytics is a significant system and is good at examining malware repackaging and transformations [10].

DREBIN [11]. It is an expensive static examination of permissions for malware identification in Android mobiles. It is a light-weighted technique which helps to recognize malicious applications on a mobile phone. It performs static investigation and collects as many as possible features of applications. In a joint vector space, highlights are en-captured featuring that malware can be easily distinguished. Among 5560 malware samples and 123,453 applications, 94% of the malware cases are found. It is observed that the checking time is 10 s. For dissecting applications which are downloaded, this strategy is appropriate, and explanations provided are used to reveal detection's relevant properties of malware. The nature of results delivered by DREBIN depends on the openness of malignant and generous applications. It has the absence of dynamic examination and utilization of AI [11].

Risk Ranker [12]. It spots zero-day Android malware. Without depending on malware tests and outcomes of tests, it evaluates potential security dangers which occur by some unauthorized applications. It builds a framework which is automated and analyzes whether any application displays risky conduct. In order to deliver decreased applications rundown, above obtained yield is used, and 118,318 Android applications are inspected which are gathered from various Android markets. Utilizing Risk Ranker, it is proven that preparing this enormous number of applications will take less than four days. Among 3281 suspicious applications, it reports 718 malevolent applications, 322 of them being zero-day.

APK Auditor [13]. It categorizes and classifies Android applications as either benign or malicious using static analysis. It consists of three components—Firstly, to store the extracted information of applications and its analysis results, a signature database is used. Secondly, an Android client that end-users use for granting applications requests, and for the communication between signature database and smartphone client, a central server is used. It uses 8762 applications in total and analyzed by the developed system. It detects the most well-known malware with 88% accuracy.

Permission-Based Android Malware Detection System Using Feature Selection with Genetic Algorithm [14]. It uses a machine learning method to detect Android malware using the genetic algorithm. Using GA with three different classifiers, different feature subsets are selected and used to analyze and detect Android malware. 1740 samples are tested using GA for feature selection and for classification of Android malware, SVM is implemented with along selected 16 permission. The result gives an accuracy of 98.45% and performs more efficiently with fewer number of permissions. Here, we have lower level API which is used to include permission features.

3 Conclusion

Android users have been extremely increasing over the years. This growth of Android users have been taken as an advantage by malware authors to harm many users. This paper surveyed various types of Android malwares and their detection approaches. The observation says that detecting malicious apps using static approach is comparatively less efficient than dynamic approach. Dynamic approach follows frequent monitoring of applications but it sometimes fails to detect some parts of malicious code that is not executed. Static approaches are generally faster and efficient in providing detection accuracy. Building up a hybrid technique can also give better results. We can come to the conclusion that no single methodology is sufficient to make a framework secure and no single machine learning algorithm is sufficient to give required effectiveness.

References

1. C. Liu, Z. Zhang, S. Wang, An android malware detection approach using Bayesian inference, in *2016 IEEE International Conference on Computer and Information Technology (CIT)*, Nadi (2016), pp. 476–483
2. Krishna Sugunan, T. Gireesh Kumar, K.A. Dhanya, Static and dynamic analysis for android malware detection, in *Advances in Big Data and Cloud Computing*. 645ISBN: 978-981-10-7199-Lookout app genome report (2018). https://www.mylookout.com/appgenome,2011
3. J. Lopes, C. Serrão, L. Nunes, A. Almeida, J. Oliveira, Overview of machine learning methods for android malware identification, in *2019 7th International Symposium on Digital Forensics and Security (ISDFS)*, Barcelos, Portugal (2019), pp. 1–6
4. A. Utku, I.A. DoGru, M.A. Akcayol, Permission based android malware detection with multilayer perceptron, in *2018 26th Signal Processing and Communications Applications Conference (SIU)*, Izmir, (2018), pp. 1–4

5. R. Zachariah, K. Akash, M.S. Yousef, A.M. Chacko, Android malware detection a survey, in *2017 IEEE International Conference on Circuits and Systems (ICCS)*, Thiruvananthapuram (2017), pp. 238–244
6. L. Sun, Z. Li, Q. Yan, W. Srisa-an, Y. Pan, SigPID: significant permission identification for android malware detection, in *2016 11th International Conference on Malicious and Unwanted Software (MALWARE)*, Fajardo (2016), pp. 1–8
7. P. Faruki, A. Bharmal, M.S. Gaur, V. Laxmi, V. Ganmoor, Androsimilar: robust statistical feature signature for android malware detection, in *Proceedings of the 6th International Conference on Security of Information and Networks* (2013), pp. 152–159
8. D. Wu, C. Mao, T. Wei, H. Lee, K. Wu, DroidMat: android malware detection through manifest and API calls tracing, in *2012 Seventh Asia Joint Conference on Information Security*, Tokyo (2012), pp. 62–69
9. M. Zheng, J.C.S. Lui, M. Sun, Droid analytics: a signature based analytic system to collect, extract, analyze and associate android malware, in *12th IEEE International Conference on Trust, Security and Privacy in Computing and Communications (TrustCom)* (2013), pp. 163–171
10. D. Arp et al., DREBIN: effective and explainable detection of android malware in your pocket. NDSS (2014)
11. M. Grace, Y. Zhou, Q. Zhang, S. Zou, X. Jiang, RiskRanker: scalable and accurate zero-day android, in *Proceedings of the 10th International Conference on Mobile Systems, Applications* (2012). https://doi.org/10.1145/2307636.2307663
12. A.T. Kabakus, I. Doğru, A. Çetin, APK auditor: permission-based android malware detection system. Digit. Invest. **13** (2015). https://doi.org/10.1016/j.diin.2015.01.001
13. O. Yildiz, İ. Doğru, Permission-based android malware detection system using feature selection with genetic algorithm. Int. J. Softw. Eng. Knowl. Eng. **29**, 245–262 (2019). https://doi.org/10.1142/s0218194019500116
14. Z. Xiaoyan, F. Juan, W. Xiujuan, Android malware detection based on permissions, in *2014 International Conference on Information and Communications Technologies (ICT 2014)*, Nanjing, China (2014), pp. 1–5

Comparative Analysis of Prevalent Disease by Preprocessing Techniques Using Big Data and Machine Learning: An Extensive Review

Bandi Vamsi, Bhanu Prakash Doppala, N. Thirupathi Rao, and Debnath Bhattacharyya

Abstract Nowadays, in communities like healthcare and biomedical data, there has been a tremendous growth. The healthcare industry maintains a vast quantity of treatments that should be given to patients by analyzing the diseases that once has already occurred among patients, for further references and methodologies in curing the diseases by maintaining this vast track of history that has already been saved in the healthcare industry. In view of big data progress in biomedical data and healthcare communities, veracious study and predictive analysis of this methodologies related to medical data account to early disease recognition, patient care, and community services. When the trait of these methodologies is incomplete, the promptitude of study becomes economized. Furthermore, at particular regions, they show an uncommon trait of these certain differential regional diseases, which may cause the outcomes in diminishing the prediction of disease outbreaks. A prior task is that how the data can be accessed and how the information could be available for particular disease from these vast data saving machines. On the other hand, some machines develop techniques that are applied by providing realistic time sequence data, statistical analysis, and innovative data analytics in terms of patient's family history, laboratory reports, impact of disease, and blood pressure. The proposed work is to identify the problem in patient earlier by producing the exact treatment

B. Vamsi (✉) · B. P. Doppala · N. Thirupathi Rao
Department of Computer Science and Engineering, Vignan's Institute of Information Technology, Visakhapatnam 530049, India
e-mail: vamsib111@gmail.com

B. P. Doppala
e-mail: doppalabhanuprakash@gmail.com

N. Thirupathi Rao
e-mail: nakkathiru@gmail.com

D. Bhattacharyya
Department of Computer Science and Engineering, K L Deemed to be University, KLEF, Guntur 522502, India
e-mail: debnathb@kluniversity.in

D. Bhattacharyya and N. Thirupathi Rao (eds.), *Machine Intelligence and Soft Computing*, Advances in Intelligent Systems and Computing 1280,
https://doi.org/10.1007/978-981-15-9516-5_3

in advance before the disease attacks the patient completely, which may save the patients' life in reducing the complexities.

Keywords Health care · Machine learning · Predictive analysis · Big data

1 Introduction

1.1 Big Data

The big data is mentioned itself to be the bigger quantity of data which is structured, unstructured, and semi-structured. This statistics gathered so far from different number of means that which includes e-mails, smart devices, applications, databases, servers, and by other means which are varying in the day to day [1]. Three Vs are three defining properties or dimensions of big data which are density, assortment, and speed [2].

The main zeal of this huge quantity of this large data architecture involves in handling the processing methodologies, ingestion, and deconstruction of patients data, which is too convoluted for existed database systems [3]. In Fig. 1, data sources are repositories of large volume of data. Examples are various data stores such as real-time data sources, relational databases through IoT devices. The data storage relies in different format of this complex data which involves in tracking of the batch processing operations. Batch data processing deals with large amount of data or data sources only. The tasks in batch processing includes such as performing aggregate functions, i.e., finding the mean, maximum, or minimum of a set interval of data. In real-time message ingestion, the data is imported as it is produced by the source, and ingesting data in batches represents in importing discrete blocks of data at particular intervals. In stream processing, after gathering all real-time data, filtering should be

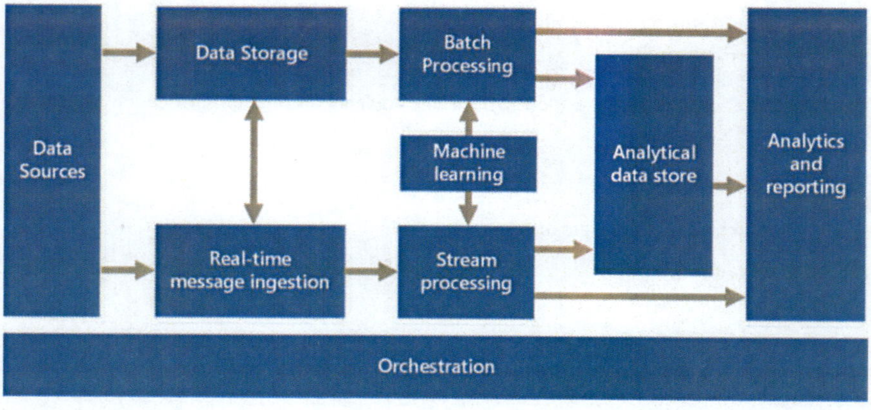

Fig. 1 Big data architecture

done for the collected process, aggregating and, on the other hand, for estimating the data [4]. In analytical, the data can be stored in many large data solutions which are combined for analysis and then provide the processed data in structured format which can be acquired by analytical tools. The perception of providing data which acts as major big data solutions can be progressed through analyzing and reporting [5]. Data scientists or data analysts can also be represented in the form of data exploration through analyzing and reporting. The most of big data consist of duplication of data processing summarized in work flows. To analyze these work flows, one can also use orchestration technology [6].

Big statistical applications are creating a new era in every industry such as health care, retail, manufacturing, education and agriculture and farming. In this work, we are focusing more on health care [7]. The levels of data generated within healthcare systems are not trivial. With the including maintenance of medical health (m-health), e-health, and wearable technologies, these bulks of data will continue to accession. This constitutes electronic (e-health) record data, sensor data, patient generated data, and other forms too complex to process the data [5]. The usage of this large statistical information in health care for individual rights, privacy and autonomy, to transparency and trust which has become a challenging significance that has been evolved depending upon the risk.

1.2 Machine Learning

The term machine learning itself deals with branch of computer science which is provided by computers the ability to understand in making determinations on given data. The decisions of machine learning are based on algorithms built on data to build an algorithm in machine learning which takes feature as an input and gives a prediction for output. One of the most using examples of machine learning in our daily life is to anticipate whether an e-mail is spam or not [3].

The machine learning algorithms are categorized into supervised and unsupervised learning algorithms as shown in Fig. 2. The supervised attainment works with labeled data. The aim of supervised attainment is to build a model which is able to pre-assume specified variable [8]. Supervised problems are divided into two types which are regression problem and classification problem. In classification problem, target variable consists of categories, whereas in backsliding problem assumed variable is continuous. There are many applications in supervised learning such as spam detection, pattern recognition, speech recognition, and many more [9]. On the other hand, unsupervised learning deals with unlabeled data. Unsupervised learning is again divided into clustering and association rule. Clustering is used to combine the similar data into subgroups. Association rule can be used to find the relations between the large variables in database [10]. The applications of unsupervised learning are anomaly detection, visual recognition, robotics, etc.

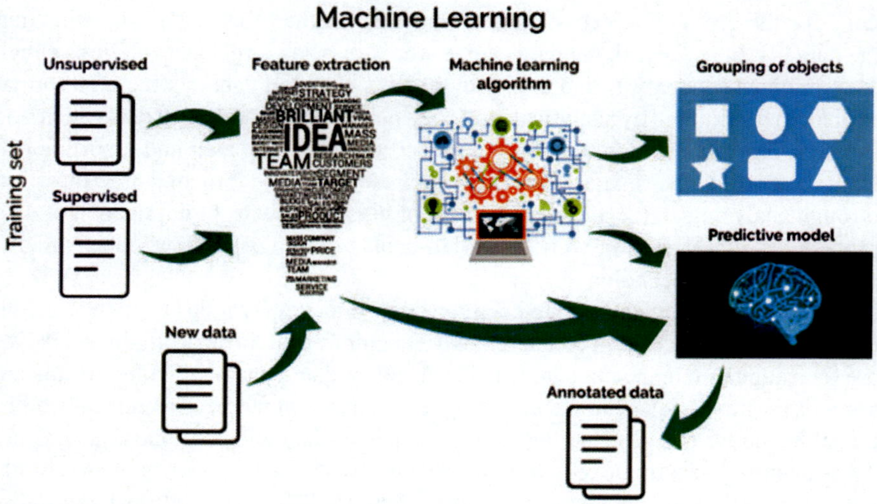

Fig. 2 Architecture of machine learning

1.3 Health Science Computing

In this modern era, big data and machine learning algorithms are having great influence in healthcare industry. There is a huge growth in biological data and healthcare industries [7]. Nowadays, the healthcare industries are producing huge amount of patient's clinical and physical data which is difficult for understanding by existing means of data processing [3]. On an alternative, big data may be processed sometimes by machine learning algorithms. Huge amount of health record data generated by hospitals or healthcare organizations might help a doctor to treat a patient in a better way with the help of records already existing in the repository [2]. Electronic health care is one of the global applications of big data in medicine in which each patient has their unique digital record which consists of medical history, demographics, etc. Predictive analysis in health care is one of the enormous applications of big data where it helps the doctors to take data-driven decision within short time which helps to improve the treatment [11].

The term bioinformatics itself is made up of two parts, bio means biology and informatics means information. It is an interdisciplinary field mainly consists of molecular biology and genetics, computer science, mathematics, and statistics to analyze biological data [5]. Databases are essential for bioinformatics research and applications. There are many applications of bioinformatics such as molecular medicine, microbial genome application, and agriculture. The fields of medicine applications of bioinformatics include personal medicine, preventive medicine, etc.

There are many diseases which are affected in recent years and now let us discuss how this large data is used to presume and cure them. Cancer, as we know cancer is one the dangerous disease which has no cure [5]. These large records can be utilized by the medical researchers which account for the latest ways and treatments which are mainly used in the recovery of cancer patients which ranked in the highest priority in this existing society [12]. For instance, assume that the researchers can determine the tumor instances in biobanks that which are related with patient's medical records in Fig. 3. With the use of these data researchers, one can estimate how certain mutations and cancer proteins interact with various treatments and find the ways that will lead to patient outcomes. Heart attack is one the deadliest diseases which is also known as acute myocardial infarction (AMI). The main cause for often heart attacks is due to which the flow of blood in one of the coronary arteries gets hold which results in blockage of the artery. Most of the examines include big data analytical health care, prognosis of heart attack, and other curing technologies which were verified in these theories that use national and international databases in big data and solitude concerns [7]. The main priority in which the big data analysis is used was in presuming the heart attack, and the technologies used in big data play a crucial role to management

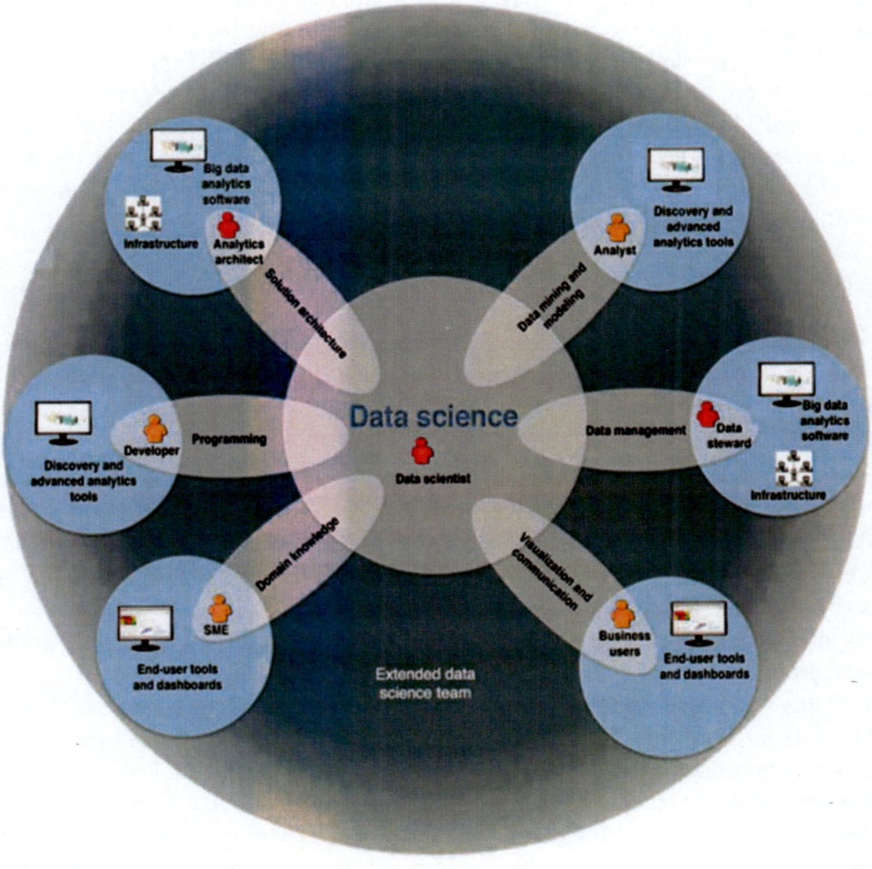

Fig. 3 Architecture of health science

and monitoring of treatment for various cardiovascular diseases [5]. Effective and maintained methodologies of medical treatment might be advanced by the use of these technologies. Today many people in the world are affected by heart diseases. Big data plays an important role in order to save patients health and to reduce the death of heart patients. Smart phones and sensors can identify and transmit different types of health data [11]. Nowadays, few modern devices like wrist bands or watches have been designed as heart attack revelation devices which are used in identifying heart condition, detect heart attack and to immediately seek for rescue [8].

For example, the common widespread disease which is attacking most of the people was dengue, in which the infection is carried out through the female mosquitoes. In this context, these are most usually found in the climatic conditions mainly in the hottest regions. Long back, experts are finding a way in sorting out the different types of patients who have been infected by this disease for bringing out the best results for the treatment that they have to carry out for different types of infections caused by these mosquitoes [6]. Past beyond few years ago in the contemporary study, dengue disease quarried Pakistan. Based on the nature and performance for comparing, this dengue fever has been explained by dividing them under certain techniques. For accurately categorizing, our available dataset and different classification techniques are used. These techniques are REP tree, Naïve Bayesian, random tree, SMO, and J48. The data mining tools like WEKA are used for data classification. Evaluation of the performance of above techniques separately can be done by using graphs and tables depending upon dataset and finally by comparing the performance of all the techniques.

2 Introduction

Ives [5] have proposed their work on "Big Data and Machine Learning Meet the Health Sciences." This work defines how to handle large size of data that can be recorded in various forms using multiple ways. The family health history of patients can be used to solve future generation problems and using this we can control the spreading of diseases. The big data tools and machine learning algorithms together are used to predict the clinical information using healthcare systems. The model is used to maintain continuous monitoring of patients health and can be used for early diagnosis. The clinical calculators are used to support the machine learning results to improve the therapy.

Georga [11] have proposed their work on "Artificial Intelligence and Data Mining Methods for Cardiovascular Risk Prediction." In this work, the model focused on how to use big data analytics to predict the stroke-level artery disease in the regions of Europe. As per the records given by World Health Organization, the death caused by cardiovascular strokes is 20% out of 45% total deaths. The progression factors depend on age, weight, diabetics, etc. The problem can be solved by binary classification. The imaging of stroke can be detected by magnetic resonance imaging (MRI) which is a rational model and to produce results it takes more than 24 h but in non-imaging machine learning model can produce the results by classification techniques. The input data can be divided by clustering way to identify the similarities of patients to find the true positive and true negative.

Venkatesh [6] have proposed their work on "Development of Big Data Predictive Analytics Model for Disease Prediction using Machine learning Technique." In this work, the model focused on heart diseases caused around the world. The machine learning technique is having better decision-making algorithms to predict the health parameters of patients. The feature extraction can be done by Naive Bayes algorithm to get the best outcome. The accuracy of the model about 97% is to predict the rate of disease. The model can divide the data into two classes are: true negative and negative–positive.

Beulah [12] have proposed their on "Improving the accuracy of prediction of heart disease risk based on ensemble classification techniques." In this work, the model can predict the patterns from the available data and it applies to an unavailable dataset to get the accurate outcome. By using ensemble classification, the model can improve the accuracy of weak algorithms with multiple classifiers. The parameters work with family history, age, cholesterol, obesity, etc. By comparing the various techniques, the model produces best accuracy with 92.1% with SVM. The *K*-means clustering algorithm can be used to extract the data from the master dataset using the MAFIA algorithm to predict the stroke with different stages.

Junfie [10] have proposed their work on "A survey of machine learning for big data processing." In this work, very large datasets are collected on research areas of biomedical sciences, social media, and social sciences. These large datasets are very difficult to handle with traditional architectures and algorithms, so the solution is machine learning through big data done through signal processing techniques. The classification and learning techniques are going to input to big data learning techniques and these are given to signal processing techniques to get the conclusion.

Francisco [7] have proposed their on "Advanced Machine Learning and Big Analytics in Remote Sensing for Natural Hazards Management." In this work, the datasets are collected from 116 metropolises of topic like floods, landslides, earthquakes, soil, etc., for analyzing the natural hazards. The big data techniques are used to extract the high-resolution satellite images for data visualization and patterns.

Ives [3] have proposed their work on "Machine learning and big data analytics in bipolar disorder: A position paper from the International Society for Bipolar Disorders Big Data Task Force." In this work, the big data and machine learning algorithms are used to predict the treatment decisions on bipolar disorder of individual patients. This can be done by ROC-AUC curve. The expected outcome can be done through the mapping of X to Y for predicting the suicide attempts of patients with mental disorders. The machine learning model helps to get the conclusion in various steps of gathering the patient data, selection of features, selecting the model and tuning, testing and knowledge extraction.

Robb [13] have proposed their work on "Machine learning and big data in psychiatry: toward clinical applications." In this work, the mental disorders of the patients can be analyzed by approaches of computational and machine learning with high-dimensional mechanisms. The analysis of resting EEG after medication is given to classifier performance to get the required parameters. The performance can be identified through AUC using reinforcement learning.

3 System Architecture

In Fig. 4, the big data characteristics are the features for machine learning model for data processing. To extract the useful information from large volume of datasets, the feature extraction can be useful to get the transfer learning.

3.1 AUC-ROC Curve

Area under curve studies the area under a receiver operating characteristics (ROC curve). It is a performance measurement of a classification problem at a various

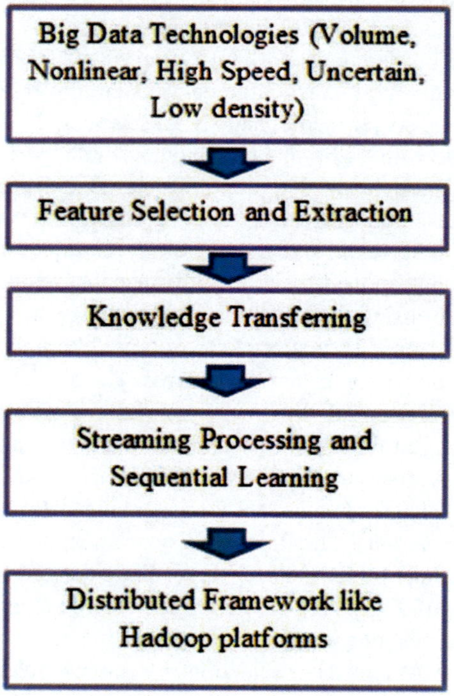

Fig. 4 Architecture of machine learning model for big data processing

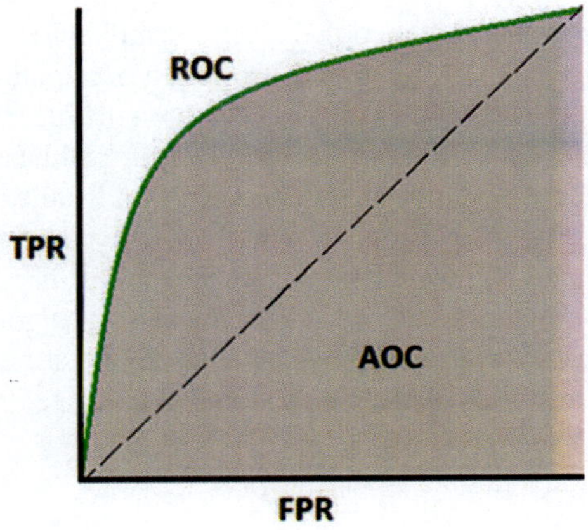

Fig. 5 AUC-ROC curve

threshold settings. ROC curve is a graph which shows the performance of classification model at all classification thresholds. There are two parameters which are plotted by ROC which are true positive rate (TPR) and false positive rate (FPR).

True Positive Rate (TPR)

$$\mathrm{TPR} = \frac{\mathrm{TP}}{\mathrm{TP} + \mathrm{FN}}$$

False Positive Rate (FPR)

$$\mathrm{FPR} = \frac{\mathrm{FP}}{\mathrm{FP} + \mathrm{TN}}$$

The categorization of thresholds related to production on a cluster measure is given by AUC. On a graph, AUC ranges from 0 to 1 where the prognosis is 100% false whose AUC is 0.0 and the prognosis is 100% true where AUC is 1.0. AUC advantageous of two causes which are AUC is a scale-invariant and AUC is classification-threshold-invariant (Fig. 5).

3.2 *Data Preprocessing*

It is an essential step of data mining whose goal is to transform the raw data into understandable format and it improves the prediction accuracy. Data cleaning—the data which is taken from the real world is hardly clean and complete, especially the healthcare field. The data cleaning step deals with the missing values inconsistencies in dataset. Missing value is defined as the value which is not present in the cell of a particular column. The reason behind this missing value in the context of health care can be the human exclusion, not recorded electronically by senor and others. These missing values can be handled either by discarding it or imputing the missing data.

Data reduction—the data is generating rapidly through different electronic devices which is very huge and also unstructured, especially in health care. It is showed that more than 75% of data reduction task is concerned with feature selection about 15% is concerned with feature extraction and less than 10% concerned with discretization method. Discretization means the process of converting quantitative data into qualitative data. Discretization became essential where the algorithm works well on nominal data like Naive Bayes and decision tree. The limitation of this technique is information loss but it simplifies the data and makes accurate.

4 Results and Discussion

In Table 2, we can observe the main heart pain scale that can be occurred based on the age and cholesterol factors. The range of pain scale is more in old age patients based on the cholesterol range more than 200 (Tables 1, 3, 4 and 5).

Table 1 Sample dataset of heart disease

Age	Sex	Pain scale	Trestpbs (in mm)	Cholesterol	Fbs	Restecg
65	M	3.0	119	219	0	2
45	M	4.0	123	268	0	1
40	F	4.0	120	309	0	2
55	M	2.0	134	163	0	0
63	M	1.0	103	101	0	1
72	F	3.0	138	293	0	2
82	M	4.0	140	263	0	2
50	F	3.0	137	282	0	2

Table 2 Classification report

Sex	Precision	Recall	*F*-score	Support
M	0.78	0.73	0.81	864
F	0.45	0.40	0.35	272
Average	0.80	0.82	0.81	1132

Table 3 Comparison of boosting and bagging time in sec

	Without ensembling	With bagging	With boosting
Naive Bayes	0.02	0.05	0.23
Bayes net	0.03	0.10	0.15
C 4.5	0.04	0.50	0.34
Multilayer perception	2.5	8.01	15.20
PART	0.07	0.35	0.83

Table 4 Accuracy of bagging with feature selection

Algorithm	Bagging accuracy	Improvement accuracy	Feature dataset (FDS)
C 4.5	73.7	75.18	FDS-1
C 4.5	81.9	83.3	FDS-5
Random forest	80.3	81.2	FDS-6
Random forest	80.5	80.9	FDS-2
Multilayer perceptron	80.2	81.3	FDS-4
Multilayer perceptron	81.5	82.6	FDS-6
Multilayer perceptron	80.5	81.7	FDS-3
Bayes net	85.3	85.9	FDS-1
Naïve Bayes	85.1	86.3	FDS-6

Table 5 Accuracy of boosting with feature selection

Algorithm	Boosting accuracy	Improvement accuracy	Feature dataset (FDS)
C 4.5	75.9	79.7	FDS-6
C 4.5	72.9	73.4	FDS-3
C 4.5	70.3	74.1	FDS-1
C 4.5	80.3	83.4	FDS-4
Random forest	76.3	80.1	FDS-6
Random forest	76.3	81.3	FDS-3
Random forest	79.4	85.9	FDS-6
Multilayer perceptron	75.3	76.9	FDS-5
Multilayer perceptron	74.6	75.8	FDS-4
Naïve Bayes	85.3	85.9	FDS-6
Naïve Bayes	86.4	87.1	FDS-2

5 Conclusion

Big data analytics and machine learning algorithms play a huge and essential role in healthcare industry. In this work, we discussed how big data is being generated and how do big data deals with that huge amount of data, how machine learning techniques are used to predict the risks and provide life-saving outcomes. We have compared some prevalent disease using big data and machine learning techniques. The discussion about the AUC-ROC curve is used to evaluate and how the models performance and also the study of different preprocessing techniques. This work causes the effect and importance of big data and machine learning models on the healthcare predictions.

References

1. F. Shoayee, Sehaa: a big data analytics tool for healthcare symptoms and diseases detection using twitter, apache spark, and machine learning. Appl. Sci. **10**(4), 1398–1427 (2020)
2. F. David, The basics of data, big data, and machine learning in clinical practice. Clin. Rheumatol. 1–3 (2020)
3. F. Ives, Machine learning and big data analytics in bipolar disorder: a position paper from the international society for bipolar disorders (ISBD) big data task force. Bipolar Disord. **21**(7), 1–13 (2019)
4. F. Ngiam, Big data and machine learning algorithms for health-care delivery. Lancet Oncol. **20**, 262–273 (2019)
5. F. Ives, Big data and machine learning meet the health sciences: big data analytics in mental health (Springer Nature Switzerland AG, Berlin, 2019), pp. 1–3
6. F. Venkatesh, Development of big data predictive analytics model for disease prediction using machine learning technique. J. Med. Syst. **43**(8), 272 (2019)
7. F. Francisco, Advanced machine learning and big analytics in remote sensing for natural hazards management. Remote Sens. **2**(2), 301–303 (2020)
8. F. Roh, A survey on data collection for machine learning: a big data—AI integration perspective. IEEE Trans. Knowl. Data Eng. **1**(3), 1 (2019)
9. F. Tai, Machine learning and big data: implications for disease modelling and therapeutic discovery in psychiatry. Artif. Intell. Med. **99**, 1–11 (2019)
10. F. Junfie, A survey of machine learning for big data processing. EURASIP J. Adv. Signal Process. **2016**(1), 1–16 (2016)
11. E.I. Georga, F. Nikolaos, Artificial intelligence and data mining methods for cardiovascular risk prediction, in *Cardiovascular Computing—Methodologies and Clinical Applications* (Springer, Berlin, 2019), pp. 279–301
12. F. Beulah, Improving the accuracy of prediction of heart disease risk based on ensemble classification techniques. Inf. Med. Unlocked **16**, 100203 (2019)
13. F. Robb, Machine learning and big data in psychiatry: toward clinical applications. Curr. Opin. Neurobiol. **55**, 152–159 (2019)

Prediction of Diabetes Using Ensemble Learning Model

Sapna Singh and Sonali Gupta

Abstract Diabetes is an illness which may cause blood sugar, by producing insufficient amount of insulin. Untreated diabetes can increase the danger of heart attack, cancer, liver problem and other disorders. All over the world, billions of people are affected by diabetes disease. Early treatment of diabetes is very important to maintain a healthy lifestyle and culture. Diabetes is a major reason of global concern as the cases of diabetic patients are increasing rapidly day by day. Machine learning is an advance technical growing field which helps to understand the meaning of real data. It is widely used in healthcare community for detecting and analyzing serious and complex situations. Machine learning is a mathematical computational model which is used for learning new knowledge from its old knowledge and improves model efficiency to make more accurate results. Machine learning is the process of gathering, selecting, analyzing, and designing large real-time dataset. This research work using a special class of machine learning technique called bagging and boosting which is also known ensemble learning techniques. In this work, ensemble learning techniques are experimented on Pima Indians dataset to analyze and identify diabetic patient with highest risk factors using Python language. To classify a diabetic patient and non-diabetic patient, five different predictive models, namely random forest (RF), light gradient boost (LG Boost), extreme gradient boost (XGBoost), gradient boost (GB), and adaptive boost (AdaBoost) have been used for better prediction of diabetes.

Keywords Healthcare industry · Diabetes prediction · Machine learning · Ensemble learning · Predictive analytics

---

S. Singh (✉) · S. Gupta
J.C. Bose University of Science and Technology, YMCA, Faridabad, India
e-mail: sapna4882@gmail.com

D. Bhattacharyya and N. Thirupathi Rao (eds.), *Machine Intelligence and Soft Computing*, Advances in Intelligent Systems and Computing 1280,
https://doi.org/10.1007/978-981-15-9516-5_4

1 Introduction

Diabetes is a metabolic disorder which may caused by producing insufficient amount of insulin, and due to this, glucose levels in the blood are going to be rise [1]. According to the recent World Health Organization report, diabetes kills 1.6 million people each year across the globe. Diabetes affects the different parts of human body like eyes, kidney, heart, liver, and nerves. Usually, onset of type 2 diabetes happens in adult people and sometimes in old people. But nowadays, children less than 15 years are also affected by diabetes. There are lots of reasons for developing diabetes in a human body such as extended genes from their parents, daily lifestyle activities, body weight, craving junk food, and physical inactivity.

Untreated and unidentified diabetes may increase the risk of heart diseases, including coronary artery disease, heart attack, stroke, and narrowing of the arteries. So, early detection and treatment of diabetes are very important to improve the quality of patient's life [2–4].

Nowadays, healthcare industry generates huge amount of unstructured data which is very difficult to handle and store. In the era of technology, machine learning is concerned with handling huge amount of structured and unstructured data. Machine learning allows building models to analyze data and serve better results, leveraging both past and real-time data. In machine learning, "learning" means that the system is capable to learn new knowledge from input raw data, so that system can make effective decisions and prediction results based on its previous learning knowledge [5]. Different types of machine learning algorithms have been used for disease prediction depicted in Fig. 1.

Machine learning algorithms are also used for classification, pattern prediction, image recognition, object detection, and pattern recognition, in which classification techniques have been widely used for prediction. Predictive analytics is part of supervised machine learning often used for finding patterns to predict future behavior of object based on the existing data and previous experience. Predictive analytic techniques are often applied in various medical fields like disease identification and diagnosis, drug discovery and manufacturing, medical imaging, personalized medicine treatment, smart health records, and disease prediction [6]. With

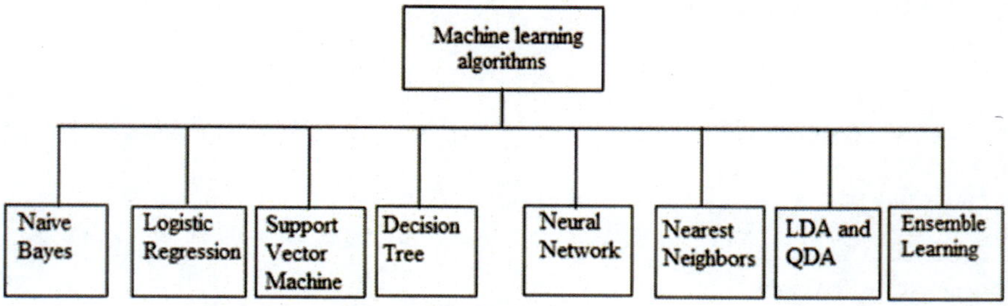

Fig. 1 Machine learning algorithms used for prediction

machine learning techniques, doctors can make better decisions on patient's diagnoses and treatment options, which lead to the improvement of healthcare services. Some machine learning algorithms, such that support vector machine, Naïve Bayes, logistic regression, decision tree, support vector machine, nearest neighbor and neural network, were previously widely used techniques for the prediction of diabetes [7, 8]. In this research work, ensemble learning techniques used for better prediction results and identify correct patterns. Ensemble learning model can give better performance than individual model. The proposed model is implemented using five different ensemble learning techniques on Pima Indians dataset (PID) to understand newly discovered patterns form real unstructured dataset. This diabetes dataset contains 768 patient records of Pima Indians women with 9 independent numerical attribute and one dependent categorical attribute. The presented work has elaborated the performance of five different models, namely random forest (RF), light gradient boost (LG Boost), extreme gradient boost (XGBoost), gradient boost (GB), and adaptive boost (AdaBoost) used for predict the presence of diabetic patients. The performance of each model has been evaluated by accuracy, misclassification rate, precision, recall, F1-score, and roc_auc_score.

2 Literature Work

In recent years, many researchers around the globe worked with big data and predictive modeling in healthcare industry and other medical domains, to predict or forecast about the future challenges and opportunities.

Sarwar et al. [9] in their work machine learning algorithms have been discussed for prediction of diabetes. The researcher has been provided a basic overview of all six machine learning algorithms based on performance and accuracy and find the best suitable algorithm for diabetes prediction. For research and experiment purpose, the dataset has been downloaded from UCI machine repository. Their dataset contains 768 instances of Pima Indians women with 9 attributes. Six machine learning algorithms are Naïve Bayes (NB), *K*-nearest neighbor (KNN), support vector machine (SVM), logistic regression (LR), decision tree (DT), and random forest (RF). In all algorithms, SVM and KNN achieved highest accuracy 77% (Fig. 2).

Kumar et al. [10] presented a diabetes predictive model for diabetes diagnosis. Different machine learning algorithms, namely decision tree, random forest, Naïve Bayes, and gradient boosting, have been used for making better prediction and observed that preprocessing technique enhances the performance of the designed model. For research experiment conduct, the medical dataset has been collected from University of California, Irvine machine learning repository. From all experimented model, the best performance was achieved by random forest method as 99.5% (Fig. 3).

Bano and Khan [11] proposed a classification predictive model to predict and identify the diabetic patient. In their work, two predictive classifiers, *k*-nearest neighbor

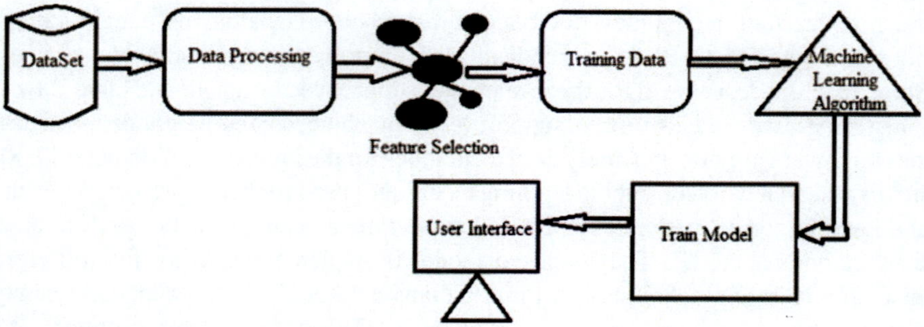

Fig. 2 Machine learning model to predict diabetes

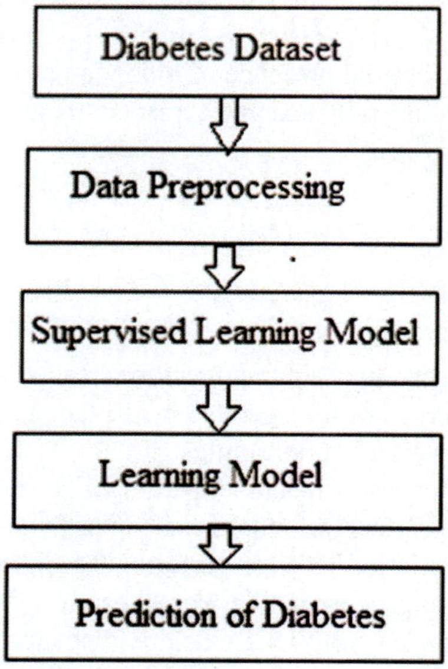

Fig. 3 A diabetes diagnosis system using machine learning

and support vector machine, have been used. The dataset Pima Indians has been used. Accuracy achieved by KNN is 85% and by SVM is 80.92% (Fig. 4).

Nidhi and Kakkar [12] designed a model to classifying the diabetic patient whether they have tested positive and negative. The diabetes dataset used for classification modeling was PIMA dataset which is further collected by the National Institute of Diabetes and Digestive and Kidney Diseases. The dataset contains 768 records having eight attributes. Four classification algorithms such as decision tree J48, PART, multilayer perceptron, and Naïve Bayes have been used and found that Naïve Bayes achieved highest accuracy 76.3% (Fig. 5).

Sai Prasanna Kumar Reddy et al. [13] proposed an effective diabetes prediction model using artificial intelligence. The dataset has been used for experiment is

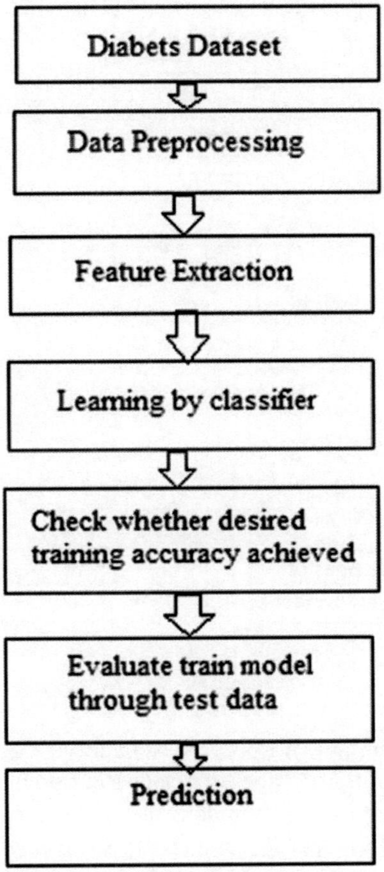

Fig. 4 Classification model to predict diabetes in patient

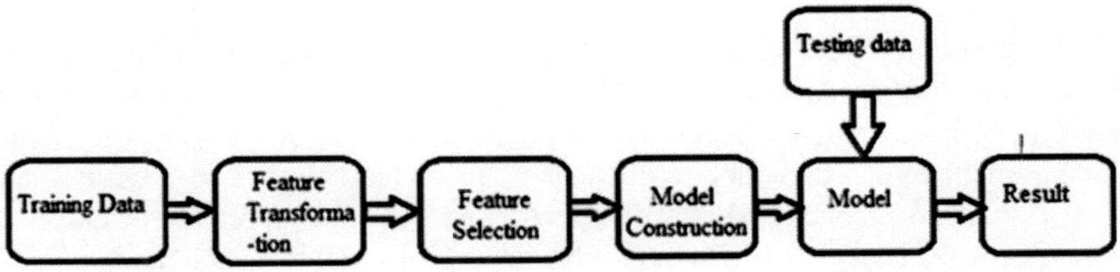

Fig. 5 Model to predict diabetes patient using machine learning

Pima Indians diabetes dataset which is a collection of 768 patients' health records. Convolution neural network technique has been used for disease prediction. It has a unique approach in regularization and convenes more complex patterns from existing hierarchical patterns. CNN algorithm has been achieved 84.4% accuracy (Fig. 6).

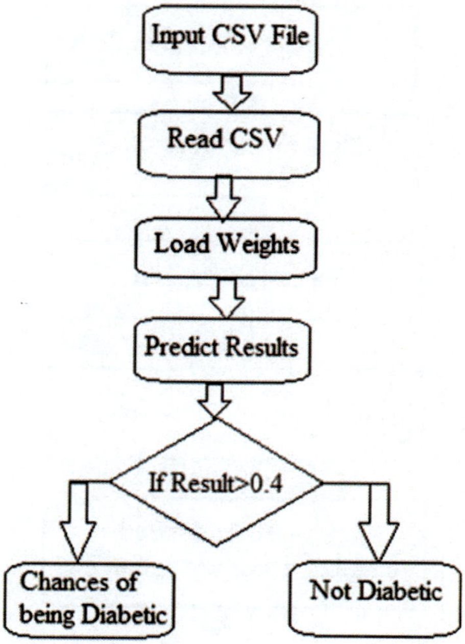

Fig. 6 A model to predict diabetes using artificial intelligence

Jakka and Vakula Rani [14] the purpose of their work is to assess the classifiers that can predict the highest probability of diabetes in patients with the greatest precision and accuracy. The work has been done using classification algorithms such as k-nearest neighbor (KNN), decision tree (DT), Naive Bayes (NB), support vector machine (SVM), logistic regression (LR) and random forest (RF) on Pima Indians diabetes dataset using nine attributes which are available online on UCI repository. It has been observed that logistic regression performs better than with 77.6% accuracy (Fig. 7).

Karun et al. [15] explained the application of different machine learning techniques for better prediction of diabetes. The performance of each model is calculated with or without feature extraction process. For the research purpose, Pima Indians diabetes dataset with 8 attribute and 1 predicted class has been used. Their final results work concluded that with feature extraction, support vector machine (SVM), *K*-means nearest neighbor (KNN), neural network (NN), and random decision forest (RDF) show better prediction accuracy, while logistic regression (LR), Naïve Bayes (NB), and decision tree (DT) show better prediction accuracy without feature extraction (Fig. 8).

Kaur and Kumari [16] developed a predictive analytic model for predict diabetic mellitus using predictive learning. Dataset of female patients with minimum twenty one year age of Pima Indians population has been taken from UCI machine learning repository. For prediction whether a patient is diabetic or not, five different algorithms have been used. The machine learning algorithms such as linear kernel and

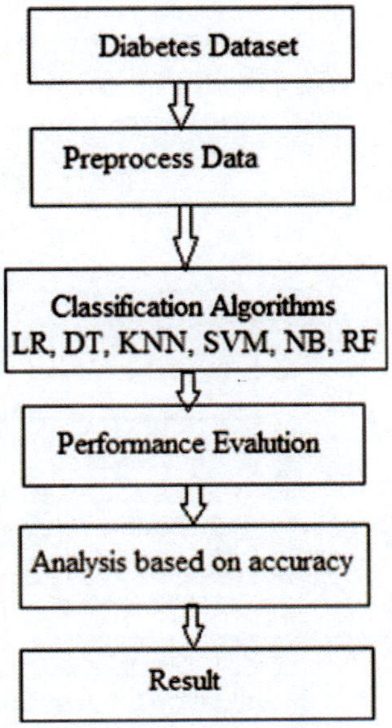

Fig. 7 Diabetes diagnosis using classification algorithms

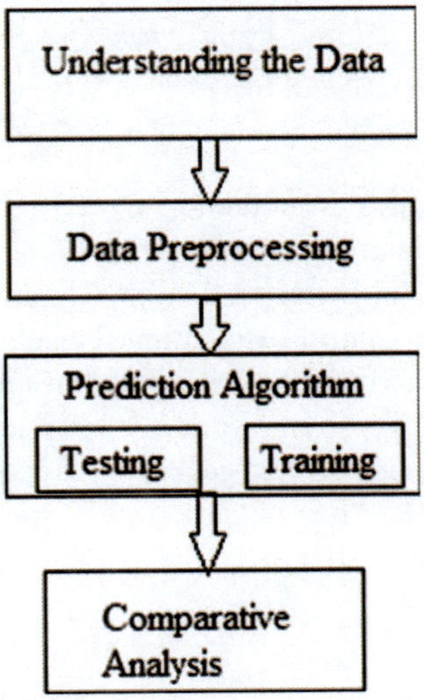

Fig. 8 Methodology used for diabetes prediction

radial basis function (RBF) kernel support vector machine (SVM), *k*-nearest neighbor (*k*NN), artificial neural network (ANN), and multifactor dimensionality reduction (MDR) have used in their predictive model. Final results concluded that kernel support vector machine (SVM linear) and kNN are two best classifiers to find whether patient is diabetic or not (Fig. 9).

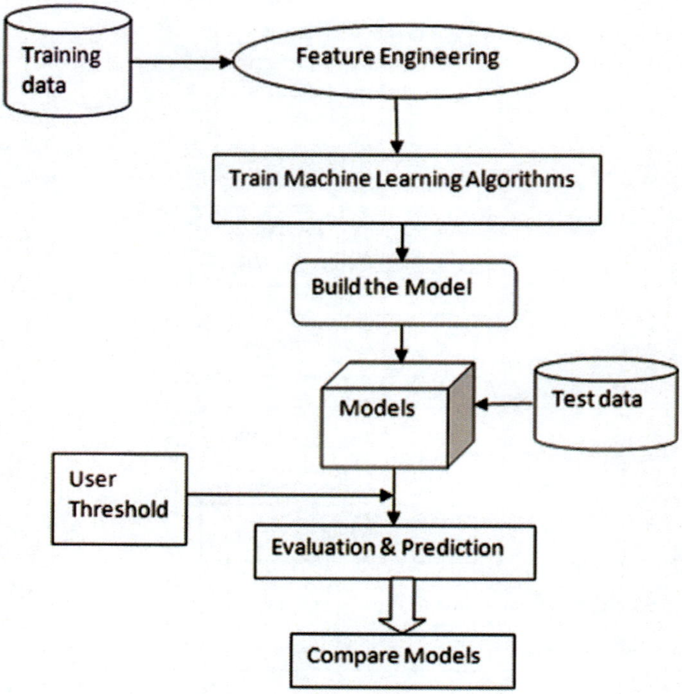

Fig. 9 Framework of evaluating predictive model

Priyadarshini [17] designed a prediction model to predict diabetic metabolic mellitus using ensemble learning algorithm. In their work, XGBoost gradient boosting algorithm is used for better prediction. The model is a regularized model and has been formalized to control over-fitting for better performance. The model is trained by ensemble method which is composed of multiple trained weak models to make one single model. The framework has been used to build the model was WinPython environment with XGBoost package. After the initial iteration, the accuracy of the model was 77%. After much iteration, the accuracy keeps increasing gradually from 77 to 90% (Fig. 10).

3 Learning Approaches Used for Prediction

Ensemble Learning

Ensemble learning is a supervised machine learning technique that can be used to improve the accuracy of a classifier. Machine learning built a predictive model using classifiers which can be used to classify new test samples, but it is difficult to achieve high accuracy with a single classifier. Also, it cannot be applied to all datasets to solve different problems. This leads to the generation of ensemble learning models.

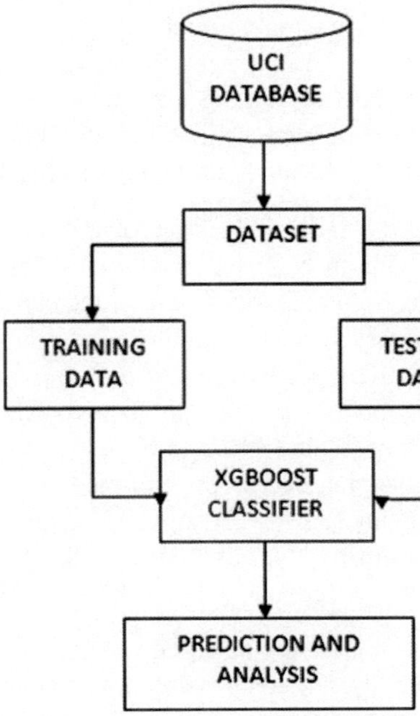

Fig. 10 Block diagram of diabetes prediction system using XGBoost

Ensemble learning models are widely used production models in the industry recently because of the combination of the stacked models which has the capability of learning the whole dataset well without over-fitting. Ensemble learning technique combines weak classifiers with strong classifiers to improve the efficiency of the weak classifier. Basically, ensemble learning is a tree-like approach where multiple decision trees are combining to make a strong model. Figure 11 represents different bagging and boosting techniques used for diabetes prediction in this work.

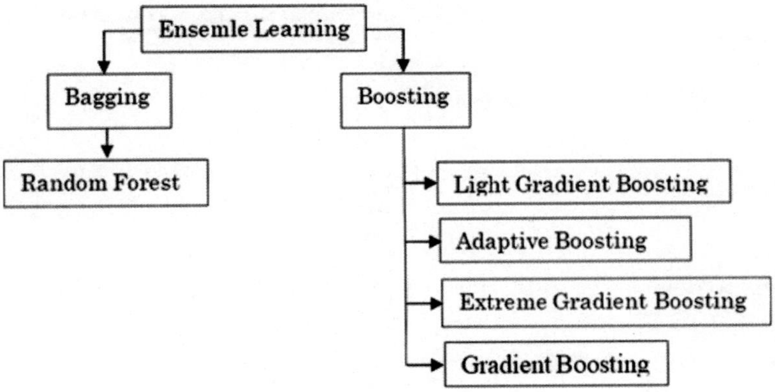

Fig. 11 Ensemble learning techniques

Boosting Boosting is an ensemble learning method combining multiple models in order to improve classification results in terms of stability and accuracy. In this method, the initial model is built by training data, then creating a second model which attempts to correct the errors from the first model. Models are added until the training set is predicted correctly or a maximum number of models are created.

Bagging Bagging is composed of two keywords: bootstrap and aggregation. In bootstrapping, data samples are chosen from the processed dataset, and the model is trained with each sample. Bagging randomly selects some patterns from the training set with replacement. The newly created training set will have the same number of patterns as the original training set with a few omissions and repetitions. The new training set is known as bootstrap replicate. The voting from each model is combined, and the classification result is selected based on majority voting or averaging. Research shows that bagging can be used to increase the performance of a weak classifier optimally. Bagging is generally applied by using decision tree models; however, bagging can be used with any type of models. In this research work, bagging is applied by using J4.8 decision tree model.

Random Forest Random forest is a bagging algorithm which uses a decision tree as its weak classifier. The random forest method is used for regression and classification to create multiple decision trees at training time and outputs the class that is the mode of the classification or regression of the individual trees. Decision tree is a graph that appears like a tree or decision model with the possible results. In the beginning process, a random sample of data is chosen up by each tree, and each tree is trained separately independent of each other. This level is at the row level. After this, the data is trained at the column level. Again each decision tree gets some set of columns. The results from each tree are taken, and majority ones are selected based on majority voting. Random forest method decreases the chance of over-fitting of the data by decreasing the variance since so many decision trees are in work. Random forest classifier builds several decision trees by applying bootstrap aggregation and incorporates them to get the best result. For a given data, $X = \{x_1, x_2, x_3, \ldots, x_I\}$ with responses $Y = \{y_1, y_2, y_3, \ldots, y_I\}$ which repeats the bagging from $i = 1$ to I. I represent total samples of the dataset. The unseen sample x' is made by averaging the predictions.

$\sum_{i=1}^{I} F_i(x')$ From every individual tree on:

$$j = \frac{1}{I}\sum_{i=1}^{I} F_i(x') \tag{1}$$

The uncertainty of prediction on these trees is made through its standard deviation "σ"

$$\sigma = \sqrt{\frac{\sum_{i=1}^{I}(F_i(x') - \widehat{F})^2}{I-1}} \tag{2}$$

AdaBoost Adaptive boosting (AdaBoost) initially assigns equal weights to each training instance. It uses multiple weak models and assigns higher weights to those observations for which were misclassified by model. As it uses multiple weak models, it combines the result of the decision boundaries obtained during multiple iterations, the accuracy of the misclassified instances is increased, and hence, the accuracy of the overall model is also improved. The weak models are evaluated using the error rate as given in (3).

$$\varepsilon_t = pr_{i\sim Dt}\left\lfloor ht_{(x_i)} \neq y_i \right\rfloor \equiv \sum_{t:ht_{(x_i)}\neq y_i} Dt(x_i) \tag{3}$$

where ε_t is the weighted error estimate, $pr_{i\sim Dt}$ is the probability of the random example i to the distribution Dt, ht are the hypotheses of the weak learner, x_i is the training observation, y_i is the target variable, t is the iteration number. The loss function error is 1 if the prediction is incorrect and 0 if the prediction is correct.

Gradient Boost Gradient boost (GB) is a boosting technique which sequentially creates new models from weak models with the idea that each new model can minimize the loss function. This loss function is calculated by gradient descent method. With the use of the loss function, each new model fits more accurately with the observations, and thus, the overall accuracy of the model is improved. However, minimization of loss function needs to be stopped; otherwise, the model will move toward over-fitting. The stopping criteria can be a threshold on the accuracy of predictions or reached maximum number of iterations.

Gradient boosting algorithm:

- Initialize $F_0(x) = \arg\min\rho \sum_{i=1}^{I} L(y_i, \rho)$
- For $t = 1$ to T do: //t be the number of iterations from 1 to T
- Step 1. Compute the negative gradient

$$\hat{y}_i = -\left[\frac{\partial L(y_i, F(x_i))}{\partial F(x_i)}\right]$$

- Step 2. Fit a model

$$\alpha_t = \arg\min_{\alpha,\beta} \sum_{i=1}^{I} \left[\hat{y} - \beta h(x_i; \alpha_t)\right]^2$$

- Step 3. Choose a gradient descent step size as

$$\rho_t = \arg\min_{\alpha,\beta} \sum_{i=1}^{I} L(y_i, F_{t-1}(x_i) + \rho h(x_i; \alpha))$$

- Step 4. Update the estimation of $F_t(x_i)$

$$F_t(x_i) = F_{t-1}(x_i) + \rho_t h(x; \alpha_i)$$

- End for loop
- Output the final regression function $F_t(x_i)$

ρ be the parameter chosen for alignment accuracy residual.

Light Gradient Boost (LG Boost) Light gradient boost is a gradient boosting technique that uses tree-based learning algorithms. LG Boost grows tree vertically means trees grows leaf wise While other algorithms grow trees level wise. The term "light" refers to the high speed which means large amounts of data. It can handle large amounts of data and take less memory to process the data. It uses histogram-based algorithm where each continuous feature is bucked into discrete bins. Now in order to compute the split, it needs to iterate the number of bins instead of number of points. LG Boost provides a variety of parameters setting and adjusts the different parameters to get the optimal model. The objective function is shown in Eqs. (4) and (5).

$$y^t = y^{t+1} + F_t(x_i) \tag{4}$$

$$\text{obj}^{(t)} = \sum_{i=1}^{I} L\left(y_i, y_i^t\right) + \sum_{i=1}^{I} \omega(F_i) \tag{5}$$

$\omega(F_t)$ is a regular term, F_i is decision tree, "L" be the loss function, I be the total samples of the data, t be the number of iterations and i be the sample from 1 to I.

Extreme Gradient Boost (XGBoost) XGBoost is an optimized version of a gradient boosting tree which creates a decision tree in a sequential manner. XGBoost is widely used for its performance in modeling new attributes and classification labels. This algorithm is identified as a helpful approach to optimize the gradient boosting algorithm by handling missing values, eliminating over-fitting issues using parallel processing. So the system optimization is achieved by implementing parallelization. The XGBoost algorithm also supports L2 regularization to obtain its objective function.

$$\begin{aligned}&\text{Objective function} = \text{training loss} + \text{regularizationa}\\&\text{i.e., obj} = L + \theta\end{aligned} \tag{6}$$

where L is training loss and θ is regularization.

In XGBoost, the objective function is optimized by the gradient descent.

$$\text{obj}^{(t)} = \sum_{i=1}^{l}\left[g_i F_t(x_i) + \frac{1}{2} g_i F_t^2(x_i) + \theta(F_t)\right] \tag{7}$$

4 Materials and Methods

Dataset Description The Pima Indian diabetes dataset is downloaded from online UCI machine repository which is for pregnant women aged 21 and above. This dataset includes a total of 768 instances with nine numerical attributes and one Boolean predictive class. Out of 768 instances, 500 instances belong to class "True" which indicates that diabetic patient and 268 instances belong to class "False" means non-diabetic patient. The description the dataset is shown in Table 1.

Proposed Model The aim of this research work is to enhance performance of ensemble model with feature extraction to predict whether a woman is diabetic patient or not. For experimental purpose, Pima Indian diabetes dataset is taken from online UCI machine learning repository [18]. There are various steps involved to design a predictive model. These steps are explained below, and diagram of proposed model is depicted in Fig. 12.

Step 1: Data Collection Data is collected from the UCI machine repository. Figures 13, 14, and 15 show description of the dataset and nature of the attribute.

Step 2: Data Preprocessing Before training and testing the model, data must be filtered and cleaned. Extracting information from raw dataset is very difficult. Raw

Table 1 Feature information of the diabetes dataset

Attribute	Description	Range
num_Pregnant	Frequency of being pregnant	0–17
plas_Glucose	Oral glucose tolerance test	0–199
diastolic_bp	Diastolic blood pressure	0–122
Thickness	Thickness	0–99
Insulin	2 h serum insulin	0–846
BMI	Body mass index	0–67.1
Dia_pedigree	Diabetes pedigree	0.078–2.42
Age	Age	21–81
Skin	Skin	0–99
Target	Positive and negative	True–False

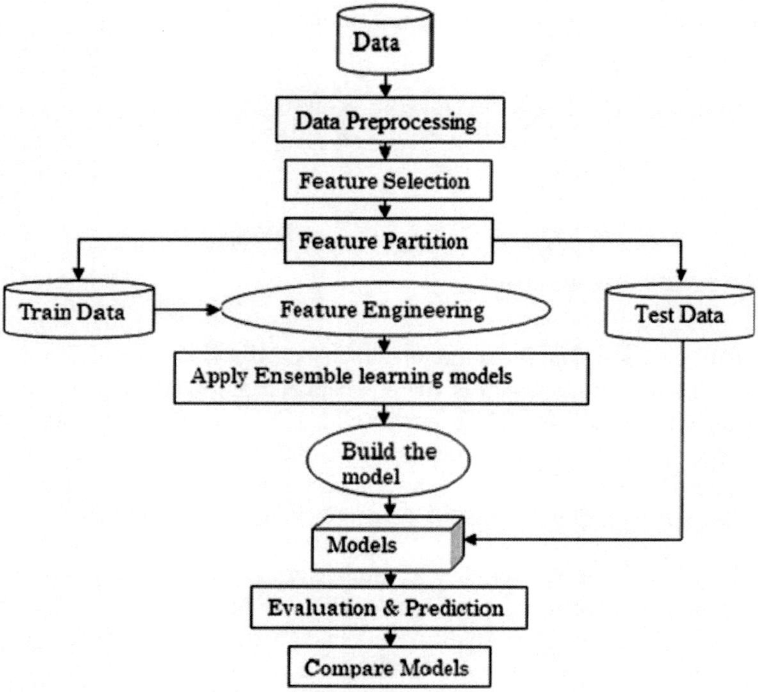

Fig. 12 Diabetes prediction model based on ensemble learning

	num_preg	glucose_conc	diastolic_bp	thickness	insulin	bmi	diab_pred	age	skin	target
0	6	148	72	35.0	NaN	33.6	0.627	50	1.3790	True
1	1	85	66	29.0	0.0	26.6	0.351	31	1.1426	False
2	8	183	64	0.0	0.0	23.3	0.672	32	0.0000	True
3	1	89	66	23.0	94.0	28.1	0.167	21	0.9062	False
4	0	137	40	35.0	168.0	43.1	2.288	33	1.3790	True

Fig. 13 Sample view of data

	num_preg	glucose_conc	diastolic_bp	thickness	insulin	bmi	diab_pred	age	skin
count	768.000000	768.000000	768.000000	767.000000	766.000000	768.000000	768.000000	768.000000	767.000000
mean	3.845052	120.894531	69.105469	20.563233	80.007833	31.992578	0.471876	33.240885	0.810191
std	3.369578	31.972618	19.355807	15.945349	115.322203	7.884160	0.331329	11.760232	0.628247
min	0.000000	0.000000	0.000000	0.000000	0.000000	0.000000	0.078000	21.000000	0.000000
25%	1.000000	99.000000	62.000000	0.000000	0.000000	27.300000	0.243750	24.000000	0.000000
50%	3.000000	117.000000	72.000000	23.000000	34.000000	32.000000	0.372500	29.000000	0.906200
75%	6.000000	140.250000	80.000000	32.000000	127.750000	36.600000	0.626250	41.000000	1.260800
max	17.000000	199.000000	122.000000	99.000000	846.000000	67.100000	2.420000	81.000000	3.900600

Fig. 14 Description of dataset

```
<class 'pandas.core.frame.DataFrame'>
RangeIndex: 768 entries, 0 to 767
Data columns (total 10 columns):
num_preg        768 non-null int64
glucose_conc    768 non-null int64
diastolic_bp    768 non-null int64
thickness       767 non-null float64
insulin         766 non-null float64
bmi             768 non-null float64
diab_pred       768 non-null float64
age             768 non-null int64
skin            767 non-null float64
target          768 non-null bool
dtypes: bool(1), float64(5), int64(4)
memory usage: 54.9 KB
```

Fig. 15 Dataset-type information

data in unstructured format may contain null value, irrelevant and inconsistent information which may not be useful for pattern prediction. The idea of preprocessing is to improve and augment the reliability of the chosen data. The cleansing process removes noisy value and missing value. In this work, data missing values are replaced with the mean value of the attribute. In the given dataset, glucose_conc, diastolic_bp, BMI, thickness, and insulin contain missing. So these missing values are replaced with these attribute mean values (Fig. 16).

Step 3: Feature Selection Feature selection is the process of selecting out the most relevant features from given dataset and improve the performance of model. It enables an algorithm to train faster and reduce the complexity of a model which makes it easier to interpret. A feature selection process also reduces over-fitting and improve

```
rows in dataframe 768
rows missing glucose_conc: 5
rows missing diastolic_bp: 35
rows missing thickness: 226
rows missing insulin: 372
rows missing bmi: 11
rows missing diab_pred: 0
rows missing age: 0
```

Fig. 16 Dataset contains missing values

```
Training feature shape: (614, 8)
Testing feature shape: (154, 8)
Training Label shape: (614,)
Testing Label shape: (154,)
```

Fig. 17 Training and testing feature shapes

the performance of a model by choosing is the right subset of attributes. In this work, correlation-based feature selection (CFS) method is used to get rid of redundant features. It is a statistical approach which provides a correlation score. Correlation score infers that how much linear dependency exists between two features. Higher correlation score between two features infers more linear dependency between the features, and low correlation score infers low liner dependency between the features. Therefore, low correlated features are eventually retained in the feature set, and highly correlated features are deleted from the feature set to reduce the redundancy and dependency among the features.

Step 4: Feature Partition Feature partition is a process of splitting data into train sets and test sets. The training set partition is used to build the model, and test set partition is used to validate the model or to test the model. In this research work, the dataset is divided into two parts (80 and 20%, training/testing) to avoid any bias in training and testing. Of the data, 80% data is used to train the ensemble model, and the remaining 20% data is used for testing the performance of the model. The idea of partition of data into 80:20 ratio is that more training data is good thing because it makes the prediction model better, while more test data makes the error estimation more accurately. The shape of training and testing feature sets is depicted in Fig. 17.

Step 5: Apply Ensemble Learning Models Five ensemble learning models have been run over the preprocessed dataset, and appropriate results have been shown in next section.

Step 6: Model Evaluation After model creation, final model is evaluated based on some parameters. The evaluation of the model is performed with the confusion matrix. Totally, four outcomes are generated by confusion matrix, namely TP (True Positive), TN (True Negative), FP (False Positive), and FN (False Negative). The expressions to calculate Accuracy are provided in Eq. (8).

$$\text{Accuracy} = (\text{TN} + \text{TP})/(\text{TN} + \text{TP} + \text{FN} + \text{FP}) \tag{8}$$

The expressions to calculate precision, recall, and $F1$-score are provided in Eqs. (9), (10), and (11). Precision provides a measure of how accurate a model is in predicting the actual positives out of the total positives. Recall provides the number of actual positives captured by a model by classifying these as True Positive.

$$\text{Precision} = \text{TP}/\text{TP} + \text{FP} \times 100 \tag{9}$$

$$\text{Recall} = \text{TP}/\text{TP} + \text{FN} \times 100 \quad (10)$$

$$F1\text{-Score} = 2 * (\text{Precision} * \text{Recall})/\text{Precision} + \text{Recall} \quad (11)$$

where TP—True Positive, FP—False Positive, FN—False Negative.

5 Results and Discussion

To diagnose diabetes for Pima Indian population, performances of all the five different models are evaluated upon parameters like accuracy, precision, recall, and $F1$-score in Table 3. In order to avoid problem of over-fitting and under-fitting, data is divided into 80:20 ratio. Accuracy defines how often our model is correct to diagnosis a diabetic patient. Table 2 represents the confusion matrix of used ensemble learning models using 80-20-sized dataset.

Table 3 is observed that first random forest (RF) with accuracy 82.46% and next Naive XGBoost with 81.16 showing the maximum accuracy, and gradient boost is showing minimum accuracy of 77.92%. Moreover, the LGBoost and ADA Boost classifiers also perform same achieving the overall performance of 80.51%. The accuracy comparison of five ensemble learning model to classify diabetic is depicted in Fig. 18.

Table 2 Confusion matrix of used model with 80-20 split dataset

Model	Train				Test			
	TP	TN	FP	FN	TP	TN	FP	FN
RF	393	0	0	221	95	15	12	32
ADA	351	76	42	145	93	16	14	31
LGBM	392	1	1	220	90	13	17	34
XGBM	363	46	30	175	92	14	15	33
GBM	393	0	0	221	87	14	20	23

Table 3 Results of different models

Method	Accuracy %	Recall	Precision	$F1$-score	Misclassification rate	ROC
Random forest	82.46	74.8	88.7	81.15	17.5	78.43
AdaBoost	80.51	75	86.9	80.51	19.4	76.4
Light gradient boost	80.51	72.5	84.1	77.87	19.48	78.22
XGBoost	81.16	73.6	85.9	79.27	18.8	78.09
Gradient boost	77.92	72.5	81.3	76.66	22.07	75.76

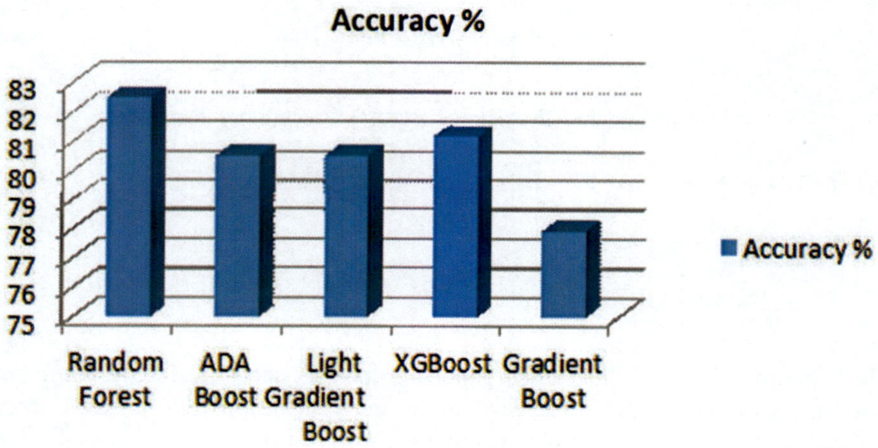

Fig. 18 Accuracy comparison of models

Precision has been used to decide model ability to provide correctly positive predictions of diabetes. Recall learns the proportion of actual positive cases of diabetes which are correctly recognized by the used model. $F1$-score can be calculated to by the weighted mean value of precision and recall. The classification accuracy, misclassification rate, precision, recall, error rate, and $F1$ score are used to measure the performance of the models and are presented in the Table 2. Figures 18 and 20 depict the comparison of accuracy and misclassification rate of the used models. Figure 19 shows the comparison of precision, recall, and $F1$-score.

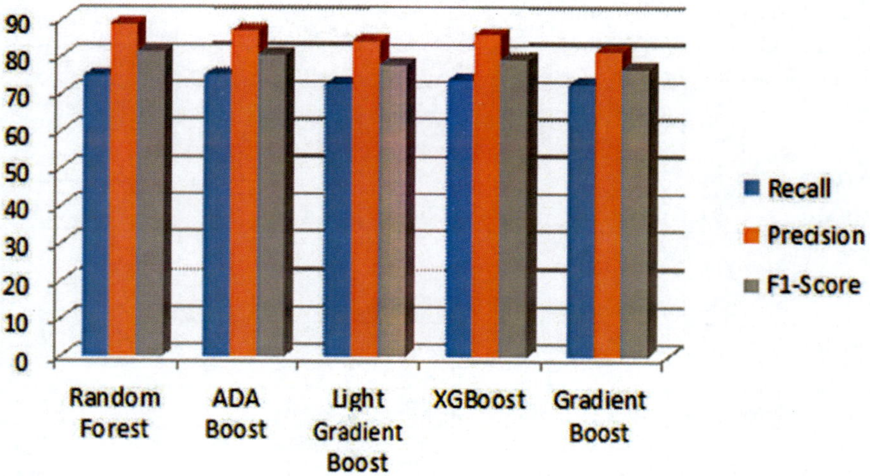

Fig. 19 Comparison of precision, recall and $F1$ score

Misclassification rate gives the information that a model is able to classify a correct class for prediction. Misclassification rate refers to the frequency of misclassified instances occurred during prediction. As the misclassification rate increases, performance of model decreases. The comparison of misclassification rate is depicted in Fig. 20.

ROC is plot of True Positive rate against False Positive rate as the threshold for assigning observations is varied to a particular class. Figure 21 shows ROC score of different classifiers used for the prediction. In Fig. 21, RF, light boost, XGBoost, AdaBoost, and gradient boost are found to be 78.73, 78.22, 78.09, 76.4, and 75.76 ROC score, respectively. So, from the above research studies, it can be said that on the basis of all the evaluated parameters, RF and XGBoost are two best models to find whether a patient is diabetic or not. Further, it can be seen that accuracy, precision, and $F1$-score of RF model are higher in comparison to other models, while recall value AdaBoost model is higher than RF model.

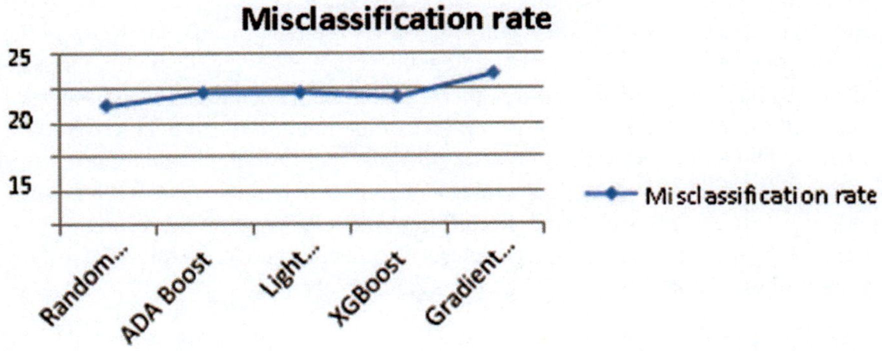

Fig. 20 Comparison of misclassification rate

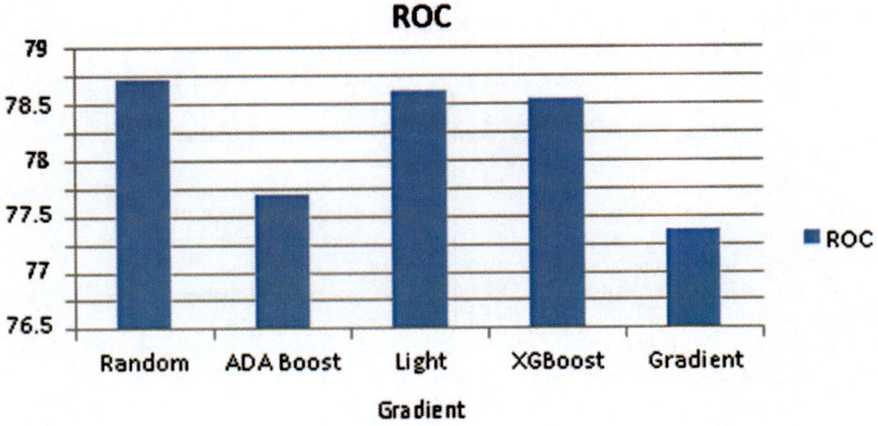

Fig. 21 ROC score of different algorithms

So, from the above research studies, it can be said that on the basis of all the evaluated parameters RF and XGBoost are two best models to find whether a patient is diabetic or not. Further, it can be seen that accuracy, precision, and $F1$-score of RF model is higher in comparison to other models while recall value AdaBoost model is higher than RF model.

6 Conclusion

This study provides a fair and unbiased performance analysis of the ensemble learning model for diabetes prediction. The study also investigates the performance of newly developed light gradient boost classifiers and how well they perform with feature selection. All the models are evaluated on the basis of different parameters—accuracy, misclassification rate, ROC, recall, precision, and F1-score. The experimental results suggested that all the models achieve good results; random forest model provides best accuracy of 82.46% with 88.7%. On the other hand, AdaBoost model provided best recall with 75%. The gradient boost model having maximum misclassification rate 22.07 which indicates that gradient boost achieves lowest accuracy among all used model. Finally, it is concluded that random forest achieved highest accuracy. The reduction of feature set implies the real-time implementation as this is directly connected with the computational complexity of the system. Our future work will focus on integration of two different models for better prediction results. Then, testing these models with large dataset having minimum or no missing attribute values will give more insights and better prediction accuracy.

References

1. M. Alehegn, R. Joshi, P. Mulay, Analysis and prediction of diabetes mellitus using machine learning algorithm. Int. J. Pure Appl. Math. **118**(9), 871–878 (2018)
2. A.C. Jamgade, S.D. Zade, Disease prediction using machine learning. Int. J. Res. Eng. Technol. **06**(05) (2019)
3. N. Sneha, T. Gangil, Analysis of diabetes mellitus for early prediction using optimal features selection. J. Big Data, Article 13 (2019)
4. S. Park, D. Choi, M. Kim, W. Cha, C. Kim, I.C. Moon, Identifying prescription patterns with a topic model of diseases and medications. J. Biomed. Inf. **75**, 35–47 (2017)
5. N.A. Farooqual, Ritika, A. Tyagi, Prediction model for diabetes mellitus using machine learning technique. Int. J. Comput. Sci. Eng. **06**(03) (2018)
6. B. Tamilvanan, V. Murali Bhaskaran, An experimental study of diabetes prediction system using classification techniques. IOSR J. Comput. Eng. **19**(01), Version 04 (2017)
7. D.K. Harini, M. Natesh, Prediction of probability of disease based on symptoms using machine learning algorithm. Int. Res. J. Eng. Technol. **05**(05) (2018)
8. M. Shilpa, C. Nandini, M. Anushka, R. Niharika, P. Singh, P.R. Raj, Heart disease and diabetes diagnosis using predictive data mining. Int. Res. J. Eng. Technol. **05**(09) (2018)

9. M.A. Sarwar, N. Kamal, W. Hamid, M.A. Shah, Prediction of diabetes using machine learning algorithms in healthcare, in *24th International Conference on Automation and Computing (ICAC)*, 8748992, New Castle upon Tyne, United Kingdom (2018)
10. C. Kumar, N. Singh, J. Singh, Prediction of diabetes using data mining algorithm. Int. J. Res. Appl. Sci. Eng. Technol. **07**(02) (2019)
11. S. Bano, M.N.A. Khan, A framework to improve diabetes prediction using *k*-NN and SVM. Int. J. Comput. Sci. Inf. Secur. (IJCSIS) **14**(11), 3–10 (2016)
12. M.K. Nidhi, L. Kakkar, Classification of diabetes patient by using data mining techniques. Int. J. Res. Eng. Appl. Manage. **04**(05) (2018)
13. K. Sai Prasanna Kumar Reddy, G. Mohan Seshu, K. Akhil Reddy, P. Raja Rajeswari, An efficient intelligent diabetes disease prediction using AI techniques. Int. J. Recent Technol. Eng. **08**(04) (2019)
14. A. Jakka, J. Vakula Rani, Performance evaluation of machine learning models for diabetes prediction. Int. J. Innov. Technol. Exploring Eng. **08**(11) (2019)
15. S. Karun, A. Raj, G. Attigeri, Comparative analysis of prediction algorithms for diabetes, in *International Conference on Computer, Communication and Computational Sciences, IC4S*, vol. 759, Kathu, Thailand (2019)
16. H. Kaur, V. Kumari, Predictive modelling and analytics for diabetes using a machine learning approach. Appl. Comput. Inf. 329807137 (2018)
17. P. Priyadarshini, Prediction of diabetes mellitus using XGBoost Gradient Boosting. Int. J. Adv. Sci. Eng. Technol. **05**(04) (2017)
18. Pima Indians Diabetes Dataset. https://archive.ics.uci.edu/ml/datasets/Pima+Indians+Diabetes+Dataset (2017)

Improve K-Mean Clustering Algorithm in Large-Scale Data for Accuracy Improvement

Maulik Dhamecha

Abstract If you want to classify object and you do not have any specific labels, then how can you classify that objects? Clustering is the best way to classify this kind of object. Whenever we are talking about large-scale data relating to variety of fields, beginning of new gen techniques for concentrate collection of data was resulted. Typical database query processing is not able to find exact information from large number of data, and hence, clustering is important analysis method for large-scale data. Among these many clustering algorithms, k-means and k-medoids are superiorly utilized for large-scale database. Initially, select centroids in k-mean algorithm and medoids for k-medoids algorithm for batter quality of the resulting clusters. Key highlight of this algorithm is that, whenever number of iterations increase, it will also increase the computational time. Proposed k-means algorithm initially finds the starting level centroids "k" as per requirements and returns with better compare to previous, effective and very stabilized cluster. Here proposed algorithm consumes very less time for execution as it segregates unnecessary computational distance because it uses the previous iteration of cycle. On the basis on initial centroids, here proposed algorithm systematically selects initial k-medoids. And thus, it will produce stable clusters for efficiency improvement.

Keywords Cluster analysis · Centroid · Medoid · Data sets · Euclidian distance-means · K-medoids · Partitioning method

1 Introduction

"A cluster is a collection of data objects that are similar to one another within the same cluster and are dissimilar to the objects in other clusters" [1]. The main requirements that a clustering algorithm should satisfy are [2]:

M. Dhamecha (✉)
Department of Computer Engineering, VVP Engineering College, Rajkot, Gujarat, India
e-mail: mvdhamecha@gmail.com

D. Bhattacharyya and N. Thirupathi Rao (eds.), *Machine Intelligence and Soft Computing*, Advances in Intelligent Systems and Computing 1280,
https://doi.org/10.1007/978-981-15-9516-5_5

(1) Domain knowledge
(2) Scalability for large database
(3) Ace of use and interpretability
(4) High dimensionality
(5) Clean noisy data
(6) Insensitivity to order of complex input
(7) Deal with outliers' noise and outliers.

Several problems with clustering are also viewed. Some problems are listed below [3, 4]:

(1) For all available requirements, current journal techniques of clustering will not be so effective.
(2) Because of time complexity issue, there may be problem to deal multidimensional data and many data items.
(3) Whenever we talk about distance-based clustering, the method's effectiveness depends on the criteria of "distance."
(4) For multidimensional spaces, suppose regular distance criteria was not defined, we must define it and that is not simple task.

There are several clustering algorithms, which fail to handle the main requirements. The partition method is widely used and is one of the most efficient clustering methods which can handle most of the issues related to clustering. [5] The clusters generated by this method are sensitive to noise; outliers and accuracy depend on initial centroids and medoids. However, the complexity of this method increases with size of data objects [5, 6]. For this work, a new clustering algorithm is proposed, which is based on obtaining the initial centroids and efficient enhanced k-mean clustering algorithm. Generally, initial centroids cannot generate a good and sustained cluster always and that is why in standard partition method, we have to randomly select "k" objects from the given population [7, 8]. To obtain better clustering, first stage is to find initial consistent centroids among the data in proposed algorithm. A simple idea of second stage is to enhance the efficiency of algorithm, with no miss of clustering quality. To measure total time to be taken and distance calculation's magnitude, we proposed the improved algorithm [9, 10].

2 Preprocessing of Data

In recent time, advance technology creates data collection easier and faster. With the use of that, it handles large as well as more complex attributes with many dimensions and facts. Now, the attributes of datasets behave like varied as they are larger, need to adopt quality based and robust clustering algorithms [11, 12]. Recently, all traditional clustering methods are adequate for all the dimensions of the attribute's input to learn about relevant object which was described into the database. Many of the attributes are frequently extraneous whenever we are talking about high-dimensional data.

These extraneous attributes can generate confusion in clustering as they hiding noisy data in clusters. Now whenever we are talking about completely masking of the clusters, we have to deal with high dimensions of attributes which are common for all available attributes in available datasets which are situated as equidistant from each attributes. As we know the real-world data, the actual data in organization are tend to be noisy, inconsistent and incomplete. And thus with the use of data preprocessing, we can improve the standard of the result [13–15]. This preprocessing technique will help to improve the accuracy and efficiency in result which we can find through mining techniques. If we want quality decisions, we have to apply preprocessing on quality data. Finding data anomalies, redefining them in early stages and reducing data for the analyzing can generate huge effect on decision-making system. Mostly, preprocessing of the dataset is application dependent [16, 17].

2.1 Step of Preprocessing

Now, as we already know the importance of preprocessing which helps to improve the efficiency, scalability and accuracy of the clustering, for that following are some steps which may be applicable to the attributes [12, 18–20]. Data cleaning: As we all know that the real-world data tend to be noisy, that is why we have to apply data cleaning method at initial level to clear the data. This step makes data more suitable for future processing. Different techniques are applied to handle the different types of attributes in database.

Handling the missing values of continuous attribute: We can handle it by deleting that missing value, by putting mean value, by putting median value, by putting mode value or with the user-specified value. Handling the missing values of nominal attribute: For nominal attribute, first we find all distinct values for a given attribute as value set, and then, missing value of it is replaced by most frequent item from those attributes.

Data transformation: Basically, we have to transform type of the data and consolidated into one common format of data. Normalization: Generally, we can apply normalization for distance measurement to improve the accuracy and efficiency of data mining methods like clustering. Different normalization techniques are applied to handle the different types of attributes. For this attribute consideration, the following method of normalization is used. Normalization with decimal scaling: As name says, this technique depends on moving parameters of the decimal point of values which are relevant to their data. The number of moves of the decimal points depends on the max value of attribute's absolute value.

Dimensionality reduction: In real-world datasets which come for analysis that may have thousands of data, in that database many attributes may be irrelevant or redundant with each other. Keeping irrelevant attribute slows down the process of clustering. Such a way this technique reduces the size of database by eliminating such attributes from the database.

Discretization: With dividing the data range of the given attribute into some specific data intervals, discretization is used to reduce the value size for available continuous attribute range. Interval data labels may generate in place of actual data attributes.

3 Categorization of Major Clustering Methods

Generally, clustering techniques are available as following way [19–22].

Partitioning technique: For n objects data tuples or any database, k partitions of the data constructed in partitioning technique, where each partition known as a cluster.

Hierarchical technique: A hierarchical segregation of data objects is created in hierarchical technique. On the basis of data segregation, there are two types of this method—agglomerative method and divisive method. The agglomerative method is much known as the bottom-up method. In this, each and every object forms the individual group of object. In this, closed objects and attributes are merged into one another group which have similar objects until all the other available groups can also generate the group or they can reach up to the termination condition. The divisive method is much known as the top-down method. In this, all the available objects lie in one common group in each and every iteration cycle, and every time it splits into other small group until a termination condition initiated.

Density-based technique: Generally, partitioning method of clustering is based on the distance between the dataset attributes. With the use of this technique, we can find only spherical-shaped clusters. With the use of this technique, we also face the difficulty whenever we deal with the arbitrary shapes of clusters. Many clustering algorithms also can be deal with the notion of density of attributes. General idea behind this is to steady growth of the given cluster up to the density in the "neighborhood" exceeds the threshold level.

4 Applied Concept

Get initial centroids:

Let's take U be a data point set. We can find initial centroids by following below mention steps [23, 24]

1. For U, find out the distances of each and every data point.
2. Generate set of data points A1 in which shortest distance of set of data points are available, and delete those data points from U.
3. For next, find every other data point's distance from A1, and find that which data point is nearer to A1. Also delete those data points from U.
4. Decide any threshold for termination the process.

5. Do again step 3 until you reached up to threshold level.
6. At the end of this process, after taking average of every data point, we can find the initial centroid.

Let's take set of data points "U" in that two-dimensional vector data points are available, in which suppose that "U" contains 12 data. Now let's take that "U" is divided into two classes. In this scenario, one thing is very obvious that the distance between "a" and "b" is shortest for given data points. So, now we have to remove "a" and "b" data points from "U" and create a new data point set "A1" which have these new data points in their set. Now let's take any data point "c" which is closest to "A1" data point set. At that time, we have to remove it from "U" and add these data points into "A1" dataset. Now, if the total data points which are available in "A1" are not more than 4, we have to generate new data point "d" which is also included into "A1" data point set. Continue with above process, and we can generate different set of data point "A2" which have "g", "h", "I" and "j". At the end, we can get two initial centroids which come from the averaging of available data points in "A1" data set and "A2" data set. Whatever centroids we find with above process are obviously reliable with the distribution system of data points of dataset [21, 22].

Similarity Here we can use Euclidean distance to find similarity. As per data, the distance between one vector "*X*" = "(*x*l, *x*2,..., *xn*)" and the other vector "*Y*" = "(*y*l, *y*2, ..., *yn*)" is described as below.

$$\mathrm{d}(X, Y) = \text{sqrt}\ ((x1 - y1) + (x2 - y2) + \cdots + (xn - yn))$$

Eliminate unnecessary distance computation.

If we want to create k-means as a powerful, especially whenever any dataset which have large number of clusters in their pan, we have to limit for unnecessary calculation steps for distance. For large amount of datasets, calculation cost is very high as distance between every data points is calculated in k-means algorithm [25, 26]. To improve the efficiency, it is better choice to take advantage of previous iteration result which is available from k-means algorithm. We can arrange every single cluster to that nearest distance cluster. We can calculate previous nearest cluster's distance in next iteration. Point lies in same cluster if distance of cluster and previous cluster is less or equal, and with this process, we can reduce the time for calculation of $k - 1$ cluster center's distance [27, 28].

5 Proposed Algorithm

5.1 Proposed Algorithm for k-means

Input:

$M = \{m1, m2, \ldots, mn\}$ // collection of n data items set.

k // Number of expected clusters from data.

Output:

A set of data points which have k clusters.

Steps:

Phase 1: Consider the initial centroids of the given clusters from the data:

The starting phase of centroids is considered systematically so they can produce clusters which have better accuracy in result.

Algorithm: Calculating for the initial centroids into the cluster.

Input:

$M = \{m1, m2,..., mn\}$ // Collection of n data items.
where K = Number of expected clusters from data.

Output:

A set of k primary centroids $Nm = \{N1, N2 \ldots Nk\}$ // set of centroids = k.

Step:

1. Define $a = 1$
2. Distance calculation of every single data point in dataset M.
3. Search the nearest pair from the set M for all available data points, and generate a dataset Xa of data points which contain these two types of data points. After that, remove it from main set M of data points.
4. Search the data point in set D that is closest to the Am data point set, and add it to data point set Am and delete it from data set point D.
5. Continue with step 4, and reach up to the data points in Am which becomes 0.75*(n/k) for data point set.
6. Whenever $a < k$ happen, then take $a = a + 1$, calculate from M data points from another pair of data point set which have smallest distance, generate separate set of data points Xa and after that delete it from main dataset M.
7. Now for every set Xa of data points, calculate the mean of the vectors of Xd and declare it as starting centroids of available data point set.

Phase 2: Allocate each data point of set to its relevant clusters:

From the first centroid, it finds the relative distance from every data points to create initial clusters. With the use of heuristic approach, these clusters were adjusted to improve efficiency.

Algorithm: Allocate data points to relevant clusters.

Input:

$M = \{m1, m2,..., mn\}$ // set of n data items
$N = \{n1, n2,......, nk\}$ // set of k centroids of cluster

Output:

X set of *k* available clusters.

Steps:

1. Calculate the possible distance between every data set point "*mi*" where ($1 \le i \le n$) with all available centroids "*nj*" where ($1 \le j \le k$) as *m*(*mi*, *mj*).
2. Search the nearest available centroid "*nj*" from the cluster and assign "*mi*" to cluster *j*.
3. Define cluster "Id[*i*] = *j*"; where "j:Id" of nearest cluster data points.
4. Define Closest_Dist[*i*] = *m*(*mi*, *nj*)
5. Regenerate the cluster's centroids for every cluster "*j*" where *j* is ($1 \le j \le k$).
6. Go to step-1
7. Finalized result for data point "*di*" from available datasets.

   7.1. Calculate centroid distance from the closest cluster.

   7.2. Keep data points into similar cluster when we find that distance is less compared to current distance
   Else

   7.2.1. Compute distance *m*(*mi*, *ni*), for each centroid where *cj*($1 \le j \le k$).

   End for

   7.2.2. For closest centroid *Nj*, define data point mi to the cluster.

   7.2.3. Define cluster Id[*i*] = *j* for data points

   7.2.4. Define closest_Dist[*i*] = *m* (*mi*, *nj*);

   End for

8. Regenerate centroid till every cluster *j* ($1 \le j \le k$).

5.2 Proposed Algorithm for k-medoids

Input: $M = \{m1, m2,\ldots,mn\}$ // set in cluster of *n* data points.

K = total number of expected cluster from the current data sets.

Output: X is a set of some k which is initial centroids $Nm = \{N1, N2, \ldots Nk\}$Steps:

Set $a = 1$.

1. For set M, calculate the distance for every data point.
2. Generate set of data points "*Xa*" where ($1 \le c \le k$), and search for the nearest pair of data set points which lie in set M.
3. Remove these similar data set points from the set M.
4. Search the nearest data set points in "*Xa*" which lies in M.
5. Remove these data points from *M* and add in *Xa*.
6. Continue with step 4 until the data points of *Xa* reached up to 0.75*(*n*/*k*).
7. Find another set of data points in set *M* if $a < k$, continue with step 4.
8. For generate the mean value of *Na*(($1 \le a \le k$) of data set points in "Xm" for every data set point of *Xa* ($1 \le a \le k$) and declare this value as initial centroids.

Algorithm

Input:

- Data points of P objects.
- Initial centroids $Na = \{N1, N2, \ldots Nk\}$ of k cluster.

Output:
Data point Set of k clusters from the current data sets.
Steps:

1. Find nearest centroids {N1,N2,...Nk} of data points for initial cluster.
2. Combined every data point with nearest medoid.
3. To compute total configuration cost, swap m and p of every medoid.
4. Find out lower cost configuration.
5. Continue with step 2 until not find any difference.

6 Conclusion

Whenever we are talking about large-scale database, first clustering algorithm which comes in our mind is "k-means." Whenever we are talking about standard "k-means" clustering algorithm, final cluster's accuracy depends on initial selection of centroid which is not given always good result. Here, the proposed "k-medoid algorithm" behaves as it working like "K-means clustering algorithm." In this proposed clustering algorithm, we can suggest systematic selection of the initial medoids. We can say that, depend on initial selection of medoids, the performance of the algorithm may fluctuate. And thus, we can say that this algorithm is efficient when compared to the existing k-medoid clustering algorithm.

Reference

1. D. Pi, X. Qin, Q. Wang, Fuzzy clustering algorithm based on tree for association rules. Int. J. Inf. Technol. **12**(3) (2006)
2. M. Dhamecha, A.G. Ganatra, C.K. Bhensdadiya, Comprehensive study of hierarchical clustering algorithm and comparison with different clustering algorithms, in *CiiT* (2011)
3. G. Godhani, M. Dhamecha, A study on movie recommendation system using parallel MapReduce technology. IJEDR (2017)
4. D. Vekariya, N. Limbasiya, A novel approach for semantic similarity measurement for high quality answer selection in question answering using deep learning methods, in *ICACCS* (2020)
5. N. Limbasiya, P. Agrawal, *Bidirectional Long Short-Term Memory-Based Spatio-Temporal in Community Question Answering* (Springer, 2020)
6. O. Beaumont, T. Lambert, L. Marchal, B. Thomas, Data-locality aware dynamic schedulers for independent tasks with replicated inputs, in *IEEE International Parallel and Distributed Processing Symposium Workshops* (2018)

7. M. Dhamecha, T. Patalia, Scheduling issue for dynamic load balancing of mapreduce in large scale data (big data). J. Xidian Univ. (2020)
8. M. Dhamecha, K. Dobaria, T. Patalia, A survey on recommendation system for bigdata using mapreduce technology (IEEE, 2019)
9. S. Garg, R.C. Jain, Variation of k-mean algorithm: a study for high dimensional large data sets. Inf. Technol. J. **5**(6), 1132–1135 (2006)
10. M. Dhamecha, T. Patalia, MapReduce Foundation of Big data with Hadoop environment, ELSEVIER—SSRN (2018)
11. A.M. Fahim, A.M. Salem, Efficient enhanced k-means clustering algorithm. J. Zhejiang Univ. Sci., 1626–1633 (2006)
12. F. Yuag, Z. HuiMeng, A New Algorithm to get initial centroid, in *Third International Conference on Machine Learning and Cybernetics*, Shanghai, 26–29 August 2004
13. J. MacQueen, Some method for classification and analysis of multi varite observation, University of California, Los Angeles, pp. 281–297 (2015).
14. M. Dhamecha, T. Patalia, Comparative study of dynamic load balancing algorithm in large scale data (Big data). IJAST (2020)
15. R. Xu, D. Wunsch, Survey of clustering Algorithm. IEEE Trans. Neural Netw. **16**(3) (2005)
16. D. Chandarana, M. Dhamecha, A survey for different approaches of outlier detection in data mining (IEEE, 2015)
17. K. Parmar, N. Limbasiya, M. Dhamecha, Feature based composite approach for sarcasm detection using MapReduce (IEEE, 2018)
18. L. Parsons, E. Haque, H. Liu, Subspace clustering for high dimensional data: a review. SIGKDD Explor. Newsletter **6**, 90–105 (2004)
19. Z. Huang, A fast clustering algorithm to cluster very large categorical data sets in data mining (2017)
20. M. Dhamecha, A. Ganatra, C.K. Bhensadadiya, Comprehensive study of hierarchical clustering algorithm and comparison with different clustering algorithms, in *CiiT* (2011)
21. A.N. Nandakumar, Y. Nandita, A survey on data mining algorithms on Apache Hadoop platform. Int. J. Emerg. Technol. Adv. Eng. (2014)
22. D. Miner, A. Shook, *MapReduce Design Patterns: Building Effective Algorithms and Analytics for Hadoop and Other Systems* (O'Reilly Media, Sebastopol, 2012).
23. Z. Matei, D. Borthakur, S.J. Sarma, K. Elmeleegy, Delay scheduling a simple technique for achieving locality and fairness in cluster scheduling, in *Proceedings of the 15th European Conference on Computer Systems* (2010)
24. M. Dhamecha, T. Patalia, Fundamental survey of map reduce in bigdata with Hadoop environment, in *Spinger—CCIS* (2018)
25. N. Limbasiya, P. Agrawal, *Semantic Textual Similarity and Factorization Machine Model for Retrieval of Question-Answering* (Springer, 2019)
26. I. Polato, R. Ré, A. Goldman, F. Kon, A comprehensive view of Hadoop research—a systematic literature review. J. Netw. Comput. Appl. (2014)
27. X. Bu, J. Rao, C.Z. Xu, Interference and locality-aware task scheduling for MapReduce applications in virtual clusters, *in International Symposium on High-Performance Parallel and Distributed Computing* (2013)
28. N. Thirupathi Rao, P. Aleemullah Khan, D. Bhattacharyya, *Prediction of Cricket Players Performance Using Machine Learning, LNNS,* vol. 105, pp. 155–162 (2020)

A Novel Approach to Predict Cardiovascular Diseases Using Machine Learning

Bhanu Prakash Doppala, Midhunchakkravarthy, and Debnath Bhattacharyya

Abstract Heart is considered to be one of the major organs which plays a vital role in the human body functioning. In recent days, due to many health hazards and living style, most of the people are suffering from heart-based diseases. Among many, one of the problems is cardiomegaly which is an enlargement of heart size. That can be affected by any age group person based on the health conditions and daily routine. Majorly, this can be noticed during the delivery time for a pregnant woman, athlete during his running habits, etc. Our proposed ensemble mechanism identified cardiovascular diseases with an accuracy of 85.24% which is better when compared with existing machine learning techniques.

Keywords Cardiovascular disease · Naive Bayes · SVM · Ensembles · Machine learning

B. P. Doppala (✉) · Midhunchakkravarthy
Department of Computer Science and Multimedia, Lincoln University College, Petaling Jaya, Malaysia
e-mail: bhanu.doppala@lincoln.edu.my

Midhunchakkravarthy
e-mail: midhun@lincoln.edu.my

D. Bhattacharyya
Department of Computer Science and Engineering, K L Deemed to be University, KLEF, Guntur 522502, India
e-mail: debnathb@kluniversity.in

D. Bhattacharyya and N. Thirupathi Rao (eds.), *Machine Intelligence and Soft Computing*, Advances in Intelligent Systems and Computing 1280,
https://doi.org/10.1007/978-981-15-9516-5_6

1 Introduction

Today's healthy life is entirely dependent on the efficient functioning of the heart. Heart-based diseases mainly cause due to the malfunction of the heart vessel system. Heart disease-based deaths stand as the leading number in the world from the past few years.

Usual risk factors like stress, abnormal weight and improper intake of food cause severe damage to the heart. In most of the cases, disease identification related to the heart may not be possible at the early stage due to the unavailability of decision-making systems. Many hospitals were majorly building their information management systems and inventory management systems for generating statistical data. A few hospitals are majorly concentrating on decision-making systems which are very limited in the number that can mainly focus on answering queries related to success ratio of operations. Middle age group patients are affected and suffering from the disease. For many medical practitioners, better classification of heart disease can be more valuable in saving patients life.

The cardiovascular diseases can occur due to improper functioning or sudden changes in the major parameters like as blood pressure, temperature, humidity and heartbeat value which is extremely susceptible and variant.

2 Literature Review

Baad et al. [1] conducted a study of various techniques administered by the analysts to forecast the heart disease coming from a patient's historical data. Chala Beyene et al. 2018 worked out with a couple of machine algorithms in anticipating the incident of cardiovascular disease with help from help angle maker, decision plant, Naïve Bayes, k-nearest neighbor as well as artificial neural network [2].

Corsetti et al. [3] presented Bayesian choices in treatment in addition to Bayesian system subgraphs (origin nodes to result from occasions) made use of as the particulars from which PAI-2-associated impact process was obtained. Goldman et al. introduced relevant information based on the analytical methods of the investigation, which is a population-based study [4].

Christalin et al. [5] exercised a technique called ensemble classification for improving the accuracy of fragile algorithms by integrating several classifiers. Tryouts of this particular device were executed, making use of a heart problem dataset. Leiherer et al. offered extra information on the organization of betatrophin along with its power to forecast cardio fatality in coronary clients. Besides, it displayed the functionality of betatrophin as a biomarker using c data [6].

Lizabeth et al. intended to use data mining category choices in approaches, particularly selection vegetation, Naïve Bayes and semantic network, alongside weighted affiliation Apriori formula as well as MAFIA algorithm in heart disease forecast [7]. Marimuthu et al. provided an insight into the existing algorithm, and it can offer an

Table 1 Existing algorithms performance on heart disease dataset

Name of the algorithm	Accuracy
Decision tree	77.80
Random forest	78.68
Logistic regression	81.96
Naïve Bayes	81.14
Gradient boosting	81.20
Extreme gradient boosting	79.50

overall ECG test summary of the present work [8]. Nyaga et al. [9] summarized available data on predictors, rates, treatment, aetiologies and the prevalence of mortality due to heart failure in SSA.

Prasad, Reddy et al. proposed a mechanism toward the prediction of heart problem utilizing artificial intelligence methods by recapping minority current investigates [10]. Wu et al. [11] proposed a new cardiovascular disease forecast system that incorporates all methods into one singular protocol, phoned hybridization. The result validates an accurate diagnosis by utilizing a mixed style coming from all approaches.

The main objective of this research work is to propose a new approach in machine learning algorithms to increase the accuracy rate towards predicting the heart disease. The following table depicts the performance of existing machine learning algorithms on heart disease dataset with 14 attributes. Performance of existing models is represented in Table 1.

We are living in an information world where, for every given period, we have been receiving a lot of data from different sources around the globe. In the same manner, information in the medical field is also getting generated very rapidly.

Data is available as a huge base, but we need to purify the required information with the help of different pre-processing techniques.

3 Proposed System

In this work, we have taken heart disease dataset from Cleveland repository with 14 attributed having 303 instances and pre-processed the collected data to eliminate the unwanted data with missing values. We have been providing the dataset to different classifiers and evaluating their performance, and in this process to increase the accuracy of weak classifiers, we are using ensembles classifiers by a voting method. Finally, proposed system produced results toward the increased accuracy (Fig. 1).

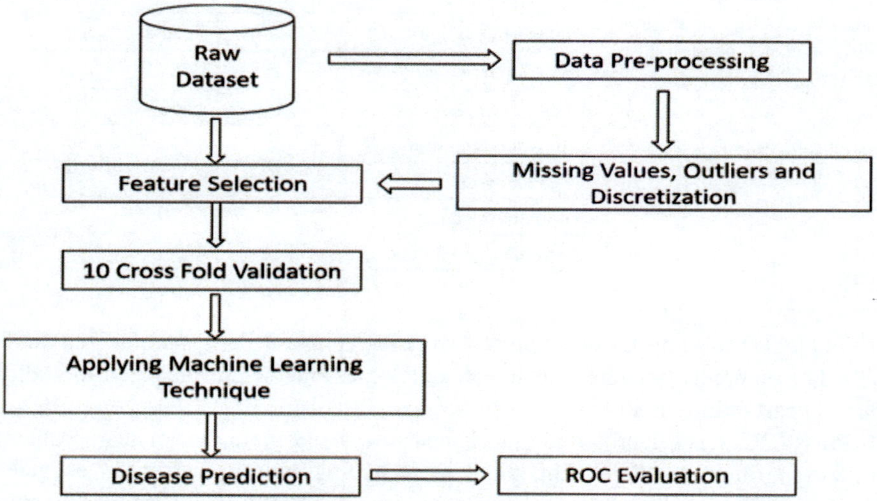

Fig. 1 Proposed system architecture

3.1 Dataset Description

We have taken heart disease UCI dataset from Cleveland repository for this paper. This catalog encloses 14 characteristics represented in Fig. 2.

Dataset was obtained from the UCI repository [12], which has 303 records with fourteen attributes. We used the WEKA tool for data analysis of this dataset. Here, attribute **target** is the class attribute that is used by this prediction system to identify heart disease.

3.2 Information Pre-processing

Processing of data plays a important key role for efficient representation of the data as well as performance of the classifier. Minimax scalar has been applied to the dataset which shifts the data in such a way all the features are between 0 and 1. Missing values feature will be deleted from the dataset.

3.3 Feature Selection

Feature assortment is actually the process where you instantly or even manually select those features which add very most to your prediction variable or even outcome in

S.No	Name	Description
1	age	age in years (28 - 77)
2	sex	0- female , 1- male
3	cp	chest pain type 1-typical angina, 2-atypical angina, 3-non-angina pain, 4-asymptomatic
4	trestbps	resting blood pressure (mm/Hg) (0 - 200)
5	chol	serum cholesterol (mg/dl) (0 - 603)
6	fbs	fasting blood sugar 0-false(< 120 mg/dl) , 1-true (> 120 mg/dl)
7	restecg	resting electrocardiographic results 0-normal , 1- having ST-T wave abnormality 2- showing probable or definite left ventricular hypertrophy by Estes' criteria
8	thalach	maximum heart rate achieved (60 - 202)
9	exang	exercise induced angina 0-no 1-yes
10	oldpeak	ST depression induced by exercise relative to rest (mm) (-2.6 – 6.2)
11	slope	the slope of the peak exercise ST segment 1-upsloping , 2-flat , 3-downsloping
12	ca	number of major vessels colored by fluoroscopy (0-3)
13	thal	3-normal 6-fixed defect 7-reversable defect
14	target	diagnosis of heart disease (angiographic disease status) – target attribute 1. - negative diagnosis (absence) 2. - 4 (from least serious most serious - presence)

Fig. 2 Feature information of the Cleveland dataset [12]

which you have an interest in [13]. Sample process of feature extraction is represented in Fig. 3.

3.4 Cross-Validation

We commonly divided available dataset into test and train collections when we are developing a learning model utilizing some information. The training set is actually utilized to teach the design, and the validation/test collection is utilized to verify it

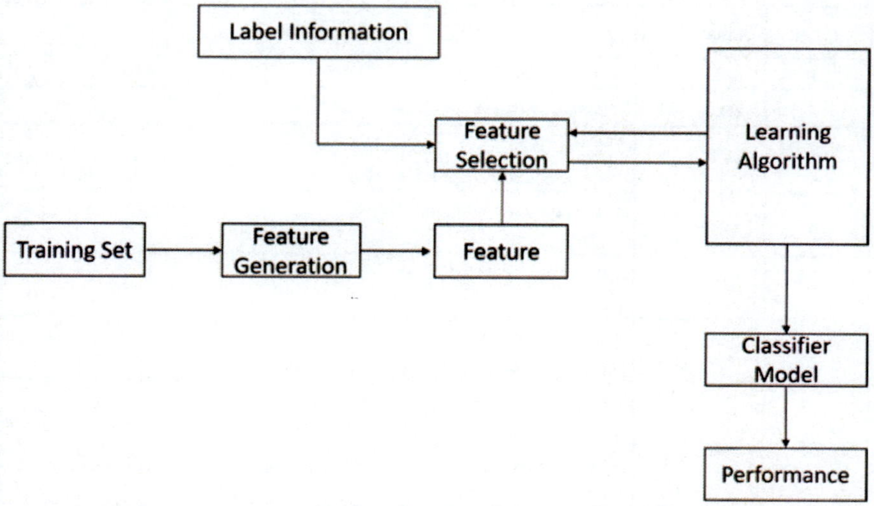

Fig. 3 Process of feature selection [14]

on data it has never ever observed just before. In general, we use 80–20% split, and occasionally with different market values like 70–30% or 90–10% will be considered. In cross-validation, we perform more than one split. Our experts can possibly do 3, 5, 10 or any kind of K lot of splits. Those divides are referred to as folds, and also there are actually numerous strategies our team can easily produce these folds with. Sample model of four folds is represented in Fig. 4.

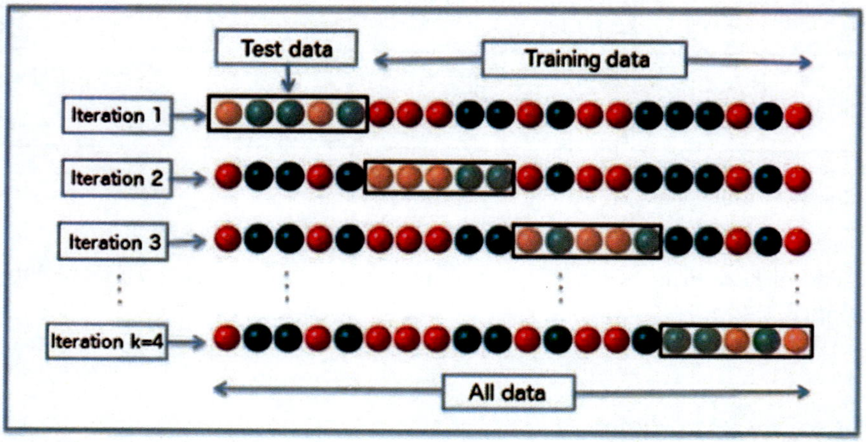

Fig. 4 Sample diagram for cross-validation with $k = 4$ [15]

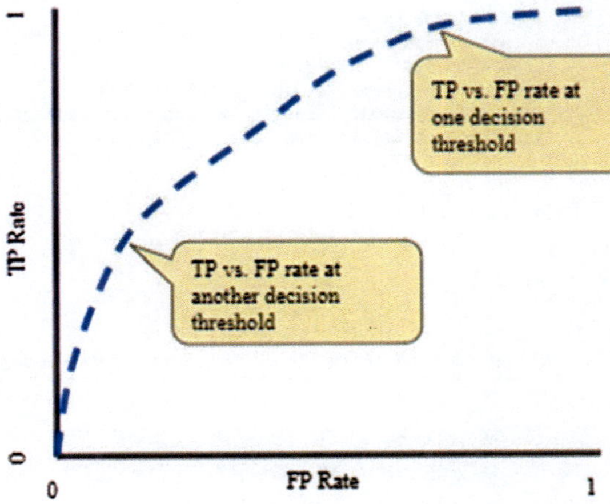

Fig. 5 Sample ROC curve [16]

3.5 *Classification*

It is simpler to recognize the correct class, and the outcome would be more precise than with the clustering method. With the end goal of the comparative investigation, we used three machine learning calculations. The diverse machine learning (ML) algorithms are decision tree, Gaussian Naive Bayesian and SVC. The motivation to pick these calculations depends on their popularity.

3.6 *Receiver Operating Characteristic (ROC) Curve Evaluation*

This graph presents the staging of a category style at all classification thresholds. This curve will be plotted using true and false positive rate. Figure 5 explains typical ROC curve.

4 Results

In this exploration work, determination of coronary illness test was acquired from the UCI information vault. The dataset comprises of fourteen (14) features which contain 303 samples (Fig. 6).

Here, we represented the detailed distribution of the dataset attributes in Fig. 7.

Table 2 exhibits the attributes correlation. The accuracy of the proposed system is 85.24% for 14 characteristics out of all benchmark algorithms our proposed system performed better in terms of efficiency (Fig. 8).

```
In [1]: import numpy as np
        import pandas as pd
        import os,sys
        from sklearn.preprocessing import MinMaxScaler
        from sklearn.model_selection import train_test_split
        from sklearn.metrics import accuracy_score

In [2]: df = pd.read_csv('heart.csv')

In [3]: df.head()
```

Out[3]:

	age	sex	cp	trestbps	chol	fbs	restecg	thalach	exang	oldpeak	slope	ca	thal	target
0	63	1	3	145	233	1	0	150	0	2.3	0	0	1	1
1	37	1	2	130	250	0	1	187	0	3.5	0	0	2	1
2	41	0	1	130	204	0	0	172	0	1.4	2	0	2	1
3	56	1	1	120	236	0	1	178	0	0.8	2	0	2	1
4	57	0	0	120	354	0	1	163	1	0.6	2	0	2	1

Fig. 6 Dataset description

```
In [8]: plt.figure(figsize=(14,8))
        sns.heatmap(df.corr(), annot = True, cmap='coolwarm',linewidths=.1)
        plt.show()
```

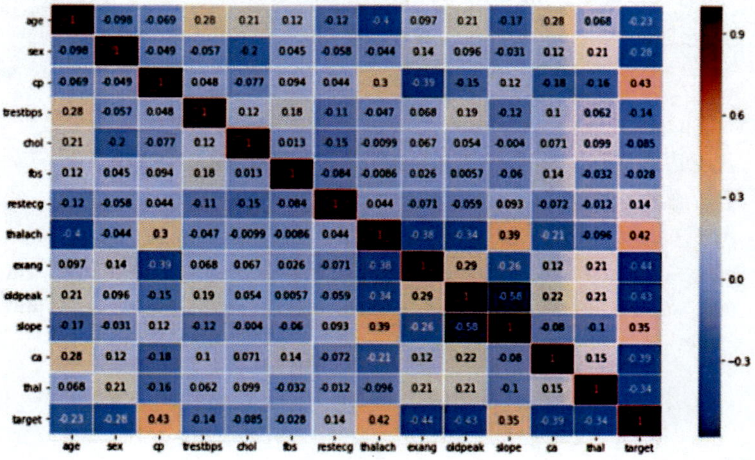

Fig. 7 Distribution of attributes in the dataset

A noticeable change in the accuracy value is obtained with the proposed model when compared with existing machine learning techniques. We analyzed the result with the help of the ROC curve in Fig. 9.

Table 2 Model comparison for accuracy

Name of the model	Accuracy
Decision tree	77.80
Random forest	78.68
Logistic regression	81.96
Gradient boosting	81.20
Extreme gradient boosting	79.50
Proposed model	85.24

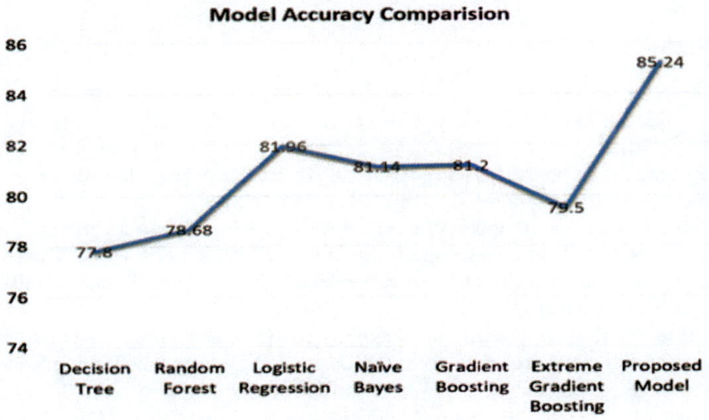

Fig. 8 Comparison of accuracy between models

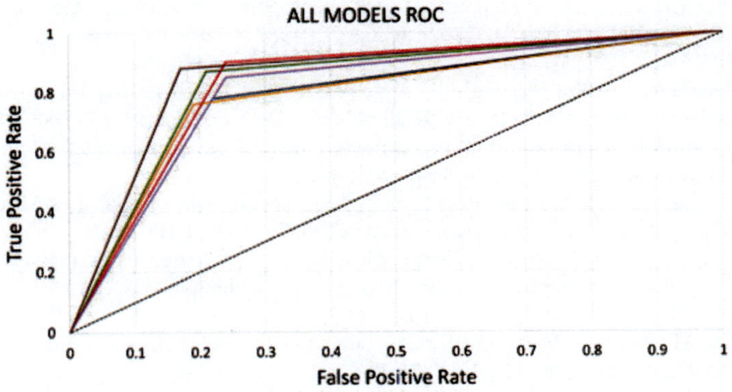

Fig. 9 ROC curve representation for all the models

5 Conclusion and Future Work

Cardiovascular disease is actually complicated, and also yearly great deals of people are perishing with this disease. Taking this into consideration, we have developed an ensemble-based algorithm, which is made use of combination machine learning protocol to emphasize even more accuracy of prediction over existing machine learning classification mechanisms.

The primary intention of this job is the accurate forecast of heart problem, along with a higher cost of accuracy. For predicting heart problem, our company may utilize logistic regression protocol, Naive Bayes, sklearn in artificial intelligence. The potential range of the proposed work can be implemented in different datasets available.

References

1. B. Baad, Heart disease prediction and detection. Int. J. Res. Appl. Sci. Eng. Technol. **7**(4), 2293–2299 (2019)
2. C. Beyene, Survey on prediction and analysis the occurrence of heart disease using data mining techniques. Int. J. Pure Appl. Math. **118**(8), 165–74 (2018)
3. J.P. Corsetti, et al., Data in support of a central role of plasminogen activator inhibitor-2 polymorphism in recurrent cardiovascular disease risk in the setting of high HDL cholesterol and C-reactive protein using bayesian network modeling. Data in Brief **8**, 98–104 (2016). https://doi.org/10.1016/j.dib.2016.05.026
4. A. Goldman, H. Hod, A. Chetrit, R. Dankner, Data for a population based cohort study on abnormal findings of electrocardiograms (ECG), recorded during follow-up periodic examinations, and their association with long-term cardiovascular morbidity and all-cause mortality. Data in Brief **26**, 104474 (2019). https://doi.org/10.1016/j.dib.2019.104474
5. C.B.C. Latha, S. Carolin Jeeva, Improving the accuracy of prediction of heart disease risk based on ensemble classification techniques. Informatics in Medicine Unlocked 16(November 2018), 100203 (2019). https://doi.org/10.1016/j.imu.2019.100203
6. A. Leiherer et al., Data on the power of high betatrophin to predict cardiovascular deaths in coronary patients. Data in Brief **28**(December), 104989 (2020). https://doi.org/10.1016/j.dib.2019.104989
7. M.E. Lizabeth, , B Rickner, and A L Ange., "C ONGENITAL H EART D ISEASE IN A DULTS First of Two Parts." Review Article Medical Progress 256: 53–59 (2000).
8. M. Marimuthu et al., A review on heart disease prediction using machine learning and data analytics approach. Int. J. Comput. Appl. **181**(18), 20–25 (2018)
9. U.F. Nyaga, et al., Data on the epidemiology of heart failure in sub-Saharan Africa. Data in Brief **17,** 1218–39 (2018). https://doi.org/10.1016/j.dib.2018.01.100
10. R. Prasad, P. Anjali, S. Adil, N. Deepa, Heart disease prediction using logistic regression algorithm using machine learning. Int. J. Eng. Adv. Technol. **8**(3 Special Issue), 659–662 (2019)
11. C.S. Wu, M. Badshah, V. Bhagwat, Heart disease prediction using data mining techniques. ACM Int. Conf. Proc. Ser. **3**(7), 7–11 (2019)
12. Cleveland heart disease dataset, UCI Repository (1988). https://archive.ics.uci.edu/ml/datasets/Heart+Disease. Last Accessed 25 May 2020
13. S. Biswas, T.H. Kim, D. Bhattacharyya, Features extraction and verification of signature image using clustering technique. Int. J. Smart Home **4**(3), 43–56 (2010)
14. N.F.L Mohd Rosely, R. Salleh, A.M. Zain, Overview feature selection using fish swarm algorithm. J. Phys. Conf. Ser. **1192**(1). https://doi.org/10.1088/1742-6596/1192/1/012068
15. https://towardsdatascience.com/5-reasons-why-you-should-use-cross-validation-in-your-data-science-project-8163311a1e79. Last Accessed on 5 June 2020
16. https://developers.google.com/machine-learning/crash-course/classification/roc-and-auc. Last Accessed on 5 June 2020

Comparative Analysis of Machine Learning Models on Loan Risk Analysis

M. Srinivasa Rao, Ch. Sekhar, and Debnath Bhattacharyya

Abstract Financial institutions suffer from the risk of losing money from bad customers, specifically banking sectors, where the risk of losing money is higher, due to bad loans. This causes an economic slowdown of the nation. The banking industry has a significant action of lending cash to individuals who are needing cash. In order to payback, the principal borrowed from the depositor bank collects the interest made by the principal borrowers. Credit risk investigation is turning into a significant field in financial risk management. Many credit risk analysis strategies are utilized for the assessment of credit risk of the client dataset. In this paper, we designed a model which takes loan data of the customers who applied for a loan from a bank and predicted to give the credit of the client or reject the utilization of the client. The proposed model takes the factors which affect the loan status of a person, thus providing accurate results for issuing credit to the client or reject the utilization of the client by considering all possibilities.

Keywords ML algorithms · SVM · Random forest · Logistic regression · Naive Bayes classifier · Decision tree

1 Introduction

The banking industry faces several challenges in the competitive world. The banks have to change their strategies. The customers are also facing difficulties of a lengthy process, delay in processing while applying for a loan and sanctioning of the loan. The

M. Srinivasa Rao (✉) · Ch. Sekhar
Department of CSE, Vignan's Institute of Information Technology, Visakhapatnam, India
e-mail: srinivas.marada22@gmail.com

Ch. Sekhar
e-mail: sekhar1203@gmail.com

D. Bhattacharyya
Department of CSE, K L Deemed to be University, KLEF, Guntur, AP, India
e-mail: debnathb@kluniversity.in

D. Bhattacharyya and N. Thirupathi Rao (eds.), *Machine Intelligence and Soft Computing*, Advances in Intelligent Systems and Computing 1280,
https://doi.org/10.1007/978-981-15-9516-5_7

repayment capacity of the borrowers has depended on their liabilities, depended on family members, a loan from other sources individual age, increase in future income, etc. These all are the challenges faced while doing traditional banking, but in this new era, everything is automated, and machine learning is taking place in every sector of finance, health care, business and many more. Using machine learning technology has become a necessity rather than a trend. This helps the customer to consume less time while processing the loan application and the organizations can predict the good customers and bad customers by gathering the complete data that is required for the processing of the loan application. By using machine learning technologies, the application will perform accurately, productively and are cost-efficient.

Machine learning is the usage of human-made intellectual competence that activates systems to usually take in and improve for a reality without being explicitly altered. ML pivots the development of PC programs that can get to information and use it to learn for them. "The direction toward learning begins with affirmations or information, for instance, methods, direct understanding, or heading, to scan for plans in data and choose better resolutions succeeding on subject to the models that we give. The key point is to enable the PCs to adjust this way without human mediation or help, and change rehearses suitably" [1]. Budgetary establishments experience the ill effects of the danger of losing cash from awful clients and explicitly banking segments where the danger of losing cash is greater, because of awful credits. This causes financial stoppage of the country.

1.1 Detailed ML Algorithms

Naive Bayes Classifier Algorithm. Naive Bayes classifier depends on Bayes' hypothesis and arranges each incentive as free of some other esteem. It enables us to foresee a class/classification, because of a given arrangement of highlights, utilizing likelihood. Regardless of its effortlessness, the classifier does shockingly well and is regularly utilized because of the reality it beats increasingly advanced grouping techniques [7, 11, 14].

K-Means Algorithm. The k-means is unsupervised clustering technique. Data samples do not have any labels to classify the available data as different classes. Data instance is grouped based on the similar behavior instance formed as a cluster. In k-means method, k indicates the no. of clusters and each case in the cluster having similar means with the centroid [7, 17].

SVM Algorithm. Support vector machine algorithms are regulated to learning models that break down information utilized for characterization and relapse examination. They basically channel information into classifications, which is accomplished by giving much preparing precedents, each set apart as having a place with either of the two classes [8, 13].

Logistic Regression. It centers on evaluating the likelihood of an occasion happening dependent on the past information given. It is utilized to cover a double needy erratic that is the place just two qualities, either zero or one, speak to results [8, 12].

Decision Trees. This is a stream diagram resembling tree structure that uses a spreading strategy to show each likely result of choice. Every hub inside the tree speaks to an experiment on a particular erratic, and each division is the result of that experiment [10, 15].

Random Forests. It is a group learning procedure, joining numerous calculations to turn out enhanced outcomes for arrangement, relapse and different undertakings. Every individual classifier is powerless, yet when joined with others can deliver incredible outcomes. The calculation begins with a choice tree, and info is entered at the top. It at that point goes down the tree, with information being divided into littler and littler sets, in view of explicit variables [9, 16].

1.2 Problem Statement

The economic growth of the nation decreases due to the bad customers who applied for a loan. This can be solved by analyzing the customer's data and identifies bad customers by rejecting their loan application. This makes automated, more accurate, productive and less costly concerning employee time.

1.3 Existing System

Previously, the traditional banking system conquered the finance sector. Where, when the customer applies for a loan, there is a lengthy of work to be done, longer wait time for verification of documents, screening process and should wait for the response from the bank. All this process takes time and slows down the process, and this traditional banking system has to maintain the commissions for the maintenance of transactions where sometimes these commissions are expensive. Among various fields in finance, loan process has garnered a great deal of attention in banking industry.

1.4 Proposed System

The proposed system acts as a better customer service and will prove to be an added advantage where this provides a digital platform by reducing customer time and cost. Already existing data is used for predictions.

1.5 *Benefits of Proposed System*

High performance and accuracy rate are compared to existing system. It can analyze large amounts of data. Prediction is done by analyzing the entire dataset which is utilized to prepare the machine learning model.

2 Literature Review

Hamid and Ahmed proposed to manufacture a prescient approaches that could be utilized to anticipate and group the uses of advances that acquainted by the clients with fortunate or unfortunate advance by exploring client rehearses and past compensation back credit [1].

Dr. Kavitha proposed a system for chance assessment dependent on k-means grouping procedures. Client information is extricated, and the significant characteristics are chosen utilizing gain hypothesis. Rule gauge is executed for every credit type subject to the predefined criteria. Perceived and pardoned up-and-comers are considered as "material" and "non-pertinent" credits in like way primer results indicated that advanced methodology predicts better exactness and eats up less time than the current procedure [2].

Abhijit et al. proposed a framework utilizing information digging for advanced default hazard examination which empowers the bank to diminish the manual mistakes engaged with the equivalent. From the trial, it was directed that out of the five information mining calculations, i.e., Bayesian algorithm, decision tree, boosting, bagging and random forest applied on various datasets of various size, random forest algorithm is generally predictable and has most elevated exactness contrasted with others [3].

Huang et al. has utilized two informational collections for Taiwan money-related organizations and US business banks as examination tried. The outcomes indicated that help vector machines accomplished exactness equivalent to that of backpropagation neural systems [5].

Kala expresses a system for chance assessment, where mass volumes of client information are caused and chance appraisal in addition to assessment is done dependent on the data mining procedure. The client information is separated for highlighting determination of the important properties. The characteristics are chosen utilizing gain hypothesis. Rule estimate is developed for each credit type. Hazard evaluation is acted in two levels, principal and colleague to be unequivocal. An edge regard is defined, with the objective that the credit competitor underneath the edge regard is excused and remaining credits are approved [4].

3 Dataset

The dataset is collected from KAGGLE Web site [6]. The dataset consists of 100,000 cases of data which demonstrates the loan status of the customers.

Variables in the dataset are:

Loan ID: Loan identification number of the applicant.
Customer ID: Customer identification number in the bank.
Loan Status: It describes the loan status of the customer (**Approved or Rejected**).
Current Loan Amount: It describes the total amount took by the customer from the bank.
Term: It describes the loan term, i.e., **Short or Long** term of the customer.
Credit assessment: It is a number running from 300 to 850 that delineates a customer's financial soundness.
Annual Income: It describes the total income earned by the customer in a year.
Years in Current Job: It describes how many years the customer has been working.
Monthly Debt: It describes the amount that customer has to pay back to the bank.
Years of Credit History: It is a record of a borrower's capability of reimbursement of obligations.
Number Of Credit Problems: It describes the customers who have a history of not paying their bills on time.
Current Credit Balance: It is the amount that you have available to spend.
Maximum Open Credit: A credit limit is the most extreme measure of credit that a monetary organization or other loan specialists will reach out to a borrower for a specific credit extension.

4 Case Study

See Tables 1 and 2.

5 Results

Figure 1 shows the percentage of the output that is the loan status approved (75.4%) or not approved (24.6%) based on the comparison between loan status and count, where the values are taken from the dataset which we had used.

Logistic regression predicts categorical outcomes. The results of logistic regression range between 0 and 1. Figure 2 shows the probability values of the approval and disapproval of the loan.

Figure 3 shows the two-group classification values of approval and disapproval of loan using support vector machine (SVM).

Table 1 Unit test case-1

S. No.	1
Name of Test	Check for the case of rejection of loan application
Sample input	Current loan amount = 123,456
	Credit score = 678
	Term = 1
	Annual income = 1,234,654
	Current credit balance = 563,245
	Year of credit history = 14
	Number of credit problems = 1
Expected output	SORRY!YOUR LOAN IS REJECTED
Actual output	SORRY!YOUR LOAN IS REJECTED
Remarks	Pass

Table 2 Unit test case -2

S. No.	2
Name of test	Check for the case of acceptance of loan application
Sample input	Current loan amount = 564,789
	Credit score = 810
	Term = 1
	Annual income = 123,456
	Current credit balance = 56,789,
	Year of credit history = 12,
	Number of credit problems = 0
Expected output	CONGRATULATIONS! YOUR LOAN IS APPROVED
Actual output	CONGRATULATIONS! YOUR LOAN IS APPROVED
Remarks	Pass

Figure 4 shows the probability values of approval and disapproval of the loan.

Figure 5 shows the probability values of approval and disapproval of loan. Random forest is a tree-based machine learning algorithm that uses the intensity of numerous choice trees for deciding.

Figure 6 gives detailed knowledge of the accuracy values of different machine learning models. Finally, we conclude that Naïve Bayes classifier gives high accuracy (75%) when compared to remaining machine learning models for the taken dataset.

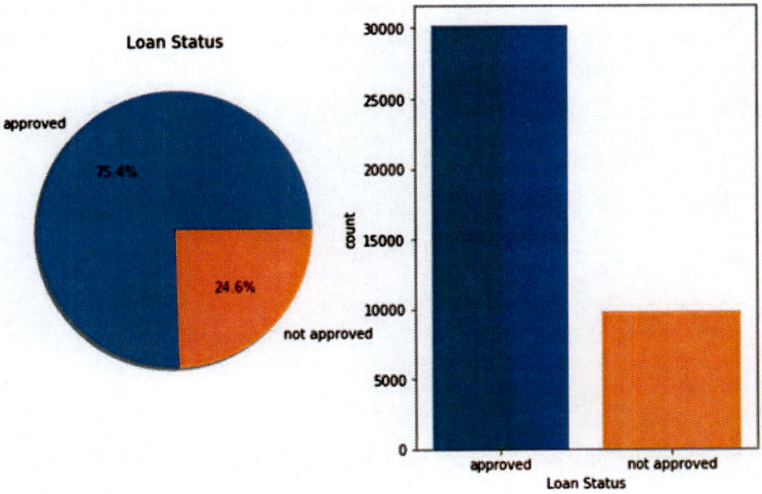

Fig. 1 Display of bar plot and pie chart

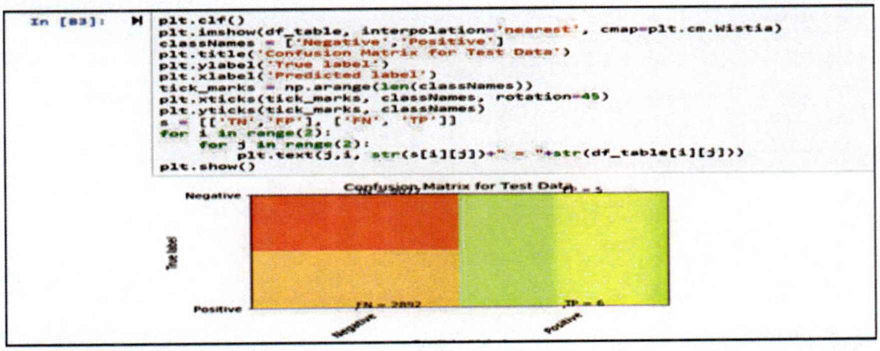

Fig. 2 Model-1 logistic regression

6 Conclusion

The scope of this paper is to provide an added advantage for the banks to easily detect bad customers by rejecting their loan application. We have used five algorithms to develop a loan risk analysis system, which includes the models built using the above five algorithms. Of all the models built, the Naïve Bayes model yields the most accurate results. This paper opens the door for future research by adding the advanced classifiers to the existing system in future to yield more accurate results and efficient model by helping the organizations in the finance sector.

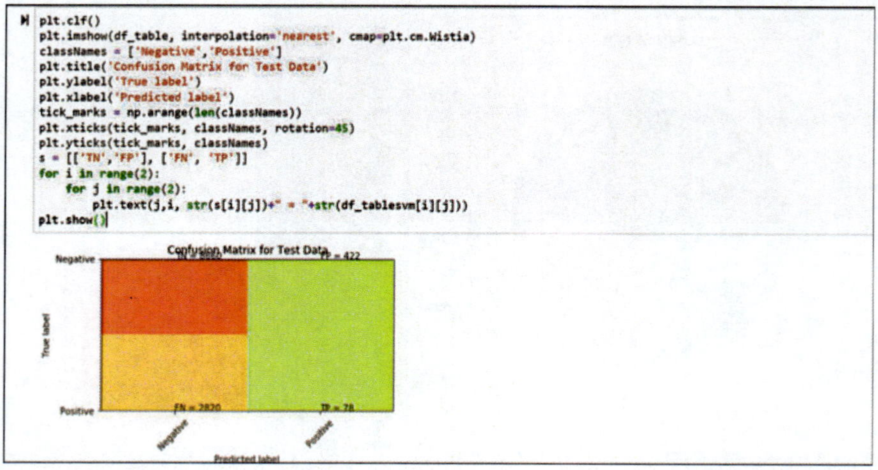

Fig. 3 Model-2 support vector machine

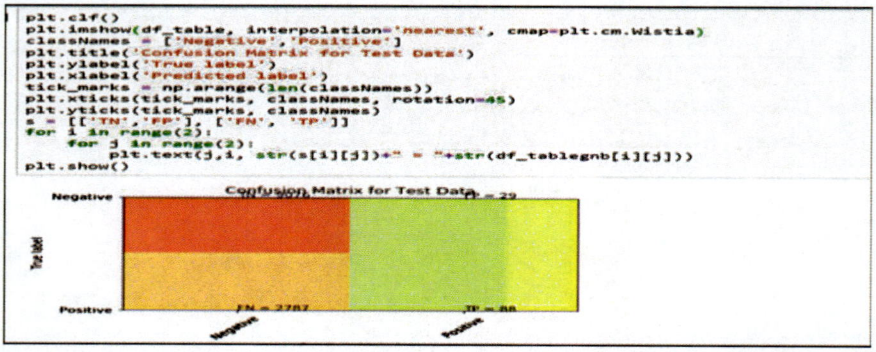

Fig. 4 Model-3 Naive Bayes

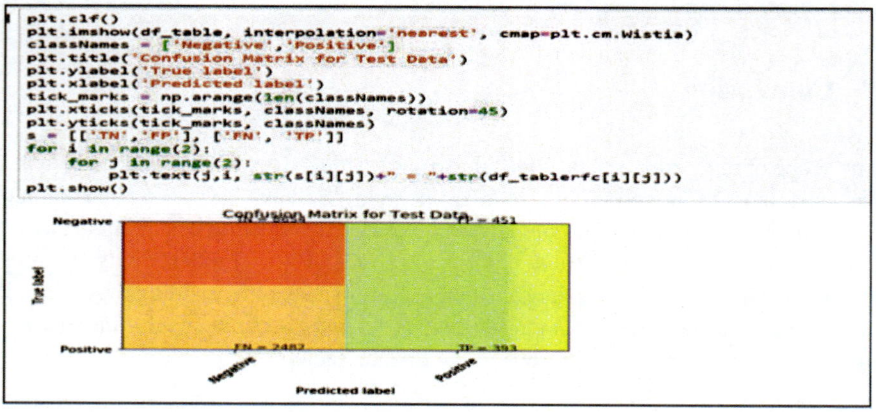

Fig. 5 Model-5 random forest classifier

Table 3 Accuracy values of machine learning models

S. No.	Name of the algorithm/model	Accuracy
1	Logistic regression	74.691
2	Support vector machine	74.796
3	Naïve Bayes	75.317
4	Decision tree	68.021
5	Random forest	74.023

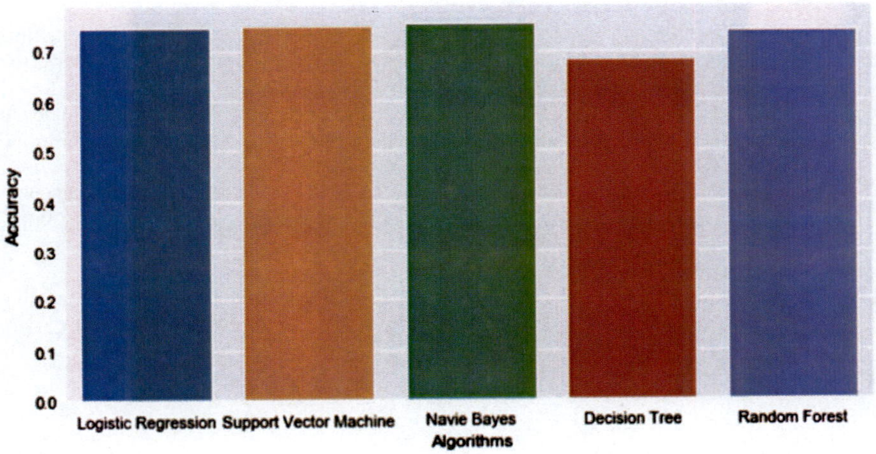

Fig. 6 Model accuracy graph

References

1. A.J. Hamid, et al., Developing prediction model of loan risk in banks using data mining. Mach. Learn. Appl. Int. J. (M LAIJ) **3** (2016). https://doi.org/10.5121/mlaij.2016.3101
2. K. Kavitha, Clustering loan applicants based on risk percentage using k-means clustering techniques. Int. J. Adv. Res. Comput. Sci. Softw. Eng. **6**,162–166 (2016)
3. A.A. Sawant et al., Comparison of data mining techniques used for financial data analysis. Int. J. Emerg. Technol. Adv. Eng. (2013)
4. K. Kala, A customized approach for risk evaluation and prediction based on data mining technique. Int. J. Eng. Res. Technol. **3**, 20–25 (2018)
5. J. Huang et al., Credit rating analysis with SVM and neural network: a market comparative study. Decis. Support Syst. **37**, 543–558 (2004)
6. https://www.kaggle.com/omkar5/dataset-for-bank-loan-prediction
7. T.N. Pandey, et al., Credit risk analysis using machine learning classifiers, in *International Conference on Energy, Communication, Data Analytics and Soft Computing (ICECDS)* (2018). https://doi.org/10.1109/ICECDS.2017.8389769
8. Attigeri, et al., Credit risk assessment using machine learning algorithms. J. Comput. Theor. Nanosci. **23**, 3649–3653 (2017). https://doi.org/10.1166/asl.2017.9018
9. S. Kalayci, M. Kamasak, S. Arslan, Credit risk analysis using machine learning algorithms, in *2018 26th Signal Processing and Communications Applications Conference (SIU)*, July 2018. https://doi.org/10.1109/SIU.2018.8404353

10. M. G. Saragih et al., Machine learning methods for analysis fraud credit card transaction. Int. J. Eng. Adv. Technol. **8**, 870–874 (2019)
11. D. Varmedja, et al. Credit card fraud detection—machine learning methods, in *2019 18th International Symposium INFOTEH-JAHORINA (INFOTEH)*, May 2019. https://doi.org/10.1109/INFOTEH.2019.8717766
12. A. Bhanusri et al., Credit card fraud detectionusing Machine learning algorithms. J. Res. Humanities Soc. Sci. **8**, 4–11 (2020)
13. R. Melendez, Credit Risk analysis applying machine learning classification models, in *Intelligent Computing. CompCom 2019. Advances in Intelligent Systems and Computing*, 997, ed. by K. Arai, R. Bhatia, S. Kapoor (Springer, Cham, 2009). https://doi.org/10.1007/978-3-030-22871-2_57
14. P.M. Addo, D. Guegan, B. Hassani, Credit risk analysis using machine and deep learning models," Documents de travail du Centre d'Economie de la Sorbonne 18003, Universite Pantheon-Sorbonne(Paris 1), Centre d'Economie de la Sorbonne (2018)
15. S.Z.H. Shoumo et al., Application of machine learning in credit risk assessment: a prelude to smart banking, in *TENCON 2019–2019 IEEE Region 10 Conference (TENCON)*. https://doi.org/10.1109/TENCON.2019.8929527
16. Ch. Sekhar, M. Srinivasa Rao, K. Venkata Rao, A.S. Keethi Nayani, Role of machine learning concepts in disease prediction using patient's health history. Int. J. Adv. Sci. Technol. **29**(8s), 4476–4482 (2020). Retrieved from https://sersc.org/journals/index.php/IJAST/article/view/25502
17. M. Srinivasa Rao, C.R. Pattanaik, A. Sarkar, M. Ilayaraja, R. Pandi Selvam, Machine learning models for heart disease prediction. Int. J. Adv. Sci. Technol. **29**(2), 4567–4580 (2020). Retrieved from https://sersc.org/journals/index.php/IJAST/article/view/28464

Compact MIMO Antenna for Evolving 5G Applications With Two/Four Elements

Srinivasa Naik Kethavathu, Sourav Roy, and Aruna Singam

Abstract In this article, we designed a low-cost MIMO antenna for 5G-based IoT applications. The frequency range of the designed antenna is 25.8–29.4 GHz. The gain was approximately 5.7 dBi at 28 GHz. The antenna has been designed, simulated and evaluated on ANSYS HFSS. The efficiency of ECC and DG performance of the MIMO two/four-element antenna is studied and both results have been found within this range are useful. The different MIMO antenna configurations are studied. The antenna maximum dimension is $9.48 \times 7.36\ \text{mm}^2$.

Keywords Compact MIMO · IoT · ECC · Directive gain · 5G applications

1 Introduction

Today, the world is running toward technology, and each one carrying at least two antennas with them. Every ten years, a new generation is coming into mobile communication. Present 4G technology does not have better channel capacity, so research is going on 5G to improve capacity, latency and mobility. From generation to generation, frequency is going to increase, which results in the reduction in the antenna size due to this system, complexity increases. Most of the other 5G are 27.5–29.5 GHz, 33.4–36 GHz, 37–40.5 GHz, 42–45 GHz, 47–50.2 GHz, 50.4–52.6 GHz and 59.3–71 GHz, respectively. One of the essential applications of 5G is IoT for connecting the number of devices at a time.

S. N. Kethavathu (✉) · S. Roy
Vignan's Institute of Information Technology, Visakhapatnam, India
e-mail: nivas97033205@gmail.com

S. Roy
e-mail: sourav31roy@gmail.com

A. Singam
Andhra University College of Engineering, Visakhapatnam, India
e-mail: aruna9490564519@gmail.com

D. Bhattacharyya and N. Thirupathi Rao (eds.), *Machine Intelligence and Soft Computing*, Advances in Intelligent Systems and Computing 1280,
https://doi.org/10.1007/978-981-15-9516-5_8

For stable communication and a higher transmission rate, multiple inputs multiple outputs (MIMO) technology is useful. It uses multiple antennas at the transmitting side and various antennas at the receiving side. Because of this, it increases the channel capacity with the same transmitting power. It is known that MIMO operations can make the wireless communications networks even more efficient by reliable linking, power and high data rate, and MIMO is going to be used in the 5G devices.

The revolution in mobile networking is introduced by cellular network technology from 0G up to 4G [1]. There are changes and improvements in every generation [2]. Even if 4G has too many features, it cannot solve some problems like high energy consumption, crowded channels, inadequate coverage and low quality of service (QoS) [3]. Therefore, mobile communication updated to 5G to resolve the disadvantages of 4G and to increase the data rate [4]. 5G has so many frequency bands [5], but the unused or underutilized broadband spectrum exists at 28 GHz. This spectrum has low atmospheric absorption, low path loss and better propagation conditions [6]. Kaeib et al. designed a 28 GHz antenna with slots on the patch that antenna covered 26.81–29.29 GHz frequency band with VSWR 1.02 and returned loss −39.3 dB [7].

As the number of users increasing day by day traffic of the spectrum increases, and there is a possibility of data corruption. But MIMO increases the channel capacity with the high data rate, low BER and support multi-users because of added features of diversity and multiplexing [8]. A 28 GHz MIMO antenna designed for UWB applications, and it covers a band of 13 GHz, but the gain is only 2.39 dB [9].

This article describes the two/four elements MIMO antenna for upcoming 5G application. The design and simulation antennas have done using 3D electromagnetic ANSYS HFSS [10] software. This antenna is constructed on a low-cost FR-4 substrate in the frequency range of 25.8–29.4 GHz.

2 Single Element Antenna Design

The antenna has a dielectric permittivity (ε_r) = 4.4, thickness (h) = 1.6 mm and loss tangent (δ) = 0.09 on the FR-4 substrate The antenna geometry is shown in Fig. 1. The parameters are $L_1 = 3.53$ mm, $L_2 = 1.67$ mm, $L_3 = 1.462$ mm, $W_1 = 5.12$ mm, $W_2 = 0.93$ mm, $W_3 = 3.26$ mm, $W_4 = 1.18$ mm and $W_5 = 0.9$ mm. The dimension of the antenna is 5.12×3.53 mm^2.

The simulated S_{11} parameter is shown in Fig. 2a. The antenna covers 26.9–29.5 GHz frequency bands. The antenna has also demonstrated a good gain in the frequency band in that particular band. The maximum gain is found near about 5.9 dB at 28 GHz is shown in Fig. 2b. The 3D polar gain plot is referred in Fig. 3. The polar plot showed that the antenna has a dipole radiation pattern.

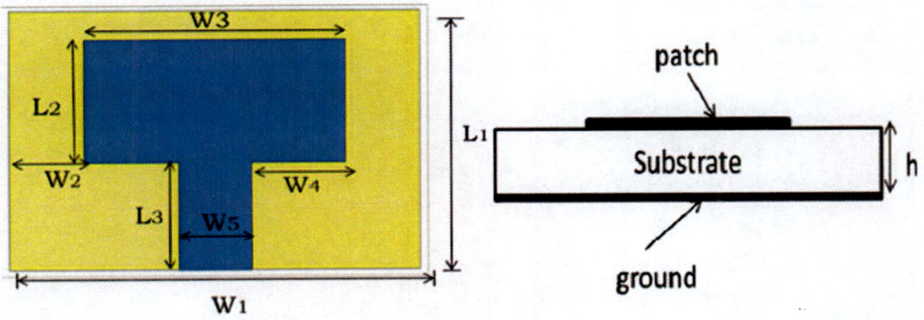

Fig. 1 Antenna 1 design geometry

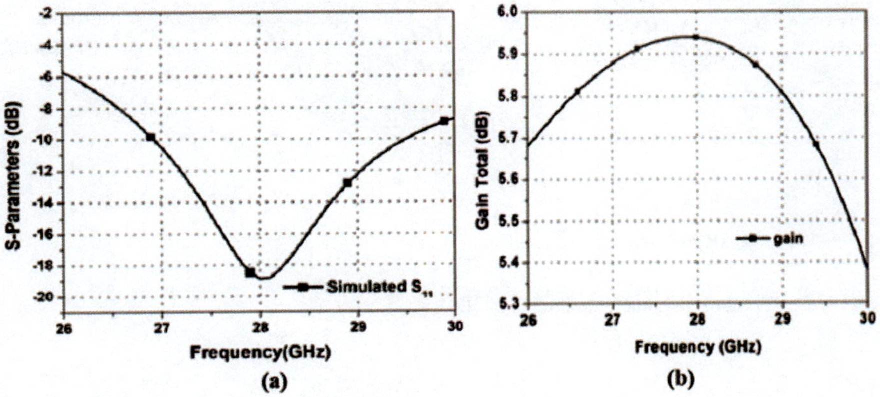

Fig. 2 Simulated **a** S_{11} (dB), **b** gain versus frequency plot of antenna 1

3 Two-Element MIMO Antenna Analysis

Two-element MIMO antenna is depicted in Fig. 4. The space between two antenna elements taken near about $W_7 = 1.2$ mm and width of the antenna is $W_6 = 9.58$ mm. The S-parameters of the simulated antenna are shown in Fig. 5. The antenna occupies a 27.2–30.3 GHz frequency band. The antenna covers a frequency band from 27.2 to 30.3 GHz. The S_{11} and S_{22} values are nearly same in the frequency ranges. The S_{12} is found near about ≤ -12 dB at 28 GHz frequency.

The MIMO antenna's two basic parameters are the envelope correlation coefficient (ECC) and the diversity gain (DG). By using radiation pattern method, ECC and DG are calculated. In the desired frequency range, ECC is nearly <0.05 and DG is nearly about >9.98 by considering port 1 and 2 as shown in Fig. 6.

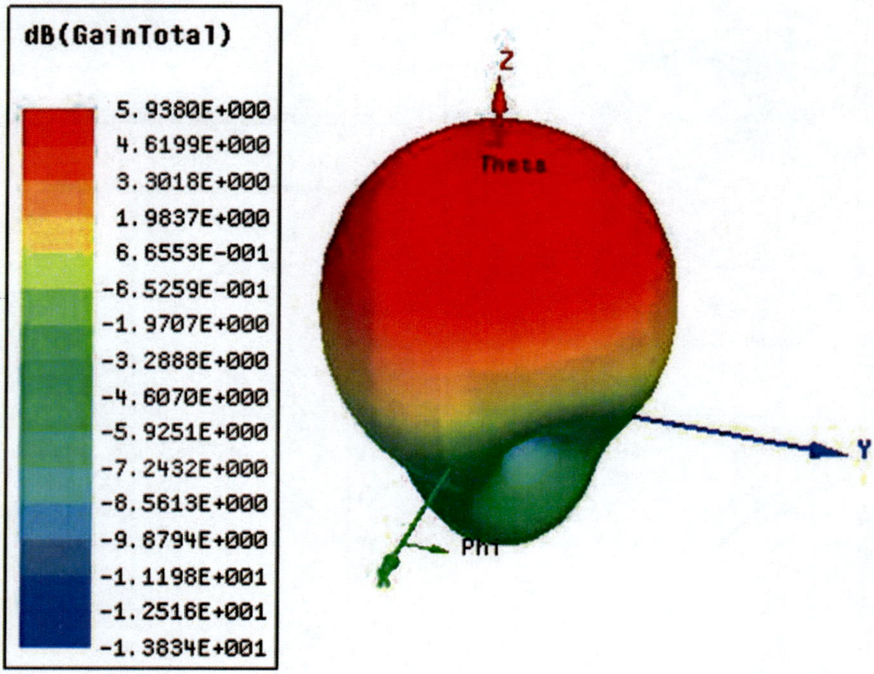

Fig. 3 Gain plot of antenna 1 at 28 GHz in 3D

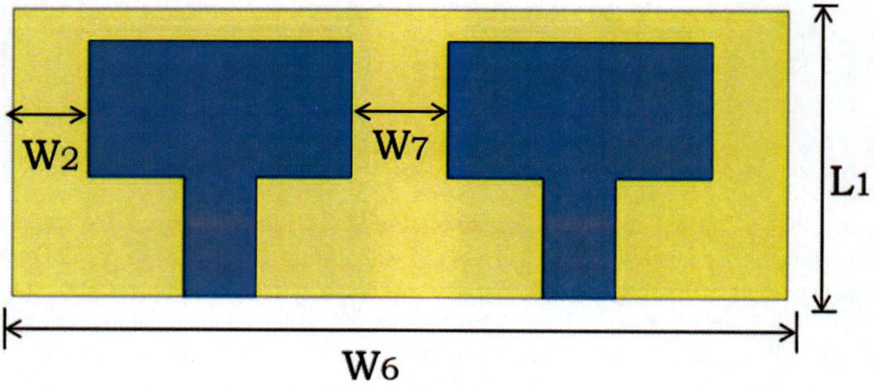

Fig. 4 Geometry of MIMO antenna with two elements

4 Four-Element MIMO Antenna Analysis

The four-element MIMO antenna is designed and simulated with help of ANSYS HFSS software. Two separate configurations are used to analyze the four-element MIMO antenna parameters.

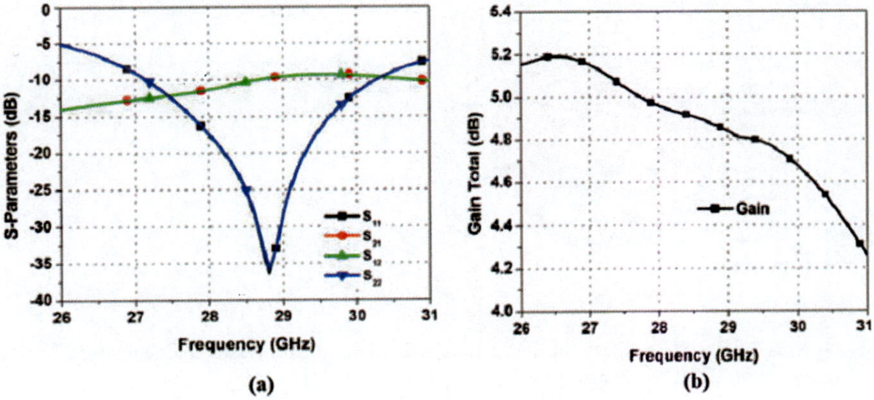

Fig. 5 Simulated **a** S-parameters, **b** gain versus frequency plot

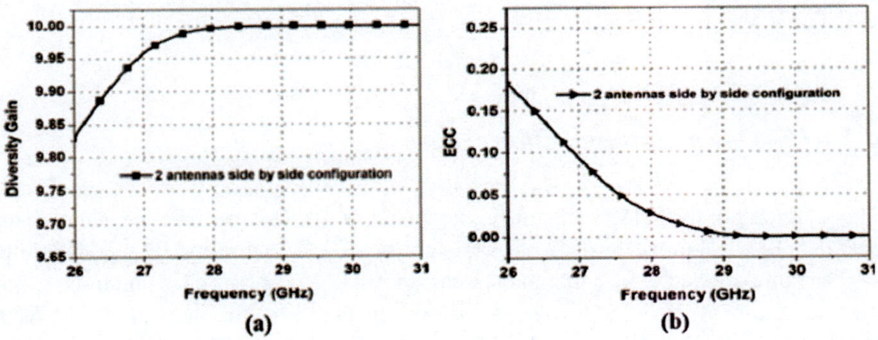

Fig. 6 Simulated **a** ECC, **b** DG parameter of the two-element MIMO antenna

4.1 Side by Side Configuration

The design geometry of the four-element side by side configuration is shown in Fig. 7. The space between antennas is $W_9 = 1.3$ mm, and the width of the overall

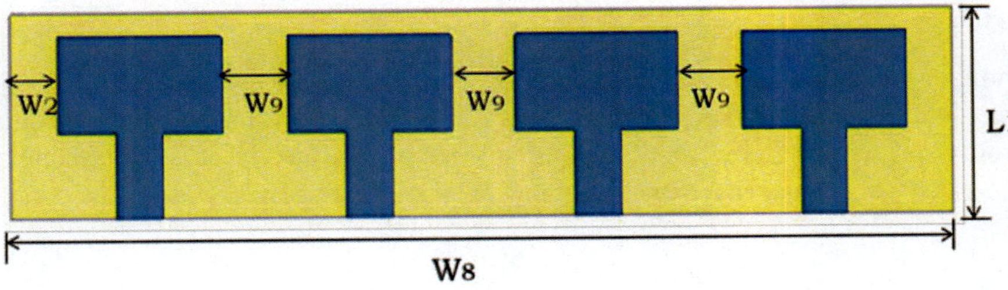

Fig. 7 Geometry of side by side configuration

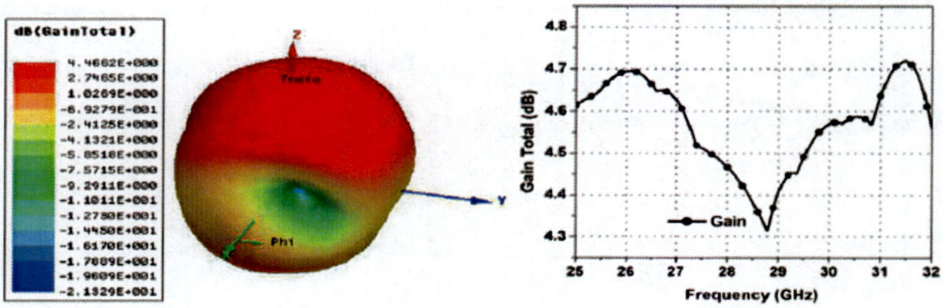

(a) (b)

Fig. 8 Simulated **a** 3D gain plot of the antenna at 28 GHz, **b** gain versus frequency side by side configuration of MIMO antenna

antenna is $W_8 = 18.8$ mm. The antenna S-parameters are shown in Fig. 9. This antenna occupies the 27.2–30.3 GHz frequency range. As shown in Fig. 8, the gain almost 4.4 dB in 28 GHz.

4.2 Up-Down Configuration

The geometry of the MIMO antenna is shown in Fig. 10. The space between antennas is $W_{11} = L_6 = 1.1$ mm. The design parameters are $L_5 = 7.36$ mm and $W_{10} = 9.48$ mm.

The simulated antenna S-parameters are shown in Fig. 11a–d. The antenna occupies spectrum of 25.8–29.4 GHz. As shown in Fig. 12b, the gain of the MIMO antenna is 5.68 dB at 28 GHz. The 3D gain plot as shown in Fig. 12a depicted that antenna radiation performance is affected due to MIMO.

The comparisons plot of the ECC of both configurations is mentioned in Fig. 13a, b. In both cases, the ECC found near about <0.05 in the desired frequency ranges. The DG also found near about >9.95 in both cases. So as per our application requirements, we can use the antenna.

5 Conclusion

In this paper, a low-cost and compact two/four-element MIMO antenna was developed and simulated for the 5G applications. This antenna covers the frequency spectra from 25.82 to 29.4 GHz. The ECC of the antenna is <0.1, and the DG of the antenna is >9.0. This antenna covers the upcoming 5G application bands.

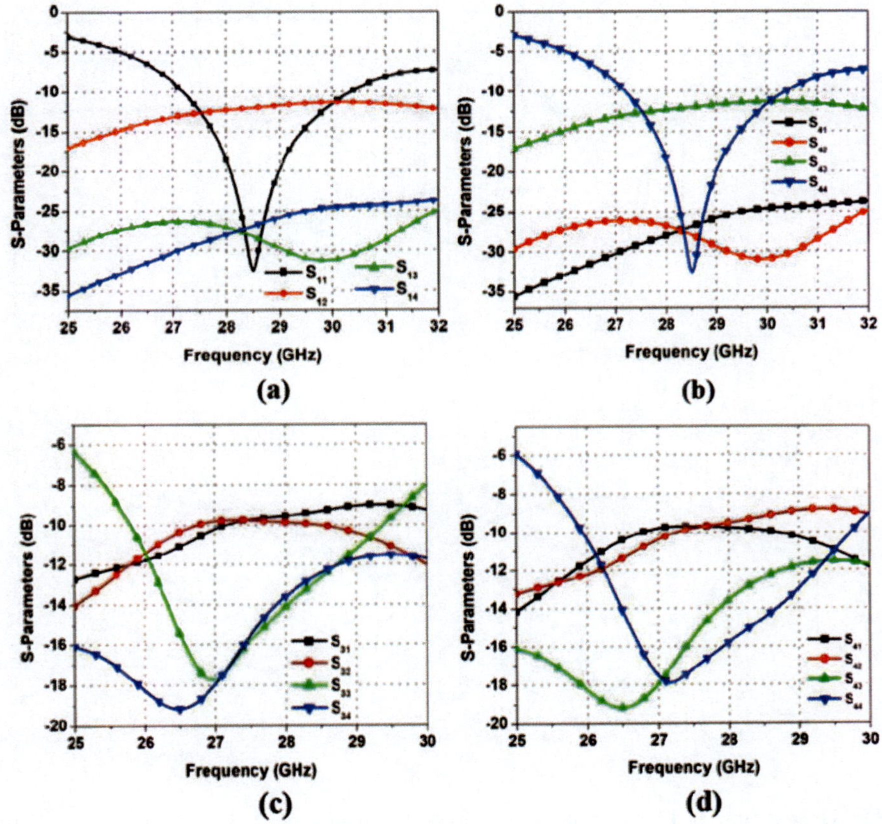

Fig. 9 **a–d** Simulated S-parameters side by side configuration

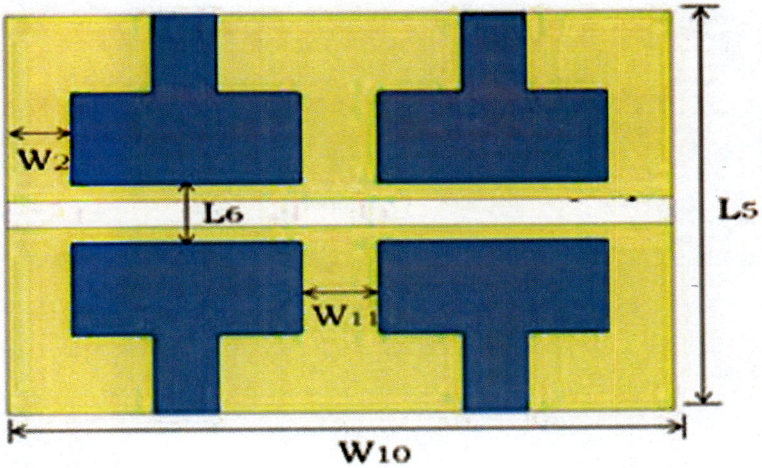

Fig. 10 Design geometry of four-element up-down configuration MIMO antenna

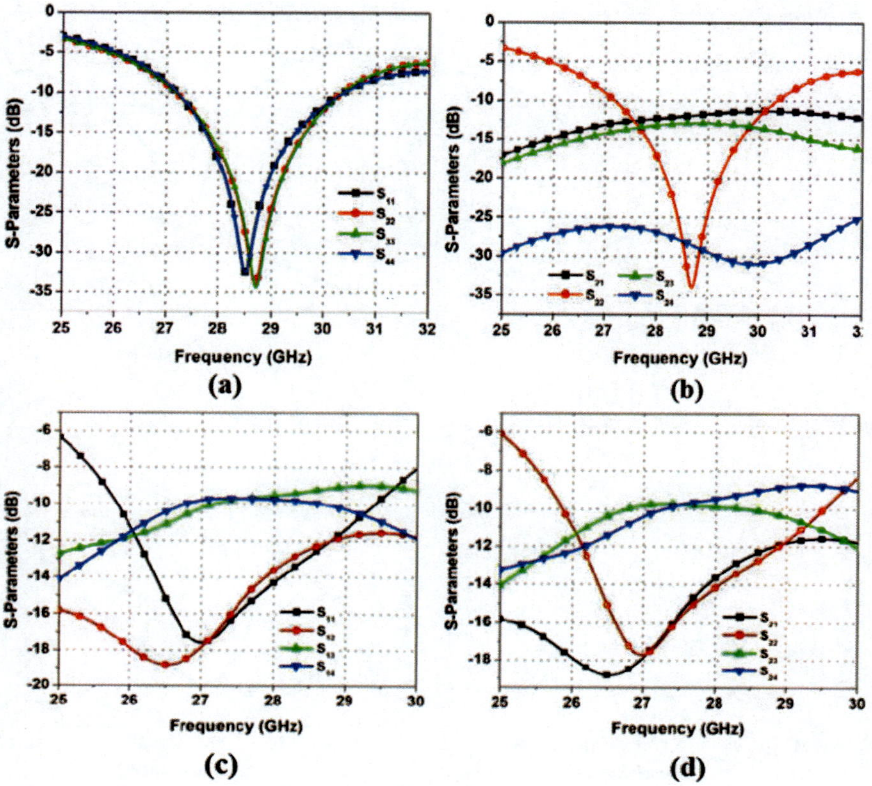

Fig. 11 a–d Simulated S-parameters of the up-down configuration

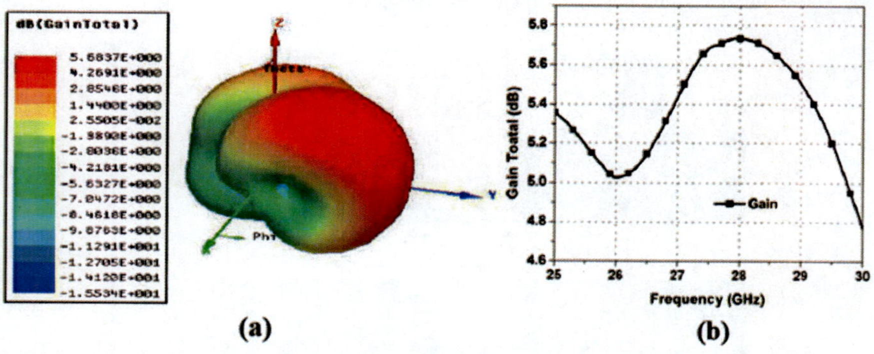

Fig. 12 Simulated a 3D gain plot of the antenna, b gain versus frequency plot of up-down configuration

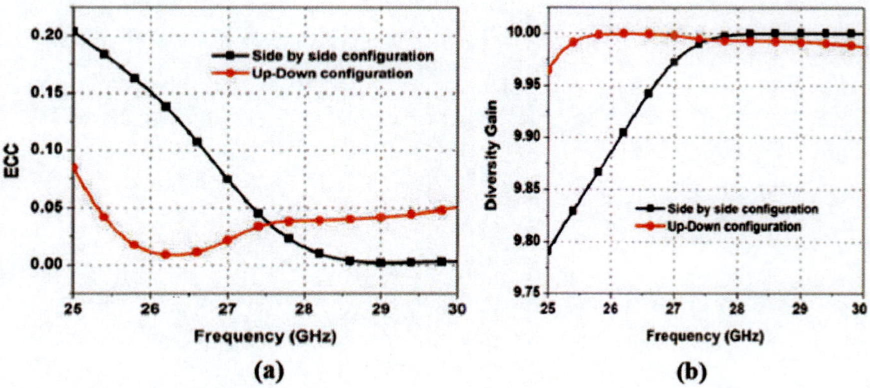

Fig. 13 Comparison of **a** ECC, **b** DG of MIMO antenna in side by side and up-down configuration

Acknowledgements This research is sponsored by the DST Science & Engineering Research Board (SERB) with File No: EEQ/2016/000391.

References

1. I. Ahmad, T. Kumar, M. Liyanage, J. Okwuibe, M. Ylianttila, A. Gurtov, Overview of 5G security challenges and solutions. IEEE Commun. Stand. Mag. **2**(1), 36–43 (2018)
2. S. Hakimi, S.K.A. Rahim, Millimeter-wave microstrip bent line grid array antenna for 5G mobile communication networks, in *2014 Asia-Pacific Microwave Conference* (IEEE, 2014), pp. 622–624
3. A.K. Jain, R. Acharya, S. Jakhar, T. Mishra, Fifth generation (5G) wireless technology revolution in telecommunication, in *2018 Second International Conference on Inventive Communication and Computational Technologies (ICICCT)* (IEEE, 2018), pp. 1867–1872
4. Y. Cao, K.S. Chin, W. Che, W. Yang, E.S. Li, A compact 38 GHz multi-beam antenna array with a multi-folded butler matrix for 5G applications. IEEE Antennas Wirel. Propag. Lett. **16**, 2996–2999 (2017)
5. A. Kumar, M. Gupta, A review of activities of the fifth-generation mobile communication system. Alexandria Eng. J **57**(2), 1125–1135 (2018)
6. T.S. Rappaport, Y. Xing, G.R. MacCartney, A.F. Molisch, E. Mellios, J. Zhang, Overview of millimeter wave communications for fifth-generation (5G) wireless networks—with a focus on propagation models. IEEE Trans. Antennas Propag. **65**(12), 6213–6230 (2017)
7. A.F. Kaeib, N.M. Shebani, A.R. Zarek, Design and analysis of a slotted microstrip antenna for 5G communication networks at 28 GHz, in *2019 19th International Conference on Sciences and Techniques of Automatic Control and Computer Engineering (STA)* (2019), pp. 648–653
8. A. Kumar, K. Mishra, A. Mukherjee, A.K. Chaudhary, Channel capacity enhancement using MIMO technology, in *IEEE International Conference on Advances in Engineering, Science and Management (ICAESM-2012)* (2012), pp. 10–15
9. M. Tiwari, P.K. Singh, A. Sehgal, D.K. Singh, Compact MIMO antenna with improved isolation for 5 g/ultra-wideband communications, in *2018 International Conference on Automation and Computational Engineering(ICACE)* (2018), pp. 229–233
10. HFSS ver. 17. Ansoft Corporation, Pittsburgh, PA

Accurate Prediction of Fake Job Offers Using Machine Learning

Bodduru Keerthana, Anumala Reethika Reddy, and Avantika Tiwari

Abstract The recent growth of online recruitment and candidate management system has established yet another media for fraudsters on the Internet. People who need job are getting being scammed by fake advertisements by fraudsters on popular websites. These fraudsters are enlarging their scams by motivating job seekers as full-time roles with basic minimum qualifications like B.Tech. and degree. Machine learning classification techniques are essential tools in extracting invisible knowledge from a huge dataset to increase the accuracy and efficiency in predictions. In this paper, we analysed fake job posting dataset and used machine learning techniques to classify job advertisements which are fraudulent or real. Various machine learning algorithms and performance metrics are used in this paper. The benefaction of this research is one encompassing of contextual capabilities space, which relieved absorbing improvements of accuracy, precision and recall.

Keywords Candidate management systems · Extract knowledge · Classification · Machine learning · Accuracy · Precision · Recall

1 Introduction

Job scam is fake online job advertising, targeting job seekers with the aim of stealing personal information or money. In 2016, over 3 lakh students were targeted; half of them handled over cash being unaware that they had been scammed. Majorly, students are being targeted because they want to financially support themselves since. It is a

B. Keerthana (✉) · A. R. Reddy · A. Tiwari
Department of CSE, Vignan's Institute of Information Technology, Visakhapatnam, India
e-mail: keerthana150828@gmail.com

A. R. Reddy
e-mail: anumala.reethu34@gmail.com

A. Tiwari
e-mail: avantika.tiwari92@gmail.com

D. Bhattacharyya and N. Thirupathi Rao (eds.), *Machine Intelligence and Soft Computing*, Advances in Intelligent Systems and Computing 1280,
https://doi.org/10.1007/978-981-15-9516-5_9

tough and competitive job market, so they are more interested in being a part with cash for securing their job [1]. Frauds are becoming extremely enlightened, with fraudsters increasing their strategies, making it really difficult to identify what is genuine or fake. According to the information, we have to classify which advertisement is fake or real and then make a forecast on the required data. This study involves extracting of job advertising data and analysing through exploratory data analysis [2]. For this, we used different machine learning techniques to predict the advertisement frauds. This approach is useful to several industries and consultancies that show interest in predicting the job advertising data.

In this paper, we are concerned with predicting the fake or real of job postings. The dataset consists of 18K job advertisements; out of which, about 800 are a fraud. The data consists of both textual information and meta-information about the jobs. This dataset is mostly used for classification models which can predict the job reports which are fraudulent. We need to perform data pre-processing on a dataset to check whether the data is properly loaded or not and missing values or NA values, etc. Drop all the missing rows among all those features. Now, apply exploratory data analysis to visualize the graphical representations [3] between input and the target features. Now, apply machine learning classification algorithms to predict whether the job is real or fake. Finally, to obtain better performance, we have to apply possible machine learning algorithms which give us the best result.

Machine learning algorithms are of three types; they are supervised learning, unsupervised learning, and reinforcement learning. In this paper, we have used only supervised classification algorithms such as baseline, logistic regression using TFIDF data, logistic regression using Count Vectorizer data, KNN, SVC, random forest, neural network—MLPClassifier w/"lbfgs", neural network—MLPClassifier w/"adam" [4]. In supervised learning, there will be input and output features, and we apply a machine learning technique to obtain a mapping function from input to the target feature. We also use performance metrics such as accuracy, precision, recall. To perform all these functionalities, we have used Python scripting language where Python has a lot of library functions like **NumPy**, **scipy**, **pandas**, **Matplotlib**, **seaborn**, **model_selection**, etc. [5], which have to be imported to perform exploratory data analysis, null value identification, visualization, and machine learning model functions.

The major objective of this research work is to classify job posting which is fraud or real by using machine learning algorithms. In the upcoming sections, we prepare the review of the related work, methodologies with detailed descriptions, prediction outcome, results and discussions. Hereby, the paper ends with a conclusion and future enhancement.

2 Literature Review

The authors, "Fernando Cardoso Durier da Silva", Rafael Vieira da Costa Alves, proposed, "Can Machines Learn to Detect Fake News? A Survey Focused on Social Media", to map the state of the art of fake news detection, defining fake news and finding the most popular machine learning techniques to apply [6]. They finalized that the regularly used method for automated fake news recognition is not just one classical machine learning technique, but it is a combination of classic techniques correlated by a neural network. They also identified the usage of a domain system that would combine different expressions and definitions of the fake news domain.

The authors "Murat Goksu, Nadire Cavus" proposed "Fake News Detection on Social Networks with Artificial Intelligence Tools: Systematic Literature Review". In this study, we found the latest article, which talks about identification of hoax news by using inclusive and reputable electronic databases [7]. The main aim is to reveal the sake of artificial intelligence tools used in recognition of fake news and their victory levels in different applications.

The authors "Xinyi Zhou, Reza Zafarani", proposed, "Fake News: A Survey of Research, Detection Methods, and Opportunities", to increase the integrative research of fake news, and they scrutinized about fake news research efficiently and also recognize and describe the fundamental hypothesis [8]. These studies mostly target on the dummy news; according to the false knowledge, it conveys writing style, diffusion patterns, and the dependability of its designers and spreaders. They underlined some prospective research tasks at the end of the survey by evaluating the quality of fake news and open issues on fake news.

"Naveed Hussain, Hamid Turab Mirza", the authors proposed, "Spam Review Detection Techniques: A Systematic Literature Review". In this study, overall, 76 existing studies were discussed and examined. The scientists estimate the studies based on how characteristics are extracted from review datasets and non-identical methods and strategies that are used to solve the review spam detection issue. Besides, this study inspects different measures that are used for the ranking of the review of spam detection techniques [9]. This research recognized two major feature extraction methods and two different perspectives to review spam detection. In addition to that, the study has associated various performance measures that are frequently used to estimate the accuracy of review spam detection replicas. Finally, this work presents an inclusive discussion about dissimilar feature extraction methods from review datasets, the proposed terminology of spam review detection methods, evaluation metrics, and publicly available review datasets.

"Sokratis Vidros, Constantinos Kolias", the authors proposed "Automatic Detection of Online Recruitment Frauds: Characteristics, Methods, and a Public Dataset". According to this study, they determine and narrate the attributes of the severe and timely novel cybersecurity research topic. It presents and extracts the knowledge on a publicly available dataset of job advertisements; recover from the use of a real-life system [10]. In this study, they explored the possible characteristics of employment hustle, an undiscovered, up to now, research area that calls for further analysis, and established EMSCAD.

The authors "Hadeer Ahmed, Issa Traore" proposed "Detecting opinion spams and fake news using text classification". In this paper, they present a new n-gram model to expose frequently fake contents on spam evaluations and fake news [11]. They considered and contrasted various features of extraction methods and machine learning classification algorithms. Exploratory evaluation is done by using the existing public datasets, and a newly inaugurate spam news dataset recommended very encouraging, and it increases accomplishment compared to the state-of-the-art methods.

3 Usage of Python With Machine Learning

Python is scripting and high-level programming language and allows us to focus on the main functionality of the application by using common programming tasks. Python is also used for expanding composite technological and numeric implementations and provides a feature to smooth data analysis and visualization in an efficient manner. Machine learning highly covers most of the algorithms, and Python makes it in a convenient way for inventors in testing [12]. Python recommends almost all the skill sets that are required for machine learning problems. Machine learning with Python provides solidity, elasticity, effectiveness, and a huge number of tools. Python helps inventors to be fertile and assured about the product that they are implementing, from the starting point of implementation to deployment and till the conservation level.

A library is a group of modules specified by several origins like PyPi which contains a pre-written piece of code that acknowledge [5] users to acquire and reach some services to carry out different operations. Machine learning requires libraries for pre-processing, visualization, to handle and transform the data. Some of the Python libraries needed for machine learning are:

Scikit-learn: This library handles machine learning classification, regression, and clustering algorithms.
Pandas: This library handles data structures and analysis. It is also used for reading datasets from various sources like csv, excel files, etc.
NLTK: used to handle natural language recognition and processing.
Matplotlib: used to create 2D plots, histograms, bar plots for visualization.
Sklearn: used to handle sklearn plugins, model selections, and evaluation metrics.

4 Methodology

In machine learning, there is a step-by-step procedure to perform pre-processing, analysing the data and model evaluation to make predictions (Fig. 1).

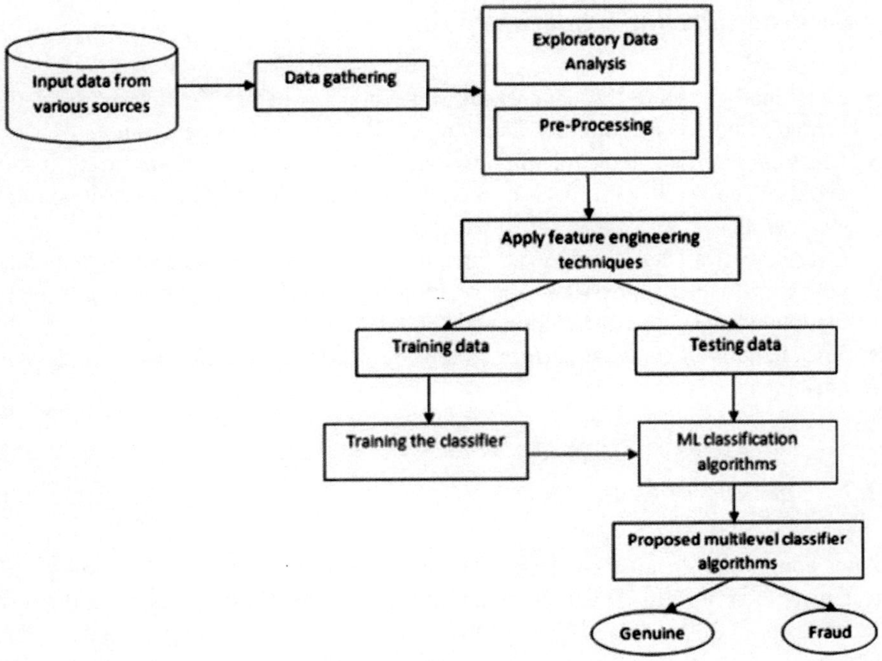

Fig. 1 System architecture for model evaluation

4.1 Description About Dataset

In this paper, we have used publicly available fake job advertisements dataset that is retrieved from the historical time series data. This dataset consists of 17,880 job descriptions and 18 features; out of which, about 800 are fake, and it consists of both textual and meta-information about jobs. Now, check whether the dataset had missing values or dummy values. If the data column contains a huge amount of data loss, then drop that column because it will decrease the model evaluation. If the data column has less amount of loss data, then fill the null values or missing values place by calculating any one of the central tendency measures like mean, median, and mode values.

After making changes, we have to apply visualization techniques to know how much the relationship exists between the input and output variables. After applying the correlation matrix, we determine that the "fraudulent" column is a target variable. The target variable is used to classify whether the job advertisement is real or fake; this can be represented in binary value like 0 and 1. Finally, apply classification algorithms to get the best results.

4.2 Procedure for Exploratory Data Analysis

- Statistically explore the data: check dimensions, whether the data is perfectly loaded or not, rows, columns, data types, and unique values per column.
- Cleaning the data: look for missing data [13]; if missing data occurred, try to avoid or replace it with mean or median or mode values to increase the model evaluation.
- Overview of a data: check head, tails, information of data to see complete data which is loaded, understand the relationship between the columns and their mappings, and check correlation and chi-square [14].
- Visualization of the data: perform graphical representation techniques on dataset attributes.

4.3 Splitting of Data

After completion of all these steps, we have to split the dataset into training and testing with split ratio 70 and 30. Training set consists of 70% of data, and a test set consists of 30% of data. This dataset splitting is done by using train_test_split library function imported from sklearn model_selection package [3]. In this paper, we have used different classification machine learning algorithms such baseline, Logistic Regression using TFIDF data, logistic regression using Count Vectorizer data, KNN, SVC, random forest, neural network—MLPClassifier w/"lbfgs", neural network—MLPClassifier w/"adam". Now, apply these algorithms on training data to train the model and classify the advertisement which is fake or real in order to perform predictions on testing data.

4.4 Feature Engineering Techniques

Feature engineering is a process of generating new input characteristics from the existing data via data mining methodologies [15]. Their features can increase the accomplishment of modelling algorithms. Generally, we imagine that cleaning is a procedure of deletion, and feature engineering is a procedure of inclusion.

In this paper, we apply feature engineering techniques like "one hot encode, TFIDF Vectorizer, and count Vectorizer" on textual data columns to see if those changes make anything to the model. In order to that, we fit a PCA model to reduce computational time.

One-hot encode: We applied this technique on "employee_type, required_experiance, required_education, industry, function, title, location" columns to split the values in a column to multiple flag columns and assign 0, 1 to them [16]. From Fig. 2, if we consider employee_type column, contains different categories

	title	location	description	telecommuting	has_company_logo	has_questions	employment_type	required_experience	required_educat
0	Marketing Intern	US, NY, New York	Food52, a fast-growing, James Beard Award-winn...	0	1	0	Other	Internship	Bachelor's Deg
1	Customer Service - Cloud Video Production	NZ, , Auckland	Organised - Focused - Vibrant - Awesome!Do you...	0	1	0	Full-time	Not Applicable	Bachelor's Deg
2	Commissioning Machinery Assistant (CMA)	US, IA, Wever	Our client, located in Houston, is actively se...	0	1	0	Full-time	Mid-Senior level	Bachelor's Deg
3	Account Executive - Washington DC	US, DC, Washington	THE COMPANY: ESRI – Environmental Systems Rese...	0	1	0	Full-time	Mid-Senior level	Bachelor's Deg
4	Bill Review Manager	US, FL, Fort Worth	JOB TITLE: Itemization Review ManagerLOCATION:...	0	1	1	Full-time	Mid-Senior level	Bachelor's Deg
6	Head of Content (m/f)	DE, BE, Berlin	Your Responsibilities: Manage the English-spea...	0	1	1	Full-time	Mid-Senior level	Master's Deg

Fig. 2 Before one-hot encode, TFIDF, Count Vectorizer

telecommuting	has_company_logo	has_questions	fraudulent	Contract	Full-time	Other	Part-time	Temporary	...	Public Relations	Purchasing	Quality Assurance	Research	Sales
0	1	0	0	0	0	1	0	0	...	0	0	0	0	0
0	1	0	0	0	1	0	0	0	...	0	0	0	0	0
0	1	0	0	0	1	0	0	0	...	0	0	0	0	1
0	1	0	0	0	1	0	0	0	...	0	0	0	0	1
0	1	1	0	0	1	0	0	0	...	0	0	0	0	0

Fig. 3 After one-hot encode

like full time, part-time, contract, others, etc. These will be placed as separate columns instead of "employee_type" column. The following figure shows the output before and after the one-hot encoding (Fig. 2 and Fig. 3).

TFIDF Vectorizer: Term frequency–inverse document frequency is a statistical measure which is evaluated by multiplying two metrics: how many number of times a word appears in a document, and the inverse document frequency of the word in a set of documents [16]. So that means if a word appears very common in many documents, then the number will approach to 0; otherwise, the number will be 1. The following figure represents the output of TFIDF Vectorizer (Fig. 4).

	telecommuting	has_company_logo	has_questions	fraudulent	Contract	Full-time	Other	Part-time	Temporary	Associate	...	write	writing	written	w communic
0	0.0	1.0	0.0	0.0	0.0	0.0	1.0	0.0	0.0	0.0	...	0.000000	0.0	0.000000	
1	0.0	1.0	0.0	0.0	0.0	1.0	0.0	0.0	0.0	0.0	...	0.000000	0.0	0.000000	
2	0.0	1.0	0.0	0.0	0.0	1.0	0.0	0.0	0.0	0.0	...	0.000000	0.0	0.000000	
3	0.0	1.0	0.0	0.0	0.0	1.0	0.0	0.0	0.0	0.0	...	0.063101	0.0	0.038383	

4 rows × 688 columns

Fig. 4 After TFIDF

	telecommuting	has_company_logo	has_questions	fraudulent	Contract	Full-time	Other	Part-time	Temporary	Associate	...	write	writing	written	written communication
0	0.0	1.0	0.0	0.0	0.0	0.0	1.0	0.0	0.0	0.0	...	0.0	0.0	0.0	0.0
1	0.0	1.0	0.0	0.0	0.0	1.0	0.0	0.0	0.0	0.0	...	0.0	0.0	0.0	0.0
2	0.0	1.0	0.0	0.0	0.0	1.0	0.0	0.0	0.0	0.0	...	0.0	0.0	0.0	0.0
3	0.0	1.0	0.0	0.0	0.0	1.0	0.0	0.0	0.0	0.0	...	1.0	0.0	1.0	0.0
4	0.0	1.0	1.0	0.0	0.0	1.0	0.0	0.0	0.0	0.0	...	0.0	0.0	0.0	0.0

5 rows × 688 columns

Fig. 5 After Count Vectorizer

Count Vectorizer: it is used to convert a block of text documents to a vector of term counts. It also enables pre-processing of text data before generating the vector representation [17]. Here, we apply the same procedure to Count Vectorizer, which is done before for TFIDF Vectorizer (Fig. 5).

Finally, we got two different data frames those are vectorized. Now, apply the logistic regression algorithm with TFIDF data and count Vectorizer data to determine which model is best. Similarly, apply to the remaining algorithms.

4.5 Prediction Algorithms

Logistic Regression with TFIDF: Take TFIDF data, store the input columns into X and target column into Y variables, and split them into train and test sets. Now, apply logistic regression model and then make predictions [18]. We obtain 59% accuracy by calculating the performance metrics on the predictions [19].

Logistic Regression with Count Vectorizer: Here, the same process is repeated as before but apply logistic regression on Count Vectorizer data and then make predictions [19]. Now, we obtain 55% accuracy. Count Vectorizer is not accurate than the previous model, and it is decreased by four percentage points. Thus, we use only TFIDF data to other models.

***K*-Nearest Neighbours:** KNN is supervised machine learning used for both classification and regression problems. It uses the data to classify into new data points based on comparison measures [20]. The data is assigned to the new class, which has the nearest neighbours [21]. As you increase the number of nearest neighbours and the value of *k*, the accuracy might increase. Similarly, we apply this algorithm on TFIDF data and obtain 20 neighbours, and the accuracy is 58%.

Support Vector Classifier: SVC is used to find a hyperplane in an n-dimensional space that classifies the data points and tries to fit the best line among the predefined error value [22]. SVC has few important keywords such as the kernel, hyperplane, boundary line, and support vector. In this algorithm, we have used a linear kernel. We got 53% accuracy.

Random Forest: Random forest is also used for classification problems [23], creates decision trees on arbitrarily chosen data samples, obtains projections from each tree, and picks the finest solution by means of voting. It also provides a pretty good indicator of the feature importance [24]. In this paper, we got 200 random estimators after applying random forest classifier and also obtain 52% accuracy.

Neural Nets—MLPClassifier with "lbfgs": It is a multi-layer perceptron classifier. This model minimizes the log-loss function using LBFGS [25]. LBFGS is an optimizer in the group of quasi-Newton methods. MLPClassifier trains repetitively at each step, and the incomplete derivatives of the loss method with respect to the model variables are calculated to upgrade the parameters [26]. It can also have a standardization term added to the loss function that decreases model parameters to stop overfitting. For this methodology, we obtain 69% accuracy.

Neural Nets—MLPClassifier with "adam": Adam is the value for solver attribute in MLPclassifier function () and works well on huge datasets in terms of both training time and validation score. "adam" refers to a random gradient-based optimizer. For larger datasets, "adam" is used, which gives fast and better performance [27]. For this algorithm, we obtain 71% accuracy.

5 Prediction Outcome

After applying all these algorithms on TFIDF dataset, we obtain different accuracy values for each algorithm. Among all of these algorithms, neural nets MLPClassifier with solver "adam" model got the highest accuracy with the percentage of 71, and this is the best model when compared to other models.

6 Results and Discussions

To assess the accuracy of all machine learning algorithms, classification performance metrics were used to compare one model with another [28]. To achieve the performance values, we have two formulas. The formulas for the calculation of performance metrics are given here [29] (Table 1):

$$\text{Accuracy} = (\text{TP} + \text{TN})/(\text{TP} + \text{TN} + \text{FP} + \text{FN}) \tag{1}$$

$$\text{Recall} = (\text{TP})/(\text{TP} + \text{FN}) \tag{2}$$

$$\text{Precision} = (\text{TP})/(\text{TP} + \text{FP}) \tag{3}$$

Table 1 Model accuracy score

S. No	Model	Accuracy score
1.	Logistic regression with TFIDF data	0.59
2.	Logistic regression with Count Vectorizer data	0.55
3.	*K*-nearest neighbours	0.58
4.	Support vector classifier	0.53
5.	Random forest	0.52
6.	Neural network—MLPclassifier with "lbfgs"	0.69
7.	Neural network—MLPclassifier with "adam"	0.71

```
0.7117329081434238
              precision    recall  f1-score   support

         0.0       0.98      0.98      0.98      1196
         1.0       0.46      0.44      0.45        43

    accuracy                           0.96      1239
   macro avg       0.72      0.71      0.72      1239
weighted avg       0.96      0.96      0.96      1239
```

Fig. 6 Neural network—MLPclassifier with "adam" output

$$F - \text{measure} = (2 * \text{Recall} + \text{Precision})/(\text{Recall} + \text{Precision}) \quad (4)$$

where

TP is "True Positive"
TN is "True Negative"
FP is "False Positive"
FN is "False Negative"

Final Result:
Figure 6

7 Conclusion and Future Enhancement

Fake news is disinformation spread through the Internet or traditional news media. It is complicated for student or jobseekers to identify which advertisement is fraud or real. To avoid such conflict, we make predictions on the dataset to know which job post is legitimate or fake. Finally, we conclude that the prediction of fake job advertisements has done and we observed how the data is classified as well as the job is spam or not. We had worked on feature engineering techniques like one-hot encode,

TFIDF Vectorizer and count Vectorizer to improve the efficiency of the model. For predicting fake job advertisements, we applied classification algorithms. Among these entire algorithms, neural network—MLPClassifier—with "adam" model gave us the best accurate result with 71%. For future enhancement, in order to reduce the system time for the prediction process and apply more efficient algorithms to reduce the error rate and increase the accuracy score.

References

1. C.J. Awati, R. More, A review on credit card fraud detection using machine learning, (IJSTR, 2019). ISSN: 2277-8616
2. I. Sadgali, N. Sael, Performance of machine learning techniques in the detection of financial frauds, (ICDS, 2019). https://doi.org/10.1016/j.procs.2019.01.007
3. M. Schoenberger, Improving university e-learning with exploratory data analysis and web log mining, (IEEE, 2011). https://doi.org/10.1109/tassp.1979.1163294
4. D. Varmedja, M. Karanovic, Credit card fraud detection—machine learning methods, (IEEE, 2019). https://doi.org/10.1109/infoteh.2019.8717766
5. Y.C. Huei, Benefits and introduction to python programming for freshmore students using inexpensive robots, (IEEE, 2014). https://doi.org/10.1109/tale.2014.7062611
6. F. Cardoso Durier da Silva, R. V. da Costa Alves, Can machines learn to detect fake news? a survey focused on social media, in *HICSS* (2019), ISBN: 978-0-9981331-2-6
7. M. Goksu, N. Cavus, Fake news detection on social networks with artificial intelligence tools: systematic literature review, in *AISC* (2019)
8. X. Zhou, R. Zafarani, Fake news: a survey of research, detection methods, and opportunities, (ACM, 2018). arXiv:1812.00315
9. N. Hussain, H.T. Mirza, Spam review detection techniques: a systematic literature review, Appl. Sci. **9**, 987 (2019). https://doi.org/10.3390/app905098
10. S. Vidros, C. Kolias, Automatic detection of online recruitment frauds: characteristics, methods, and a public dataset, Future Internet **9**(6) (2017). https://doi.org/10.3390/fi9010006
11. H. Ahmed, I. Traore, Detecting opinion spams and fake news using text classification, full-10.1002-spy2.9 (2017)
12. Z. Dobesova, Programming language python for data processing, (IEEE, 2011). https://doi.org/10.1109/iceceng.2011.6057428
13. A. Nasser, D. Hamad, C. Nasr, Visualization methods for exploratory data analysis, (IEEE, 2006). https://doi.org/10.1109/ictta.2006.1684582
14. S. Kaski, Learning metrics for exploratory data analysis, (IEEE, 2001). https://doi.org/10.1109/nnsp.2001.943110
15. D. Deepa, Raaji, Sentiment analysis using feature extraction and dictionary-based approaches, (IEEE, 2019), https://doi.org/10.1109/i-smac47947.2019.9032456
16. F. Haque, M.M.H. Manik, Opinion mining from bangla and phonetic bangla reviews using vectorisation methods, (IEEE, 2019). https://doi.org/10.1109/eict48899.2019.9068834
17. A. Humeau-Heurtier, Texture feature extraction methods: a survey, (IEEE, 2019). https://doi.org/10.1109/access.2018.2890743
18. P. Rao, J. Manikandan, Design and evaluation of logistic regression model for pattern recognition systems, (IEEE, 2016). https://doi.org/10.1109/indicon.2016.7839010
19. T. Haifley, Linear logistic regression: an introduction, (IEEE, 2002). https://doi.org/10.1109/irws.2002.1194264
20. A. Moldagulova, R.B. Sulaiman, Using KNN algorithm for classification of textual documents, (IEEE, 2017). https://doi.org/10.1109/icitech.2017.8079924

21. S. Taneja, C. Gupta, K. Goyal, D. Gureja, An enhanced *k*-nearest neighbor algorithm using information gain and clustering, (IEEE, 2014). https://doi.org/10.1109/acct.2014.22
22. S. Mahdevari, K. Shahriar, S. Yagiz et al., A support vector regression model for predicting tunnel boring machine penetration rates. Int. J. Rock Mech. Min. Sci. **72**, 214–229 (2014)
23. A. Paul, D.P. Mukherjee, P. Das, Improved random forest for classification, (IEEE, 2018). https://doi.org/10.1109/tip.2018.2834830
24. S.V. Patel, V.N. Jokhakar, A random forest based machine learning approach for mild steel defect diagnosis, (IEEE, 2016). https://doi.org/10.1109/iccic.2016.7919549
25. E.C. Popovici, O.G. Guta, MLP neural network for keystroke-based user identification system, (IEEE, 2013). https://doi.org/10.1109/telsks.2013.6704912
26. G. Singh, M. Sachan, Multi-layer perceptron (MLP) neural network technique for offline handwritten Gurmukhi character recognition, (IEEE, 2014). https://doi.org/10.1109/iccic.2014.7238334
27. S.B. Mohod, V.N. Ghate, MLP-neural network based detection and classification of power quality disturbances, (IEEE, 2015). https://doi.org/10.1109/icesa.2015.7503325
28. M. Fatourechi, R.K. Ward, S.G. Mason, J. Huggins, A. Schlögl, Comparison of evaluation metrics in classification applications with imbalanced datasets, (IEEE, 2008). https://doi.org/10.1109/icmla.2008.34
29. P. Maillard, D.A. Clausi, Comparing classification metrics for labeling segmented remote sensing images, (IEEE, 2005). https://doi.org/10.1109/crv.2005.28

Emotion Recognition Through Human Conversation Using Machine Learning Techniques

Ch. Sekhar, M. Srinivasa Rao, A. S. Keerthi Nayani, and Debnath Bhattacharyya

Abstract Emotion recognition will perform a hopeful role in the field of artificial intelligence, uniquely in the case of human–machine interface development. It is the process of recognizing and analyzing the emotion of chat and text, i.e., moods of the people can be easily found, and this process can be used in various social networking websites and various business-oriented applications. The mood of the person will be confirmed by making proper observations, i.e., by asking multiple questions until his/her situation is correctly recognized. Based on his/her answers, it tries to refresh his/her mind if he/she is in a bad mood (mild) by providing the refreshments based on the interests of the person that were gathered initially. The proposed system goes about as a choice emotionally supportive network and will demonstrate to be a guide for the doctors with the analysis. The user expresses his or her feelings, and the Chatbot replies accordingly. Using Python packages, NLTK and Flair, we analyze the intensity of the emotion.

Keywords Emotion recognition · Naïve Bayes classifier · NLTK vader · TextBlob · Flair · DeepMoji

Ch. Sekhar (✉) · M. S. Rao
Department of CSE, Vignan's Institute of Information Technology, Visakhapatnam, India
e-mail: sekhar1203@gmail.com

M. S. Rao
e-mail: srinivas.marada22@gmail.com

A. S. K. Nayani
Department of ECE, Matrusri Engineering College, Hyderabad, India
e-mail: naskeerthi@gmail.com

D. Bhattacharyya
Department of CSE, KL Deemed to be University, Guntur, Andhra Pradesh, India
e-mail: debnathb@kluniversity.in

D. Bhattacharyya and N. Thirupathi Rao (eds.), *Machine Intelligence and Soft Computing*, Advances in Intelligent Systems and Computing 1280,
https://doi.org/10.1007/978-981-15-9516-5_10

1 Introduction

The human-made intellectual approach is in terms of artificial intelligence as a superset for machine and deep learning techniques. The machine learning that activates systems to usually take in and improve for a reality without being explicitly altered. ML revolves around the headway of computer applications that can get to data and use it discover for themselves. The knowledge begins with identification or information, for instance, points of reference, direct understanding or direction, to scan for models in data and choose better decisions later on subject to the models that we give [1, 2].

The primary point is to allow the PCs to adjust without human intervention or help regularly and change exercises as constraints. Not all the time your loved ones will be present with you to share your feelings or thoughts. So, here is a Chatbot that can interact with you just like a human friend when you feel bored or anxious or angry and you need to badly pour out your thoughts feelings you can use this bot. The movie name called HER realized in the year 2013. The theme of the movie was showing how the person is interested female voice assistant of artificial intelligence application, how he loved to overcome his loneliness, depression and a boring job. At that time film look like friction. Nowadays, AI provides various applications in the medical diagnosis. The main objective of this paper developed a Chatbot, to recognize and analyze the emotions of persons, when he/she interacted with Chatbot. All the emotions are classified in the emotion classifier, and each emotion is given a respective emoji and a color. Chatbot that can interact with he/she just like a human friend whenever you feel bored or anxious or angry and you need to badly pour out your thoughts or feelings, you can use this bot.

Acheampong et al., study on emotion from texts and highlight the main strategies selected by researchers in the perspective of text-based emotion detection systems. Figure 1 shows the survey of various papers from 2010 to 2020 [3].

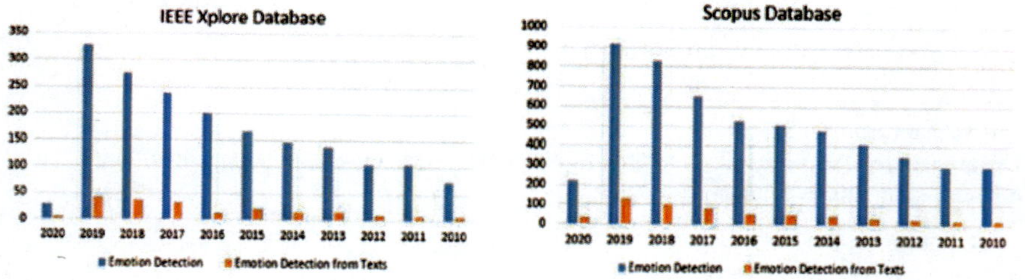

Fig. 1 Visualization of research in emotion detection and emotion detection from texts various research databases [3]

2 Literature Review

Recognizing the sensitive nature of a character by examining text studies composed by him/her appears challenging but also required numerous occasions. Most of the times, textual representations do not solely be straight applying excitement messages but also result from the formation of the meaning of concepts and interaction of ideas described in the text document. Classifying the emotion of the text plays a vital part in human–computer intercommunication. A person's talk may signify emotions, facial appearance and written text known as communication, facial and text-based sentiment [4].

Many researchers are done to achieve speech recognition and facial emotion recognition, and they succeeded. In the case of text-based emotion, identification needs improvement. Computational linguistics, the detection of human emotions in the text, is becoming increasingly important from an applicative point of view—the sentiment expressed as happiness, sorrow, excitement, wonder, hatred, worry any many more. Since there is not any standard feeling word chain of command, the emphasis is on the related research about feeling in subjective brain science space. In 2001, Parrot composed a book named "Feelings in Social Psychology," in which he defined the mood frame and correctly discriminated the personal feelings through a feeling series of importance in six groups at a fundamental level which are love, joy, anger, sadness, fear and surprise [5, 6]. Certain different words additionally fall in optional and tertiary levels. Bearings to enhance the capacities of current strategies for content-based feeling identification are recommended in this paper.

Naive Bayes Classifier Algorithm: The Naive Bayes classifier depends on Bayes' hypothesis and arranges each an incentive as free of some other esteem. It enables us to foresee a class/classification, in view of a given arrangement of highlights, utilizing likelihood. Regardless of its effortlessness, the classifier does shockingly well and is regularly utilized because of the reality it beats increasingly advanced grouping techniques.

NLTK Vader: A sentiment analysis tool is able to analyze data on social media. A combination of lexical words with labeled is as per the semantic orientation, either positive or negative. This tool is able to identify how far the lexical toward positive or negative sentiment [7, 8].

TextBlob: Data type is in Python library and performs the basic NLP tasks using API. It placed on top of the NLTK module.

Flair: It is open-source NLP package developed by Zalando research and is available with PyTorch, useful framework for deep learning applications. The Zalando Research team has also delivered several pre-trained prototypes for NLP jobs. To identify whether a word depicts a person, location or names in the text with NER. Classifying the text based on the pattern or labels is using text classification [8].

Complete Research Objective Lifecycle

ML undertakings are exceptionally repetitive through the ML lifecycle. You will end up repeating on a segment until achieving a tasteful dimension of execution, at

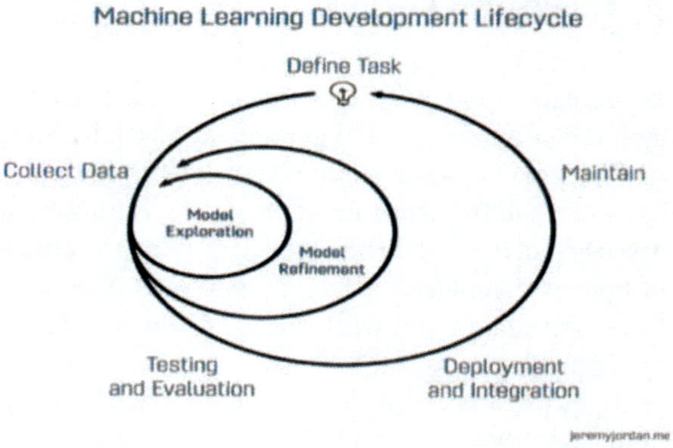

Fig. 2 Machine learning development lifecycle

that point continuing forward to the following errand (which might hover back to a considerably prior advance). In addition, a venture is not finished after you send the main form; you get criticism from true collaborations and reclassify the objectives for the following emphasis of organization [9].

In Fig. 2, it is showing that machine learning life cycle development process during the predicting the any dataset. It contains major stages like planning, collecting of data requirement, how the exploration of model done, later redefine model with proper hyperparameter and testing of the developed model. After successful completion of the model going deploy on the different platform using REST API2 [2].

3 Research Problem Analysis

Research Problem: Emotion recognition plays a promising role in the field of artificial intelligence, especially in the case of human–machine interaction interface development [9]. Using Chatbot, we can analyze the emotion of chat text, in terms of personal behavior, how he/she is giving response to the other end person. Based on the feelings, he/she expressed emotion in the form of text. This application can be used in various areas toward social interaction Web sites, business applications and the banking sector to respond to the customer for most common data. The type of emotion needs to identify and respond as per that. Hence, we need classification-based machine learning algorithms to classify the text using NLP.

Existing System: Emotion identification is the method of recognizing individual sentiment, most typically from facial expressions as well as from verbal expressions. At present, we are able to analyze the emotion of a person based on speech, and facial expression APIs and tools are available based on various machine learning methods like Bayesian networks and Gaussian mixture models.

Proposed System: The proposed system goes about as a choice emotionally supportive network and will demonstrate to be a guide for the doctors with the

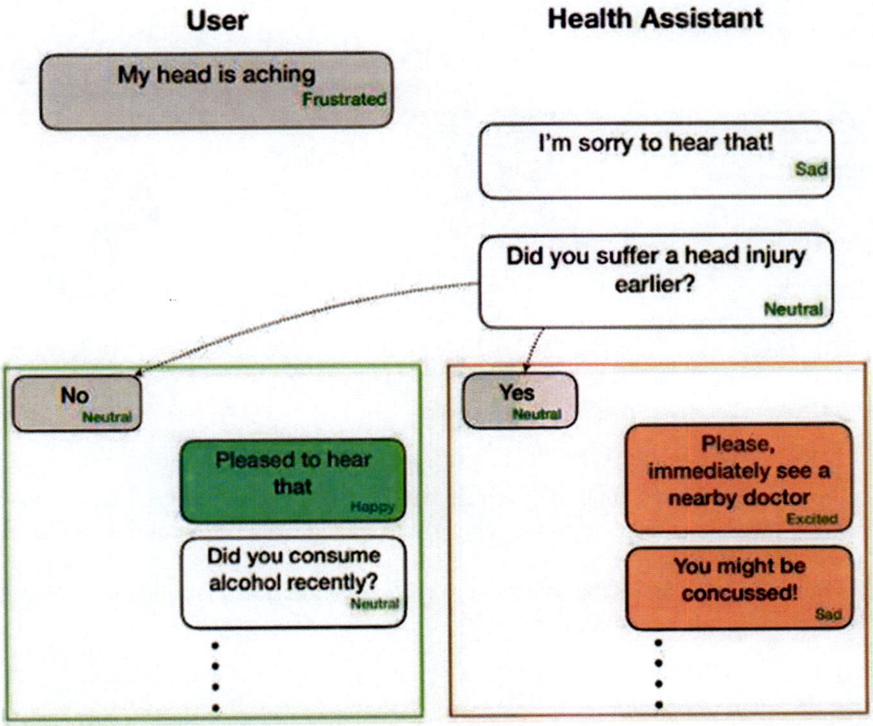

Fig. 3 Example of a health assistant chatbot

analysis. The user expresses his or her feelings, and the Chatbot replies accordingly. For analyzing the emotion of the user, we are using some Python packages which includes TextBlob, Flair, NLTK [10], and DeepMoji model to analyze the intensity of the emotion. The text entered by the user is divided based on keyword-based technique. The recognized keyword is compared with the emotions present in the emotion classifier [10]. The classifier then categorized this with the emoji and color present in the classifier.

Figure 3 shows the basic conversation of Chatbot application for hospitals.

4 Methodology

To find the emotion based on chat information between the person and Chatbot, we need to get the chat data based on the data needed to determine the score for each word based on the parameters of the previous stage [1]. The score can be determined with the frequency of the word and inversely proportional to its depth in the ontology. Based on the scores of each word, the higher the score word considered

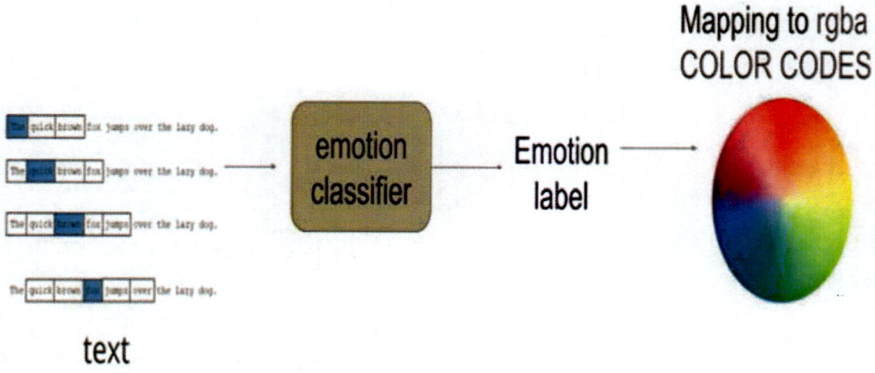

Fig. 4 Text to color classification of emotion state

as the emotional state. Input is given as text, the text is divided based on keyword splitting method, and the keyword is given a label and mapped to color codes based on emotion classifier.

Figure 4 above shows the basic text to color emotion state classification, reading text and make relevant color to the text.

Step 1: Train the chatbot with predefine tokens based on the conversation.
Step 2: Load chat dataset.
Step 3: Tokenization of data using lexical analyzer.
Step 4: Word frequency calculation.
Step 5: Train Naïve Bayes algorithm.
Step 6: Classify the positive and negative polarization words.
Step 7: Calculate the sentence emotion based on polarization value.
Step 8: Predict the emotion outcome.

Figure 5 shows the training of emotion identification step wise process.

Figure 6 shows that training of Chatbot with various tokens that can be able to respond to the user during the conversation.

Experimental Results

Figures 7 and 8 above show the experimental results of the user conversation with Chatbot. The number of words frequently occurrence is shown as histogram based on the emotion score and performance pair wise plot emotion recognition each parameter. Figure 9 shows the sentence polarity calculations.

5 Conclusion

The main scope of this paper is the Chatbot named, and we developed a Chatbot called "My Friend Bot," designed to recognize and analyze the emotions of a person.

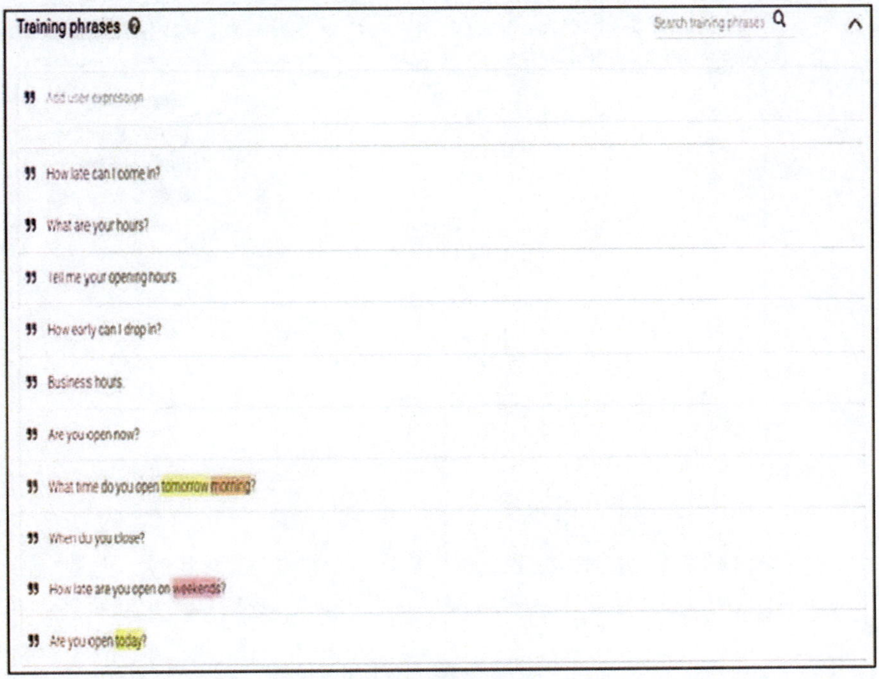

Fig. 5 Step wise process of emotion identification

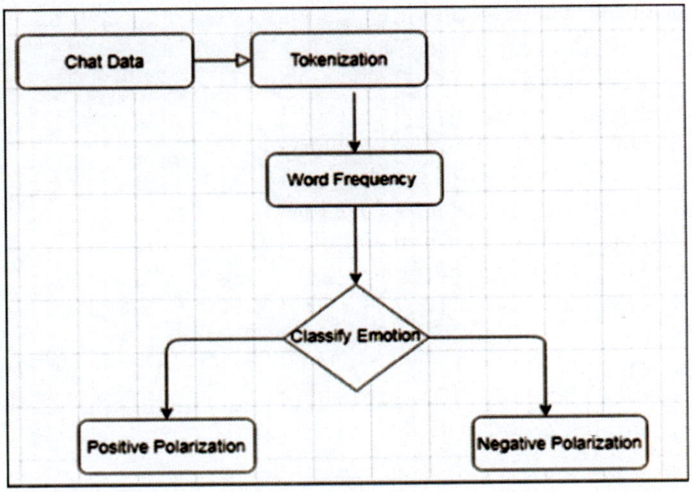

Fig. 6 Training of chatbot

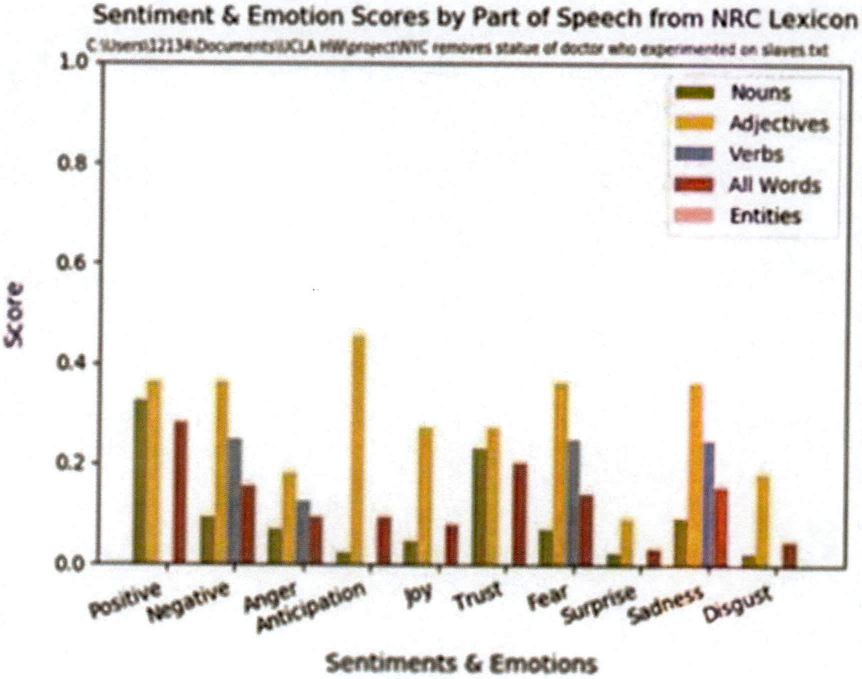

Fig. 7 Display histogram showing various emotion scores

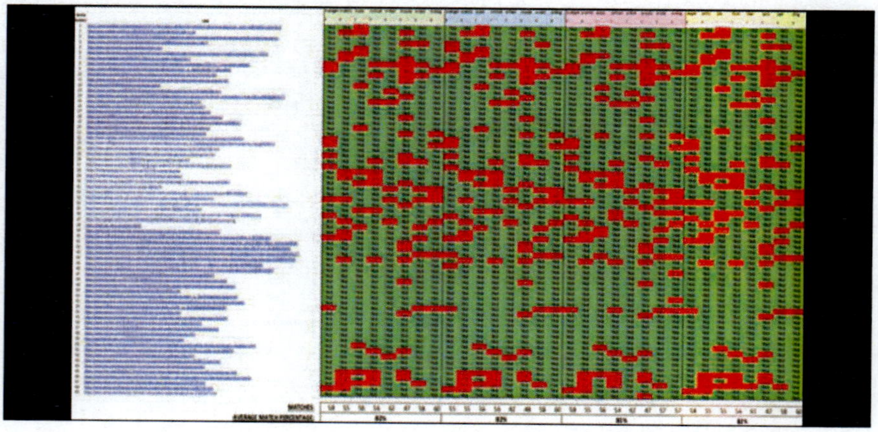

Fig. 8 Display pair wise plot emotion recognition each parameter

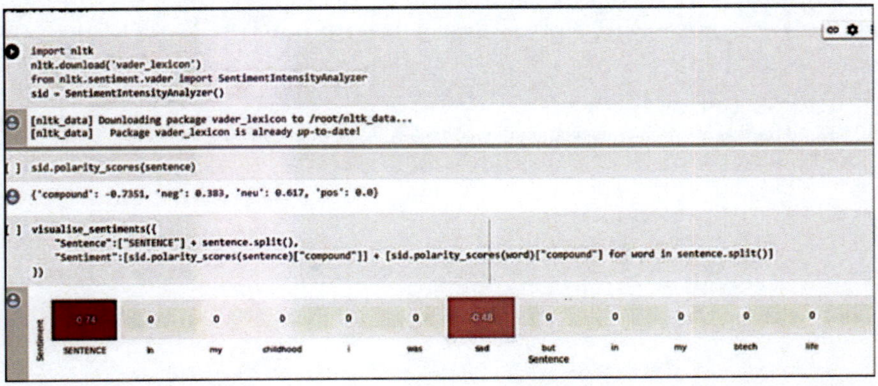

Fig. 9 Determine the sentence polarity

All the emotions are classified in the emotion classifier, and each emotion is given a respective emoji and a color. Emotion can be of in any form, as joy, sorrow, irritation, shock, hate and suspicion. We cannot say it is in standard level. Some people genuinely dislike human interaction.

Whenever they are forced to socialize or go to events that involve lots of people, they feel detached and awkward. So, here is a Chatbot that can interact with you just like a human friend whenever you feel bored or anxious or angry and you need to badly pour out your thoughts or feelings you can use this bot. The future work, in this article, we worked out with Naive Bayes classifier and achieved 76% emotion detection rate based on Chatbot conversation. To improve performance, we can use shallow-based machine learning methods.

References

1. S.N. Shivhare, S. Khethawat, Emotion detection from text. https://doi.org/10.5121/csit.2012.2237. May, 2012
2. E. Batbaatar, M. Li, K.H. Ryu, Semantic-emotion neural network for emotion recognition from text. IEEE Acc. (2019)
3. F.A. Acheampong, C. Wenyu, H. Nunoo-Mensah, Text-based emotion detection: advances, challenges, and opportunities. https://doi.org/10.1002/eng2.12189
4. S. Moharreri, N.J. Dabanloo, K. Maghooli, Detection of emotions induced by colors in compare of two nonlinear mapping of heart rate variability signal: triangle and parabolic phase space (TPSM, PPSM). J. Med. Biol. Eng. **39**, 665–681 (2019). https://doi.org/10.1007/s40846-018-0458-y
5. R. Cowie, E. Douglas-Cowie, N. Tsapatsoulis, G. Votsis, S. Kollias, Emotion recognition in human-computer interaction. IEEE Sig. Process. Mag. **18**(1), 32–80 (2001). https://doi.org/10.1109/79.911197
6. Physchology research reference emotion. http://psychology.iresearchnet.com/socialpsychology/emotions/#:~:text=A%20parallel%20point%20is%20that,that%20are%20often%20explicitly%20emotional
7. F. Calefato, F. Lanubile, N. Novielli, EmoTxt: a toolkit for emotion recognition from text

8. A. Akbik, T. Bergmann, D. Blythe, K. Rasul, S. Schweter, R. Vollgraf, FLAIR: an easy-to-use framework for state-of-the-art NLP. https://doi.org/10.18653/v1/n19-4010
9. V.V. Ramalingam, A. Pandian, A. Jaiswal, N. Bhatia, Emotion detection from text, J. Phy.: Conf. Ser. (2018)
10. Build natural and rich conversational experiences. https://dialogflow.com/

Intelligent Assistive Algorithm for Detection of Osteoarthritis in Wrist X-Ray Images Based on JSW Measurement

Anil K. Bharodiya and Atul M. Gonsai

Abstract Osteoarthritis is a common disease in human having age more than 40 years. It happens in bone joints such as knee, hips, hand, wrist. In osteoarthritis, joint cartilage is ruptured, and as a result, bone rubs against bones with a severe pain. Orthopedic doctors are advising patients to take an X-ray image of joints to diagnose osteoarthritis. After physical X-ray printing, a doctor or physician studies it physically to diagnose joint space width (JSW). But, it is not exactly possible to measure accurate JSW because of visual interpretation and JSW may be different from orthopedician to orthopedician. In this research paper, we have presented an intelligent algorithm to assist an orthopedician to measure JSW of wrist X-ray images automatically and to diagnose whether patient suffers from wrist osteoarthritis or not based on JSW measurement using digital image processing. We have given name to this algorithm as WODJSW (wrist osteoarthritis detection using JSW). The algorithm is divided into steps such as inputting X-ray images of human hand, ROI detection, edge detection, measurement of vector height, conversion of vector height into millimeter, comparison of JSW with standard wrist joint width and making decision. To diagnose osteoarthritis, we have conducted experiments on 75 X-ray images of human hand collected from openly available dataset and achieved an average accuracy of 96%.

Keywords Osteoarthritis · Orthopedics · ROI · JSW · X-ray · Image processing

---

A. K. Bharodiya (✉)
BCA Department, UCCC & SPBCBA & SDHG College of BCA and I.T., Surat, Gujarat, India
e-mail: anilbharodiya@gmail.com

A. M. Gonsai
Department of Computer Science, Saurashtra University, Rajkot, Gujarat, India
e-mail: atul.gosai@gmail.com

D. Bhattacharyya and N. Thirupathi Rao (eds.), *Machine Intelligence and Soft Computing*, Advances in Intelligent Systems and Computing 1280,
https://doi.org/10.1007/978-981-15-9516-5_11

1 Introduction

Human body is collection of bones, muscles, blood, water and many other components. Bone has a great importance in physical working of human body. Joint makes a connection between bones. This joint provides bending and flexibility of human hands, legs, wrist or other parts of human body. The joint contains cartilage or slick which is one type of cushioning surface on the ends of bones. When the bone rubs against bone in the joint area, it decreases the strength of the cartilage, and it results in joint pain, swelling, stiffness and sometime leads to deformity of joints in human body. This is called arthritis which is musculoskeletal condition that often causes disability and loss of joint functions especially in the wrists, fingers, feet and knees joints [1, 2].

Osteoarthritis and rheumatoid arthritis are two major types of osteoarthritis. Osteoarthritis is a common arthritis in humans of many countries under various joints such as knee, hips, hand and wrist. According to World Health Organization (WHO), 18.0% of women and 9.6% of men age over 60 years have symptoms of osteoarthritis throughout the world. Main problem found in 80% of the patients of osteoarthritis is difficulty in physical movement, while 25% of them are not able to perform their routine activities [3]. Arthritis can occur in 23% of the adult population it means 54 million people approximately in USA only, and it is more common disease that results into disability for the last 15 years. [1] And in India, it is 22% to 39% [3]. Women are more suffering from OA as compared to men. Women having the age of 65 years contain major symptoms and 70% of women those over 65 years contain any of the radiological evidence of osteoarthritis.

In modern medical treatment, CADD has its vital role to play. Osteoarthritis can be detected by radiography, ECG, computed tomography, magnetic resonance imaging, US image, etc. Among these modalities, X-ray is the best solution to detect. JSW that is joint space width is an important measurement to detect arthritis from radiological images that is X-ray. JSW is used to measure the space between two bones in joint cartilage, and it helps to analyze the severity of osteoarthritis [4].

In this research paper, we have discussed an intelligent assistive algorithm to detect osteoarthritis from X-ray images of human wrist using image processing. We have achieved an average accuracy of 96% to detect wrist osteoarthritis detection. The result of our research paper will help to physician or orthopedics to identify wrist arthritis in early stage.

The structure of this research paper is organized as follows: Sect. 2 explains related literature work. Section 3 explains on methodology. Section 4 explains proposed algorithm to detect wrist osteoarthritis. Section 5 details out the analysis of algorithm. Finally, the research paper is concluded in Sect. 6.

2 Related Works

Many researchers have shown their interest to detect wrist osteoarthritis using X-ray images of human arm. The following text is discussed on the work done by researchers in terms of algorithms, methods, tools and devices.

Gornale et al. [5] have introduced a survey on many medical imaging techniques for the osteoarthritis assessment. This paper also discovers the X-ray imaging and magnetic resonance imaging (MRI) techniques for detection and classification of osteoarthritis in comparative and descriptive manner. They have concluded that analysis of MRI and X-ray images is done manually by the doctor or physician which is tedious and time-consuming and unpredictable. There should be an automatized method to detect osteoarthritis.

Wagaj and Patil [6] developed an algorithm on pixel-based classification and SVM classifier. This includes steps such as input image, segmentation, feature extraction, classification and analysis. They have achieved overall accuracy of 93.75%. However, they have not considered dataset of abnormal images.

Bhavyashree and Rao [7] have worked on pixel thickness to identify osteoarthritis using human X-ray images. They used Bezier spline to measure a discrete dimension from image. They have tested 31 images and achieved sensitivity of 92%, specificity of 88%, precision of 87% and accuracy of 92%. They have not specified much on dataset sources.

Chokkalingam and Komathy [8] have developed an intelligent method for diagnosis of rheumatoid arthritis. They have used gray-level co-occurrence matrix. They have trained system with the use of neural network. However, they have not specified accuracy and size of the dataset.

In Paper [9], authors have developed a technology to quantify osteoarthritis from 20 knee X-ray image-based support vector machine. An X-ray image is divided into nine dissimilar blocks, selection of 4, 5 and 6 image blocks and finally, feature extractions and classifications to detect osteoarthritis. They have achieved 80% and 86.67% accuracy in normal and impacted X-ray images, respectively. The dataset size is very small.

Hegadi et al. [10] have worked on detection of osteoarthritis from knee X-ray images using artificial neural network. The proposed method extracts synovial cavity region from knee X-ray images based on global thresholding and curvature values like standard deviation, mean, range and skewness. They have claimed 100% accuracy of proposed method. However, images are not classified into different grades.

Kale and Bhisikar [11] have presented an automated framework to detect and quantify joint space width from X-ray images and classified into three stages—abnormal, normal and sever. The accuracy of proposed method was 92%, 95%, 70% and 100% in joint location detection, normal stage classification, abnormal stage classification and severe stage classification, respectively. They have used only 60 test images.

Chan and Dittakan [12] have classified knee X-ray image into six different stages to detect osteoarthritis. They have used 130 images as a dataset and concluded that

ROC curve gives good performance in medial tibia and logistic regress algorithm is the suitable classification algorithm. Large dataset with deep learning may perform a better result.

Stachowiak et al. [13] have developed a DSS to detect osteoarthritis from hand and knee joints based on the fractal analysis of X-ray images. They have selected different types of classifiers for discrimination of ROI. Texture of TB had the prediction accuracy of 0.77 and 0.75 areas under the curve (AUC), respectively. Accuracies of 90.51% and 80% for the detection of radiographic OA and prediction of radiographic OA progression, respectively, are achieved in the DMC classification. These outcomes can be improved in future.

Subramoniam [14] has developed a non-invasive method for arthritis analysis based on segmentation of the image. The developed algorithm detects abnormal and normal from the bone joints. This method is semi-automated. Accuracy of the algorithm has not been analyzed.

Huo et al. [15] have presented the method to detect the joint space width of 3 wrist joints that are least affected by bone overlapping and are frequently involved in RA. It was noticed that 90% of the joints had a JSW deviating less than 20% from the mean JSW of manual indications, with the mean JSW error less than 10%.

Bear et al. [16] have presented a method to correlate joint space height (JSH) with arthroscopic grading of wrist arthritis. They have concluded that JSH ratio accurately grades radio scaphoid arthritis and detects early radio lunate arthritis. However, the percentage of accuracy of proposed method was not measured.

Sharp et al. [17] have developed a computer-oriented method for calculating JSW and estimating volume of erosion to detect arthritis from finger and wrist X-ray images. Estimates of erosion volume in two experiments showed a greater disparity between gold and placebo therapies than erosion scores in the test set ($P = 0.049$ and $P = 0.016$ versus $P = 0.27$). Authors have analyzed results by comparing with the software of the NIH.

Schenk et al. [18] have presented a technique to validate the accuracy of an automated JSW measurement mechanism. The method includes a systematic evaluation of the sources of errors, from image acquisition to automated measurements. In the two series of radiographs, the system could automatically locate and measure 1003/1088 (92.2%) and 1143/1200 (95.3%) individual joints, respectively. The proposed method should also be further validated for normal and RA joints in order to establish reference values.

Pandey et al. [19] presented an approach to detect osteoarthritis by using knee X-ray images using image processing method, extraction of JSW and comparison with Ahlback grading. They have considered 100 knee normal and abnormal X-ray images. The automated system worked well on clear images of knee, excluding the damaged images.

Banerjee et al. [20] presented a technique to detect hand osteoarthritis using X-ray images based on CNN. Authors have experimented on nine images of hand and achieved 90% accuracy to detect osteoarthritis. This work can be extended to improve accuracy and to develop same algorithm for knee and hip joints by considering large dataset.

Shamir et al. [21] have described an automated method to detect radiographic osteoarthritis (OA) using knee X-ray images. This is based on the Kellgren–Lawrence (KL) classification grades, which relate to dissimilar grades of OA seriousness. They have achieved accuracy of 91.50% from dataset of 350 X-ray images.

Lim et al. [22] have developed a method to detect osteoarthritis using statistical data based on deep neural network. They have conducted study on 5749 subjects, and their experiments resulted that the proposed method using DNN with scaled PCA resulted in 76.8% of AUC and minimized the effort to generate features. This method can also be developed for health behavior dataset to detect osteoarthritis.

Kurniasih and Pratiwi [23] have proposed a method to detect osteoarthritis using self organizing map based on Ossa Manus X-ray. The method contains steps such as image contrast repair phase, grayscale conversion, thresholding, histogram processing, training and testing on images. They have used 42 X-ray images and achieved 92.8% accuracy from the result of testing. The dataset size can be increased to measure the accuracy.

Kiselev et al. [24] have developed a method to detect osteoarthritis using acoustic emission analysis. Authors have examined 24 patients having arthritis. Sensitivity was 0.92, and the specificity was up to 0.78. They have considered less number of participants.

According to Hunter and Bierma-Zeinstra [25], policy-makers and healthcare providers have to understand that the worldwide population will be suffered with osteoarthritis in the future, and they have to decide whether it will lead to the world's most serious disease. There should be such mechanism to prevent this disease from the entire world. It is the ethical role of not only researchers but also imaging technologies.

Above presented broad related work guarantees that no method/technique/algorithm is 100% perfect and achieved accuracy exactly 100%. Each and every method achieves different accuracy, precision, specificity or sensitivity and also suffers from one or more negative points. It provides us persistently inspiration to work in the field of osteoarthritis detection from human X-ray image to increase the accuracy of clinical diagnostic based on the JSW measurement. The ultimate goal of this research paper is to discuss an intelligent assistive algorithm or technique to detect osteoarthritis using wrist X-ray images.

3 Methodology

In this division, we have presented a comprehensive methodology that we have used to detect osteoarthritis using X-ray images of human arm. The methodology is divided into two sections, viz. flowchart which is used to explain proposed algorithm and evaluation metrics which are used to measure the effectiveness of the proposed algorithm. Figure 1 depicts the step-by-step flowchart of proposed method/algorithm to represent the flow of an algorithm.

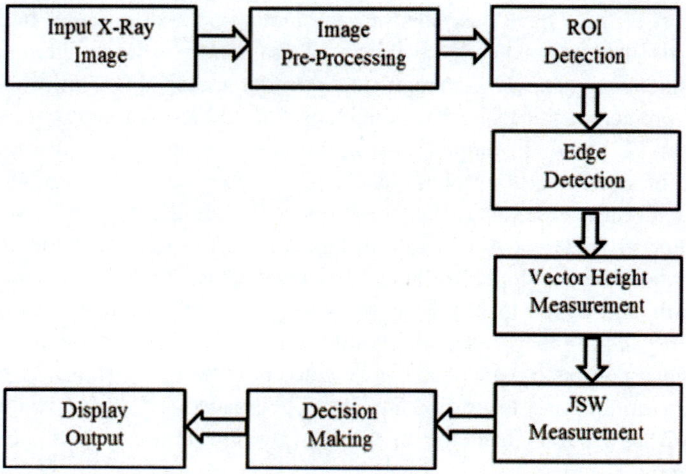

Fig. 1 Flowchart to detect wrist osteoarthritis

3.1 Input X-Ray Images

We have gathered X-ray images of human hand from National Health Portal of India, National Institute of Health (NIH) and selected hospitals in person. The dataset comprises 75 such X-ray images to conduct experiments for detection of wrist osteoarthritis. These images are fed to the developed algorithm as an input in JPG format having resolution of 512 * 512 pixels.

3.2 Image Pre-processing

X-ray image may usually contain improper blurring, noise or out of focus, and hence, image is not displayed clearly. If image is not perfect, it becomes difficult to interpret and process further. Image enhancement is the technique used to pre-process or enhance the quality of the image by removing distortions and noise. An X-ray image mostly contains salt and pepper noise. According to papers [26, 27], to remove salt and pepper noise, median filter is the best solution and it also improves the quality of the image. Inputted images are cropped into 512 * 512 pixel size after enhancement. Here, we have used median filter for image enhancement. Image contrast also plays a dominant role for image processing. As per paper [28], contrast limited histogram equalization (CLAHE) method can be used to improve contrast of any image. Here, we have also used CLAHE for contrast and blur improvement.

3.3 ROI Detection

This is very important phase of the algorithm in which location of the wrist is extracted which is our region of interest. In this phase, ROI is extracted based on maximum connected pixels in inputted X-Ray image. Here, our ROI is the wrist part of the hand X-ray image. Wrist part of the X-ray is in the form of small arc so that it can be detected by eight connected neighbor pixels of the middle. The ROI image is stored as a separate image for further processing.

3.4 Edge Detection

Edge is the part of the image boundary where pixel intensity changes abruptly. As per our paper [29], edge detection through statistical range gives efficient edge detection. Here, we have used algorithm given in our paper. ROI image is transformed into binary format after edge detection for further process.

3.5 Vector Height Measurement

In binary ROI image, there are two lines or arcs like top arc and bottom arc. In this phase, distance between top and bottom arcs is measured for each column in terms of number of pixels. Distance measurement starts from first 1′s pixel to last 1′s pixel of starting from first column to last column of the ROI binary image. The distance (in terms of pixels) of each column is stored in a vector of matrix. Finally, the mean of distance vector is calculated for further process.

3.6 JSW Measurement

Wrist is one type of joint in the human arm. In this phase, mean of the vector which is calculated in the above step, is transformed into millimeter by using a formula 1 pixel = 0.264583333 mm [30]. The result that we get after conversion of mean vector pixel into millimeter is termed as joint space width which is used in further process.

3.7 Decision Making

In this, phase decision is made about osteoarthritis. As per [31, 32], normal JSW in man and woman without osteoarthritis is 1 to 2 mm. Based on these criteria, we have classified osteoarthritis detection result into three different classes such as "Abnormal" if wrist JSW is between 0 and 0.49 mm, "Moderate" if wrist JSW is between 0.50 and 0.85 mm and "Normal" if wrist JSW is greater than 0.85 mm.

3.8 Display Output

In this final phase, we have displayed the result for orthopedics doctor. The output can be any one from three options of type of osteoarthritis such as Abnormal, Moderate and Normal. Orthopedics doctor will get help from this algorithm and prescribe appropriate advice to the patient suffering from wrist osteoarthritis.

Metrics are measurement or evaluators for measuring the performance of any method, technique or algorithm. We have considered accuracy as measure of the performance [33]. The algorithm is easy to implement through certain software tool for machine learning purpose and high diagnosis accuracy can be achieved [34].

4 Proposed Algorithm

In this part, we have presented an intelligent assistive algorithm for detection of osteoarthritis in wrist X-ray images based on joint space width measurement. The proposed algorithm contains systematic steps to diagnose osteoarthritis. The algorithm is programmed in Scilab 5.5 (open-source image processing software). We have conducted experiments on 75 X-ray images of human hand to detect osteoarthritis.

Algorithm: Wrist arthritis detection from X-ray images of human arm

Input: X-ray image of human wrist

Output: An image with wrist arc (ROI) and JSW to decide osteoarthritis type

Step-1: Browse X-ray image of human wrist
Step-2: Median filter (image enhancement to remove salt and pepper noise)
Step-3: Contrast stretching and contrast limited adaptive histogram equalization (CLAHE)
Step-4: ROI detection (based on longest arc of binary image)
Step-5: Convert ROI into binary
Step-6: Edge detection based on statistical range edge detector
Step-7: Measure height of each column and store into vector (starts form first 1 of each column to last 1 of each columns of above image got in step 6)

Step-8: Take mean of vectors derived in step-7
Step-9: Convert result of step-8 into mm (joint space) 1 Pixel = 0.264583333 mm
Step-10: Take mean of vectors derived in step-7
Step-11: Check joint space to make decision for Abnormal, Moderate or Normal type of arthritis (Abnormal if WJSW between 0 and 0.49 mm, Moderate if WJSW between 0.50 and 0.85 mm and Normal if WJSW > 0.85 mm)
Step-12: Display output.

Above algorithm can be converted into pseudo-code as follows. If pseudo-code is available, then it will provide a great help and support for the enthusiastic programmer to implement proposed intelligent assistive algorithm using real programming language code.

Pseudo-code: Wrist arthritis detection from X-ray images of human arm

SELECT an X-ray image
CONVERT an inputted image into grayscale image
DO Median filter on grayscale image
DO CLAHE for contrast improvement
DETECT ROI based on longest arc of binary image
PERFORM edge detection based on statistical range edge detector
MEASURE height of each column starting from first 1's to last 1's of respective column and store into a vector
CALCULATE mean of the vector
CONVERT mean of the vector millimeter (1 Pixel = 0.264583333 mm) and mark as wrist joint space width (WJSW)
F WJSW between 0 and 0.49 mm, then print Abnormal osteoarthritis in wrist, ELSE IF WJSW between 0.50 to 0.85 mm, then, print Moderate osteoarthritis in wrist, ELSE IF WJSW > 0.85 mm, then, print Normal wrist, ELSE print undefined.

5 Experimental Result & Discussion

Many experiments have been conducted to detect osteoarthritis from 75 X-ray images of human arm. As the X-ray images are in different size, resolution or format, they are fixed in specific size, for example, 512*512 after pre-processing in JPG format. Scilab 5.5 which is open-source image processing software has been used for interface and coding.

Above Fig. 2 depicts the X-ray image of human arm. This type of image is passed as an input to the developed algorithm for wrist osteoarthritis detection. Figure 3 shows detection of region of interest from inputted X-ray image to measure JSW of the wrist. On the basis of JSW, the algorithm decides the class of the wrist osteoarthritis out of three classes, viz. abnormal, moderate and normal. We have inputted 75 collected X-ray images one by one in developed algorithm and got following classification matrix populated in Table 1.

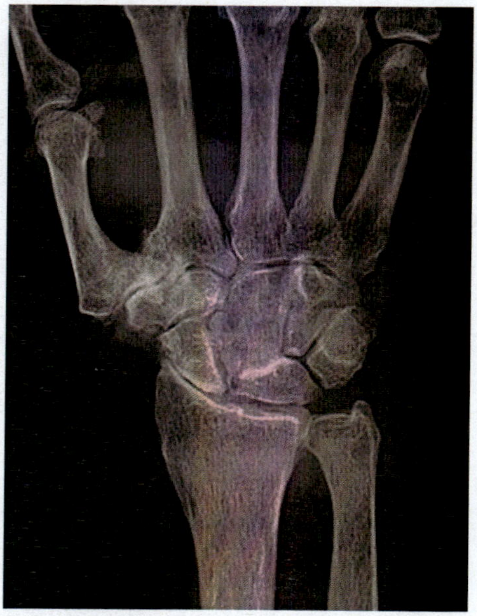

Fig. 2 X-ray image of human arm

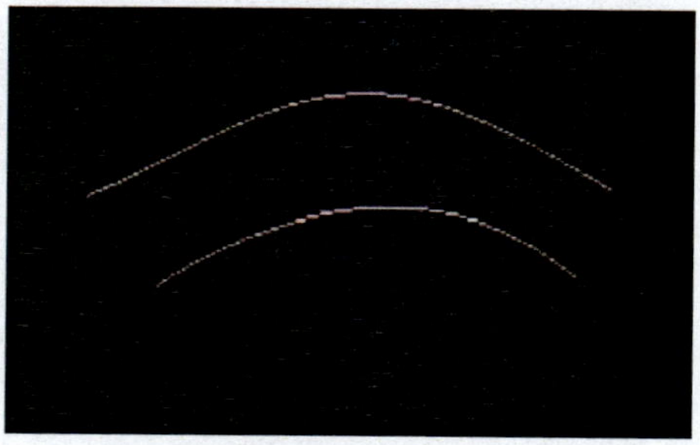

Fig. 3 ROI detection and measurement of JSW

Table 1 Classification rate of osteoarthritis

Osteoarthritis	Correctly classified	In orrectly classified	Ground truth	Accuracy rate (%)
Abnormal	38	1	39	97.30
Moderate	21	2	23	90.00
Normal	13	0	13	100.00
Total	72	3	75	96.00%

Table 1 shows classification of wrist osteoarthritis into three different classes like Abnormal, Moderate and Normal. This classification is derived from experiments conducted on an intelligent assistive wrist osteoarthritis detection algorithm which is developed in Scilab 5.5. We have used collected 75 X-ray images of human arm for this purpose. The algorithm has classified 38 images correctly and 1 image incorrectly out of 39 abnormal X-ray images of wrist osteoarthritis. The accuracy rate of abnormal classification is 97.30%. In moderate osteoarthritis, the algorithm has classified 21 images correctly and 2 image incorrectly out of 23 moderate X-ray images of wrist osteoarthritis. The accuracy rate of moderate classification is 90.00%. The algorithm has classified all normal X-ray images correctly. The accuracy rate of normal classification is 100.00%. The overall accuracy rate of classification is 96.00%. It means that algorithm has classified 72 images correctly, and only 3 images incorrectly out 75 X-ray images of different types of wrist osteoarthritis that is abnormal, moderate and normal or no wrist osteoarthritis.

Figure 4 shows graphical representation of success rate of different classes. It is very clear from Table 1 and Fig. 2 that proposed intelligent assistive algorithm achieves the overall accuracy rate 96.00% which is very efficient and better than algorithms/methods specified in literature for detection of wrist osteoarthritis from X-ray images of human arm.

We also compared result of our algorithm WODJSW with three latest osteoarthritis detection algorithms. The comparison is given in Table 2.

Table 2 shows comparison of accuracy of proposed WODJSW algorithm with some efficient algorithms. It reveals from the above table that our algorithm achieved highest accuracy, i.e., 96% to detect wrist osteoarthritis from human arm X-ray

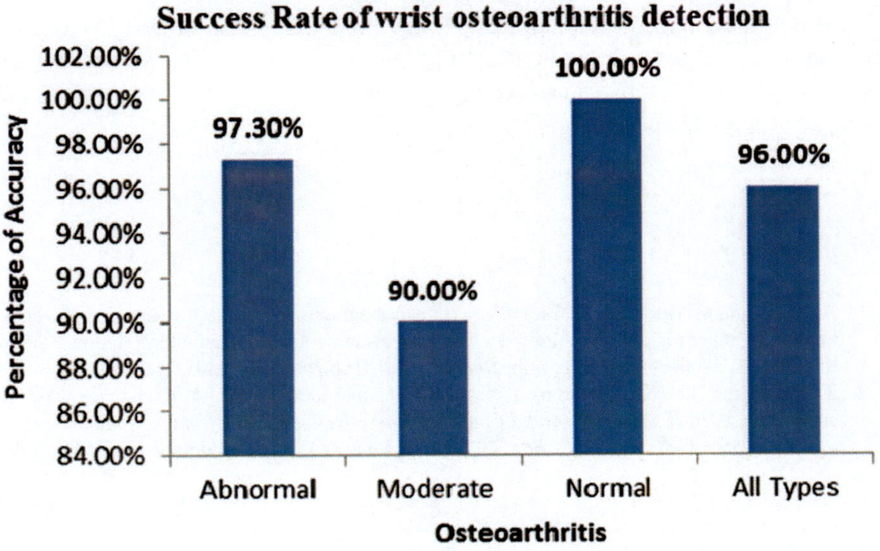

Fig. 4 Success rate of wrist osteoarthritis detection

Table 2 Comparison of proposed algorithm

No	Algorithm name	Accuracy (%)	Dataset
1	Machine vision-based approach [4]	87.92	200
2	Class analogy based on CNN [20]	90.00	9
3	Radial basis function [11]	95.00	60
4	ReadMyXray [13]	77.00	40
5	Self-organizing map [23]	92.80	42
6	**Proposed WODJSW**	**96.00**	**75**

images. Our dataset comprises 75 human arm X-ray images which is higher than all datasets given in Table 2 except machine vision-based algorithm.

6 Conclusion and Future Scope

This research paper presents an intelligent assistive method or algorithm to detect wrist osteoarthritis using X-ray of human arm based on joint space width measurement. We have considered accuracy of wrist osteoarthritis detection as an evaluation metric. The proposed algorithm achieves the accuracy rate of 96.00% in detection of wrist osteoarthritis which is almost higher than those discussed in related work section of this paper. Here, we have also compared our algorithm WODJSW with latest five methods to detect osteoarthritis in terms of accuracy and dataset. The proposed algorithm can be used by orthopedic doctor to detect wrist osteoarthritis. It can also be used in medical fraternity to design and develop DSS or CADD based on human X-ray images to detect wrist osteoarthritis from X-ray images of human arm. Researcher can add more features, comparisons and parameters to further evolve and analyze proposed algorithm.

References

1. A. Bieleckia, M. Korkoszb, B. Zielinskic, Hand radiographs preprocessing, image representation in the finger regions and joint space width measurements for image interpretation. Pattern Recogn. **41**, 3786–3798 (2008). https://doi.org/10.1016/j.patcog.2008.05.032
2. S.A. Bhisikar, S.N. Kale, Automatic joint detection and measurement of joint space width in arthritis, in *2016 IEEE International Conference on Advances in Electronics, Communication and Computer Technology (ICAECCT)*, Rajarshi Shahu College of Engineering, Pune, India (2016)
3. National Health Portal, https://www.nhp.gov.in/disease/musculo-skeletal-bone-joints-/osteoarthritis. Last accessed 1 Dec 2019
4. S.S. Gornale, P.U. Patravali, R.R. Manza, Detection of osteoarthritis using knee X-ray image analyses: a machine vision based approach. Int. J. Comput. Appl. **145**(1), 0975–8887 (2016)

5. S.S. Gornale, A survey on exploration and classification of osteoarthritis using image processing techniques. Int. J. Sci. Eng. Res. **7**(6), 334–355 (2016). https://www.ijser.org
6. B.L. Wagaj, M.M. Patil, Osteoarthritis disease diagnosis with the help of pixel based segmentation and SVM classifier. Int. J. Adv. Sci. Eng. Technol. **3**(4), 136–138 (2015)
7. K.G. Bhavyashree, S.N. Rao, Determination and analysis of arthritis using digital image processing techniques. Int. J. Electr. Electron. Data Commun. **2**(9), 46–49 (2014)
8. S.P. Chokkalingam, K. Komathy, Intelligent assistive methods for diagnosis of rheumatoid arthritis using histogram smoothing and feature extraction of bone images. Int. J. Comput. Inf. Eng. **8**(5), 905–914 (2014)
9. D.I. Navale, R.S. Hegadi, N. Mendgudli, Block based texture analysis approach for knee osteoarthritis identification using SVM, in *2015 IEEE International WIE Conference on Electrical and Computer Engineering (WIECON-ECE)*, Dhaka, Bangladesh (2015). https://doi.org/10.1109/WIECON-ECE.2015.7443932
10. R.S. Hegadi, D.I. Navale, T.D. Pawar, D.D. Ruikar, Osteoarthritis detection and classification from knee X-ray images based on artificial neural network, in Recent Trends in Image Processing and Pattern Recognition. RTIP2R 2018. Communications in Computer and Information Science. ed. by K. Santosh, R. Hegadi, vol. 1036 (Springer, Singapore, 2015), 97–105. https://doi.org/10.1007/978-981-13-9184-2_8
11. S.A. Bhisikar, S.N. Kale, Classification of rheumatoid arthritis based on image processing technique, in *Recent Trends in Image Processing and Pattern Recognition. RTIP2R 2018. Communications in Computer and Information Science*. ed. by K. Santosh, R. Hegadi, vol. 1036 (Springer, Singapore, 2019), pp. 163–173. https://doi.org/10.1007/978-981-13-9184-2_15
12. S. Chan, K. Dittakan, Osteoarthritis stages classification to human joint imagery using texture analysis: a comparative study on ten texture descriptors, in *Recent Trends in Image Processing and Pattern Recognition. RTIP2R 2018. Communications in Computer and Information Science*. ed. by K. Santosh, R. Hegadi (eds), vol. 1036 (Springer, Singapore, 2015), pp. 209–225. https://doi.org/10.1007/978-981-13-9184-2_19
13. G.W. Stachowiak, M. Wolski, T. Woloszynski, P. Podsiadlo, Detection and prediction of osteoarthritis in knee and hand joints based on the X-ray image analysis. Biosurface Biotribol. **2**(4), 162–172 (2016). https://doi.org/10.1016/j.bsbt.2016.11.004
14. M. Subramoniam, A non-invasive method for analysis of arthritis inflammations by using image segmentation algorithm, in *2015 IEEE International Conference on Circuits, Power and Computing Technologies [ICCPCT-2015]*, Nagercoil, India (2015). https://doi.org/10.1109/ICCPCT.2015.7159337
15. H. Yinghe, L.K. Vincken, D.V. Heijde, M.J.H. De Hair, F.P. Lafeber, M.A. Viergever, Automatic quantification of radiographic wrist joint space width of patients with rheumatoid arthritis. IEEE Trans. Biomed. Eng. **64**(11), 2695–2703 (2017). https://doi.org/10.1109/TBME.2017.2659223
16. D.M. Bear, G. Moloney, R.J. Goitz, M.L. Balk, J.E. Imbriglia, Joint space height correlates with arthroscopic grading of wrist arthritis. Hand (N Y) **8**(3), 296–301 (2013). https://doi.org/10.1007/s11552-013-9522-9
17. J.T. Sharp, J.C. Gardner, E.M. Bennett, Computer-based methods for measuring joint space and estimating erosion volume in the finger and wrist joints of patients with rheumatoid arthritis. Arthritis Rheum. **43**(6), 1378–1386 (2000). https://doi.org/10.1002/1529-0131(200006)43:6%3C1378::AID-ANR23%3E3.0.CO;2-H
18. O. Schenk, Y. Huo, K.L. Vincken , M.A. Laar, I. Kuper, K.C. Slump, F.P. Lafeber, H.J. Moens, Validation of automatic joint space width measurements in hand radiographs in rheumatoid arthritis. J. Med. Imag. **3**(4), 044502-1–044502-8 (2016). https://doi.org/10.1117/1.JMI.3.4.044502
19. M.S. Pandey, B. Rajitha, S. Agarwal, Computer assisted automated detection of knee osteoarthritis using X-ray images. Sci. Technol. **1**(2), 74–79 (2015)
20. S. Banerjee, S. Bhunia, G. Schaefer, Osteophyte detection for hand osteoarthritis identification in x-ray images using CNNs, in *2011 Annual International Conference of the IEEE Engineering in Medicine and Biology Society*, Boston, MA, USA (2011). https://doi.org/10.1109/IEMBS.2011.6091530

21. L. Shamir, S. M. Ling, W.W. Scott, A. Bos, N.Orlov, T.J. Macura, D. M. Eckley, L.Ferrucci, and I.G. Goldberg, Knee X-Ray Image Analysis Method for Automated Detection of Osteoarthritis. IEEE Transactions on Biomedical Engineering 56(2) (2009). https://doi.org/10.1109/TBME.2008.2006025
22. J. Lim, J. Kim, S. Cheon, A deep neural network-based method for early detection of osteoarthritis using statistical data. Int. J. Environ. Res. Public Health **16**(1281), 1–12 (2019). https://doi.org/10.3390/ijerph16071281
23. P. Kurniasih, D. Pratiwi, Osteoarthritis disease detection system using self organizing maps method based on Ossa Manus X-Ray. Int. J. Comput. Appl. **173**(3), L 42–47 (2017). https://doi.org/10.5120/ijca2017915278
24. J. Kiselev, B. Ziegler, H.J. Schwalbe, R.P. Franke, U. Wolf, Detection of osteoarthritis using acoustic emission analysis. Med. Eng. Phys. **65**, 57–60 (2019). https://doi.org/10.1016/j.medengphy.2019.01.002
25. D.J. Hunter, S. Bierma-Zeinstra, Osteoarthritis. The Lancet **393**(10182), 1745–1759 (2019). https://doi.org/10.1016/S0140-6736(19)30417-9
26. E.J. Leavline, D. Antony, Int. J. Signal Proces. Image Process. Pattern Recogn. **6**(5), 343–352 (2013). https://doi.org/10.14257/ijsip.2013.6.5.30
27. U. Erkan, L. Gokrem, S. Enginoglu, Different applied median filter in salt and pepper noise. Comput. Electr. Eng. **70**, 789–798 (2018). https://doi.org/10.1016/j.compeleceng.2018.01.019
28. L. Liangliang, S. Yujuan, J. Zhenhong, Medical image enhancement based on CLAHE and unsharp masking in NSCT domain. J. Med. Imag. Health Inf. **8**(3), 431–438 (2015). https://doi.org/10.1166/jmihi.2018.2328
29. A.K. Bharodiya, A.M. Gonsai, An improved edge detection algorithm for X-Ray images based on the statistical range. Heliyon **5**(e02743), 1–9 (2019). https://doi.org/10.1016/j.heliyon.2019.e02743
30. A.J. Patil, P. Jain, A. Pachpande, Automatic brain tumor detection using K-means and RFLICM. Int. J. Adv. Res. Electr. Electron. Instrum. Eng. **3**(12), 13896–13903 (2014). https://www.ijareeie.com/upload/2014/december/57_AUTOMATIC.pdf
31. W. Lin, B. Do, M. Nguyen, A Radiologist's Guide to wrist alignment: the good, bad, and ugly (2016). https://xrayhead.com/rsna2016.pdf. Last accessed 17 Dec 2019
32. Stanford MSK MRI Atlas, https://xrayhead.com. Last accessed 17 Dec 2019
33. S. Gil, O. Luciano, P. Matheus, Automatic segmenting teeth in X-ray images: trends, a novel data set, benchmarking and future perspectives. Expert Syst. Appl. **107**, 1–38 (2016). https://doi.org/10.1016/j.eswa.2018.04.001
34. A. Jafar, N. Anand and K. Akshi, Machine learning from theory to algorithms: an overview. J. Phys. Conf. Ser. **1142**(1), 012–018 (2018). https://doi.org/10.1088/1742-6596/1142/1/012012.

Blockchain Embedded Congestion Control Model for Improving Packet Delivery Rate in Ad Hoc Networks

V. Lakshman Narayana and Divya Midhunchakkaravarthy

Abstract The blockchain, with its own attributes, has got a lot of consideration toward the start of its introduction to the world and been applied in numerous fields. Simultaneously, its security issues are uncovered continually and digital attacks have caused huge troubles in it. At present, there is little concern and research in the field of system security of the blockchain. This paper presents the uses of blockchain in wireless systems which deliberately dissect the security of each layer of the blockchain and possible digital attacks, expose the difficulties brought by the blockchain to arrange oversight, and sum up exploring the progress in providing security to the ad hoc networks. The proposed model portrays the group model of blockchain administrations; their unwavering quality is affirmed based on test information. Data communication is known to be an inefficient activity as far as message multifaceted nature, and overhead is observed. When data transmission is initiated and all nodes using the same available route, congestion in the network is increased. Notwithstanding the effect on the framework execution, inefficient model may have serious impacts on results with respect to security and affability of the blockchain. In the proposed work, an efficient model is designed that can be used in ad hoc networks to reduce the overhead of the network. As the overhead or congestion in the network is more, the performance of the ad hoc system will be decreased. To avoid such circumstances and to improve the network performance, the proposed Congestion Control Reduction with Blocks Linking (CCRBL) for generating blockchain is introduced. The proposed method is compared with the traditional congestion control systems in terms of congestion levels, same route usage levels, congestion reduction levels and throughput, and the results show that the proposed method is more efficient in congestion reduction.

V. L. Narayana (✉) · D. Midhunchakkaravarthy
Lincoln University College, Wisma Lincoln, No. 12-18, Jalan SS 6/12, 47301 Petaling Jaya, Selangor Darul Ehsan, Malaysia
e-mail: lakshmanv58@gmail.com

D. Midhunchakkaravarthy
e-mail: divya@lincoln.edu.my

D. Bhattacharyya and N. Thirupathi Rao (eds.), *Machine Intelligence and Soft Computing*, Advances in Intelligent Systems and Computing 1280,
https://doi.org/10.1007/978-981-15-9516-5_12

Keywords Blockchain · Congestion control · Data transmission · Blocks linking · ad hoc systems · System performance

1 Introduction

Self-setup and dynamic routing make ad hoc networks the best way and easiest way for data communication. Because of the expanding interest for interchanges in different fields from content information to sight and sensitive information at a similar rate, the transmission rate likewise expanded. Expanding the information rate causes congestion in the system and simultaneously diminishing the rate fast transmission, decreased throughput and nature of administration, which is an exploration challenge in the ad hoc systems [1]. The ad hoc systems are progressively complex because of dynamic nature, resource limitations, and such systems completely depend upon intermediate nodes. Blockage is frequently brought about by unreasonable contribution of the traffic rate into restricted system; transfer speed and congestion control techniques limit this traffic rate to forestall congestions [2].

The developing utilization of the remote system challenges the basic components of remote structure to help the present rapid assistance and future applications. Fixed infrastructure systems face many issues in route management, data transmission and identification of attacks [3]. To overcome the drawbacks of the wired systems and using the advancements of ad hoc systems and cost-sparing patterns, nodes can have diverse system interfaces, for example, "Ethernet" or "Wi-Fi" to get to the distinctive application administrations arrangements, the vast majority of which are known as wireless ad hoc network data transmission. The term data transmission is as often as possible depicted as a transmission among the transmitter and the receiver with various connections at the proportionate time [4]. It can upgrade the quantity of various system measures as far as unwavering quality, efficiency, and transmission by structuring proper route identification models [5].

Contrasting with the fast advancement of blockchain innovation, important standards and guidelines on it are as yet deficient [6]. The principal distinct collection on the blockchain is the "Bitcoin: A Peer-to-Peer Electronic Cash System" composed by Nakamoto, in which blocks and chains are depicted as an information structure recording the chronicled information of the bitcoin exchange accounts [7]. A timestamp server works by taking a hash of a block of things to be time stamped and generally distributing the hash [8]. The blockchain is additionally called the Internet of significant worth, which is a dispersed record database for a shared system.

Blockchain can be utilized for proprietorship and copyright of the executives and transactions (Fig. 1). It incorporates exchanges of assets, for example, data, memory, hardware gadgets just as including computerized distributions and advanced assets that can be labeled. At present, by arranging application cases, they can be isolated into three classifications, "Reusing Box", "Dull Box" and "Sandbox". The application cases in every classification bring numerous difficulties for the lawful,

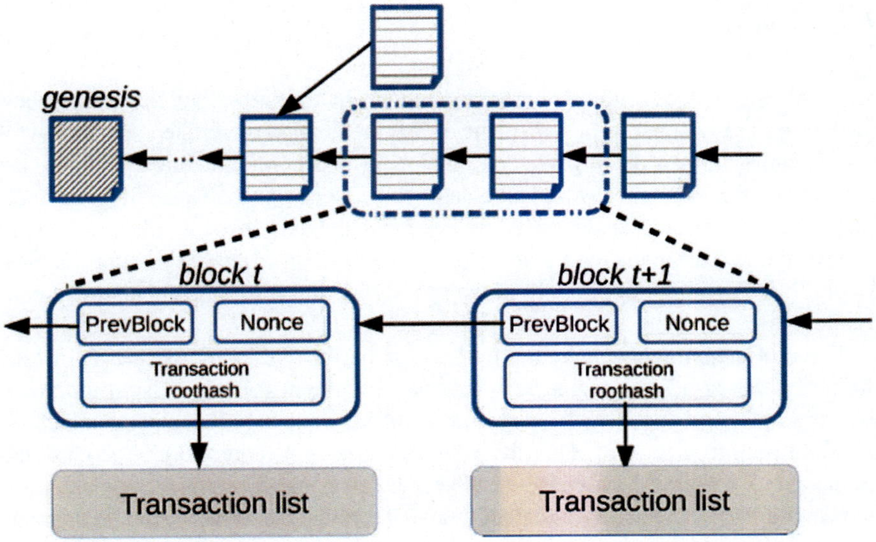

Fig. 1 Structure of blockchain

oversight and dynamic divisions [9]. The three classes are completely discussed below.

1.1 Reusing Box

Reusing boxes are those cases that endeavor to tackle industry needs that focus through blockchain arrangements in a superior, quicker, and less expensive way. Their objectives are not unlawful, and the inspiration is straightforward. During the time spent the application propelled, the system oversight specialists can actualize management just by making minor adjustments to the present oversight structure [10].

The most useful model is the interbank settlement framework created by Ripple. The installment arrangement utilizes a solitary dispersed record to associate the world's major money related organizations and cross-bank exchanges that happen between one another should be possible continuously. Contrasted and the customary technique, it not just spares a great deal of time, improves proficiency, yet in addition spares an assistance expense.

1.2 Dim Box

Cases having a place with this class, no matter what, all repudiate the present law. Such cases are various, for instance, the online medication advertise, other unlawful stock details, human dealing systems, financing and communication systems, tax evasion and tax avoidance would all be able to be named such. These illegal administrations have existed in obscurity arrangement for quite a while. These days, in light of the use of blockchain innovation, some of them resemble finding the New World. It is anything but difficult to distinguish the dim box, yet it tends to be hard to attempt to stop them.

The motivation behind why the dull box is hard to be halted is that as of late, the advanced cash has become a significant instrument for tax evasion, unlawful exchanges, and getting away from remote exchange control because of its namelessness and decentralization [11]. Advanced cash does not require a charge card and financial balance data. Lawbreakers can keep away from the oversight organizations and cannot follow the source and goal of assets through customary capital exchange records, which makes conventional management techniques malfunction [12].

1.3 Sandbox

The sandbox is one of the most energizing and cerebral pains for administrators in these three classifications, and a considerable lot of the most problematic and open conspiracy cases fall into this class. The expression sandbox was taken from an ongoing activity by the Financial Conduct Authority (FCA) called Administrative Sandbox. Application cases having a place with this class have truly important business destinations, yet the present circumstance is that because of the different attributes of the appropriated record innovation, the vast majority of these cases cannot meet the current oversight prerequisites. Their regular element is the thing that the business sought after is legitimate, yet it might cause different dangers, so the legislature will not let it proceed.

Every one of the node may have at least one remote connections interface, and they impart each other by means of radio transmissions. An ad hoc network should seriously think over of various home processing gadgets, PC, mobiles, etc. A node can convey straightforwardly to another node, on the off chance that it is in the transmission scope of another node, in any case communication occur through intermediate nodes [13]. There are various parameters that can be considered for execution assessment of a genuine ad hoc system; it implies how well a system runs, for example, accessibility, unwavering quality, reaction time, usage, throughput and packet loss. Data packet loss is acknowledged as one of the most noteworthy parameters to assess the exhibition of the portable ad hoc system [14].

Because of unique topology, ad hoc network transmission is exceptionally error inclined. Versatility, blockage, transmission errors are the significant reasons that are answerable for the data loss in ad hoc networks [15]. These issues are straightforwardly connected with the system setting. All the packets that have dropped because of the portability, congestion, transmission error and the attack are known as the packet loss [16]. The proposed model establishes a framework for reducing the congestion in the ad hoc network by monitoring the routing nodes and the data transmission levels such that the routing process is balanced to avoid congestion in the ad hoc system.

2 Literature Survey

In the present requirements, remote system administration plays a significant capacity in the upcoming age of Internet associations. Progressively new administrations are progressively accessible in business, diversion and person-to-person communication applications in every single remote system because of consistent versatility and maintaining a strategic distance from blockage brought about by the far-reaching utilization of the Internet. Notwithstanding, these present advances cannot adapt to the expanding utilization of dropping interactive media applications that raise high blockage and packet loss. For example, the information created in an emergency is the most significant, and the loss of such information may harm the objective of arranging an ad hoc network. Simultaneously, congestion control in the remote system must be put together not just with respect to the proficiency of the system yet moreover should have essential unwavering quality of utilizations.

Congestion will in general spotlight on forestalling obstructions, as it brings about a critical loss of system in states of nature of administration and asset use. Blockage anticipation systems give an answer dependent on strategies for organized packet transmission, congestion determining, congestion control and "congestion balancing." Congestion prediction is an incredibly critical way to control routing overhead. After congestion is identified, the most significant way is the manner by which it illuminates the congestion causing node [17].

The packet loss is because of the dynamic qualities of ad hoc network, for instance, "node route," "channel bit error," "man in the middle" and "route failure." Due to this exceptional nature, the packet loss rate of the remote connection is a lot higher than wired association. Routing protocol reacts to these remote data losses similarly as packet loss because of congestion. This is on the grounds that it expects to manage losses in this way. The congestion control is a significant issue in remote systems, particularly in ad hoc networks.

Gowtham et al. [1] proposed a technique to comprehend the fundamental issues those are straightforwardly identified with packet loss and indicated that AODV has more packet loss because of the versatility as contrast with congestion; henceforth, AODV is progressively delicate for portability. Shayesteh Tabatabaei et.al [4] proposed a strategy for secure channel to expel the issue of ambiguity and confirmation which assists with limiting overhead and packet drop issue.

Jan et al. [6] proposed a technique in which they utilized combined features of static and dynamic just as unique routing calculation. The proposed framework is sufficiently competent to discover the following node for conveyance of the packet. The proposed framework checks the traffic thickness by figuring the proportion of approaching sections active packets and register the traffic level, and afterward conclude whether to send the packet or not. It diminishes the chance of packet loss.

3 Proposed Method

While blockchain brings mechanical advancement, it likewise brings tremendous difficulties for ad hoc network management. To utilize the blockchain innovation and the current lawful framework to direct the use of the blockchain is one of the issues that the ad hoc networks focus on. So as to beat the issues of congestion control in ad hoc networks, blockchain model is used in the data transmission management. It is important to cross the hidden innovation and consider how to consolidate the particular instances of innovation application with oversight.

Routing table stores the data of the nodes that can involve in communication. Based on the routing table information, nodes will transfer data by using intermediate nodes. As ad hoc networks are established to data transfer in situations where fixed networks fail, many nodes will involve in communication and multiple groups are involved in using the available routes. This results in increasing congestion in the network that impacts the performance of the ad hoc network system.

To avoid congestion in the network, the proposed work considers a Congestion Control Reduction with Blocks Linking (CCRBL) by appointing a central node in the network as the Congestion-Level Verification Authority (CLVA) which initially registers all the routing information into it, number of groups using the available routes. When multiple groups are registered with the CLVA node, it continuously monitors the congestion levels in the network and a route which is having more congestion is made unavailable to remaining ad hoc network groups. The process of controlling congestion in the ad hoc system is depicted clearly.

Initially, nodes establish an ad hoc network.

A(N) $\leftarrow$ Nodes(S(i)).

AG(N,N + i) $\leftarrow$ A(N) + A(N + i).

Update AG (N).

Here A is considered as ad hoc network, AG is considered as ad hoc group of networks. A(N),A(N + i) are the groups of adjacent ad hoc groups, i is the initial node in the group. After updating the ad hoc nodes, group of ad hoc networks will search for a route and routing table is updated. From the ad hoc system, a CLVA node whose computational capabilities are high and whose power consumption is less is selected. This node continuously monitors the congestion level of the node.

For every node in AG(N).

Available Route(AG(N)) $\leftarrow$ Routing(AG(N) + A(N)).

N ε AG(i) $\leftarrow$ Route(i).

N(i + 1) → AG(i).
if (Route(A(N(i)) AND Route (AG(N(i)).
CLVA(N(i)) ← (No.of data packets received + No.of data packets sent + No.of REQ messages
Received) / Count(AG(N)).
If CLVA(N) > θ.
Route (A(N(i))) → Available Route(AG(N(i))) ! = Route(A(N(i))).
else.
Dp → AG(N(D))

DP is the data packets, D is the data. Despite the fact that the blockchain gives anonymization, it is not totally mysterious. The aggressor can even now play out certain mapping by breaking down system traffic and exchange data. Blockchain primary thought is that the client sends some bitcoin from a location and puts the bitcoin into another location so that it is hard to track down the communication between the information and yield locations of a similar client. At present, there are two fundamental kinds of strategies for blockchain security insurance: One is to add an unknown assurance instrument to a current blockchain through an innovation, for example, secure transmission. The CLVA node utilizes blockchain model to create blocks when data is transferred from one node to the other. The generated block is linked with the remaining blocks when data transferred is completed from a node to the other node. After creating a block, CLVA node can analyze which node is causing more congestion and then suggest the other ad hoc groups to use a different route rather than the existing route for reducing the congestion. The process in which a node creates a block after data transfer is depicted in Fig. 2.

The process of generating a block is clearly depicted in Fig. 3.

The generated block is validated by using CLVA node, and the process is depicted in Fig. 4.

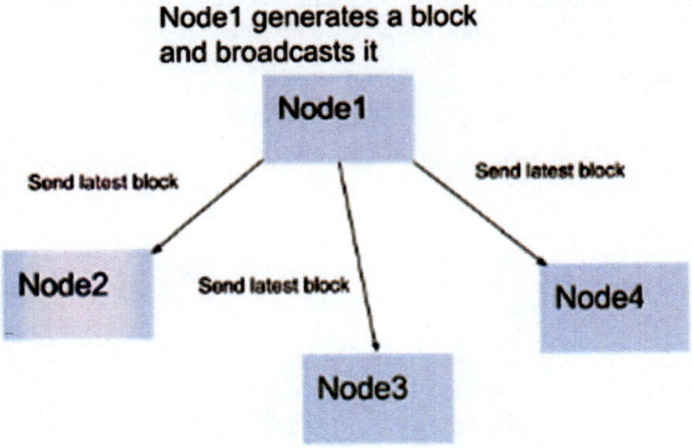

Fig. 2 Block generation after data transfer from a node

```
var initHttpServer = () => {
    var app = express();
    app.use(bodyParser.json());

    app.get('/blocks', (req, res) => res.send(JSON.stringify(blockchain)));
    app.post('/mineBlock', (req, res) => {
        var newBlock = generateNextBlock(req.body.data);
        addBlock(newBlock);
        broadcast(responseLatestMsg());
        console.log('block added: ' + JSON.stringify(newBlock));
        res.send();
    });
```

Fig. 3 Block generation module after data transmission from a node

```
def confirm_validity(block, previous_block):

        if previous_block.index + 1 != block.index:

            return False

        elif previous_block.compute_hash != block.previous_hash:

            return False

        elif block.timestamp <= previous_block.timestamp:

            return False

        return True
```

Fig. 4 Block validation process

4 Results

The proposed model is implemented in NS2 for performing simulation of nodes and data transmission. The creation of blockchain is implemented in ANACONDA SPYDER. The blocks are created in Python platform, and the block generation and block analysis are done by providing the results obtained by the NS2 simulator. The parameters for creating an ad hoc network are depicted in Table 1.

Table 1 Parameters used to create an ad hoc system

Parameter	Value
Network type	Mobile
Number of nodes	6
Network size	500 m * 600 m
Channel type	Channel/wireless channel
Radio propagation model	propagation/two way ground
Antenna type	Antenna/omni-antenna
Simulation time	1200 s
Packet size	552 bytes
Interface queue type	Queue/droptail/priqueue
Network interface type	Phy/wirelessPhy
Number of packets in IFQ	50
Adhoc routing protocol	AODV, DSDV
Application layer protocol	FTP

The congestion levels in the proposed work are always monitored by the CLVA node for attaining better outcomes. The proposed Congestion Control Reduction with Blocks Linking (CCRBL) is compared with the traditional ad hoc Distance Vector with Congestion Control (DV-CC) method, and the results show that the proposed method has less congestion than the traditional methods. The congestion levels based on the ad hoc network groups are depicted in Fig. 5.

As ad hoc networks are dynamic in nature, multiple networks use same route for completing data transmission. The usage of route and the nodes information is analyzed by the CLVA node for reducing the congestion of the system by avoiding usage of the same route by multiple ad hoc network groups. The proposed Congestion Control Reduction with Blocks Linking (CCRBL) is compared with the traditional ad hoc Distance Vector with Congestion Control (DV-CC) method, and the results show that the usage of same route levels in the proposed method is less than the traditional methods. The route usage levels in the ad hoc network groups are depicted in Fig. 6.

The performance of the ad hoc system is mainly depending on congestion level. The more the congestion level, the less is the performance. The proposed Congestion Control Reduction with Blocks Linking (CCRBL) is compared with the traditional ad hoc Distance Vector with Congestion Control (DV-CC) method, and the results show that the proposed method has less congestion than the traditional methods. The congestion levels based on the ad hoc network groups are depicted in Fig. 7.

The ad hoc network performance levels are depicted in Fig. 8.

The throughput of a system represents the overall performance of the network. The more the throughput, the better the performance of the system. The throughput of the proposed model is high when compared to traditional ad hoc Distance Vector

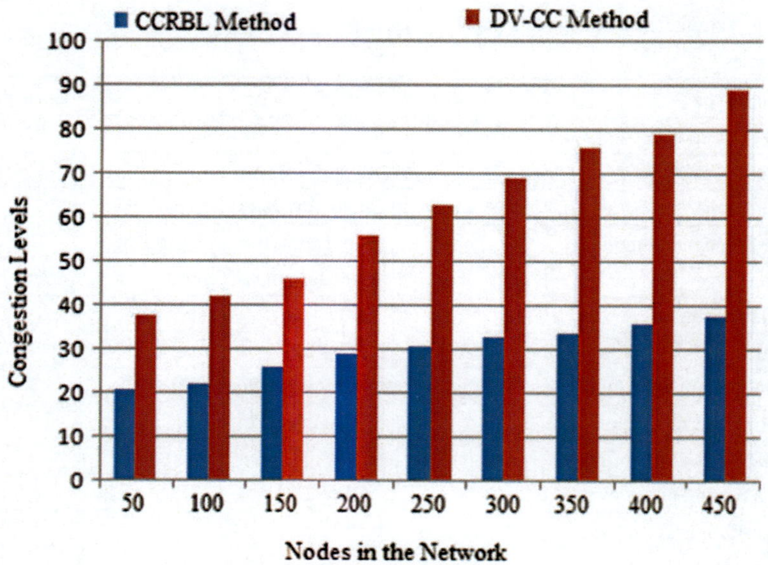

Fig. 5 Congestion level based on ad hoc network groups

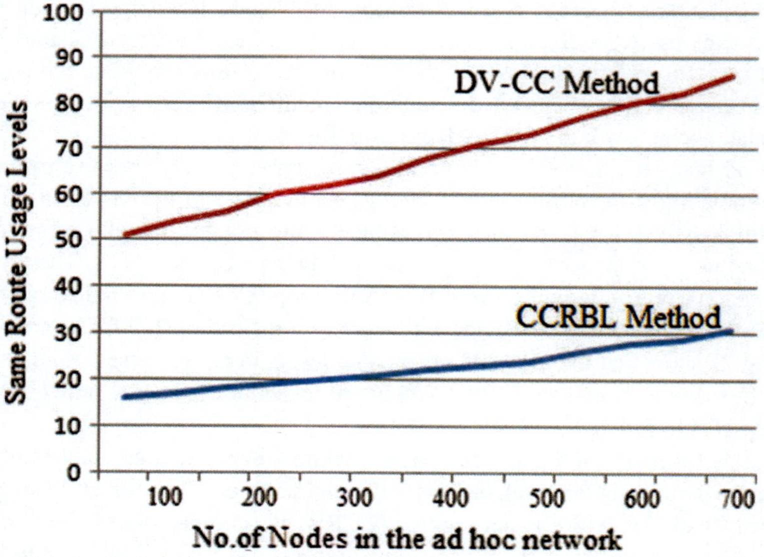

Fig. 6 Same route usage levels in ad hoc systems

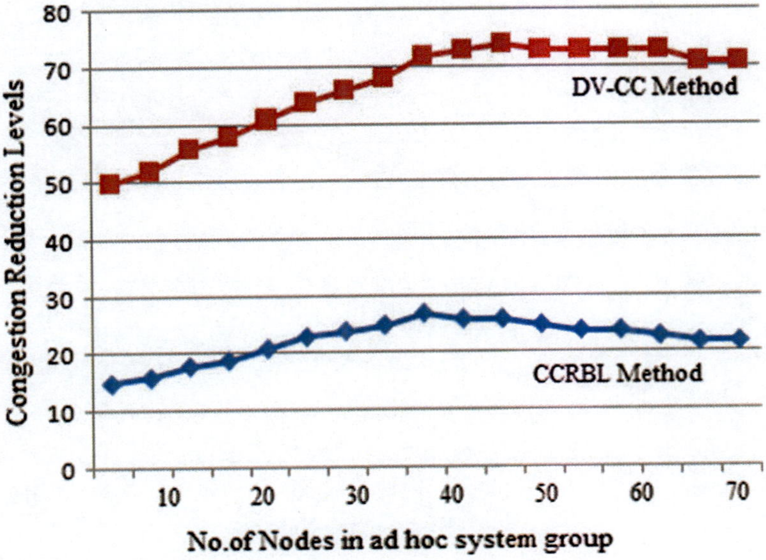

Fig. 7 Congestion reduction levels

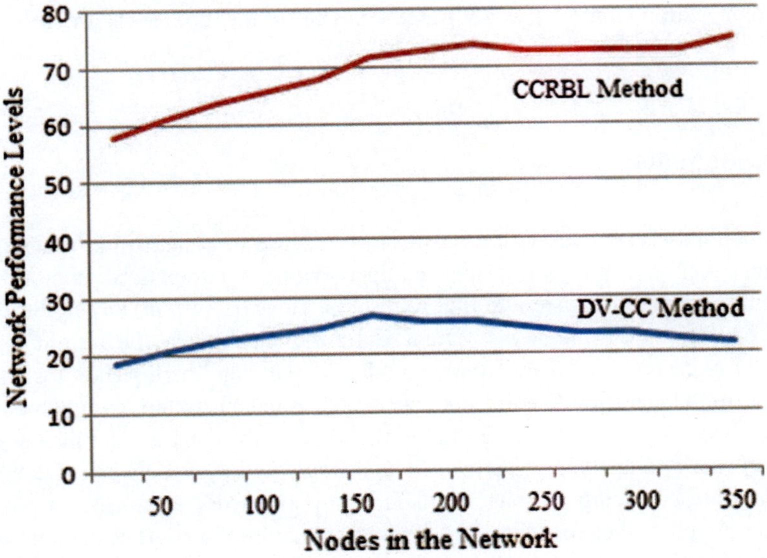

Fig. 8 Ad hoc network performance

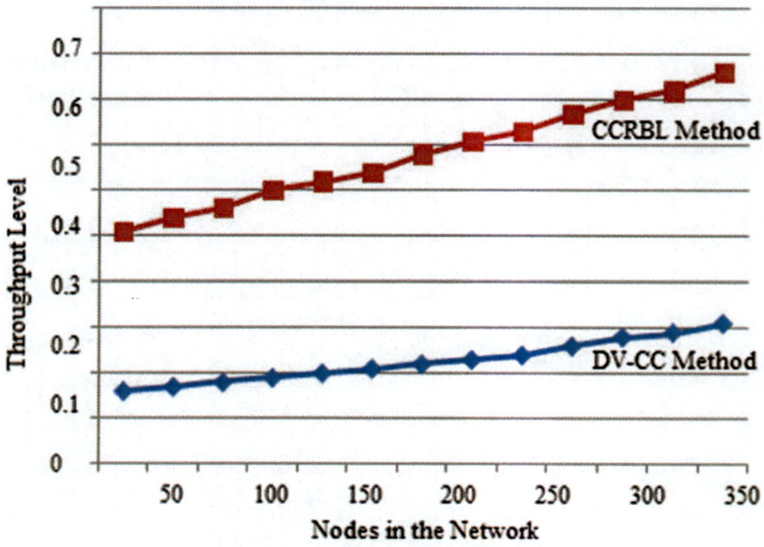

Fig. 9 Throughput

with Congestion Control (DV-CC) method. The throughput levels are depicted in Fig. 9.

5 Conclusion

In ad hoc networks, packets are mostly lost because of congestion brought about by data overflow and high transmission delay between association routes. Routing conventions flood the system to find routes that clog the system, paying little need to the condition of traffic on the system. Extra traffic can make the system progressively clogged and increment communication idleness and high packet loss. In this manuscript, Congestion Control Reduction with Blocks Linking is implemented to identify the congestion levels in the network by analyzing the data packet transmission rate and analyze the error rate and improve packet delivery rate through the dynamic balancing of congestion. The proposed work establishes a powerful exchanging controller for managing the determination of a route dependent on the routing affirmation considered by the congestion expectation controller. The congestion in the network is analyzed by analyzing the blocks and linking them to form a blockchain that is much helpful for increasing the ad hoc network performance. The proposed model achieves 92% of packet delivery rate by reducing the data loss. In future, the congestion at each and every node can be analyzed by a secured or trusted node, which is used to improve the performance of the ad hoc network.

References

1. M.S. Gowtham, K. Subramaniam, Congestion control and packet recovery for cross layer approach in MANET. Clust. Comput. **22**, 12029–12036 (2019)
2. Y. Harold Robinson, S. Balaji, E. Golden Julie, PSOBLAP: particle swarm optimization-based bandwidth and link availability prediction algorithm for multipath routing in mobile ad hoc networks. Wireless Personal Commun. **106**(4), 2261–2289 (2019)
3. Y. Harold Robinson, S. Balaji, E. Golden Julie, Design of a buffer enabled ad hoc on-demand multipath distance vector routing protocol for improving throughput in mobile ad hoc networks. Wireless Personal Commun. **106**(4), 2053–2078 (2019)
4. S. Tabatabaei, M.R. Omrani, Proposing a method for controlling congestion in wireless sensor networks using comparative fuzzy logic, in *Wireless Personal Communications*, pp. 1–18 (2018)
5. X. Yang et al., Wireless sensor network congestion control based on standard particle swarm optimization and single neuron PID. Sensors **18**(4), 10–18 (2018)
6. M.A. Jan, et al., A comprehensive analysis of congestion control protocols in wireless sensor networks, in *Mobile Networks and Applications*, pp. 1–13 (2018)
7. T.-S. Chen, C.-H. Kuo, Wu. Zheng-Xin, Adaptive load-aware congestion control protocol for wireless sensor networks. Wireless Pers. Commun. **97**(3), 3483–3502 (2017)
8. Y. Jing, et al., An energy-efficient routing strategy based on mobile agent for wireless sensor network, in *Control and Decision Conference (CCDC) 2017 29th Chinese* (2017)
9. G. Elhayatmy, N. Dey, A.S. Ashour, Internet of Things based wireless body area network in healthcare, in *Internet of Things and Big Data Analytics Toward Next-Generation Intelligence* (Springer, Cham, Switzerland, 2018), pp. 3–20 (2018)
10. J. Bae, H. Lim, Random mining group selection to prevent 51% attacks on Bitcoin, in *2018 48th Annual IEEE/IFIP International Conference on Dependable Systems and Networks Workshops (DSN-W)*, pp. 81–82 (2018)
11. K. Hasan, X.-W. Wu, K. Biswas, K. Ahmed, A novel framework for software defined wireless body area network, in *Proceedings of the 8th International Conference on Intelligent Systems, Modelling and Simulation (ISMS)*, pp. 114–119 (2018)
12. M. Mettler, Blockchain technology in healthcare: the revolution starts here, in *2016 IEEE 18th International Conference on e-Health Networking Applications and Services (Healthcom)* (2016)
13. K. Christidis, M. Devetsikiotis, Blockchains and smart contracts for the internet of things. IEEE Access **4**, 2292–2303 (2016)
14. B. Betts, Blockchain and the promise of cooperative cloud storage. Computer weekly, [online] Available: https://www.computerweekly.com/feature/Blockchain-and-the-promise-of-cooperative-cloud-storage
15. B. Mounika, P. Anusha, Use of blockchain technology in providing security during data sharing. J. Critical Rev. **7**(6), 338–343 (2020). https://doi.org/10.31838/jcr.07.06.59
16. S. Pasala, V. Pavani, G. Vidya Lakshmi, Identification of attackers using blockchain transactions using cryptography methods. J. Critical Rev. **7**(6), 368–375 (2020). https://doi.org/10.31838/jcr.07.06.65
17. P. McCorry, C.F. Shahandashti, F. Hao, A smart contract for boardroom voting with maximum voter privacy, in *International Conference on Financial Cryptography and Data Security*, pp. 357–375 (2017)

Predicting Student Admissions Rate into University Using Machine Learning Models

Ch. V. Raghavendran, Ch. Pavan Venkata Vamsi, T. Veerraju, and Ravi Kishore Veluri

Abstract Machine learning (ML) is a convention of using specified algorithms to get data, acquire some knowledge from it, and making a prediction about something in the nature. In ML, regression algorithms are used to anticipate output values basing on the input attributes from a dataset. Selecting a good ML algorithm will be helpful for predicting better outputs. It is also influenced by several parameters like data size, quality of the output, training time, etc. This paper is based on analyzing the dataset of different universities for predicting the chance of being admitted in that university. It will support the students to recognize the probabilities of their admission to a university being considered. Similarly, it will revile them in detecting the universities which are proper for their profile and also guide them with the particulars of those universities. The execution of the regression algorithms–Multi-linear regression, polynomial regression, decision tree regression, and random forest regression are examined on the dataset. The metrics used are accuracy score and mean squared error.

Keywords Machine learning · Supervised learning · Regression · Linear regression · Polynomial regression · Decision tree regression · Random forest regression

Ch. V. Raghavendran (✉)
Aditya College of Engineering & Technology, Surampalem, AP, India
e-mail: raghuchv@yahoo.com

Ch. Pavan Venkata Vamsi
Pragati Engineering College (A), Surampalem, AP, India
e-mail: pavanvamsi1507@gmail.com

T. Veerraju
Associate Professor, Aditya College of Engineering, Surampalem, AP, India
e-mail: veerarajut108@gmail.com

R. K. Veluri
Sr. Assistant Professor, Aditya Engineering College (A), Surampalem, AP, India
e-mail: ravikishorev1985@gmail.com

D. Bhattacharyya and N. Thirupathi Rao (eds.), *Machine Intelligence and Soft Computing*, Advances in Intelligent Systems and Computing 1280,
https://doi.org/10.1007/978-981-15-9516-5_13

1 Introduction

Students regularly have numerous queries concerning the courses, universities, job openings, overheads involved, etc., while preparing for higher education. Acquiring admission in their dream university is one of their main worries. It is understood that repeatedly students wish to pursue their education from universities which have international appreciation. Conferring to research, there are over 21 million students studying in more than 266 ranked universities and colleges across the USA [1].

The study we made in this paper has wide importance and applicability. In this, we focused an essential query facing nearly every student preparing for higher education. To accurately forecast whether a student wants to admit into an academic institution will admit or reject, their admissions offer can help students to manage their admission [2, 3]. For many students, admission into a ranked university or institution is an extremely important component for their higher education.

Getting entrance into a university is based on many parameters of student. Some of the parameters may be Graduate Record Examination (GRE) score, university rating, research experience, undergraduate GPA, etc. This can be predicted by using several ML algorithms which produces a good analysis on the chance of being admitted into the University [4, 5]. Using ML to solve this problem will give a clear view on which university you are capable of applying.

The remaining part of the paper consists of four sections. In Sect. 2, we have discussed about the regression algorithms, and in Sect. 3, we proposed the methodology. Section 4 consists of application of machine learning (ML) algorithms on the dataset. The paper is concluded in Sect. 5 with analysis on the results obtained.

2 Regression Algorithms

Regression is a machine learning algorithm based on supervised learning. Regression is a good statistical method that helps us to examine the relation between independent variables and dependent variables. ML provides many regression algorithms to perform regression. We cannot say which algorithm is best among all the algorithms [6–8]. The performance of the algorithm is completely based on the dataset and application. Selecting a good regression model or algorithm depends on the performance metrics such as mean square error (MSE), root mean square error (RMSE), and accuracy score [9–15]. The more the accuracy score gives the good relation between dependent variable and independents variables. Following are few regression algorithms

- Simple linear regression
- Multiple linear regression
- Least absolute selection shrinkage regression

- Multivariate regression
- Decision tree regressor
- Random forest regressor.

3 Methodology

In this section, we discuss about the data, preprocessing of the data, perform data visualization. We summarize our methodology in the form of a flowchart as shown in Fig. 1.

3.1 Data Source and Contents

The dataset used in this analysis is taken from Kaggle data science community namely "Admission_Predict.csv." The dataset is used to predict the chance of admission of a student to get admitted in the university. It contains details about many parameters like Graduate Record Examination (GRE) score, TOEFL score, university rating, Statement of Purpose (SOP), Letter of Recommendation (LOR) strength, undergraduate GPA, and research experience. Based on these independent variables, we find the dependent variable, i.e., Chance of Admit.

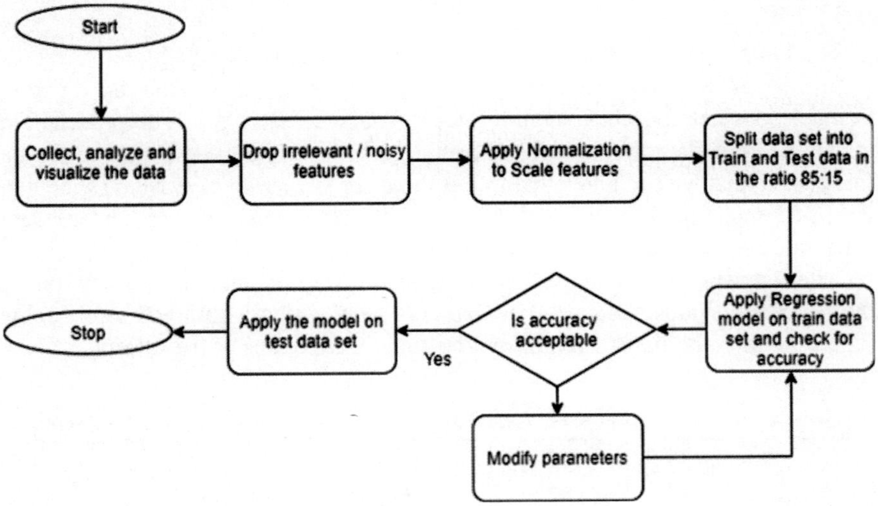

Fig. 1 Methodology of the proposal

3.2 Data Preprocessing

Data preprocessing is an important and essential step in the process of applying ML techniques. This includes the steps—*Cleaning, Integration, Transformation, and Reduction*. This step converts raw data into clean data. If quality input is given to the algorithm, then we can expect quality results and predictions.

The first thing to do before analyzing the data is to preprocess the data. Preprocessing the data includes identifying the missing data, dropping of data which is not useful for making the prediction, replacing NAN values with suitable value, etc. [16–19]. To perform preprocessing, we have to import the following libraries.

```
#importing packages
import pandas as pd
import numpy as np
import seaborn as sns
import matplotlib.pyplot as plt
```

Next, we have to load the dataset into a pandas data frame.

```
#Reading csv file
admission_df=pd.read_csv("Downloads/Admission_Predict.csv")
admission_df.head()
```

The sample output for the above command is displayed in Fig. 2.

In the dataset, Serial No. does not provide any information about the universities and does not contribute anything for the analysis. So we can drop Serial No. column/feature from the dataset. Next thing to do is finding the data types of the features, so that we can know that find if there are any string type which should be classified into numeric data, because we can perform analysis only on numeric data. Data types of the features are shown in Fig. 3.

After knowing the data types of all the features, we should check if any one of the features contains NAN value which is to be replaced with any suitable value. Figure 4 shows the statistical information about the features of the dataset.

	Serial No.	GRE Score	TOEFL Score	University Rating	SOP	LOR	CGPA	Research	Chance of Admit
0	1	337	118	4	4.5	4.5	9.65	1	0.92
1	2	324	107	4	4.0	4.5	8.87	1	0.76
2	3	316	104	3	3.0	3.5	8.00	1	0.72
3	4	322	110	3	3.5	2.5	8.67	1	0.80
4	5	314	103	2	2.0	3.0	8.21	0	0.65

Fig. 2 Contents of the dataset

```
1 admission_df.dtypes
```

```
GRE Score              int64
TOEFL Score            int64
University Rating      int64
SOP                  float64
LOR                  float64
CGPA                 float64
Research               int64
Chance of Admit      float64
dtype: object
```

Fig. 3 Data types of the features

	GRE Score	TOEFL Score	University Rating	SOP	LOR	CGPA	Research	Chance of Admit
count	400.000000	400.000000	400.000000	400.000000	400.000000	400.000000	400.000000	400.000000
mean	316.807500	107.410000	3.087500	3.400000	3.452500	8.598925	0.547500	0.724350
std	11.473646	6.069514	1.143728	1.006869	0.898478	0.596317	0.498362	0.142609
min	290.000000	92.000000	1.000000	1.000000	1.000000	6.800000	0.000000	0.340000
25%	308.000000	103.000000	2.000000	2.500000	3.000000	8.170000	0.000000	0.640000
50%	317.000000	107.000000	3.000000	3.500000	3.500000	8.610000	1.000000	0.730000
75%	325.000000	112.000000	4.000000	4.000000	4.000000	9.062500	1.000000	0.830000
max	340.000000	120.000000	5.000000	5.000000	5.000000	9.920000	1.000000	0.970000

Fig. 4 Statistical information of the features

3.3 Data Visualization

Another most important process to do before analyzing the data is to visualize the data. This process helps us to know how much a feature is important for analyzing. Finding the correlation between dependent variable and independent variables is shown in Fig. 5 in the form of heat map.

From the observation, we can conclude that the dependent feature, i.e., Chance of Admit, is completely dependent on all the features in the dataset. If the correlation vale is negative, we can drop those features from the dataset. The relation between the dependent variable and independent variable can be shown statistically using regression plot as shown in Fig. 6. Histograms are very useful tool to study the distribution of the data of a particular feature. Figure 7 shows the histograms of the range features of the dataset.

4 Data Analysis Using ML Algorithms

Before analyzing the data, we should split the dataset into training set and testing set. Training set is used while analyzing the algorithms, and testing set is used to decide which model will have a good fit to the dataset. The splitting of dataset can be done as shown in Fig. 8. Here, X_train and X_test are the subset of the dataset which

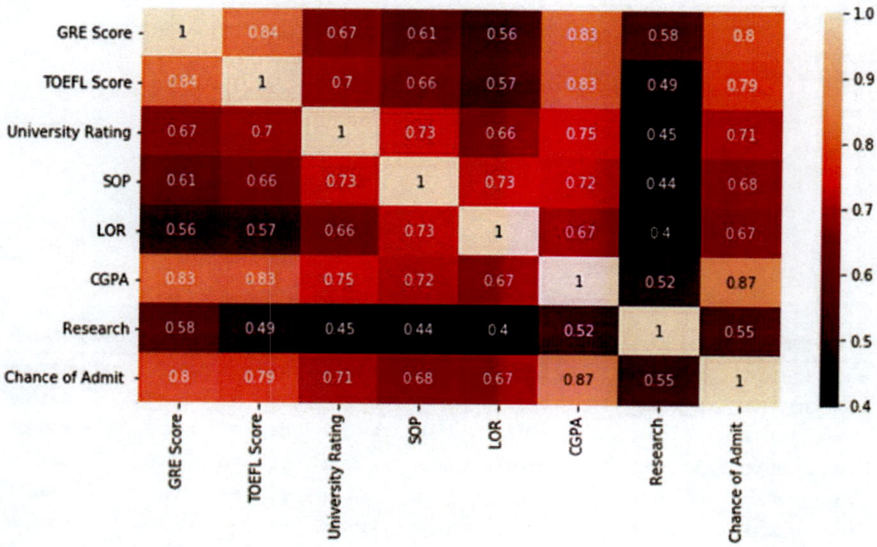

Fig. 5 Correlation among the features

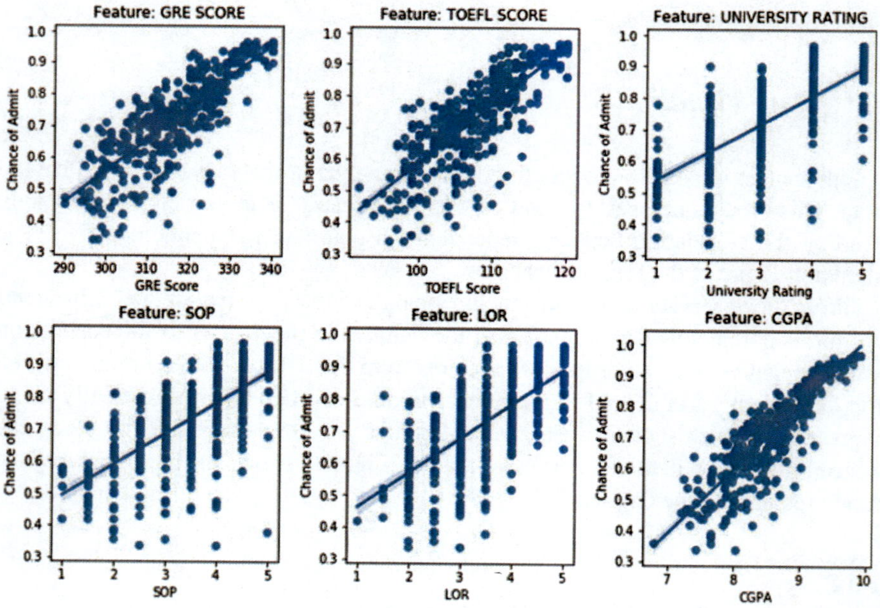

Fig. 6 Regression plots for independent versus dependent features

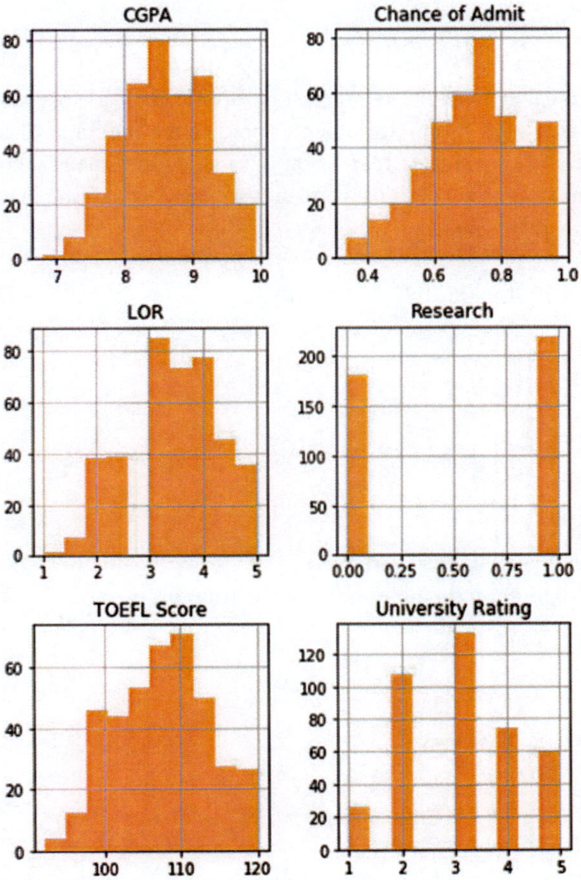

Fig. 7 Histograms of range features

```
from sklearn.model_selection import train_test_split
X=admission_df.drop(columns=['Chance of Admit '])
y=admission_df['Chance of Admit ']
X_train,X_test,y_train,y_test=train_test_split(X,y,test_size=0.15)
```

Fig. 8 Splitting of dataset

contains independent variables, and y_train and y_test contain dependent variables. X_train and y_train variables are used while training, and X_test y_test are used while testing the algorithm.

4.1 Multi-Linear Regression

Multi-linear regression is also called as multiple regression. In ML, multiple regression is a statistical analysis that uses no. of variables to predict the outcome of a response variable. The main aim of multiple regression is to show the linear relationship between independent variables and the dependent variables. Equation (1) is the formula for multiple linear regression.

$$y_i = \beta_0 + \beta_1 x_i^1 + \beta_2 x_i^2 + \cdots + \beta_p x_i^p \quad (1)$$

where

y_i = dependent variable

x_i = explanatory variables

β_0 = y-intercept (constant term)

β_p = slope coefficients for each explanatory variable.

We can perform multiple regression on the training set as follows,

```
from sklearn.linear_model import *
```

```
lr_model=LinearRegression(normalize=True)
lr_model.fit(X_train,y_train)
```

```
LinearRegression(copy_X=True, fit_intercept=True, n_jobs=None, normalize=True)
```

We performed multiple regression model on the training data, and now, we have to predict the accuracy of the model. This can be done by using mean squared error (MSE), r^2 error. To conclude that the model is good fit for the data, the mean squared error should me less, and r^2 error should be more.

```
p=lr_model.predict(X_test)
mse=mean_squared_error(y_test,p)
mse
```

```
0.22809783566727743
```

```
r2=r2_score(y_test,p)
r2
```

```
0.8048665393561916
```

From the above observation, the accuracy of the model is 74%, and the mean squared error is 0.22 which is close to zero.

4.2 Polynomial Regression

Polynomial regression is same as linear regression where the relation between independent variables and dependent variable is modeled as an nth degree polynomial. Polynomial regression is used to fit a nonlinear relationship between the dependent and independent variables. We can apply polynomial regression on dataset in the following way,

```
from sklearn.preprocessing import PolynomialFeatures
pr=PolynomialFeatures(degree=3,interaction_only=True)
X_train_pr=pr.fit_transform(X_train)
X_test_pr=pr.fit_transform(X_test)
pr.fit(X_train_pr,y_train)
```

```
PolynomialFeatures(degree=3, include_bias=True, interaction_only=
True)
```

```
lr_model.fit(X_train_pr,y_train)
```

```
LinearRegression(copy_X=True, fit_intercept=True, n_jobs=None, no
rmalize=True)
```

Now, we should predict the model depending on the mean squared error and r^2 error.

```
r2=r2_score(y_test,prediction)
r2
```

```
0.7978948587203767
```

```
mse=mean_squared_error(y_test,prediction)
mse
```

```
0.20112527627721602
```

From the above observation, the accuracy of the model is 60%, and the mean squared error is 0.34.

4.3 Decision Tree Regressor

Decision tree regression detects features of an object and trains a model in the organization of a tree to predict data in the future to produce important output [20]. This model can be applied on the dataset as follows,

```
from sklearn.tree import DecisionTreeRegressor
dt_model=DecisionTreeRegressor(criterion='mse',max_depth=5,max_features="auto",min_samples_leaf=3)
dt_model.fit(X_train,y_train)
```

```
DecisionTreeRegressor(criterion='mse', max_depth=5, max_features='auto',
           max_leaf_nodes=None, min_impurity_decrease=0.0,
           min_impurity_split=None, min_samples_leaf=3,
           min_samples_split=2, min_weight_fraction_leaf=0.0,
           presort=False, random_state=None, splitter='best')
```

To get a good accuracy of the model, we should adjust some parameters of DecisionTreeRegressor(). The adjustment of the parameter can be done by using trial and error method which means changing the values of every parameter and checking to which values of the parameters that we get a good accuracy. The accuracy for the above fit method can be done by,

```
accuracy_dt=dt_model.score(X_test,y_test)
accuracy_dt
```

```
0.774492862924727
```

The accuracy for the above model is 69% which is the max accuracy which can be obtained for the model by selecting specified values for the parameters.

4.4 Random Forest Regressor

Random forest regressor is the most accurate algorithm for most of the datasets. This model can be applied for the data as follows,

```
from sklearn.ensemble import RandomForestRegressor
rf_model=RandomForestRegressor(n_estimators=100,max_depth=3)
rf_model.fit(X_train,y_train)
```

The accuracy of the model can be increased by adjusting the parameters like we did in decision tree regressor. The accuracy of the model is,

```
accuracy_rf=rf_model.score(X_test,y_test)
accuracy_rf
```

```
0.7634116530932239
```

Table 1 Accuracy from the regression methods

Method	Accuracy (%)
Multiple regression	80.48
Polynomial regression	79.79
Decision tree regression	77.44
Random forest regression	76.34

5 Conclusion

The main purpose of this paper is to compare which regression algorithm of ML algorithms will produce more accuracy when applied on Admission_Predict.csv dataset. In this paper, we came across four regression algorithms namely multiple regression, polynomial regression, decision tree regressor, and random forest regressor. The data is divided into train set of 85% of the data and testing set of 15% of the data. The observations are given in Table 1. From this table, it is evident that polynomial regression is giving good accuracy comparing with remaining regression techniques.

References

1. MasterPortal (2020). Master Portal, URL: https://www.mastersportal.eu/countries/82/united-states.html
2. S. Mishra, S. Sahoo, A quality based automated admission system for educational domain, pp. 221–223 (2016)
3. A. Nandeshwar, S. Chaudhari, V. Sampath, A. Flagel, C. Figueroa, P. Sugrue, D. Ahlburg, M. Mcpherson, Predicting higher education enrollment in the United States: an evaluation of different modeling approaches. Int. J. Oper. Res. **19**(26), 60–67 (2014)
4. K. Basu, T. Basu, R. Buckmire, N. Lal, Predictive models of student college commitment decisions using machine learning. Data **4**, 65 (2019)
5. A.R.Y. Dalton, J. Beer, S. Kommanapalli, J.S. Lanich, Machine learning to predict college course success. SMU Data Sci. Rev. **1**(2), Article 1 (2018)
6. L.M. Abu Zohair, Prediction of student's performance by modelling small dataset size. Int. J. Educ. Technol. High Educ. **16**, 27 (2019). https://doi.org/10.1186/s41239-019-0160-3
7. J. Jamison, Applying machine learning to predict Davidson college's admissions yield, pp. 7650–7766 (2017)
8. R.V. Mane, Predicting student admission decisions by association rule mining with pattern growth approach, pp. 202–207 (2016)
9. K.N. Bhargavi, G. Jaya Suma, Quasi analysis of rainfall prediction during floods using machine learning, in *Proceedings of the 2020 8th International Conference on Communications and Broadband Networking (ICCBN '20)*. Association for Computing Machinery, New York, NY, USA, pp. 63–67 (2020). https://doi.org/10.1145/3390525.3390535
10. R. Agarwall, P. Sagar, A comparative study of supervised machine learning algorithms for fruit prediction. J. Web Dev. Web Des. **4**(1), 14–18 (2019)
11. O. Altay, T. Gurgenc, M. Ulas, et al., Prediction of wear loss quantities of ferro-alloy coating using different machine learning algorithms. Friction **8**, 107–114 (2020). https://doi.org/10.1007/s40544-018-0249-z

12. S. Baumann, U. Klingauf, Modeling of aircraft fuel consumption using machine learning algorithms. CEAS Aeronaut. J. **11**, 277–287 (2020). https://doi.org/10.1007/s13272-019-00422-0
13. J. Huang, K. Ko, M. Shu et al., Application and comparison of several machine learning algorithms and their integration models in regression problems. Neural Comput. Appl. **32**, 5461–5469 (2020). https://doi.org/10.1007/s00521-019-04644-5
14. Y. Fei, California rental price prediction using machine learning algorithms (Doctoral dissertation, UCLA) (2020)
15. S. Alawadi, D. Mera, M. Fernández-Delgado, F. Alkhabbas, C.M. Olsson, P. Davidsson, A comparison of machine learning algorithms for forecasting indoor temperature in smart buildings. Energy Syst. 1–17 (2020)
16. B.L. Sucharitha, C.V. Raghavendran, B. Venkataramana, Predicting the cost of pre-owned cars using classification techniques in machine learning, in *International Conference on Advances in Computational Intelligence and Informatics*. Springer, Singapore, 2019, pp. 253–263
17. G. Naga Satish, C.V. Raghavendran, M.D. Sugnana Rao, C. Srinivasulu, House price prediction using machine learning. Int. J. Innov. Technol. Explor. Eng. **8**(9), 717–722 (2019). https://doi.org/10.35940/ijitee.I7849.078919
18. T. Srinivasulu, S. Srujan, C.V. Raghavendran, Enhancing clinical decision support systems using supervised learning methods. Int. J. Anal. Exp. Modal Anal. **12**(5), 1188–1195 (2020)
19. C.V. Raghavendran, G.N. Satish, V. Krishna, S.M. Basha, Predicting rise and spread of COVID-19 epidemic using time series forecasting models in machine learning. Int. J. Emerg. Technol. **11**(4), 56–61 (2020)
20. M.M. Islam, N. Sultana, Comparative study on machine learning algorithms for sentiment classification. Int. J. Comput. Appl. **182**(21), 1–7 (2018)

ACP: A Deep Learning Approach for Aspect-category Sentiment Polarity Detection

Ashish Kumar, Vasundhra Dahiya, and Aditi Sharan

Abstract In recent years, sentiment analysis has gained popularity due to its wide range of application domains. With the fundamental aim of identifying the polarity of sentiment hidden in a piece of text, the ways to conduct sentiment analysis have improved with time too. Aspect-based sentiment analysis (ABSA) is one such approach. The main objective of ABSA is the extraction of sentiment polarity with respect to certain 'aspects' of the text. Semantically, an aspect itself is not a single entity but rather a representation of an aspect-category. Therefore, it makes more sense to find sentiment over an aspect-category. As the aspect-category concept is quite subjective, determining sentiment over aspect-category becomes challenging. In this paper, we developed a deep learning model that aims to perform ABSA by explicitly incorporating the aspect information into the model itself by assigning sentiment to every aspect-category. We evaluate our model on a benchmark SemEval'14 dataset and explore the state-of-the-art text-representations techniques. This paper tries to bring forward performances of these representations on the taken aspect-based model.

Keywords Aspect-based sentiment analysis · Aspect-category polarity · Long short-term memory · Opinion mining · Natural language processing

A. Kumar (✉) · V. Dahiya · A. Sharan
School of Computer and Systems Sciences, Jawaharlal Nehru University, New Delhi 110067, India
e-mail: ashish29_scs@jnu.ac.in; ashishkumar2912@gmail.com

V. Dahiya
e-mail: vasundhra.dahiya@gmail.com

A. Sharan
e-mail: aditisharan@mail.jnu.ac.in

D. Bhattacharyya and N. Thirupathi Rao (eds.), *Machine Intelligence and Soft Computing*, Advances in Intelligent Systems and Computing 1280,
https://doi.org/10.1007/978-981-15-9516-5_14

1 Introduction

In today's time, user-generated content on different social networking sites, blogs, and forms play an important role in every individual's decision-making process. To make the best use of such content, conducting opinion mining, also referred to as, sentiment analysis came into play in the early 2000s and has flourished since. In both academic and industrial fields, it has been revolutionary. Sentiment analysis is a computational method for automatically detecting the attitude, emotion, and sentiment of a speaker in a given piece of text [1], which in our case is the customer review. As reviews do not always essentially talk about just a product with only one feature, it was established that to study them with respect to different aspects would be more insightful. The sentences come from many customer reviews and represent different sentiment polarities toward varied aspects [2]; polarities mainly being: positive, negative, conflict, or neutral. Classifying the sentiment polarity of a review sentence without considering the aspect-category presented in the review sentence can be misleading and hence inaccurate.

The study of aspect-categories corresponding to every review type technically came to be known as, aspect-based sentiment analysis (ABSA). ABSA of customer reviews involves identifying the aspects of the entities being reviewed and mining the sentiment mentioned for the aspects [3]. Till yet, the ABSA problem has not been tackled fully and it has much room for improvement. Recent works done for ABSA are generally based on either identifying the sentiments by associating it with its aspect terms or using a topic modeling approach to categorize an aspect and find its corresponding sentiment polarity [4]. This approach is too crude as the explicit representation of aspect terms does not associate any sentiment with the term. Moreover, a general user is usually interested in knowing the sentiment corresponding to a certain aspect-category. Therefore, an extension of aspect term-based sentiment analysis leads to a notion of aspect-category sentiment analysis. The real challenge of ABSA starts with subjectivity associated with the aspect-category itself which makes its explicit representation difficult. Thus, finding the polarity of aspect-categories deems essential. This is termed as aspect-category polarity (ACP).

Initially, terms that were discussed more often in the review text were identified as aspect terms and extracted with their sentiment polarity [5–7]. But as many aspect terms may represent a similar concept, aggregation of these synonymic aspect terms will represent an aspect-category. For example, aspect terms like 'money,' 'bug,' 'price,' and 'cost' will belong to a single aspect-category 'price'. Hence, instead of finding sentiment toward each aspect term mentioned in the review text, it would be beneficial to extract the sentiment orientation toward the aspect-category. This task was introduced in SemEval's 2014 Conference. The purpose of this challenge was to study the aspect-categories and identify their respective polarities [8]. Let us consider an example review: "*they are often crowded on the weekends but they are efficient and accurate with their service*"; in the given snippet of a review, there are two aspect-categories: 'service' and 'ambience'. Sentiment associated with respect to the 'service' category is *positive* while it is *negative* for 'ambience' category.

Aspect-category polarity detection (ACP) is a classification problem, and generally classification in machine learning is handled in two ways, namely supervised and unsupervised learning. If supervised approaches are used, a high level of feature engineering is needed. But with changes in machine learning approaches to deep learning, handling such huge amounts of data with deep neural networks (DNNs) is preferred [9]. The success of any deep learning method depends on the representation of text in the field of natural language processing (NLP). In NLP, over time, text representations have been improved upon from the initial bag-of-word representation to one-hot representation to the latest deep neural network-based representations (like Word2Vec, GloVe, ELMo, BERT). The major contribution of this paper is as follows:

1. We have suggested a deep learning-based technique for representation of aspect-categories.
2. We have proposed a deep learning model to find out the sentiment polarities of a review corresponding to a given aspect-category.
3. We also tried to find out what text representation works best for the sentiment classification task by comparing the result of our proposed model using various pre-trained state-of-the-art text embeddings.

2 Related Work

This section sheds light on some recent deep learning approaches used for aspect-sentiment polarity detection. Wang et al., 2016 [10] used an attention mechanism called aspect-to-sentence attention that explores the prominence part of the sentence allied to the given aspect. They achieved so by proposing the following two models, attention-based LSTM (AT-LSTM) and attention-based LSTM with aspect embedding (ATAE-LSTM). The former involves concatenating aspect embeddings with the sequential output of the LSTM layer before passing it to the attention layer. The latter, however, combines aspect embeddings with each input of LSTM. These models showed an impressive performance in the ABSA task but were designed to handle only three sentiment classes: positive, negative, and neutral. Following the same architecture, Al-Smadi et al., 2018 [11] evaluated the sentiment polarity for an Arabic review dataset. Establishing a single review could be a composition of many sentiment-aspect pairs, hence, extracting opinion context regarding different aspects seemed more beneficial. So, He et al., 2018 [12] proposed an approach to enhance the potential of attention mechanism to understand sentiment from opinion's context in the review. They engraved semantic and syntactic information into an attention mechanism with the help of a dependency parser. This approach failed to distinguish the neutral sentiment, if the sentence contains an opinion word. Furthermore, Wang and Liu, 2015 [13] weighted the word-embeddings of review sentences based on aspect specific probabilistic mass calculated using the parse tree. These modified word-embeddings were used as an input to a convolution neural network (CNN)-based sentiment prediction model.

3 Background and Motivation

For a while now, deep learning has been a great means for improvement in the detection of sentiments. Hence, we highlight some deep learning models for sequential inputs, and further, some state-of-the-art techniques are to represent the textual information.

3.1 Deep Learning Models for Sequence Learning

Recurrent Neural Network. Recurrent neural networks (RNNs) were introduced because basic neural networks are incapable of extracting useful information (like position-based features) that is present in the sequential data. Fundamentally, a RNN deals with the complete input sequence one by one. It stores the information of the previous instances as *context*, technically then known as, 'memory.' This memory is actually the activation value that is combined with input value at that time step and is used to predict the output. This output (new activation value) is passed further to the next RNN cell and this process goes on until the end of the complete sequence. In this way, RNNs conserve the sequential information by capturing the relationship between the present and past input data.

Long Short-Term Memory Network. RNNs are improved significantly over the basic neural networks but failed to serve its fundamental purpose of capturing context in sequential data accurately and efficiently. This happened because of the problem of vanishing gradient causing RNNs to fail in preserving long-term dependencies, thereby originating a modification named, long short-term memory network (LSTMs) [14].

In LSTM, the concept of gates is introduced in the basic RNN cells. The main aim of LSTMs by using these gates and is to determine which information is useful in a particular time step and which is useless. Essentially, an LSTM cell keeps, updates, and forgets the information in the cell with the help of input (i), forget (f), and output (o) gates. The LSTM cell at time step t is formulated as below:

$$\hat{c}^{\langle t\rangle} = \tanh\left(W_c\left[a^{\langle t-1\rangle}, x^{\langle t\rangle}\right] + b_c\right) \tag{1}$$

$$i^{\langle t\rangle} = \sigma\left(W_i\left[a^{\langle t-1\rangle}, x^{\langle t\rangle}\right] + b_i\right) \tag{2}$$

$$f^{\langle t\rangle} = \sigma\left(W_f\left[a^{\langle t-1\rangle}, x^{\langle t\rangle}\right] + b_f\right) \tag{3}$$

$$o^{\langle t\rangle} = \sigma\left(W_o\left[a^{\langle t-1\rangle}, x^{\langle t\rangle}\right] + b_o\right) \tag{4}$$

$$c^{\langle t\rangle} = i^{\langle t\rangle} \odot \hat{c}^{\langle t\rangle} + f^{\langle t\rangle} \odot c^{\langle t-1\rangle} \tag{5}$$

$$a^{\langle t \rangle} = o^{\langle t \rangle} \odot \tanh\left(c^{\langle t \rangle}\right) \tag{6}$$

where σ is the logistic sigmoid function, $\odot$ is element-wise multiplication function, $\hat{c}$, c are memory cell states, a is activation(hidden) state, W_c, W_i, W_f, W_o are weight matrices, and b_c, b_i, b_f, b_o are bias vectors of different gates for input x.

Bi-directional Long Short-Term Memory. Another improvement to deal with sequential data is to use memory information from both past and future, i.e., previous and future sequence to preserve the context of text data. To achieve bi-directionality, two separate unidirectional LSTMs are used—one for capturing past context (forward LSTM) and another to capture future context (backward LSTM). In forward LSTM, sequential inputs are fed in the same order as they occur but in backward LSTM, and reversed inputs are fed. This way the context (both past and future) is preserved.

3.2 Word-Embeddings

Every machine learning technique works with numeric data which means all text should be converted to any numeric representation too. This representation, known as embedding, is well accepted to be crucial to compose a powerful text representation at a higher level of studying text. A good text representation preserves the semantic meaning of the text as much as possible. Over the years, researchers have tried various approaches to deal with word-embeddings. Following are some of the famous text representations:

One-hot Encoding. In this representation, each word is encoded into a vector of fixed vocabulary size. This encoded vector contains all zeroes except a single one corresponding to the index of that word in the vocabulary. This is a sparse representation and the vector size increases with the size of the vocabulary. This is the easiest representation. One-hot vectors, although fails to capture the semantic relationship between the words.

Word2Vec. Mikolov et al., 2013 [15, 16] came up with the idea of a distributed representation of words in vector space that essentially focused on capturing a semantic relationship between words rather than just their occurrence's frequency or pattern. In this representation, each word and phrase presented in the corpora are mapped to vectors as real numbers. This representation works by capturing the meaning of a word from its surrounding context. To implement this idea, two neural network-based methods were proposed. One was Skip-gram and another was contextual bag-of-words (CBOW). The training objective of the Skip-gram model is to predict the context of the given word while the objective of the CBOW model is to predict the word given some context words. Learned weights of hidden layers are finally treated as word-embeddings whose dimensions depend on the total number of neurons presented in the hidden layer. Such embeddings' value would show if two different words have the same meaning, they will have higher cosine similarity. This representation is quite effective to capture the relationship between the words and is considered to be an NLP revelation.

GloVe. This representation is almost the same as Word2Vec, but the idea varies in the way that it builds a word-context matrix [17]. Using the word-context matrix, GloVe captures the statistics of word-to-word co-occurrences. It captures global statistics in the corpus using the Skip-gram model. Since this method studies context globally, obtained vectors have been named global vectors (in short, GloVe). It uses matrix factorization to perform dimensionality reduction, which earlier was used in latent semantic analysis (LSA). These methods have a very good performance on various tasks like word analogy, word similarity, and named entity recognition.

ELMo. In embeddings from language models (ELMo), word vectors are learned functions of the internal states of a deep two-layer bi-directional language model (biLM) [18]. Here, the character-based information is passed into a convolutional neural network (CNN) to generate an initial representation of words. This representation is passed as an input to the first layer of biLM, which consists of two LSTM layers (forward and backward) to generate intermediate word representation. Further, this intermediate word representation acts as input to the next layer of biLM that also contains two LSTM layers. This layer, then again, generates an intermediate word representation. Finally, the weighted sum of initial word representation and two intermediate word representations are used to form ELMoembeddings. This is a highly effective way of word representation. As LSTMs are used to generate the embeddings, the ELMoembeddings take care of the meaning of the entire sentence. It is notable that in the case of polysemy, ELMo generates different embeddings of a single word depending on the context in which it is used. It can also handle the out-of-vocabulary (OOV) words, since initial input is formed using characters of a word.

BERT. Recently introduced bi-directional encoder representations from transformers (BERT) are built on two key ideas: (i) the transformer architecture and (ii) unsupervised pre-training [19]. This is a transformer-based model that follows a fully attentive approach. BERT's speciality is having pre-trained embeddings that are contextually learned. It functions in two phases. Firstly, it pre-trains a masked model and then works on the next sentence prediction. The strategy to pre-train is:

- Masked language model (MLM): Here, masking some percentage (originally 15%) of the input tokens is done at random and prediction of only these masked tokens is performed.
- Next sentence prediction (NSP): Here, relationships between sentences are captured, i.e., the task here is to figure out if sentence B is immediately followed by another sentence A.

Secondly, it fine-tunes the trained model rather than training embeddings from scratch for every new task.

4 Model Description

4.1 Problem Formulation

For a given pre-defined aspect-category, set $C = \{c_1, c_2, \ldots, c_k\}$ denotes the aspect-category with k possible categories. The review dataset is $R = \{r_1, r_2, \ldots, r_n\}$ contains n review sentences. The task of aspect polarity detection (APD) is formulated as a learning function $h : R, C \rightarrow S$, where S is sentiment polarity set containing different sentiment classes. For each unseen review $r \in R$ and given aspect-category $c \in C$, the aspect polarity detection function $h(\cdot)$ predicts $h(r, c) \rightarrow s$.

4.2 Dataset

We used the dataset provided by SemEval'14 for ABSA tasks. This dataset contains separate training (3041 review instances) and testing (800 review instances) data. Each review sentence is tagged with aspect-categories and its corresponding polarities. A single review might contain multiple aspect-categories also. This dataset is tagged for five pre-defined aspect-categories: 'ambience', 'food', 'price', 'service', and 'anecdotes/miscellaneous'.

4.3 Model Architecture

We used two different models in our architecture; first, to learn aspect-category representation (ACR) and another to predict aspect-category sentiment polarities (ACP). The most difficult problem in ABSA is to represent and review text. For review text representation, we have used various pre-trained text embeddings. However, the representation of aspect-category is tricky. The basic way to represent aspect-category can be to represent it as the word-embedding of an aspect-category name. For example, aspect-categories like 'food', 'service', and 'price' can be represented by word-embeddings of the words 'food', 'service', and 'price', respectively. Since categories are not a single word entity, a single category can be represented by various words. For example, the 'price' category can be represented by various other terms like cost, amount, money, etc. But an aspect-category might be domain-dependent. 'Service' category can be represented using terms like chef and delivery in the *restaurant domain,* while it can be represented by terms like guide and travel in the *tourism domain.*

So, inspired by the work of Zhou et al., 2015 [20], in the first model, we have used two-layer dense neural networks to learn category representation (Fig. 1). Different neural networks are trained for each aspect-category considering the problem as a binary classification problem. These NNs take review sentences as input and try to

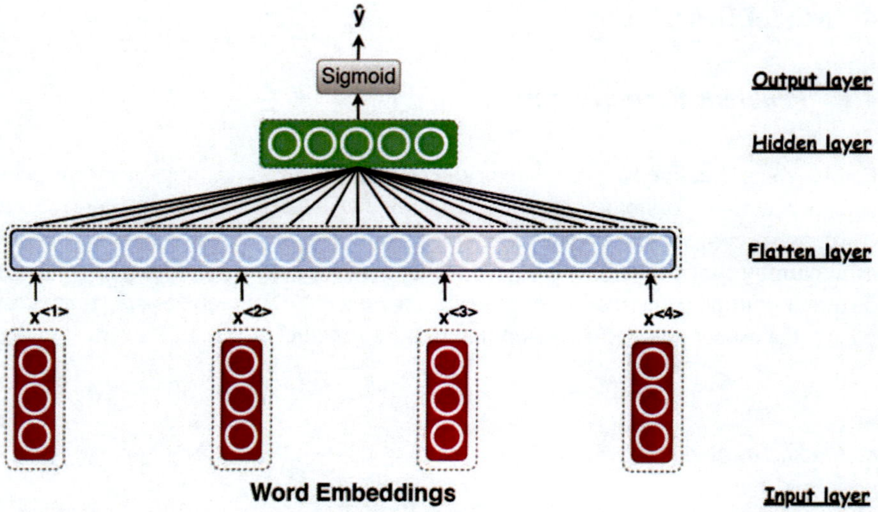

Fig. 1 Neural network model for aspect-category representation learning

predict the aspect-category in the review. After completion of training, hidden layer weights can be used to represent the given aspect-category.

In the second model, we have used LSTM layers to predict the aspect-sentiment polarities (Fig. 2). Inputs to this model are review sentences and aspect-category.

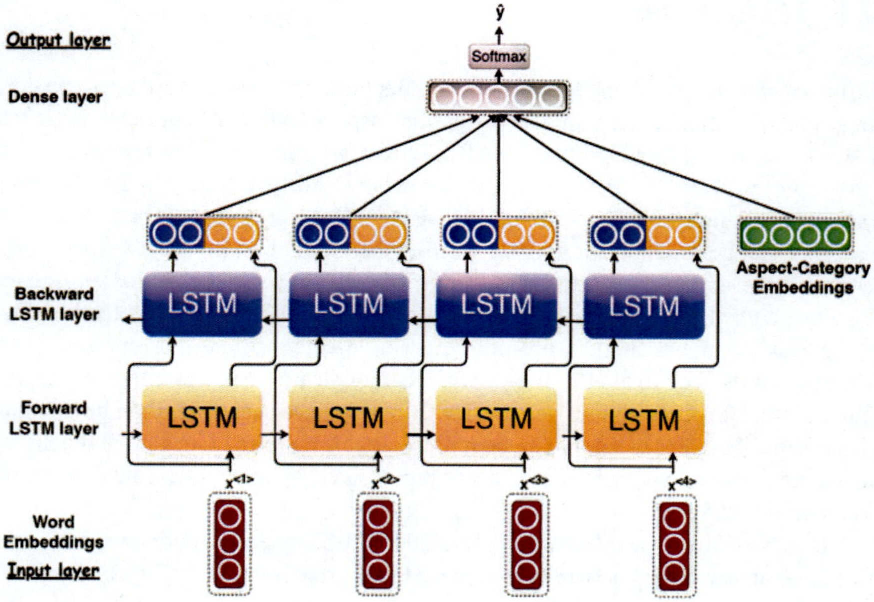

Fig. 2 Deep learning model for aspect-category sentiment prediction

5 Experiments and Results

Our proposed model was implemented on SemEval'14 dataset using previously mentioned word-embeddings like Word2Vec, GloVe, ELMo, and BERT to find out the best word-embeddings for sentiment analysis tasks, performance of proposed model, and work on SemEval's ABSA task challenge.

5.1 Experiment Phases

Experiments were conducted in two phases. The first phase was aspect-category representation and the second phase was aspect-category polarity detection.

Aspect-category Representation (ACR). In the aspect-category representation learning task, we used the two-layer neural network (hidden layer and output layer). Here, the total number of neurons in the hidden layer governs the embedding dimension of aspect-category and varies for various embeddings. As an example: If ELMoembeddings are used to represent inputs, it generates aspect-category embeddings using ELMo. Likewise, we have generated separate aspect-category embeddings for other available pre-trained embeddings too.

This experiment is done as a supervised learning method where aspect-category given in the review sentence serves as the label. For category learning, we have used a one-versus-all method and projected the problem as a binary classification problem. Once the model is trained, hidden layers' weights are used to represent aspect-category (see Fig. 1). These learned aspect-category embeddings are further used as input in the ACP model.

Aspect-Category Polarity (ACP). Reviews sentences are fed as an input to the Bi-LSTM network, where activation values for each review word are calculated by both LSTMs (forward and backward) and concatenated. Then, aspect-category embeddings are concatenated with the generated activation values of Bi-LSTM and passed to a dense layer. Further, the output of the dense layer is classified using the softmax function (see Fig. 2). Hyper-parameters used in the experiment are: LSTM unit size is set to 128, the batch size is set to 64, and for optimization function, we used RMSprop optimizer. These are general settings for LSTM-based neural networks, one can try different hyper-parameters settings also. We used the early stopping technique in all our experiments.

5.2 Effect of Word-Embeddings

We have experimented with various word-embeddings to visualize the impact of text representations for sentiment analysis. Dimensions of embeddings were 300, 300,

1024, and 768 for Word2Vec, GloVe, ELMo, and BERT, respectively. The results are documented below.

Table 1 shows a clear picture of overall performance of the proposed model. Word2Vec and GloVe almost have the same performance. ELMo performs better than GloVe and Word2Vec, which is an indication of the importance of better context capturing, yet again. Furthermore, BERT's performance is superior to any other embeddings. There is around a 4% increase in performance in BERT as compared to ELMoembeddings. This points out that BERT is rightly so, the best text representations for sentiment analysis tasks too.

The exact results of the model are shown in Fig. 3 using a graphical representation. It depicts the evaluation measures—precision, recall, and F1-score values for identified polarity of each class using ACP model with respect to different word-embeddings.

Further, Fig. 4 highlights the comparison of different word-embeddings between the confusion matrices of the ACP model. In spite of having good performance by the model (ACP_BERT), it does fail to categorize the 'conflict' class like other models (Fig. 4). This might have happened due to the less number of training instances

Table 1 ACP model performance using different word-embeddings

Model	Accuracy (%)
ACP_Word2Vec	74.93
ACP_GloVe	74.73
ACP_ELMo	77.66
ACP_BERT	81.17

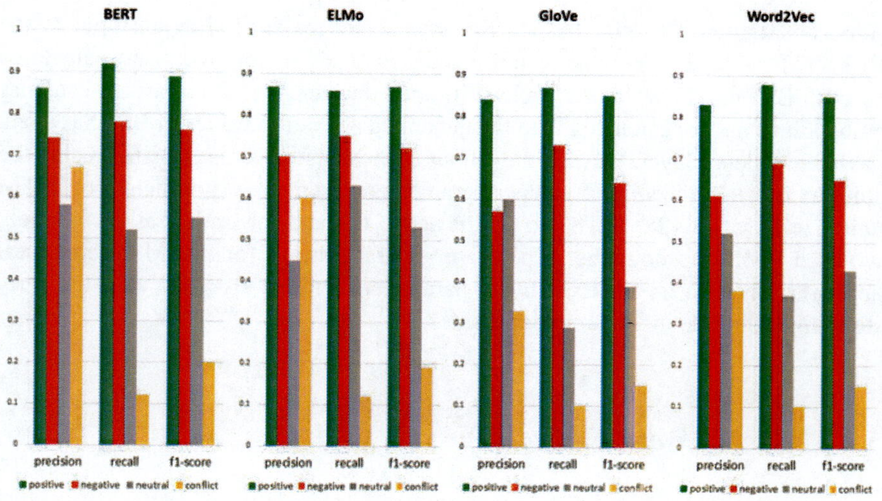

Fig. 3 Precision, recall, and F1-score values for each class using ACP model for different word-embeddings

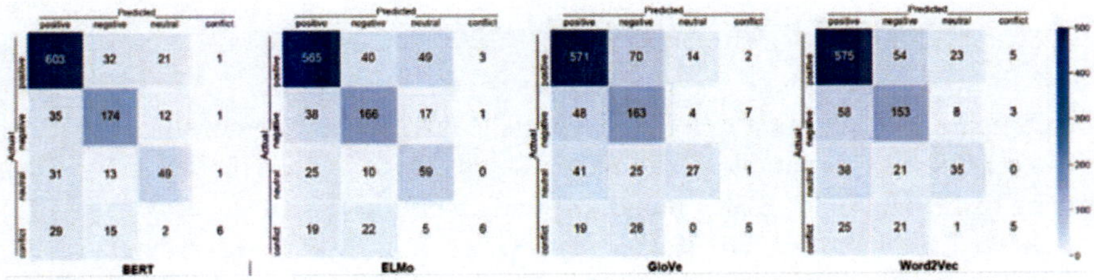

Fig. 4 Comparison between confusion matrices of ACP model using different word-embeddings

Table 2 Comparison with different SemEval'14 models

Model	Correct	All	Accuracy (%)
NRC-Canada	850	1025	82.93
ACP_BERT	**832**	**1025**	**81.17**
XRCE	801	1025	78.15
ACP_ELMo	**796**	**1025**	**77.66**
UNITOR	782	1025	76.29
SAP_RI	775	1025	75.61
ACP_Word2Vec	**768**	**1025**	**74.93**
ACP_GloVe	**766**	**1025**	**74.73**
SeemGo	765	1025	74.63
SA-UZH	749	1025	73.07
UNITOR_C	749	1025	73.07
UWB	746	1025	72.78
lsis_lif	739	1025	72.10
UBham	737	1025	71.90
EBDG	715	1025	69.76
SNAP	713	1025	69.56
COMMIT-P1WP3	694	1025	67.71
Blinov	673	1025	65.66
Baseline			65.65

Bold indicates our proposed models

belonging to 'conflict' class. Our assumption is that increasing the training instances will also improve the results.

5.3 Comparison with Different Models

On comparing our results with SemEval14's teams, Table 2 demonstrates that our method achieves comparable performance and is able to retrieve the second position among them. The ACP_BERT model correctly identifies 832 instances out of 1025 instances. Although there are 800 testing reviews, it is possible that a single review may be associated with multiple aspect-categories. Each review with a single aspect-category is treated as a single instance, thus, 800 reviews are converted to 1025 instances. Only NRC-Canada is performing better than our method. One possible reason for not performing better than NRC-Canada might be that our method does not use any external features other than word-embeddings.

Other than BERT, other embedding ELMo also achieved a significantly better performance with our proposed model. Whereas, Word2Vec and GloVe worked comparatively well. Even without pre-trained contextually learned, the reasons for their performance might be due to semantic and architectural factors, better representation for the aspect-categories, and the accommodation of better neural models.

6 Discussion and Conclusion

In this work, we proposed two deep learning approaches, one for aspect-category representation and another for aspect polarity detection. The novelty of this framework is to present an explicit exemplification technique for aspect-category representation and utilizing this information for improving the accuracy of aspect-based sentiment analysis. The model shows promising performance on SemEval'14 benchmark dataset. The use of Bi-LSTM helped in finding the contextual sentiment information from the review. Also, the incorporation of learned aspect-category representation benefits the model to focus on aspect-related sentiment context and find the correct sentiment toward the given aspect. We also explored various text-representation techniques and based on performance concluded that BERT performs better than other representations (like Word2Vec, GloVe, and ELMo) on the sentiment analysis tasks. On the basis of confusion matrices, we also conclude that the model works well for positive and negative sentiment classes but fails to achieve up-to-mark performance for neutral and conflict classes. One possible reason might be less number of training examples for these classes. In the future, we will try to integrate some data augmentation techniques for these classes. We will also consider a different outlook for the integration of aspect-category representation into the model.

References

1. B. Liu, *Sentiment Analysis: Mining Opinions, Sentiments, and Emotions* (Cambridge University Press, 2015)
2. T.A. Rana, Y.-N. Cheah, *Improving Aspect Extraction Using Aspect Frequency and Semantic Similarity-Based Approach for Aspect-Based Sentiment Analysis* (Springer International Publishing, Cham, 2018), pp. 317–326
3. J. Feng, S. Cai, X. Ma, Enhanced sentiment labeling and implicit aspect identification by integration of deep convolution neural network and sequential algorithm. Cluster Comput. 1–19 (2018)
4. K. Schouten, F. Frasincar, Survey on aspect-level sentiment analysis. IEEE Trans. Knowl. Data Eng. **28**(3), 813–830 (2016)
5. W. Bancken, D. Alfarone, J. Davis, Automatically detecting and rating product aspects from textual customer reviews, in *Proceedings of the 1st International Conference on Interactions between Data Mining and Natural Language Processing*, vol. 1202, DMNLP'14, 1–16, Aachen, DEU, 2014. CEUR-WS.org
6. M. Hu, B. Liu, Mining opinion features in customer reviews, in *Proceedings of the 19th National Conference on Artificial Intelligence, AAAI'04*. AAAI Press, 2004, pp. 755–760
7. G. Qiu, B. Liu, J. Bu, C. Chen, Expanding domain sentiment lexicon through double propagation, in *Proceedings of the 21st International Joint Conference on Artificial Intelligence, IJCAI'09*, San Francisco, CA, USA, 2009. Morgan Kaufmann Publishers Inc., pp. 1199–1204
8. M. Pontiki, D. Galanis, J. Pavlopoulos, H. Papageorgiou, I. Androutsopoulos, S. Manandhar, SemEval-2014 task 4: aspect based sentiment analysis, in *Proceedings of the 8th International Workshop on Semantic Evaluation (SemEval 2014)*, Dublin, Ireland, August 2014. Association for Computational Linguistics, pp. 27–35
9. A. Kumar, A. Sharan, Deep learning-based frameworks for aspect-based sentiment analysis, in *Algorithms for Intelligent Systems* (Springer Singapore, Singapore, 2020), pp. 139–158
10. Y. Wang, M. Huang, X. Zhu, L. Zhao, Attention-based LSTM for aspect-level sentiment classification, in *Proceedings of the 2016 Conference on Empirical Methods in Natural Language Processing*, Austin, Texas, November 2016. Association for Computational Linguistics, pp. 606–615
11. M. Al-Smadi, B. Talafha, M. Al-Ayyoub, Y. Jararweh, Using long short-term memory deep neural networks for aspect-based sentiment analysis of Arabic reviews. Int. J. Mach. Learn. Cybern. 1–13 (2018)
12. R. He, W.S. Lee, H.T. Ng, D. Dahlmeier, Effective attention modeling for aspect-level sentiment classification, in *Proceedings of the 27th International Conference on Computational Linguistics*, SantaFe, New Mexico, USA, Aug 2018. Association for Computational Linguistics, pp. 1121–1131
13. B. Wang, M. Liu, Deep learning for aspect-based sentiment analysis. Stanford University Report, 2015
14. S. Hochreiter, J. Schmidhuber, Long short-term memory. Neural Comput. **9**(8), 1735–1780 (1997)
15. T. Mikolov, K. Chen, G. Corrado, J. Dean, Efficient estimation of word representations in vector space, in 1st International Conference on Learning Representations, ICLR 2013, ed. By Y. Bengio, Y. LeCun, Scottsdale, Arizona, USA, 2–4 May 2013, Workshop Track Proceedings, 2013
16. T. Mikolov, I. Sutskever, K. Chen, G. Corrado, J. Dean, Distributed representations of words and phrases and their compositionality, in *Proceedings of the 26th International Conference on Neural Information Processing Systems*, 2, NIPS'13, Red Hook, NY, USA, 2013. Curran Associates Inc., pp. 3111–3119
17. J. Pennington, R. Socher, C. Manning, GloVe: global vectors for word representation, in *Proceedings of the 2014 Conference on Empirical Methods in Natural Language Processing (EMNLP)*, Doha, Qatar, October 2014. Association for Computational Linguistics, pp. 1532–1543

18. M. Peters, M. Neumann, M. Iyyer, M. Gardner, C. Clark, K. Lee, L. Zettlemoyer, Deep contextualized word representations, in *Proceedings of the 2018 Conference of the North American Chapter of the Association for Computational Linguistics: Human Language Technologies*, 1 (Long Papers), New Orleans, Louisiana, June 2018. Associationfor Computational Linguistics, pp. 2227–2237
19. J. Devlin, M.-W. Chang, K. Lee, K. Toutanova, BERT: pre-training of deep bidirectional transformers for language understanding, in *Proceedings of the 2019 Conference of the North American Chapter of the Association for Computational Linguistics: Human Language Technologies*, vol. 1 (Long and Short Papers), 4171–4186, Minneapolis, Minnesota, June 2019. Association for Computational Linguistics
20. X. Zhou, X. Wan, J. Xiao, Representation learning for aspect category detection in online reviews, in *Proceedings of the Twenty-Ninth AAAI Conference on Artificial Intelligence, AAAI'15* (AAAI Press, 2015), pp. 417–423

Performance Analysis of Different Classification Techniques to Design the Predictive Model for Risk Prediction and Diagnose Diabetes Mellitus at an Early Stage

Asmita Ray and Debnath Bhattacharyya

Abstract Diabetes mellitus is also known as diabetes is the chronic disease. It is basically metabolic disorder. In this type of disease, glucose level is abnormally high in blood and in consequence body unable to produce adequate insulin in order to perform the needs. Glucose is one of the most vital components of health for producing energy of cell and brain. Blood containing excessive amount of glucose can lead to several serious health complications. Designing an innovative prediction model for early recognition of diabetes using machine learning techniques is the major purpose of this research. Our proposed model will be utilized to produce closest result comparing to clinical outcomes. The proposed work is involved to conduce in the comparison of efficiency of distinct machine learning algorithms such as support vector machine (SVM), random forest, decision tree, and K-nearest neighbors (KNN). Few key factors are responsible for diabetes mellitus. Our proposed method focuses on those selective attributes for early detection of diabetes mellitus using predictive analysis method. The efficiency of each algorithm is evaluated by performance measure methods like sensitivity, specificity, positive likelihood ratio, negative likelihood ratio, disease prevalence, positive predictive value, negative predictive value, and accuracy. SVM obtained highest accuracy (77.78%) with lower error rate compared to other algorithms. The vital goal of this study is to assess the suitable classifier by comparing and analyzing the performance of four algorithms that helps the doctors and hospitals for early diagnosed of diabetes mellitus and proper plan of treatment.

Keywords Diabetes mellitus · Prediction model · Machine learning · SVM · Random forest decision tree · K-NN

A. Ray (✉)
Department of Computer Science & Engineering, Vignan's Institute of Information Technology, Visakhapatnam, A.P., India
e-mail: rayasmi.viit18@gmail.com

D. Bhattacharyya
Department of Computer Science and Engineering, K L Deemed to be University, KLEF, Guntur 522502, India
e-mail: debnathb@kluniversity.in

D. Bhattacharyya and N. Thirupathi Rao (eds.), *Machine Intelligence and Soft Computing*, Advances in Intelligent Systems and Computing 1280,
https://doi.org/10.1007/978-981-15-9516-5_15

1 Introduction

Diabetes mellitus is extremely common chronic disease that is rapidly increasing global prevalence [1, 2]. In this type of metabolic disease, pancreas is not capable to produce insulin, and as a result, blood glucose elevates effectively (known as hyperglycemia). The constant uncontrolled high level of glucose over the long term leads to serious damage of the body, especially it effects organs like eyes, kidneys, and heart [3–5].

Type 1, Type 2, and gestational are the three major types of diabetes. Type 1 diabetic usually effects children and adolescents, but there is a high chance to evolve this type of diabetes for any age group people. In this type of diabetes, pancreas produce inadequate insulin or no insulin. Insulin is a hormone that helps to allow glucose to cell that produces energy. Regulating the glucose level is too necessary for Type 1 diabetes mellitus patient, and for that purpose, daily insulin injection is too essential.

In Type 2 diabetes, middle aged or older is more likely to effect. It is one of the most common diabetes, occurs either body is unable to produce enough insulin or not capable to work efficiently [6, 7].

Gestational diabetes diagnosed during pregnancy causes high level of glucose as hormones produced at the time of pregnancy. This type of diabetes can affect the pregnancy as well as baby's health. Detection of this disease at an early stage not only reduces the medical costs but also minimizes the risks of complicated health problems of patients.

According to the report stated by World Health Organization (WHO), number of individual diabetes has increased 422 million. It is evaluated that diabetes is the main key issue of 2016 1.6 million deaths. Day by day diabetes prevalence has been risen rapidly, basically middle and low income countries. There is a significant increment in the number of individual experiencing diabetes over 18 years of age from 4.7 to 8.5% [8–10].

Diabetes is a ceaseless ailment. It has significant effect on different parts of the body that can raise the probability of heart attack, kidney failure, blindness, and so on.

If the diabetes is not properly controlled at an early stage, it develops the acute complication that leads to significant contributor to costs and poor quality of life and mortality. It is the associated with large economic burden.

2 Materials and Methods

Support vector machine is the well-known supervised learning technique [11, 12]. This most popular discriminative method used in both regression and classification. The main goal of this algorithm is to find out the hyper line. This hyper line splits the dataset into two classes. It follows mainly two steps

1. Finding out the most perfect ideal hyper line from data space.
2. Delineate the objects to the specifically defined boundaries.

In the training algorithm, new samples belong to a class play a key role to design a model.

Random Forest

This familiar supervised learning method plays a crucial role for both classification and regression [13, 14]. In this technique, in random fashion, finding the root node and splitting the feature node takes place and that makes the main difference between the decision tree and random forest. The steps to perform the random forest are as follows:

1. Load the data from dataset which consist of 'm' features.
2. In random forest, the training algorithm is known as boot strap algorithm. Random sample creation is mainly depending on the arbitrary features. Detection of OOB error is the major part of this algorithm and that is performed by trained the new samples.
3. Nodes should be split into sub-nodes. Calculation of node has been performed by the best split process.
4. This process will execute until to find n number of trees.
5. The anticipation of target is depending in the calculation of total number of votes of each tree. Final prediction of this process will be determined by the highest number of voted class.

Decision Tree

In regression and classification problems, this supervised learning method has taken a major role [15, 16]. It is basically a tree structure which consists of collection of nodes. This classification technique continuously splits the dataset into two or more sample data based on different conditions. The goal is to meet with predicted class value of the target variables. Binary and continuous both variables are used to construct the decision tree. This classification method uses the highest entropy value to find out optimal root node. Dataset consists of several attributes and instance values. In this method, dataset has given as input, and decision model will be the output [15].

K-Nearest Neighbors (KNN)

This common and familiar classification technique is mainly used for classification of new sample and that is performed based on similarity or distance measure [17–19]. Three types of distance measures are mainly used are

1. Euclidean distance
2. Manhattan
3. Minkowski
4. The steps for KNN are given below

5. In this algorithm, storing the feature of the samples and class label of training sample incorporated in training phase.
6. Classification of undefined samples depending on the value of k. Based on the similarity of features, unlabeled samples can be classified into defined class.
7. Determining the unlabeled class mostly depends on the voting classification. Different techniques can be used to determine the k-value like heuristic technique can be used for selection of k-value.

3 Dataset Description

Prediction of the diabetes mellitus at an early stage is the major aim of this study as it will help to improve the prognosis of the people. The dataset is collected from Pima Indian Diabetes (PID) dataset. The dataset comprises 768 cases and nine attributes. Descriptions of the attributes are shown in Table 1. Different samples have been used for testing and training, and 768 data items have been considered for testing the data.

Table 2 depicts the effectives of all machine learning algorithms. The effectiveness of all classifier has been estimated with respect of instances correctly and incorrectly classified and accuracy. Figure 1 shows the result of Table 2.

1. **Efficiency**

Table 1 Illustration of Dataset

S. No.	Attribute	Description	Cutoff values for Attributes
1	Age	Age of male/female	21–81
2	Pregnant	Calculate the number of pregnant women	–
3	Plasma glucose level	Concentration of glucose level	In an oral glucose tolerance test study the concentration of plasma glucose over 2 h
4	Blood pressure	Diastolic blood pressure (mm Hg)	BP < 80 = Normal BP > 80 High
5	Skin-fold thickness	Skin-fold thickness Triceps (mm)	[0–99]
6	Serum insulin	Produce and store in beta cell of pancreas	[0–846]
7	Body mass index (BMI)	Body mass divided by the square of the body height	BMI < 18.5 = Underweight 18.5–24.9 = Normal 25.0 - 29.9 = Overweight BMI > 30.0 = Obese
8	Diabetes pedigree function	It is depending on family history	[0–2.45]
9	Class (positive or negative)		0 if non-diabetic, 1 if diabetic

Table 2 Performance of classification techniques

S. No.	Classification techniques	Accuracy	Correctly classified	Incorrectly classified
1	SVM	77.78	594	174
2	Random forest	75.26	575	193
3	Decision tree	74.18	556	212
4	KNN	62.04	138	630

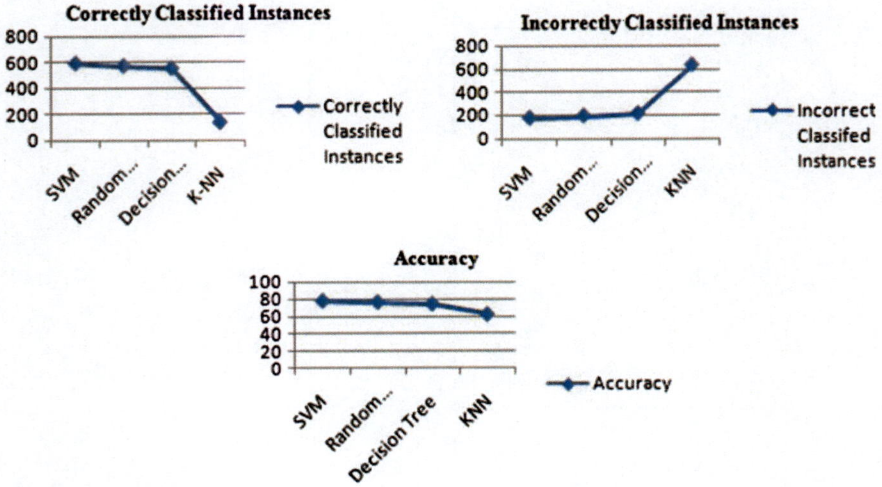

Fig. 1 Efficiency comparison graph of different classifiers

Table 3 represents the efficiency of different classification techniques. The efficiency has been measured in terms of sensitivity, specificity, positive likelihood ratio, negative likelihood ratio, disease prevalence, positive predictive value, negative predictive value, and accuracy. Figure 2 shows the corresponding graph of Table 2.

Figure 2 illustrates the analysis of different machine learning algorithms with respect of various performance measures. Nine attributes have been considered from dataset to estimate the performance of classifiers. Performance of the classification algorithm has been evaluated based on the predictive value of accuracy. Random forest and decision tree these two classification techniques obtained the highest specificity of 96.60 and 98.20%. SVM achieved a significant performance in respect of accuracy (77.78%) and low error rate.

4 Conclusion

Diabetes mellitus (DM) is the most common and widely prevalent diseases that epidemiologically and economically pose a major health challenge. Awareness plays

Table 3 Comparison of classification techniques

S. No.	Algorithm	Sensitivity (%)	Specificity (%)	Positive likelihood ratio	Negative likelihood ratio	Disease prevalence %	Positive predictive value %	Negative predictive value %	Accuracy %
1	SVM	55.09	86.80	4.10	0.52	40.90	76.21	80.43	77.78
2	Random forest	29.21	96.60	12.60	0.70	40.90	89.90	73.26	76.39
3	Decision tree	24.49	98.20	0.75	0.80	40.90	85.75	69.37	74.18
4	KNN	43.25	70.35	0.65	0.75	39.78	46.89	71.52	62.04

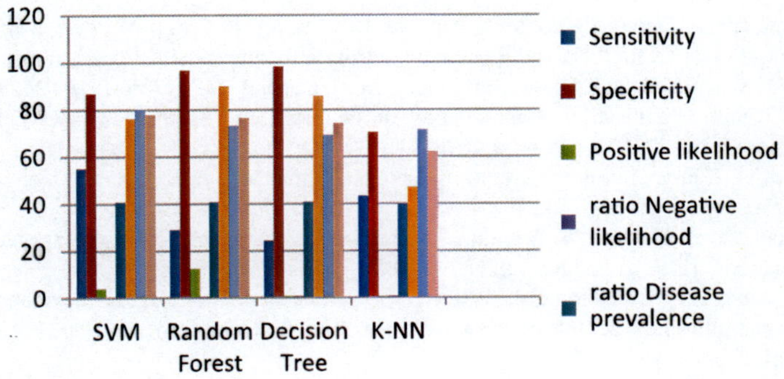

Fig. 2 Efficiency comparison of different classification techniques

a crucial role not only in prevention and control but also reduce enormous health burden. Machine learning has gained a central position due to its tremendous performance, in the field of computational medical imaging. In our proposed research work, different machine learning algorithms have been studied for detection of prediabetes, but in this paper, the main purpose is to analyze the performances of different classification methods for finding out the best accurate model that is capable to detect the diabetes mellitus at an early stage. Efficiency of the algorithms has been evaluated based on various performance measures which are played the vital role in diagnosis of this disease. SVM has acquired itself as the extensively promising algorithm due to its significant performance in early detection of diabetes mellitus prediction with regards to high accuracy and low error rate.

References

1. About diabetes. World Health Organization. Archived from the original on 31 March 2014. Retrieved 4 April 2014
2. Global Report on Diabetes 2016 by World Health Organisation. https://www.who.int/diabetes/publications/grd-2016/en/, ISBN 978 92 4 156525 7
3. Classification and Diagnosis of Diabetes: Standards of Medical Care in Diabetes—2018 American Diabetes Association Diabetes Care 2018; 41(Supplement 1): S13–S27. https://doi.org/10.2337/dc18-S002
4. K.G. Alberti, P.Z. Zimmet, Definition, diagnosis and classification of diabetes mellitus and its complications. Part 1: diagnosis and classification of diabetes mellitus provisional report of a WHO consultation. Diabet Med. **15**(7), 539–553 (1998)
5. D.G. Shoback, D. Gardner (eds.), Greenspan's basic & clinical endocrinology, 9th edn. McGraw-Hill Medical, New York, 17 (2011)
6. J.S. Kaddis, B.J. Olack, J. Sowinski, J. Cravens, J.L. Contreras, J.C. Niland, Human pancreatic islets and diabetes research. JAMA J. Am. Med. Assoc. **301**(15), 1580–1587 (2009)
7. World Health Organization. Guideline: sugars intake for adults and children. World Health Organization (2015)
8. Centres for Disease Control and Prevention. National Diabetes Statistics Report. Atlanta: Centers for Disease Control and Prevention, US Department of Health and Human Services (2017)

9. S.P. Romero, A. Garcia-Egido, M.A. Escobar, J.L. Andrey, R. Corzo, V. Perez et al., Impact of new-onset diabetes mellitus and glycemic control on the prognosis of heart failure patients: a propensity-matched study in the community. Int. J. Cardiol. **167**, 1206–1216 (2013)
10. World Health Organisation Global Report on Diabetes 2017. https://www.who.int/diabetes/publications/grd-2016/en/. ISBN 978 92 4 156525 7.
11. V. Wan, W. Campbell, Support vector machines for speaker verification and identification, in *IEEE Proceeding* (2000)
12. O. Chapelle, P. Haffner, V. Vapnik, Support vector machines for histogram-based image classification. IEEE Trans. Neural. Netw. **10**(5), 1055–1064 (1999)
13. K.Y. Yeung, R.E. Bumgarner, A.E. Raftery, Bayesian model averaging: development of an improved multi-class, gene selection and classification tool for microarray data. Bioinformatics **2005**(21), 2394–2402 (2005)
14. J.W. Lee, J.B. Lee, M. Park, S.H. Song, An extensive evaluation of recent classification tools applied to microarray data. Comput. Stat. Data Anal. **48**, 869–885 (2005)
15. S. Habibi, M. Ahmadi, S. Alizadeh, Type 2 diabetes mellitus screening and risk factors using decision tree: results of data mining. Glob. J. Health Sci. **7**(5), 304–310 (2015)
16. A. Iyer, S. Jeyalatha, R. Sumbaly, Diagnosis of diabetes using classification mining techniques. Int. J. Data Min. Knowl. Manage. Process (IJDKP) **5**(1), 1–14 (2015)
17. G. Lavanya, N. Thirupathi Rao, D. Bhattacharyya, Automatic identification of colloid cyst in brain through MRI/CT scan images, in *SMARTDSC 2019*, Visakhapatnam, LNNS, vol. 105 (Scopus), (2020), pp. 45–52
18. S.J. Griffin, P.S. Little, C.N. Hales, A.L. Kinmonth, N.J. Wareham, Diabetes risk score: towards earlier detection of type 2 diabetes in general practice. Diabetes Metab. Res. Rev. **16**, 164–171 (2000)
19. Y. Ireaneus Anna Rejani, S. Thamarai Selvi, Early detection of breast cancer using SVM classifier technique. Int. J. Comput. Sci. Eng. **1**(3) (2009)

Development of an Automated CGPBI Model Suitable for HEIs in India

Ch. Hari Govinda Rao, Bhanu Prakash Doppala, Kalam Swathi, and N. Thirupathi Rao

Abstract An appropriate performance model indeed produces good results but not to dishearten commendable faculty with biased human intervention. This paper proposed a computer-generated performance-based index (CGPBI) model for a structural performance appraisal to faculty in suitable for Indian higher educational intuitions (HEI). Using the CGPBI model is exceptionally helpful in motivating the faculty who performed superior results in all aspects in the area of academics, research, and other contributions. This model uses Python to design a centralized platform to collect data related to faculty performance. A soft-computing model is developed to direct the methodology of data collections, creates a systematized data evaluation process, and provide data indices to take performance assessment judgments. This CGPBI model delivers spirited indices for competitive performances and motivates employees with more transparent results concerning their performance. This model recommends a computer-based multi-source assessment (MSA) mechanism that suggests computer-based logical proposition for evaluation of faculty annual performance appraisal through validation impartial variables in HEIs.

Keywords Performance-based index · PBI through MSA approach in HEIs · The design of CGPBI model

Ch. Hari Govinda Rao (✉) · B. P. Doppala · K. Swathi · N. Thirupathi Rao
Vignan's Institute of Information Technology, Visakhapatnam, Andhra Pradesh 530049, India
e-mail: harigovindarao@gmail.com

B. P. Doppala
e-mail: doppalabhanuprakash@gmail.com

K. Swathi
e-mail: Swathi.kalam@gmail.com

N. Thirupathi Rao
e-mail: nakkathiru@gmail.com

D. Bhattacharyya and N. Thirupathi Rao (eds.), *Machine Intelligence and Soft Computing*, Advances in Intelligent Systems and Computing 1280,
https://doi.org/10.1007/978-981-15-9516-5_16

1 Introduction

Monetary rewards are instantaneous boosters and which generate direct results on organizational performance. Especially in private higher education intuitions (HEIs), the existence of faculty always depends on their rewards and recognition. Sometimes often an eminent faculty also moved from institutions due to lack of precise mechanism of performance appraisal. Hence, the précised performance-based incentive (PBI) approach constantly supports HEIs to retain eminent faculty and also sets benchmarks for both faculty and institutions. PBI is a merit-based incentive method that influences the growth of all individual employees and HEIs even it is big or small. Under the current psychometric studies, "incentive motivation plays an important role in the law of behavior that higher impetuses will prompt to prime execution." Hence, the structural policy of PBI does not only clarify the employee's position but also uplift their self-esteem in the organization.

At the same time, the faculty is also looking for self-development, which can be only possible through transparent policies where it rightly provides valid feedback over and above identify opportunities for career growth. The PBI are provided better pay or bonus for high performers and similarly penalized to the faculty whose lag-in behind. A systematic approach of PBI is essential in HEIs because that helps top management for deliberate discussions; besides that, it clarifies expectations by clarifying problems. It establishes a win-win platform that creates a plan for the future to construct both short-term and long-term goals of employees as well as the employer. The employees' performance in HEIs could be measured against their organizational goals, commitment, and aims; by using appropriate benchmarks and targets mark [1]. Consequently, the important stakeholders of HEIs that are faculty also require changing their methodology in the direction of institutional goals.

In fact, "the promotion of research in an enormous and varied country like India" will facilitate the country to evolve as "an information pool in the global arena." That is a reason why "the quality of research work directly translates to the quality of teaching" and also learning proposed a computer-generated performance-based incentive (CGPBI) policy to facilitate a multi-source assessment proposal for faculty appraisal in HEIs. The proposed multi-sources assessment (MSA) approach considers all set of independent variables of CGPBI policy which consists of independent variables like "subject toughness," "subject's results," and "average result of all subjects." This geometric CGPBI model produces decisive indices to take appropriate decisions to classify faculty based on performance and also motivates faculty to develop new-fangled abilities persistently.

1.1 Importance of PBI Policy in New Arena of HFIs

The modern education system is highly focused on outcome-based education (OBE), which has the need of a transform from conventional approach where highly stressed upon "result-centric." OBE remains emphasis on "skill-based and application-oriented education system. Therefore, it is necessitated to adopt HEIs should focus on the overall development of the institutions, including student progression, research and development, and overall progression of faculty intuitional goals. Inline with that the accredited agencies like NBA, NAAC, NIRF, and UGC also encourage HEIs by giving more attention toward development of core areas like "teaching-learning and evaluation," "research, consultancy, and extension," "infrastructure and learning resources," "student support and progression," "governance, leadership, and management," "innovations and best practices," etc.

Therefore, the contemporary view of HEIs has been shifted to all these parameters and forced to adopt this entire criterion. At this juncture, every HEI must adopt an appropriative administrative approach to reach organizational goals through good governance system, transparent administrative manual, systematic recruitment policy, and structural performance appraisal mechanism; the methodological approach of producing grater results through apparent stakeholder's staff and student. Because of that the importance of transparent administrative policy, this article emphasizes on the development of appropriate performance appraisal model and also make an attempt to give a solution for human intrusion in performance appraisal. The conservative approach of performance appraisal considers the academic performance of students besides the punctuality and discipline of staff. Even so often professors disagree with this approach as an "academic results always depend on the subject toughness" to a certain extent as the faculties performance.

Therefore, a traditional PBI policy sometimes considered students' feedback in conjunction with the results as a key element for assessing faculty's performance in HEIs. Although, it was also unsuccessful in implementation and it led to creating another argument that why it is not being considered "quality of student." At this juncture, the structural performance appraisal is the methodical and intermittent method to rate an employee's excellence in the parameters of matters about his/her present job role. Moreover, it is not only developing skills of an employee's but also gives confidence to employees to articulate their roles on job duties. At the same time, it also helps faculty to endorse optimum utilization of their strengths and prevents faculty accusations. It is not only helpful to faculty but also promotes students' performance in terms of placements and career opportunities.

2 Literature Review

In earlier days, faculty of HEIs has been offered consolidated perks and wages, with decided extremely by qualification and experience of faculty [2]. A performance management system can serve many vital purposes of present higher education system. Determinant factors of teacher's pay, they are not closely correlated to student performance or institutional outcomes [3]. Performance-based-incentives-based human infusion was making injustice for outstanding faculty due to implementation of misguiding politics. In this way, there are several works that have been carried out on the performance evaluation of employees in institutions. Some of the techniques of assessment used are discussed as follows.

Islam and bin MohdRasad (2006) used analytical hierarchal process (AHP) to simplify complex evaluation problems and identified, and consecutively subdivided into hierarchical levels. The weaknesses of this study are it needs professional software to measure the weights of the criteria [4].

Islam and bin MohdRasad (2006) used the analytic hierarchy process (AHP) to minimize complexity in evaluation problems into structural hierarchies. The weaknesses of the system are it requires expert choice software to compute the weights of the criteria and sub-criteria and it also requires a substantial amount of time to obtain the overall performance scores [4].

Neogi et al. (2011) proposed "fuzzy inference system" and center of gravity (COG) defuzzification method along with the fuzzy inference system (FIS) module contains five FISs sub-modules. The major weakness of the work is that an expert can easily modify the system's inputs, including the set of fuzzy rules [5].

Ghosh and Das (2013) proposed PMS models in higher educational institutions which is an important aspect of management is to monitor and assess the business performance. The weakness of this model has not covered hidden aspects of performance appraisal like mentoring and counseling significance [6].

Samuel et al. (2014) made use of fuzzy logic based on the concept of fuzzy set theory for human resource performance appraisal, but it is limited to staff under consideration and multi-users cannot access [7].

Hari Govinda Rao et al. (2017) proposed a geometric model of it highlights the performance-based incentives by using a mathematical proposition, but it is a very comprehensive model for faculty performance appraisal in latest scenario of HEIs. The main drawback is that it was a very conventional approach not provides and system base process [8].

Celik and Telceken (2018) was implemented with hypertext pre-processor (PHP) for an effective automated method that reduces sentiment in publication scoring while evaluating academic staff profiles. Although it was concentrated on the only evaluation of research metrics of staff and ignores academic performance [9].

Macwan and Sajja (2013) used fuzzy evaluation techniques to facilitate the performance appraisal process and draw definite conclusions from vague, ambiguous, or imprecise information. The evaluation parameters used were not all equally important to organization levels [10].

Jamsandekar and Mudholkar (2013) applied fuzzy inference technique in place of the traditional approach to classifying student's by scores of students' performance, and fuzzification of the input data was done by creating fuzzy inference system (FIS) subject wise. The model, however, requires the human expert to discover rule about data relationships as student's performance evaluation needs intelligent adaptive and tutoring Internet-based system [11].

Ojokoh et al. (2019) proposed "a mathematical relation approach and rough set model for effective evaluation of academic staff of an institution for promotion." The main limitation of this model is it disregards peers' comparative results and its validation [12].

Though it has been given considerable developments during the last few decades in the recorded literature, there have been enough gaps in its modality presentation and usage in higher educational institutions. At this juncture, the study attempts to review all such related review to understand the concept of PBI and its gaps in application process subjected to HEIs.

3 Performance-Based Incentives Through Multi-Sources Assessment Approach

Unless there are proper systems in place for collecting data and monitoring, analyzing, and reporting on the information, then it will not be possible to evaluate performance with any confidence [13]. Traditional PBI policy is determined objectively limited to result and student's feedback [8], whereas CGPBI policy offers fair and accurate structural provision to gather adequate information, evaluated through its developed software to deliver the output to respect stakeholders of the information. The CGPBI can be implemented through the developed portal and which can be easily implementable in private higher educational intuitions to evaluate the overall performance of faculty. This model comprises four criteria (Fig. 1). It creates internal competition spirit among faculty in terms of producing results, research output, and motivating students, respectively, that strengthen the organizational objectives. There are different ways of the procedure to evaluate performance appraisal, but GCPBI model adopts multi-sources assessment (MSA) model which mainly considers key

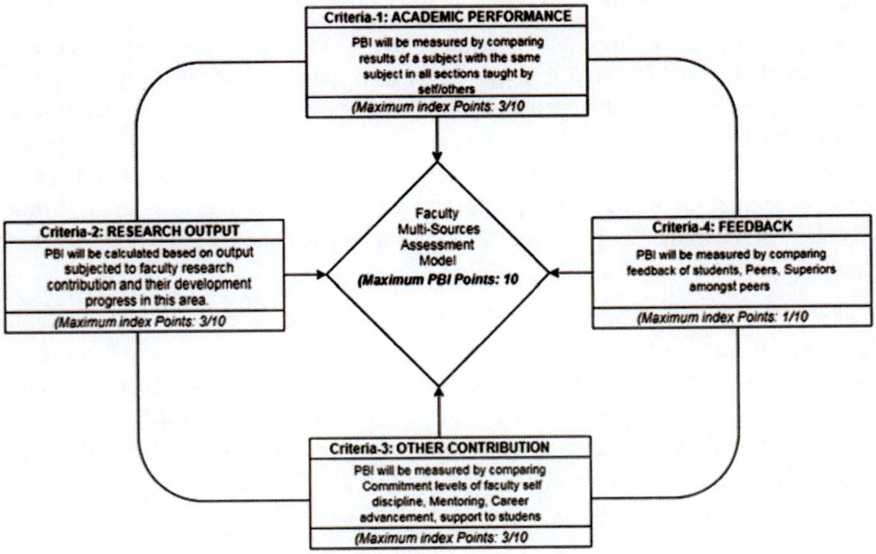

Fig. 1 Architecture of the MSA model

aspects of "cost control," "optimum utilizations of resources," and "timing saving approach." This model considers all aspects of performance appraisal like "academic performance," "research contribution," "feedback," etc.

Criteria-1: Since the days of the conventional policy, the success rate of students is to be measured based on their academic result and it is also prioritized in PBI policy, but surprisingly, it is being opposed by faculty and argued that it has a dependent variable. Sometimes, the student result is certainly influenced by their intelligence and level of paper toughness when compared to other subjects. One more argument is that "the quality of students may differ from section to section." Therefore, the MSA model pressured that "number of sections" also plays an important role in HEI where more sections and the faculty teach the same subject for different sections. To solve this problem, MSA model considers comparative result score within the same section as well as results of the other sections and evaluated PBI indices with a weighted score of three points (30% of overall PI score) for overall performance. The complete details were presented in the authors' past article [8].

Criteria-2: The next important key aspect of the MSA model is that "academic research and development" of faculty. As per as assessment agencies of HEIs is a concern, the objective of present HEI has been changed its face and "faculty research and development" played a vital role. Hence, these criteria consider 30% of weightage in overall performance while measuring PBI indices. Especially, it is being categorized it weighted score among all qualitative publications like "national and international journals, articles presented national and international conferences, conferences/workshops/symposium organized and participated" and for all these evaluations it is considered 30% of weightage in overall PBI indices. The complete details of this criterion were presented in the authors' last article [8].

Criteria-3: The next important criterion of MSA model considers three core areas of the institute like "faculty discipline in terms of punctuality, faculty participation in student counseling" and feedback of an employee from respective head of the department [HOD] and head of the institute [Principal]. Subsequently, it is evaluated for a weighted score of 30% (3 Marks) for the overall performance of PBI indices. As it was presented in the authors' preceding article [8], it will be considered "1 point for faculty discipline," 1 point for "student mentoring," and 1 point for superior's feedback.

Criteria-4: As students are a prime stakeholder in HEI, the success of any instruction is always depending on the overall development of their student. Hence, the success factors of students are sometimes measured in terms of success theory academic results, placements, entrepreneurship, and encouraging toward further studies. In line with that, CGPBI model considers students' feedback to measure PBI indices and also evaluated for 10% weightage of the overall PBI indices.

4 The Methodology of CGPBI Model Through the MSA Approach

The MSA model is a method which pooled information from various multi-sources to develop CGPBI model. MSA gathered information from all sources of stakes (self, students, peers, HOD, and head of the institute) with the assumption "ten-point scales" to appraise the performance of faculty in HEIs. As per the geometric hybrid model performance-based indices (GHMPBI) policy [8], it considers all the core areas of a technical institutional objectives like "academic result (AR) with 30% weightage, academic feedback (AF) with 10% weightage, research and development (RD) with 30% weightage, and other contributions (OC) with 30% weightage of an employee for his/her improvement as well as institutional growth," and it was subsequently measured PBI indices score three points for academic result score; three points for research and development score; three points for other contribution; and one point for students' feedback. PBI score of an Employee = [($\sum$ARS. WARS) + ($\sum$AFS. WAFS) + ($\sum$RDS.WRDS) + ($\sum$OCS. WOCS)] where ARS = academic result score; AFS = academic feedback score; RDS = research and development score; OCS = other contributions score.

Step 1: Calculation procedure of Academic Performance Score (APS)

$$\sum \text{OCS.WAP} = \frac{\sum_{i=1}^{n} (\overline{\sum \text{OCS}})_i}{n}$$

$$\text{ARS}_1 = \frac{[[1 - (x - y)^2 \times 10] \times 3] + [[1 - (z - y)^2 \times 10] \times 3]}{2}$$

$\sum \text{APS.W}_{\text{AP}}$ = Weighted average score of academic results performance.
ARS_1 = Academic result score of a subject.
x = Highest percentage of the subject result within the section; y = Percentage of the concerned subject of the individual faculty; z = Highest Percentage of the same subject results in the entire college within the semester; n = total number of papers taught by the individual faculty in the stipulated period.

Step 2: A geometric expression of measuring Research and Development Score (RDS)

$$\sum \text{RDS}.W_{\text{RDS}} = \text{TWS}.W_{\text{RDS}}$$

where TWS = Total weighted score = $(W_{\text{pci}}).(M_{\text{pci}})$;
W_{pci} = Wight of a paper category index; M_{pci} = Maximum marks of a paper in given category.

Step 3: A geometric expression of measuring other contributions Scores (AFS)

$$\text{AFS} = \sum \text{OCS.WAP} = \frac{\sum_{i=1}^{n} (\overline{\sum \text{OCS}})_i}{n}$$

Here, $\overline{\sum \text{OCS}}$ = Average contribution of faculty.
n = number of activities involved by a professor in an academic year.

Step 4: A geometric expression of measuring Academic Feedback Scores (AFS)

$$\text{AFS} = \frac{\sum_{i=1}^{n} \overline{\text{AFS}_i}}{n} \quad \text{AFS}_1 = \sum_{i=1}^{9} \text{SQ}_{i1}$$

Here,
AFS_1 = Average score of a subject feedback.
SQ_{i1} = A criterion for students' feedback.
n = number of subjects taught by a professor in an academic year.

4.1 Proposed Algorithm to Develop CGPBI Model

The CGPBI model is an automated approached to produce performance appraisal scores without any human intrusion by using soft

Step 1: Start

Step 2: Select a faculty **f**, who is teaching **n** subjects

// Calculation of ARS—Academic Result Score

Step 3: Consider a subject taught by faculty **f** which is not visited $\mathbf{sub_i}$ of section $\mathbf{sec_i}$ (If the faculty is teaching the same subject for more than one section consider them as different subjects) and mark that subject as visited

Step 4: Calculate the pass percentage $\mathbf{y_i}$ of that subject in that section

Step 5: $\mathbf{x_i}$ = max (Pass percentages of all subjects in section $\mathbf{sec_i}$)

Step 6: $\mathbf{z_i}$ = max (Pass percentages of $\mathbf{sub_i}$ from all branches and sections)

Step 7: Calculate $\mathbf{ARS_i}$:

$\mathbf{ARS_i} = [[1 - (\mathbf{x_i} - \mathbf{y_i})^2 \times 10] \times 3] + [[1 - (\mathbf{z_i} - \mathbf{y_i})^2 \times 10] \times 3]/2$

$\mathbf{total} = \mathbf{total} + \mathbf{ARS_i}$

Step 8: Repeat steps 3–7 until all subjects taught by faculty **f** are visited

Step 9: $\mathbf{ARS} = [\mathbf{total/n}]$

//Calculation of AFS—Academic Feedback Score

Step 10: For faculty **f** consider the feedback of each students $\mathbf{S_i}$

Step 11: For each student add the feedback $\mathbf{SQ_i}$for each student

$\mathbf{AFS_i} = \mathbf{\Sigma\ SQ_i}(1 <= i <= \text{no. of questions})$

Step 12: Now calculate the total Academic Feedback Score:

$\mathbf{AFS} = (\mathbf{AFS_1} + \mathbf{AFS_2} + \ldots. + \mathbf{AFS_n})/\mathbf{n}$

//Calculation of RDS - Research and Development Score

Step 13: Let **P** be the list of paper categories submitted by faculty **f**

Step 14: For each category:

get **h Index, impact Factor, other Index**

Now we need to calculate **wpci** (Weight of paper category index) and **mpci** (Maximum marks for paper category index) for these three factors

```
wpci = { }
mpci = { }
(we use hashing)
if (category == "International Journal-Un Paid (with ISSN/ISBN)") {
  mpci ["hIndex"] = 3
  mpci ["impactFactory"] = 2
  mpci ["otherIndex"] = 1.5
  if ( hIndex> 10 ){
          wpci["hIndex"] = 1
}
       else if( hIndex<= 10 and hIndex > 5){
          wpci["hIndex"] = 0.7
}
  else{
          wpci["hIndex"] = 0.4
  }

  if( impactFactor>= 3 ){
          wpci["impactFactor"] = 1
}
       else if( impactFactor< 3 and impactFactor >= 1 ){
          wpci["impactFactor"] = 0.7
}
  else{
          wpci["impactFactor"] = 0.4
  }
          wpci["otherIndex"] = 0.5
  }
```

Similarly, there are other categories based on which we need to calculate **mpci** and **wpci** as above

Step 15: Now **RDS** = **mpci["hIndex"]** x **wpci["hIndex"]** + **mpci["impactFactor"]** x **wpci["impactFactor"]** + **mpci["otherIndex"]** x **wpci["otherIndex"]**

// Calculation of OAS—Other Activity Score

Step 16:

fs = Faculty Discipline Score

Ss = Student Counselling Score

Fds = HOD, Principle Feedback Score

OAS = fs + ss + fds

//pbi

Step 17:

PBIScore = ARS + AFS + RDS + OAS

Performance-based appraisal system in this paper is implemented using complete automated mechanism. Several parameters have been considered for the calculation of PBIScore. Current work can be implemented using Python programming for the assessment of PBIScore.

5 Conclusion

This is an extensive work of authors' earlier model geometric hybrid model performance-based indices (GHMPBI) model which was a mathematical proposition that has some thrust area of further research. In this paper, we used initially MSA model which was touched up on all the key elements of institute objectives to gather necessary information of inputs variable and proposed the algorithm to develop a model called CGPBI model. To coverage the assessment area, it uses multi-sources assessment approach to evaluate the PBI indices. The CGPBI model is an automated model which generates transparent performance scores of faculties and produced more accurate and quick results. The CGPBI model is completely an automated method which saves time, produced analytical information, and helps officials who take decisions.

References

1. M.A. Camilleri, A.C. Camilleri, The performance management and appraisal in higher education, in *Driving Productivity in Uncertain and Challenging Times* (University of the West of England, 5th September), ed. by C. Cooper. British Academy of Management, UK (2018)
2. R. Richardson, Performance-related pay in schools: an assessment of the green papers: a report prepared for the National Union of Teachers. London School of Economics and Political Science (1999)
3. D. Goldhaber, The mystery of good teaching. Educ. Next **2**(1), 50–55 (2002)
4. R. Islam, S. bin MohdRasad, Employee performance evaluation by the AHP: a case study. Asia Pacific Manage. Rev. **11**(3) (2006)
5. A. Neogi, A.C. Mondal, S.K. Mandal, A cascaded fuzzy inference system for university non-teaching staff performance appraisal. J. Inf. Process. Syst. **7**(4), 595–612 (2011)
6. S. Ghosh, N. Das, A new model of performance management and measurement in the higher education sector. Management **2**(8), 1–10 (2013)
7. O.W. Samuel, M.O. Omisore, E.J. Atajeromavwo, Online fuzzy-based decision support system for human resource performance appraisal. Measurement **55**, 452–461 (2014)
8. C.H.G. Rao, P.K. Kosaraju, H.J. Kim, Performance-based incentives policy: a geometric hybrid model. Int. J. Adv. Sci. Technol. **125**, 65–79 (2017)
9. R. Cekik, S. Telceken, A new classification method based on rough sets theory. Soft. Comput. **22**(6), 1881–1889 (2018)
10. N. Macwan, D.P.S. Sajja, Performance appraisal using fuzzy evaluation methodology. Int. J. Eng. Innov. Technol. **3**(3), 324–329 (2013)
11. S.S. Jamsandekar, R.R. Mudholkar, Performance evaluation by fuzzy inference technique. Int. J. Soft Comput. Eng. **3**(2), 158–164 (2013)

12. B. Ojokoh, V. Akinsulire, F.O. Isinkaye, An automated implementation of academic staff performance evaluation system based on rough sets theory. Austr. J. Inf. Syst. **23** (2019)
13. S. Ghosh, N. Das, A new model of performance management and measurement in the higher education sector. Management **2**(8) (2013)

Range-Doppler ISAR Imaging Using SFCW and Chirp Pulse

Nagajyothi Aggala, G. V. Sai Swetha, and Anjali Reddy Pulagam

Abstract The inverse synthetic aperture radar (ISAR) produces the images of moving target in connection with the radar. This paper details the technique of ISAR imaging of target in motion with respect to the radar. The Doppler frequency shift produces a backscattered record that is received by the radar. The target in general consists of several scattering points, which are very much necessary to analyze the performance of the targets reflectivity function, i.e., target image, and is obtained from the collected complex signal with the use of chip pulse and SFCW in MATLAB simulation. The range compression is employed to every single digitized pulse by pulse compression technique, which consists of matched filtering for every returned pulse with replication of original pulse. As a result, an inverse Fourier transform (IFT) operation is accomplished with the pulse index, in order range dimensions, and also the noise at the receiver is suppressed.

Keywords Synthetic aperture radar · Inverse synthetic aperture radar · Chirp pulse · SFCW · Range-Doppler · Automatic target recognition · Cross-range · Range bins · Range profiles · Scattering points

1 Introduction

The working of any radar system is based on few specifications such as range resolution, accuracy of target detection, and its location. Other parameters incorporate its signal detection capacities during the interference from clutter echoes and some other atmospheric effects and also its tendency to note a distinction between any unintentional interfere signal from nearby unwanted signal.

Inverse synthetic aperture radar (ISAR) is a very effective signal processing technique used for the detection and classification of targets by representative distribution of scattering points known as scattering centers of target, which produce a standard

N. Aggala (✉) · G. V. Sai Swetha · A. R. Pulagam
ECE Department, VIIT, Visakhapatnam 530049, India
e-mail: aggala.jyothi@gmail.com

D. Bhattacharyya and N. Thirupathi Rao (eds.), *Machine Intelligence and Soft Computing*, Advances in Intelligent Systems and Computing 1280,
https://doi.org/10.1007/978-981-15-9516-5_17

2D ISAR image by accumulating the scattered field for various look angles and Doppler frequencies. Thus, a 2D ISAR image is a graphical figure of range profiles in one axis and cross-range profiles in other axis, keeping in mind the scattered field for various look angles and Doppler frequencies.

2 Problem Statement

ISAR image of a target is generated as a result of its rotational motion. In view of fast motion targets like jet, aircrafts, or some targets undergoing rotational motion like pitching ship, the information retrieval is done with relative motion of target [1]. One important thing to notice is the ISAR imaging that the conventional range-Doppler ISAR algorithm is not that affective when the target motion produces higher order labels in the phase of the received signal relative to each scatterer point. Figure 1 shows the 2D ISAR images for (a) pitching (b) yawing (c) rolling platform, and Fig. 2 shows the aircraft target with perfect scattering points.

The output of this non-uniformity is the distortion, which is directly proportional to the change of instantaneous range of target. Mostly, the integration time of image is just a matter of few seconds, and the rotational displacement of target is with accordance to the reference is just a few degrees [2]. The result of this is target representation visible in the ISAR image, which becomes blurred and instable and also makes the target identification and image reconstruction process very tough.

In order to settle down these complexities in ISAR imaging and create it as a viable for practical applications, a method is used based on the estimation that exists as an imaging interlude that elucidates the effect of target's motion and receiver's position with enough cross-range resolution [3]. In this paper, we represent our work on generation of a well-defined, focused image of a motion target, so as to compensate its effect on phase of echo signal received. The main problem statement of this paper is the generating of ISAR image of moving targets using chirp pulse and SFCW signal in MATLAB simulation.

3 Proposed Method

This paper explains the MATLAB modeling of the signal processing to find out moving targets. Here, the scattering centers of target are considered, as a factor of representing a 2D ISAR image of a target in motion [4]. The concerned positions and sizes of the scattering points are achieved, and the data of the range and cross-range profiles are noted in a range-Doppler matrix and an equivalent ISAR image [5]. The key objectives of this methodology are:

- To explore the real-time applications of ISAR imaging principles.

- To observe and understand the fundamentals of radar signal processing course of action, detection of moving target, imaging and Doppler processing [6, 7].
- To execute the concepts of chirp pulse and SFCW radars, pulse compression, and range resolution techniques and signal optimization [8].
- To employ proper formulation on noise and some other disorder that obstructs the signal receiving.

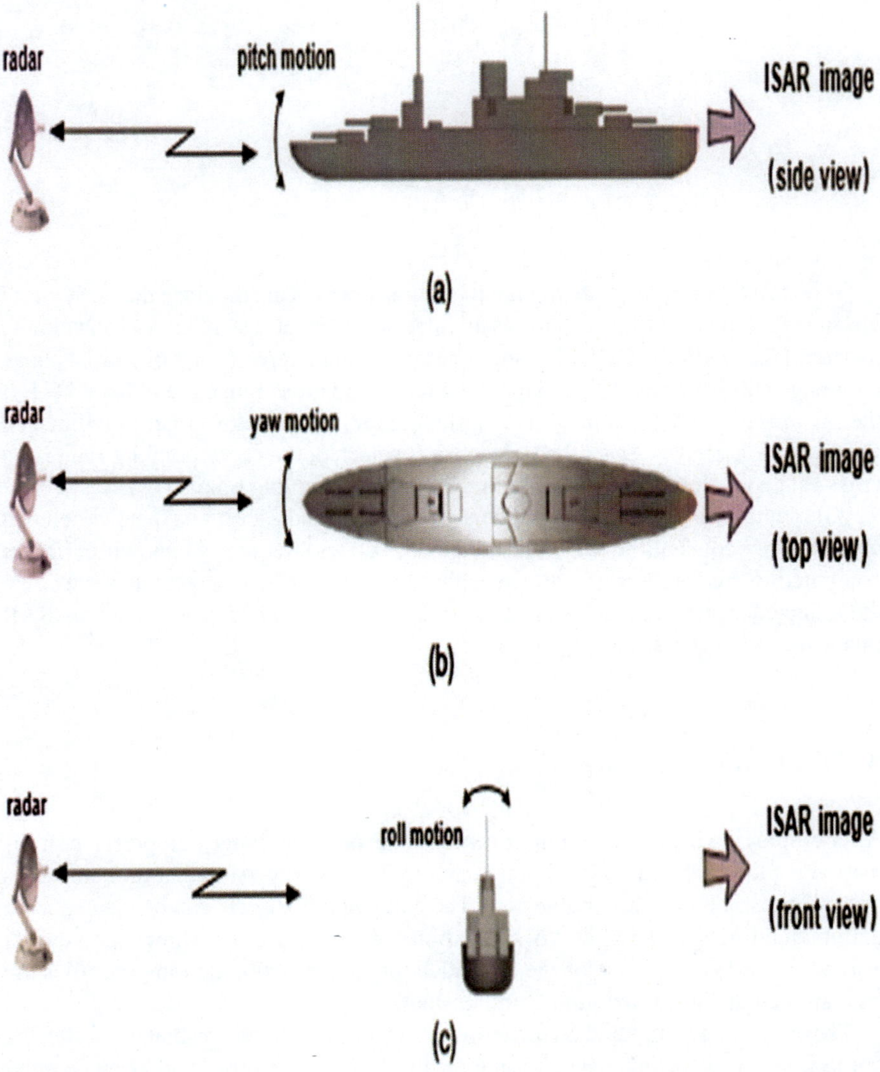

Fig. 1 Resulting 2D ISAR images for **a** pitching, **b** yawing, **c** rolling platform

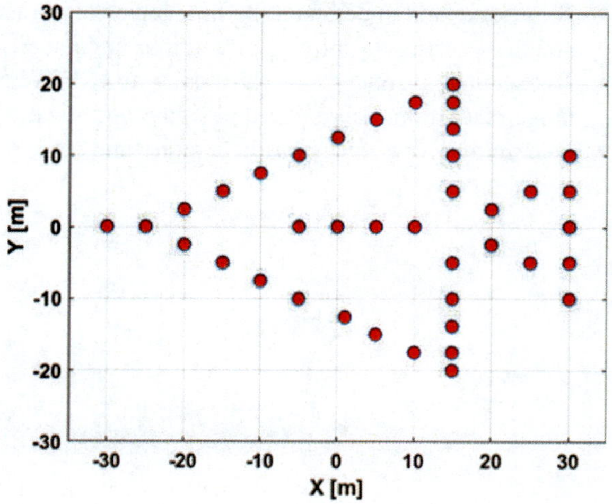

Fig. 2 Target with perfect scattering points

The very first step is study of signal circumstances. This involves the deriving of physical features of the target, such as orientation, physical size, relative size, velocity in accordance with the radar, and some different characteristics. At the initial stage, the image size is chosen [9], and then, the range and cross-range are obtained [1]. If the range is X_m and cross range is Y_m, then the relative size of target to be imaged would be as $X_m * Y_m$. Once the target size is selected, the corresponding range and cross-range resolution are drawn with theory of N sampling points [10].

After determining these, the frequency resolution and aspect ($\Delta\phi$) are concluded using Fourier concepts. The angular frequency (Ω) and frequency bandwidth (B) are computed if the frequency is at center that is f_c. Radar look angles at aspect $\Delta\phi$, the reflected scattering data of electric fields are collected and represented as ISAR image using 2D IFT [11].

4 Simulation Results

After employing the matched filter operation, the obtained range compression information is plotted in Fig. 3, in which range profile for every azimuthal time instant is observed easily at different range bins. The additive noise effect can be viewed from the obtained image in MATLAB simulation, while the noise is shown as a clutter around the image. Because of the high PRF rate about 3000, the range profiles are well aligned at various azimuthal time instants.

Therefore, an IFT method is carried out along the pulse index, by that the scattering points in cross-range dimension is found, while they are appeared in different Doppler frequency shift values, resulting an ISAR image for range-Doppler plane as shown in Fig. 4. The scattering points are framed out well in the range direction and also

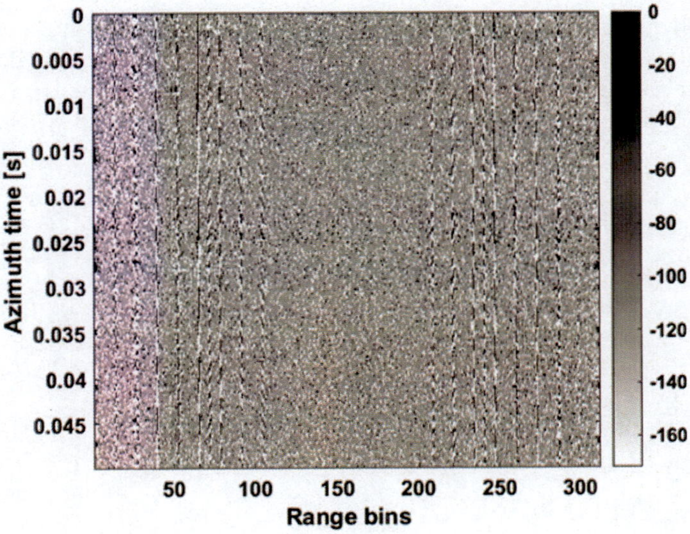

Fig. 3 Range bins with respect to azimuth time

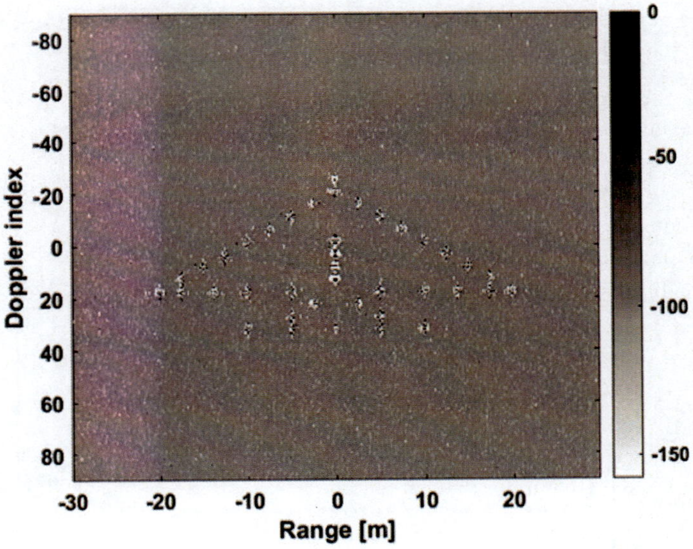

Fig. 4 Range-Doppler ISAR image of the target

in cross-range direction because of the target's finite velocity along the direction of azimuth. The matched filter suppresses the noise at the receiver.

By means of estimating the target's angular velocity, a range–cross range ISAR image is depicted as in Fig. 5. Therefore, transformation of Doppler frequency shift axis to the cross-range axis is performed.

The pitching ship is considered as shown in Fig. 1, where the target go through rotational motion, so that the concerned position of scatterers in every down range is alike in a particular integration time. Target with perfect scatter points is shown in Fig. 6. If N pulses are transmitted in a certain integration period, then there exists N

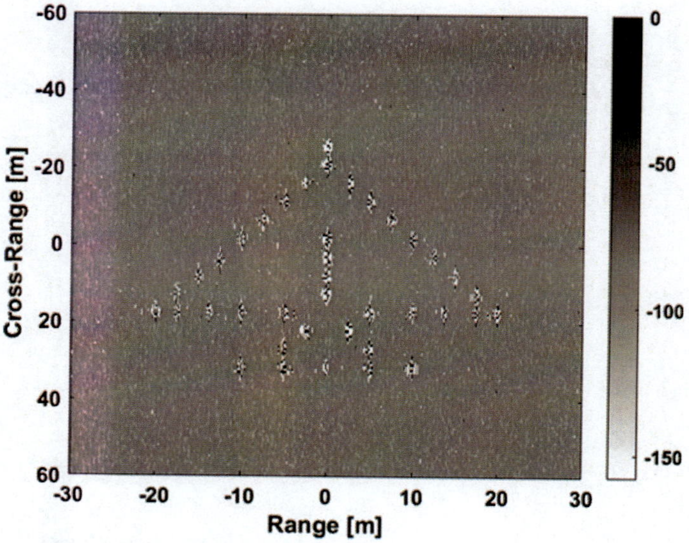

Fig. 5 Cross-range ISAR image of the target

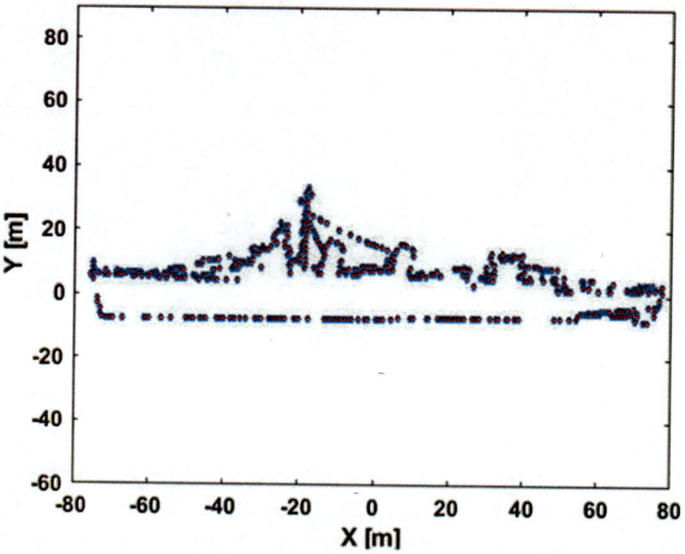

Fig. 6 Target with prefect point scatters

range profiles as shown in Fig. 7, and is shown as range bins with respect to azimuth time which can be approximated for various bursts. The range-Doppler ISAR image of the target is represented in Fig. 8.

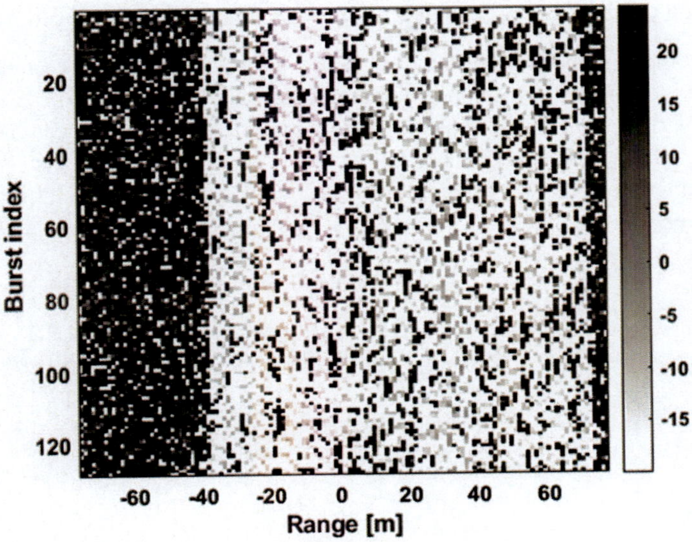

Fig. 7 Range bins with respect to azimuth time

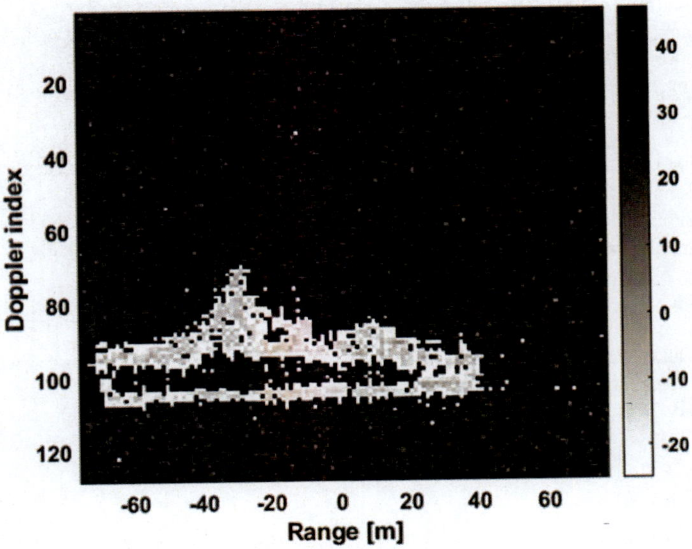

Fig. 8 Range-Doppler ISAR image of the target

5 Conclusion

In this paper, the ISAR image generation of moving target with the use of stepped frequency continuous waveform is proposed. This MATLAB simulation technique is based on the transform parameters of an assumed moving target model

- The Doppler frequency shift backscattered information is received by the radar, which is transformed to time and Doppler frequency.
- The analysis of Doppler frequency made it possible to find the scattering points along with the cross-range axis.
- The range compression is employed for every digitized pulse by using pulse compression technique and formed IFT technique and resultant is high reliable clear ISAR range-Doppler image.

Acknowledgements This work is being supported by Ministry of Science and Technology of Science and Engineering Research Board (SERB), which the Grant No.: ECR2017-000256 dated 15/07/2017.

Reference

1. M.J. Prickett, C.C. Chen, Principles of inverse synthetic aperture radar/ISAR/imaging, in *EASCON'80; Electronics and Aerospace Systems Conference*. 1980
2. M.I. Skolnik, *Introduction to Radar Systems* (McGraw Hill Book Co., New York, 1980), p. 590
3. L. Zhang, et al., Achieving higher resolution ISAR imaging with limited pulses via compressed sampling. IEEE Geosci. Remote Sens. Lett. **6**(3), 567–571 (2009)
4. M. Xing, et al., Migration through resolution cell compensation in ISAR imaging. IEEE Geosci. Remote Sens. Lett. **1**(2), 141–144 (2004)
5. S.Y. Shin, H.M. Noh, The application of motion compensation of ISAR image for a moving target in radar target recognition. Microw. Opt. Technol. Lett. **50**(6), 1673–1678 (2008)
6. M. Soumekh, Synthetic Aperture Radar Signal Processing, vol. 7 (Wiley, New York, 1999)
7. M.I. Skolnik, Radar Handbook (1970)
8. C. Ozdemir, *Inverse Synthetic Aperture Radar Imaging with MATLAB Algorithms*, vol. 210 (Wiley, New York, 2012)
9. B.R. Mahafza, *Radar Systems Analysis and Design Using MATLAB* (Chapman and Hall/CRC, 2005)
10. J.L. Walker, Range-Doppler imaging of rotating objects. IEEE Trans. Aerosp. Electron. Syst. **1**, 23–52 (1980)
11. W.M. Brown, R.J. Fredricks, Range-Doppler imaging with motion through resolution cells. IEEE Trans. Aerosp. Electron. Syst. **1**, 98–102 (1969)

Secure Communication in Internet of Things Based on Packet Analysis

V. Lakshman Narayana and A. Peda Gopi

Abstract Internet of Things (IoT) platform is projected to continue to expand to more than 80 billion connected gadgets by 2025. This flood of relevance opens up a wide range of roads to those with a noxious aim of going after the ignorant. With the anticipated development of IoT gadgets, it is inferred that a progressive way of life for your regular customer. As more gadgets coordinate our day-to-day lives, the safety of these gadgets appears to be progressively significant. In this paper, for the most part, RPL convening model is focused. Analyzing the route control messages in this protocol is performed. The proposed mechanism focuses mainly on the most problematic communicative attack called man in the middle attack. Proposed mechanism produces promising outcomes with respect to secure communication.

Keywords Secure communication · Data loss · Packet delivery · IoT analysis

1 Introduction

The Internet of Things (IoT) is a massively interconnected network of heterogeneous devices where all interchanges seem possible, even un-approved. As a result, the security specifications for such a system are specific, whereas typical standard Internet security conventions are viewed as unusable in this type of system, in particular because of certain groups of IoT gadgets with forced properties. Also, for the overwhelming amount of uses of the shrewd post, i.e., IoT, systems management innovation must be adaptable, interoperable, stable, sensible, and adaptable. New technologies are also supposed to provide a foundation for clients/gadgets, servers,

V. L. Narayana (✉)
Department of IT, Vignan's Nirula Institute of Technology & Science for Women, PedaPalakaluru, Guntur, AP, India
e-mail: lakshmanv58@gmail.com

A. P. Gopi
Department of CSE, Vignan's Nirula Institute of Technology & Science for Women, PedaPalakaluru, Guntur, AP, India
e-mail: gopiarepalli2@gmail.com

D. Bhattacharyya and N. Thirupathi Rao (eds.), *Machine Intelligence and Soft Computing*, Advances in Intelligent Systems and Computing 1280,
https://doi.org/10.1007/978-981-15-9516-5_18

and systems. RPL is vulnerable to a variety of threats. Threats may be classified as threats defining network resources, threats altering network topology, and threats related to network traffic [1]. There is a need to secure routing protocols from attacks and to provide protections for data protection [2]. As seen, RPL comes with several built-in security modes, but they were not sufficient to resolve all types of attacks, and the form of which sources the chances of the attacks that could occur is known as shown below:

(a) Threat sources: Threat sources are from other actors who threaten the network and response steps that are only recommended if we know the attack patterns, skills, and status of the attacker.
(b) Outsiders: In this case, the attackers are outsiders, they are not genuine nodes, and they were duplicate nodes that compromised the system and stole the data.
(c) Insiders: In this case, the attackers are legitimate nodes in the network, but they are engaged in some fraud activities and even make some modifications to the IoT unit in order to collect the data.

Nonetheless, there are several attacks that may take place in order to steal the data or break the topology [3]. Resource-based attacks are the one of all those whose main purpose is to absorb more node energy and also to reduce the life time of the topology. A node that is fake modifies the version number, ranks, selects the worst parent itself, and also announces false ranks to neighbor topology-related nodes, and reconstructs the entire topology that consumes a lot of energy and reduces the length of the network [4]. The version number and rank in the DIO base object, there are no methods used by the RPL to guarantee the validity of the rank and also the version number of the node to select the preferred parents. Topology reconstruction creates a rise in overhead, topology loops [5]. Past studies indicate that resource-based attacks on RPL have had a significant effect on the networks and even pre-existing studies addressed addressing these attacks and suggested solutions for certain forms of attacks, even if they involve disadvantages [6].

1.1 Man in the Middle Attack

The malicious party in this type of attack interferes with the connection between the recipient and the WAP. The intruder appears to be a legitimate person and has WAP home network exposure. After entering the WAP, as the user links to the network [7], the intruder will obtain knowledge regarding the user's MAC address. The intruder has access to both the user's WAP and MAC address details, and the intruder appears to be the consumer or the network, distributing the malware to the network or to the computer. A middle guy example occurs when a hacker reaches out and links the network and the consumer independently, causing them think they interact with one another [8] (Fig. 1).

Attackers will monitor networks and look for insecure IoT products that let them catch and manipulate state-of-the-art (malware) resources and turn them into botnets

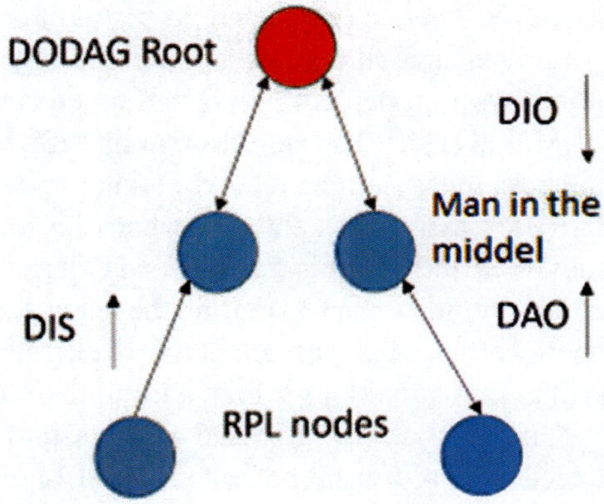

Fig. 1 Man in the middle attack in RPL protocol

for the Internet. Using management software, seize care of unsecured machines and use them for bidding [9]. Attackers will rent the user to other malicious parties for hacked computers. These attacks commonly occur in the form of distributed deni-service attacks (DDoS). The penetrator inundates servers with overcrowding demands from several botnets during DDoS attacks and makes the site inaccessible. Before that, it is no longer operated by the original owner on the attacked server [10].

We use hostapd.deny and hostapd to conduct filtering for MAC addresses on our home network firewall. Take images, please. Two different files holding MAC address lists that are to be accepted or refused are interpreted from approval and denial lists. A single MAC address is provided in each line of these data. For MAC address filtering, we use the following commands:

accept mac file (= /etc./hostapd / hostapd.acccept deny mac file).

Accept is generally sufficient to restrict exposure.

Here, in this article, we concentrate primarily on the man in the middle attack, which is a more serious attack. This triggered various types of attacks, such as packet snipping, drop, and path diversion, and rendered traffic disturbances. Here, in this paper, we suggest a method to identify and minimize the deviation induced by the man in the middle attack in the RPL routing protocol. The remainder of the paper is structured as follows, Sect. 2 offers brief information, Sect. 3 addresses the proposed study, Sect. 4 explains the experimental evaluation, and the final section concludes the article.

2 Related Work

Winter et al. [1] proposed a safety risk assessment that was carried out by the IETF RoLL work crew for RPL, where safety concerns were identified in the RPL and these issues were further addressed, and the identified hazards were further defined in four classes: security, honesty, accessibility, privacy. The report was submitted by IETF, where a completed portrayal of the RPL and the arrangement of DODAG was given, and all the relevant terminology associated with DODAG was given in a point-by-point depiction to better understand the workings of the RPL.

Murali et al. [7] proposed an energy-efficient parent choice where the node chooses the best energy-efficient parent and to reduce the loss of packets, an idea called the D-tricle timer was established where the counter was set before forwarding the DIO messages and controls the flow and determines when to send the DIOs, so that node energy could not be wasted and the results demonstrated. We will come to know why energy efficiency is important for parent selection because a node with less efficiency will flop in a topology where a lifetime of topology can be reduced.

Ghaleb et al. [8] proposed an algorithm called the Drizzle algorithm was introduced as an old-fashioned routing for LLNs. Drizzle decreases postponing problems to minimize negative impacts on transmission issues and also achieves the best results in decreasing the delay and increasing the devolution ratio of packets, additionally demonstrates the preferred results compared to ordinary RPL in terms of PDR, energy utilization, and less delay and also consumes less node electricity. This research is used to reduce RPL dealership given the fact that certain differences exist in RPL topology.

Dvir et al. [10] suggested another aspect called TRAIL, which has some benefit that computationality is less contrasted with VeRA, the first commitment VeRA, a cryptographically dependent protection story, was to be breaking up and evolving in this article. We noticed and modified the VeRA to avoid new vector attacks. When the topological nucleus of the question is retrieved, TRAIL provides our second contribution. TRAIL describes a testing methodology to refer to the routing network's actual path characteristics. The uniform solution is based on concepts of one's own choice and does not require VeRA cryptography. The fundamental cryptographic load is performed by a root node, which serves as a portal, and no node to relay neighboring node information to DODAG root is not required in TRAIL. In any case, TRAIL has the drawback that a child node may select an attacker node as its parent, which causes another hazard that is most notably the worst parent attack.

3 Proposed Work

RPL is a network in which multiple devices are connected with this program, a routing protocol is one that takes IoT into the real world, but observed se-cure is provided due to attacks on system topology, resources and traffic. In this study, another method of attack destruction has been proposed where RPL-based attacks can be totally expelled through a related node that evaluates the entire topology in order to evacuate resource-based hazards such as version number attacks, rank attacks, neighbor attacks and worst parent attacks. This associated node will be accessible to any single node in the RPL topology in order to talk to each node and choose the best preferred parent to them, and to locate the attacking node to increase the life time of the topology and reduce the energy consumption of the node, update the PDR and reduce the end-to-end latency. Tests have shown that the current model is strengthened in the sense of the defined interventions, as opposed to previous investigations. As a result, the related node strategy protects the RPL topology from various hazards and increases the ratio of packet transmission, throughput and further increases the life expectancy of the topology and the latency and energy consumption of the nodes. Mainly suggested work focuses on alleviating the man in the middle attack by considering the packet and its MAC, the IP address of the packet. By evaluating the packet address, we used the following algorithm to minimize malicious nodes.

Algorithm

1. Source node imitates the routing and checks route table
2. if (address of Destination == Available) then
   do.
   While (packets).
   PACKETcount = nodeCount.
3. Transmit PACKETcount to next node
4. PACKETcount is decremented at each node till
   PACKETcount == 0.
5. Step 5: if (DestinationMACaddress == CurrentMACaddress) then
   Packets reached its destination.
   STOP.
6. else
   Do: Send request to DODAG for node Count.
7. DODAG checks its routing table
   If (Destination Address == InRoutingTable) then.
   Do.
   Count = node Count.
8. Source node will send count
9. Routing table is updated at Source node and node Count = count
10. Repeat step 3 to 8
11. Report non-existence of destination node
12. STOP

4 Results and Discussions

Instant Contiki is a finished Contiki advancement condition running inside a Ubuntu Linux virtual machine (Ubuntu 14.04 LTS) that has every one of the compilers, improvement instruments, and simulators expected to the examination. Here, we evaluate the performance of our proposed approach with and without having the attack. And we made simulation with by varying the time of simulation.

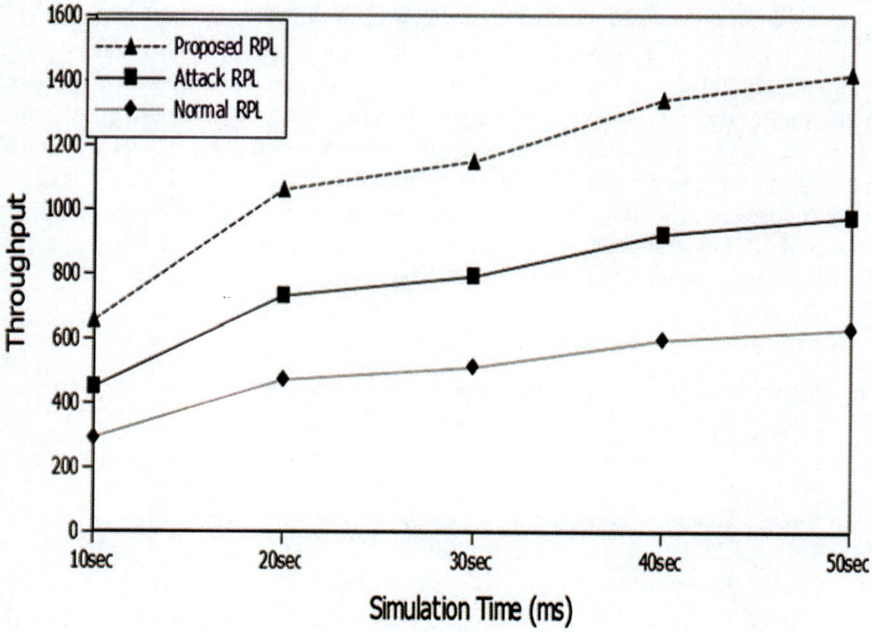

Fig. 2 Throughput

Here, Fig. 2 represents the throughput of simulation in three cases. One is normal case which does not has any attacker nodes, second case possess the simulation contains the better throughput, and finally, third case represents proposed RPL with proposed mechanism. Generally, the IOT network possess good throughput if it does not contains any attacker nodes. Once the attacker nodes increases, it starts dropping the packets, so throughput starts decreasing. By applying proposed mechanism, it starts to find the attacker nodes in the network, so throughput gets increased.

Figure 3 represents the packet delivery ratio of simulation in three cases. One is normal case which does not has any attacker nodes, second case possess the simulation contains the better packet delivery ratio, and finally, third case represents proposed RPL with proposed mechanism. Generally, the IOT network possess good packet delivery ratio if it does not contains any attacker nodes. Once the attacker nodes increases, it starts dropping the packets, so packet delivery ratio starts decreasing. By applying proposed mechanism, it starts to find the attacker nodes in the network, so PDR gets increased.

Figure 4 represents end-to-end delay of three cases, normal case, attacker case, and secure routing case. In normal scenario, packet delivery time is bit high due to route control messages. With attacker, we cannot able to determine the end-to-end delay. While an attacker present in the network, it goes fast with route control messages and establishes route after words which made more delay. While applying our mechanism, it gives better end-to-end delay.

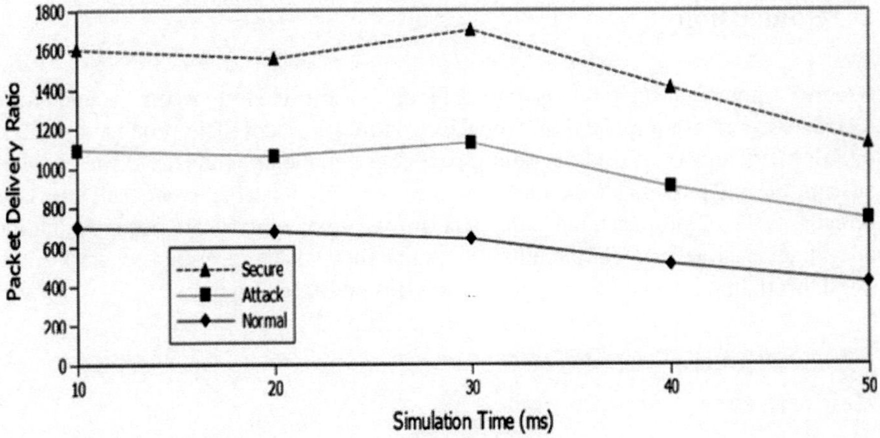

Fig. 3 Packet delivery ratio

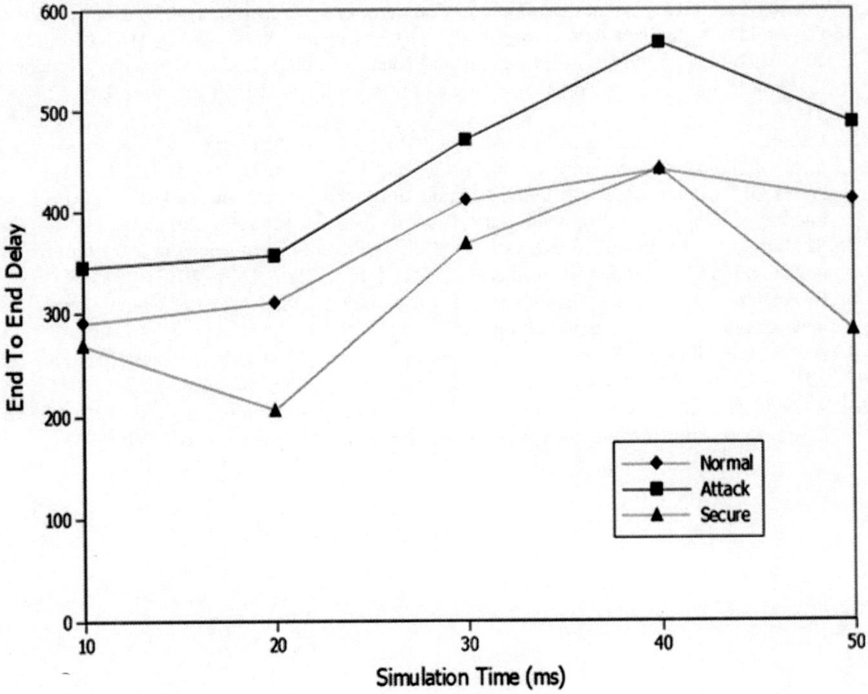

Fig. 4 End-to-end delay

5 Conclusion

Secure communication in IOT is quite difficult to achieve. Here, we made an attempt to made secure communication over RPL routing protocol. The proposed work is taking MAC and IP of each incoming packet and made an analysis. If the MAC is susceptible and possess attacker node while doing this, it is able to catch all attackers effectively. Here, proposed mechanism performs good compared to normal situation and attacker situation. The simulation results show that the proposed work gives good throughput, packet delivery ratio, and less end-to-end delay.

References

1. T. Winter, *RPL: IPv6 Routing Protocol for Low-power and Lossy Networks* (2012)
2. Kamgueu, P. Olivier, E. Nataf, T. DjotioNdie. Survey on RPL enhancements: a focus on topology, security and mobility. Comput. Commun. **120**, 10–21 (2018)
3. O. Gaddour, A. Koubâa. RPL in a nutshell: a survey. Comput. Netw. **56**(14), 3163–3178 (2012)
4. B. Ghaleb et al. A survey of limitations and enhancements of the IPv6 routing protocol for low-power and lossy networks: a focus on core operations. IEEE Commun. Surv. Tutorials **21**(2), 1607–1635 (2018)
5. I. Kechiche, I. Bousnina, A. Samet.A comparative study of RPL objective functions, in *Sixth International Conference on Communications and Networking (ComNet)*. IEEE (2017)
6. A. Raoof, A. Matrawy, C.-H. Lung, Routing attacks and mitigation methods for RPL-based internet of things. IEEE Commun. Surv. Tutorials **21**(2), 1582–1606 (2018)
7. S. Murali, A. Jamalipour, Mobility-aware energy-efficient parent selection algorithm for low power and lossy networks. IEEE Internet Things J. **6**(2), 2593–2601 (2018)
8. B. Ghaleb et al. A novel adaptive and efficient routing update scheme for low-power lossy networks in IoT. IEEE Internet Things J. **5**(6), 5177–5189 (2018)
9. A. Aris, S.F. Oktug, S. BernaOrsYalcin, RPL version number attacks: in-depth study, in*NOMS 2016–2016 IEEE/IFIP Network Operations and Management Symposium*, IEEE (2016)
10. A. Dvir, L. Buttyan.VeRA-version number and rank authentication in RPL, in *2011 IEEE Eighth International Conference on Mobile Ad-Hoc and Sensor Systems*, IEEE (2011)

Performance Investigation of Cloud Computing Applications Using Steady-State Queuing Models

Pilla Srinivas, Praveena Pillala, N. Thirupathi Rao, and Debnath Bhattacharyya

Abstract Cloud computing is the technology that was gaining the attention of most of the companies in market and utilization also increasing day to day by almost from companies to ordinary people. The working of these cloud models is effortless. A considerable number of servers are used to store the data and a vast amount of data and the service of providing data to the customers staying at remote locations too. Almost all cloud-based models are not free, and users need to pay a reasonable amount to use the services of these clouds. As the vast data is stored in these servers and the usage of this data by a vast number of customers, there is a chance of overcrowded at servers. Essential data or the hot data like the new movies, exam results or bank transactions, etc., can have the most of the crowds at various time intervals. Hence, it is required to analyse the number of customers is using the current cloud models at different intervals of time. Based on the results, the adjustments or the changes in the network model can be completed. In the current article, an attempt has been made to analyse a cloud model by considering the model working in study state and the performance was analysed for two queuing models. Several queuing models are available in research to analyse the performance of a queuing model. In the current article, the queuing models considered are M/M/1 and M/M/c models. The

P. Srinivas (✉)
Department of CSE, Dadi Institute of Engineering and Technology, Anakapalli, AP, India
e-mail: srinivasp3@gmail.com

P. Pillala
Department of Computer Science and Engineering, Centurion University of Technology and Management, Visakhapatnam, India

N. Thirupathi Rao
Department of Computer Science and Engineering, Vignan's Institute of Information Technology, Visakhapatnam, AP, India
e-mail: nakkathiru@gmail.com

D. Bhattacharyya
Department of Computer Science and Engineering, K L Deemed to be University, KLEF, Guntur, AP, India
e-mail: debnathb@kluniversity.in

D. Bhattacharyya and N. Thirupathi Rao (eds.), *Machine Intelligence and Soft Computing*, Advances in Intelligent Systems and Computing 1280,
https://doi.org/10.1007/978-981-15-9516-5_19

performance of the queuing models is analysed with various performance metrics of a network, or the cloud model is arrival rates to the model, service rates to the model, traffic density, throughput, etc. The results are displayed in the results section.

Keywords Cloud computing · Queuing models · Exponential distribution · M/M/1 · M/M/c

1 Introduction

Cloud computing is one of the interests gaining and mostly used technologies in the market today. The storage space is becoming costly day to day, most of the companies or public are looking towards the cloud centres such that to store their vast data. A cloud centre may have a considerable number of servers with a vast capacity to store data [1]. The facility was open to all types of customers. It is effortless to store the data and also to retrieve the data from a cloud centre. Almost all big companies in the market are maintaining their own data storage centres. Nowadays, the mobile phone companies like Redmi and other companies also using these cloud centres such that to give their customers to provide extra space such that to store their data more than the capacity of those mobile phones [2]. The mobile companies like the Redmi, Mi are providing such facilities to their customers. The users of these companies mobile phones can use the space allocated to them by simply login to their account and can use their space for the lifetime. It is a marketing strategy to attract the customers too. In general, the cloud service providers will charge some specific charges to store and use the space of their servers. The users who have the proper login details can use the service. The other advantage of choosing these cloud models are the damage to data due to various problems. The data stored in cloud centres cannot be damaged due to the storage of data will be in various servers at various locations. If any issue or any damage occurs to any server or cloud, the data can be retrieved from any other server that was connected in the same cloud [3].

In recent days, the utility of the Internet and its related applications had increased a lot. As the usage is increased, almost all the applications are available on the Internet. As a result, the space for storage of such applications is growing day by day. As a result, the more number of servers to be established and to be provided by service providers. The cost of establishing such large type of servers, space and maintenance of such servers is also growing day by day. As a result of these factors, the cloud utility was not made free. Nowadays, the number of customers' using these cloud centres had increased a lot. Both the private organizations and government organizations also using these cloud centres. Especially from the government pointy of view, the maintenance is the major problem for any organization, and government services are almost providing the services from various cloud centres. As a result, the network traffic to these clouds is increasing a lot during some intervals of the period. Nowadays, movies are being stored in cloud centres for more security.

The government schemes, student's results and various govt. beneficiary schemes are also being stored in these cloud centres. The traffic to such cloud centres is more during peak hours and standard at other hours. The customers entering into the cloud model and coming out from the cloud models can be symbolized as a queue line model. As a result, in the current article, the arrival of data packets to the network model is assumed to be the following queuing models and tried to analyse the performance of the model [4, 5].

1.1 Cloud Data Centres

The storing and accessing of data of a company are more cost-effectiveness than setting up a new cloud centre for any organization or a company. Hence, always thought was given on getting the service from cloud centre is more comfortable than establishing a centre and maintain such centre for any individual company. This cloud centre is a physical centre with several servers located at different locations to store the data [6]. The maintenance and the other services required to maintain such centres will be taken care of by the cloud centre people itself. Those companies have no role or no expense on such issues. These are some of the main reasons that most of the companies are looking towards cloud centres utilization rather than establishing o own cloud centre. The other benefits that they are getting by using the sources of the cloud centre are the security that was provided to these cloud servers. In typical cases, the companies need to expense a lot on providing security to these servers. Still, here the security of these centres will be taken care of by the cloud centre people [7]. As a result, the less expense on security to the centre and the security of the cloud centre will be taken care of by the intense centre people itself. The data stored in a cloud centre may not belong to only one company; several other companies are also storing their data. It is the responsibility of the cloud centre people to provide security. As a result, the customers can happily store their data and can use the services from the cloud centre [8].

1.2 Queuing Systems

In general, people used to stand in various queues to buy or purchase something or in some other cases to pay or get some items. The queue lines are formed such that to provide the service one by one such that everyone in the queue can get the service. Also, the service can be provided neat with more transparent and clarity on the type of service being provided. The queue lines follow two types of schemes like LIFO and FIFO [9–11]. The second model is used in almost all the services being provided worldwide for a queuing model. Wherever the lines are developed or proposed to provide the service, the queuing models can be considered. In the current work also, the packets being transferring from one node to the other node are in the form of a queue line. Similarly, the customers who were approaching this cloud centre are also in the form of a queue and the service by the centre will be given in the same order of arrival of requests from the customers to the centre. The data that was given as input to be considered in the form of queuing model input. Several queuing models are available in literature such that to study the performance of the model. Several authors had considered several models such that to identify the performance of the queuing models [12, 13]. In similar, in the current article, two models are considered such that the input to the current cloud model and those are M/M/1 and M/M/c models. With the help of various performances measuring metrics, an attempt has been made to analyse the performance of the cloud model with various set of inputs. The currently considered queuing model can be observed at following Fig. 1.

To fit the queuing models to the network models, several authors had considered several assumptions. The most important consideration was the flow of packets to the network, or the server is flowing in the form of first in first out. The packets which were arriving first for the service, these packets or the data need to be processed first and the other packets later in the form of the queue. All these queuing models are independent, and they follow the independent distributions and exponential distribution in some other cases [7].

2 Problem Description and Solution

In the current queuing model, the performance of the cloud data centres is analysed with the help of the queuing models. Several queuing models had considered for analysing the performance of the cloud models. In the current work, single server queuing models M/M1 and M/M/c models with different arrival rates were considered. The performance of the models is analysed concerning various parameters like the throughput, delay in packets, number of customers in queues, the number of packets in queue, etc.

3 System Design

System design includes parameters, performance measures, stability and properties considered for the data centre performance evaluation using queuing models. The detailed working flow of the model considered is as follows (Fig. 2).

System parameters are,

a. It is customary to introduce some notation for the performance measures of interest in queuing systems.

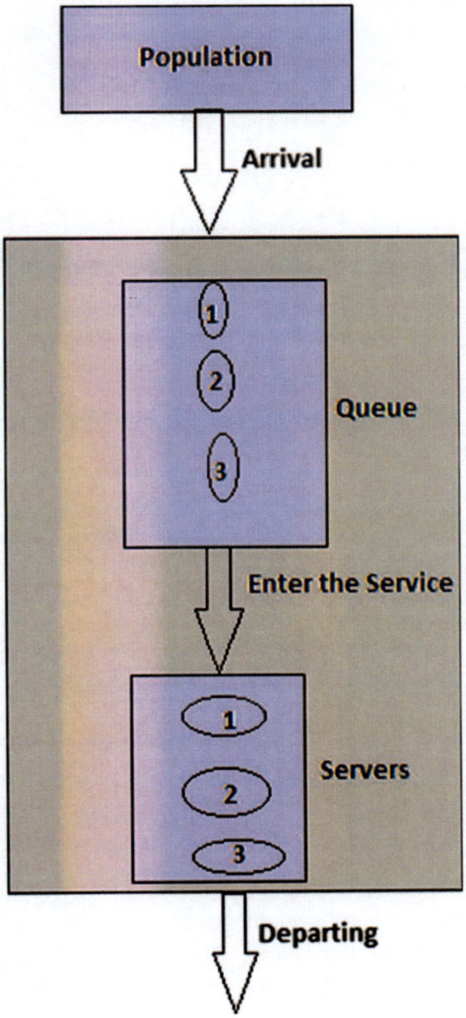

Fig. 1 Queuing system model

4 System Implementation

If any system working with no clarity or no reasonable opinion on the arrival of data to a networking system or communication system, or if any doubts or no clarity on the performance of the system or communication systems about the service times, then the excellent option for researchers to work for the queuing models. By using the queuing models with the combination of various other distribution patterns, the performance of the communication networks or the machines can be understood and realized very soon. The communication networks will provide various services;

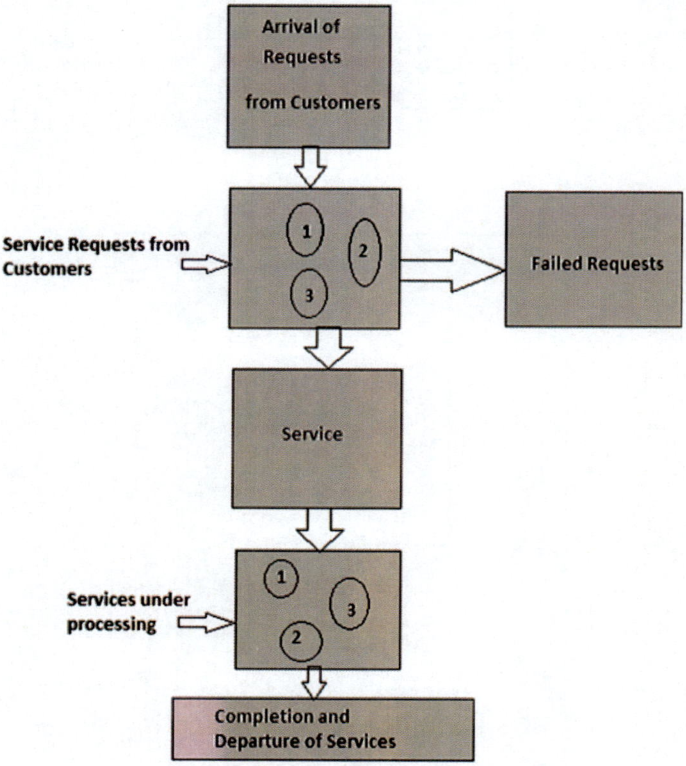

Fig. 2 Flow diagram of the considered cloud model

to achieve these performances, several performance metrics are considered. In the current work, a cloud computing model-based data centre is considered, and the arrival of packets are discussed in detail and tried to analyse the performance of such cloud model data centre. The MATLAB software is used to analyse the performance.

4.1 Queuing Models

4.1.1 Queuing Model—M/M/1

The queuing model considered in the first model is the M/M/1 model. It has interarrival rates. This interarrival also has exponentially distributed relation with the Erlang distribution.

Steady-state performance measures for the above model are:

$$L_S = \rho + \frac{(1 + \frac{1}{er})\rho^2}{2(1 - \rho)};\ L_Q = \rho + \frac{(1 + \frac{1}{er})\rho^2}{2(1 - \rho)} \quad (1)$$

$$W_S = \frac{1}{\mu} + \frac{(1+\frac{1}{er})\rho^2}{2\mu(1-\rho)};\ W_Q = \frac{(1+\frac{1}{er})\rho}{2\mu(1-\rho)} \tag{2}$$

$$Var(Ls) = \frac{\rho^3(1+er)(2+er)}{3er^2(1-\rho)} + \frac{\rho^2(1+\frac{1}{er})}{2(1-\rho)} + \rho^2(3-2\rho)\frac{(1+\frac{1}{er})}{2(1-\rho)} + \rho(1-\rho) \tag{3}$$

$$Var(Ws) = 2Wq^2 + \frac{\rho(1+er)(2+er)}{\mu^2 3er^2(1-\rho)} + \frac{(1+\frac{1}{er})}{\mu^2(1-\rho)} - Ws^2 \tag{4}$$

$$Var(Lq) = \frac{\rho^2(1+er)}{2er^2(1-\rho)}\left(1 + \frac{\rho^2}{2er(1-\rho)}\right) + \frac{2\rho(2+er)}{3er} \tag{5}$$

$$Var(Wq) = Wq^2 + \frac{\rho(1+er)(2+er)}{\mu^2 3er^2(1-\rho)} \tag{6}$$

Note: r or er is the notation for Erlang parameter.

4.1.2 Queuing Model—M/M/c

The second model considered here in the current work was the M/M/c model. It also has exponential distribution, and it has interarrival rates of transfers and arrivals. The performance of the current model is analysed with the help of various parameters and discussed below,

Steady-state performance measures for the above model are:

$$B(c, p) = \frac{\frac{\rho^c}{c!}}{1 + \frac{\rho^2}{2!} + \frac{\rho^3}{3!} \ldots + \frac{\rho^c}{c!}} \text{(Erlang } B \text{ formula)} \tag{7}$$

$$Ls = \frac{\lambda(1 - B(c,\rho))}{\mu};\ L_Q = 0 \tag{8}$$

$$W_S = \frac{1}{\mu};\ W_Q = 0 \tag{9}$$

4.1.3 QtsPlus4Calc Software (https://qtsplus4calc.sourceforge.net)

The software entitled QtsPlus4Calc was used, and the home page model of the software is shown as follows (Fig. 3).

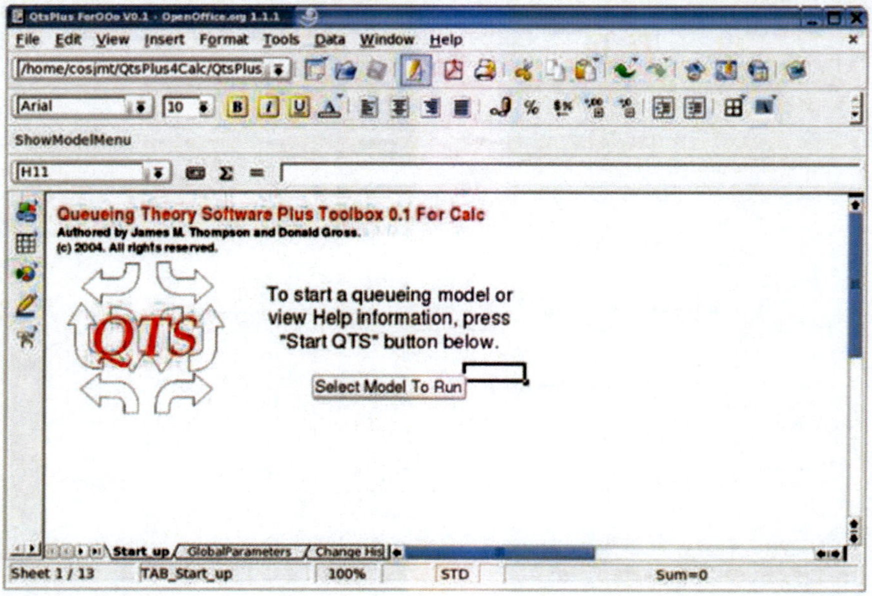

Fig. 3 QtsPlus4Calc environment screenshot

5 Results and Discussion

The individual performance can be observed from various graphical representations below. The below tables and graphical representations are as follows (Figs. 4, 5, 6 and 7; Tables 1 and 2).

The performance of both models considered here is analysed by applying various set of inputs. The time considered here is study that is the period to be considered for the execution of the model or the time is taken for the model to be complete given tasks can be taken as a study. There is no change in time value until the completion of the considered process. The performance of both models is impressive [8]. The performance can be observed in tabular formats, and for the better understanding, the results are shown under graphical representations also.

6 Conclusions

In the current work, an attempt has been made to analyse the performance of the cloud model data centre with the help of queuing models. Two queuing models have been considered here with study state. In the study state, the arrivals of the packets are following a study that is the time for the entire process is fixed and no change later on until the completion of the process. Several parameters have been studied,

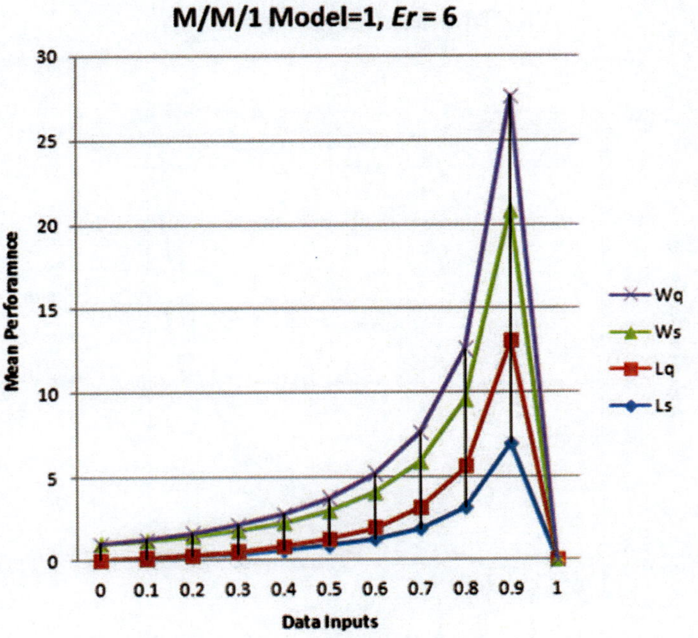

Fig. 4 Graphical representation of the first model for mean performance measures

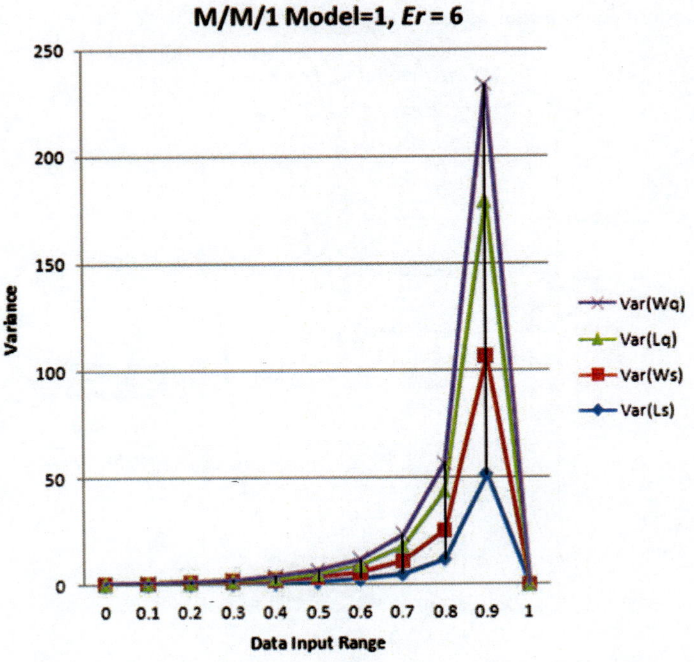

Fig. 5 Graphical representation of the first model for variance performance measures

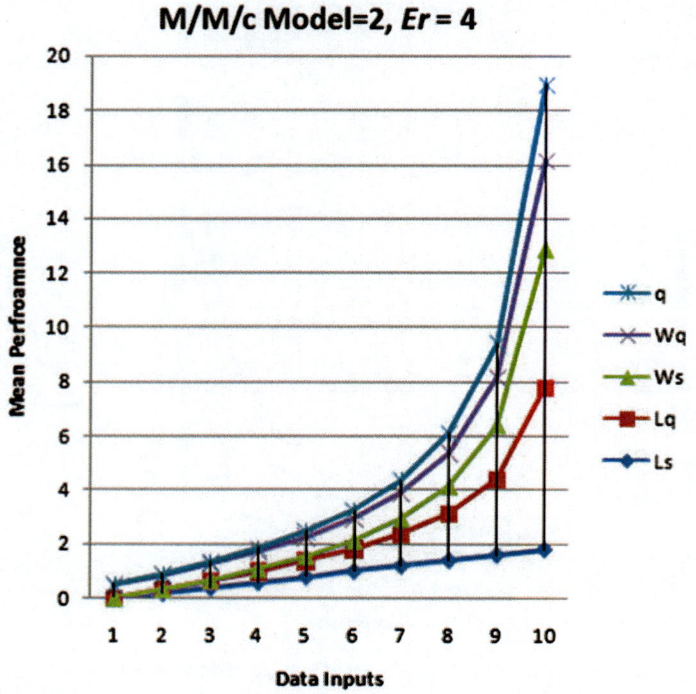

Fig. 6 Graphical representation of second model M/M/c for mean performance measures

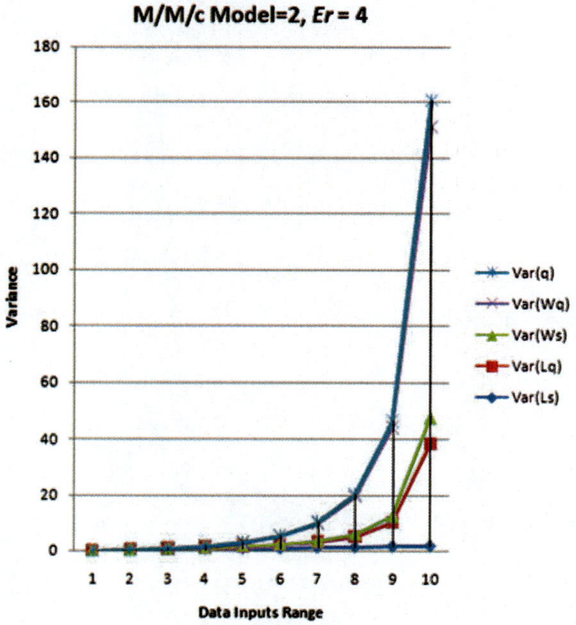

Fig. 7 Graphical representation of second model M/M/c for variance performance measures

Table 1 Performance of data centre—M/M/1, model = 1

Performance of data centre—M/M/1 model = 1, $Er = 6$											
t	Λ	ρ	Er-parameter	Input data				Variance			
				Ls	Lq	Ws	Wq	$Var(Ls)$	$Var(Ws)$	$Var(Lq)$	$Var(Wq)$
1	0	0	6	0.00	0.00	1.00	0.00	0.00	0.50	0.00	0.00
1	0.1	0.1	6	0.11	0.01	1.08	0.08	0.11	0.62	0.01	0.12
1	0.2	0.2	6	0.24	0.04	1.19	0.19	0.27	0.79	0.08	0.29
1	0.3	0.3	6	0.40	0.10	1.32	0.32	0.49	1.03	0.26	0.53
1	0.4	0.4	6	0.60	0.20	1.50	0.50	0.83	1.42	0.67	0.92
1	0.5	0.5	6	0.88	0.38	1.75	0.75	1.39	2.06	1.52	1.56
1	0.6	0.6	6	1.28	0.68	2.13	1.13	2.45	3.27	3.29	2.77
1	0.7	0.7	6	1.93	1.23	2.75	1.75	4.81	5.90	7.30	5.40
1	0.8	0.8	6	3.20	2.40	4.00	3.00	11.84	13.50	18.40	13.00
1	0.9	0.9	6	6.98	6.08	7.75	6.75	51.58	55.06	72.14	54.56
1	1	1	6	Inf	Inf	Inf	Inf	Inf	NaN	Inf	NaN

Table 2 Performance of data centre—M/M/c model = 2, *Er* = 4

t	Inputs								Variance		
	Λ	*ρ*	Er-parameter	*Ls*	*Lq*	*Ws*	*Wq*	*Var(Ls)*	*Var(Ws)*	*Var(Lq)*	*Var(Wq)*
2	0	0	4	0.00	0.00	0.50	0.00	0.00	0.06	0.00	0.00
2	0.2	0.1	4	0.11	0.01	0.53	0.03	0.11	0.08	0.02	0.02
2	0.4	0.2	4	0.23	0.03	0.58	0.08	0.25	0.11	0.13	0.05
2	0.6	0.3	4	0.38	0.08	0.63	0.13	0.43	0.15	0.47	0.08
2	0.8	0.4	4	0.57	0.17	0.71	0.21	0.70	0.21	1.26	0.15
2	1	0.5	4	0.81	0.31	0.81	0.31	1.13	0.32	2.91	0.25
2	1.2	0.6	4	1.16	0.56	0.97	0.47	1.91	0.52	6.28	0.45
2	1.4	0.7	4	1.72	1.02	1.23	0.73	3.60	0.96	13.50	0.90
2	1.6	0.8	4	2.80	2.00	1.75	1.25	8.56	2.25	31.60	2.19
2	1.8	0.9	4	5.96	5.06	3.31	2.81	36.35	9.38	103.59	9.32
2	2	1	4	Inf	Inf	Inf	Inf	Inf	NaN	Inf	NaN

and the performance was given in tabular and graphical formats. From the results, it is observed that the performance of the models is working correctly, and the results are analysed in detail.

References

1. K. Hamzeh, M. Jelena, M. Vojislav, Performance analysis of cloud computing centers using m/g/m/m+r queuing systems. IEEE Trans. Parallel Distributed Syst. **23** (2012)
2. H. Khazaei, Performance Modeling of Cloud Computing Centers, Doctoral dissertation, The University of Manitoba, Canada (2012)
3. B. Yang, F. Tan, Y. Dai, S. Guo, Performance evaluation of cloud service considering fault recovery, in *First International Conference on Cloud Computing (CloudCom) 2009* (2009)
4. I. Adan, J. Resing, *Queuing Systems* (Eindhoven University of Technology, The Netherlands, 2015)
5. J. Sztrik, *Basic Queuing Theory*. University of Debrecen Faculty of Informatics (2012)
6. Chandrakala, J. Shetty, Survey on models to investigate data center performance and QoS in cloud computing infrastructure, in *First International Conference on Recent Advances in Science & Engineering*, Netherlands (2014)
7. M. Hlynka, S. Molinaro, *Comparing Expected Wait Times of an M/M/1queue*. University of Winsor Department of Mathematics and Statistics (2010)
8. N. Khanghahi, R. Ravanmehr, Cloud computing performance evaluation: issues and challenges. Int. J. Cloud Comput. Services Archit. **3**(2), 121–130 (2013)
9. G. Rastogi, R. Sushil, Secured identity management system for preserving data privacy and transmission in cloud computing. Int. J. Future Generation Commun. Netw. NADIA **11**(1), 23–36 (2018)
10. D. Zhang, Research on collaborative filtering algorithm based on cloud computing. Int. J. Grid Distributed Comput. NADIA **9**(7), 23–32 (2018)
11. He. Kun, Research on collaborative filtering recommendation algorithm based on user interest for cloud computing. Int. J. Grid Distributed Comput. NADIA **10**(1), 255–268 (2017)

12. N. Thirupathi Rao, D. Bhattacharyya, Energy diminution methods in green cloud computing. Int. J. Cloud-Comput. Super-Comput. **6**(1), 1–8 (2019)
13. N.Thirupathi Rao, D. Bhattacharyya, S. Naga Mallik Raj, Queuing model based data centers: a review. Int. J. Adv. Sci. Technol. **123**, 11–20 (2019)

A Random Forest-Based Leaf Classification Using Multiple Features

Dipankar Hazra, Debnath Bhattacharyya, and Tai-hoon Kim

Abstract A novel method of resultant radial distances, leaf perimeter-based features, and RGB color moments-based leaf classification is proposed in this paper. In the training stage, shape features of plant leaf images are extracted by the resultant radial distances and leaf perimeter-based features; and color features are extracted using RGB color moments. Random forest is constructed with the leaf features where leaf names are the class attribute. In the testing stage, shape feature consisting of resultant radial distances and perimeter-based features, and color feature of the query image is extracted by means of same method. The query leaf image is recognized by the already created random forest in the training stage. The proposed method gives 98% recognition rate, which is similar to state-of-the-art leaf recognition methods. This is mainly due to the resultant radial distances for calculating accurate shape features. The smooth and jagged edges of the leaves are perfectly distinguished by leaf perimeter-based features. RGB moments help to distinguish different colored leaves.

Keywords Leaf recognition · Resultant radial distances · Perimeter-based features · RGB color moments · Random forest

D. Hazra (✉)
Department of Computer Science & Engineering, OmDayal Group of Institutions, Uluberia, Howrah, West Bengal 711316, India
e-mail: dipankar1998@rediffmail.com

D. Bhattacharyya
Department of Computer Science & Engineering, K L Deemed to be University, KLEF, Guntur 522502, India
e-mail: debnathb@kluniversity.in

T. Kim
School of Economics & Managements, Beijing Jiaotong University, Beijing, China
e-mail: taihoonn@diam.net

D. Bhattacharyya and N. Thirupathi Rao (eds.), *Machine Intelligence and Soft Computing*, Advances in Intelligent Systems and Computing 1280,
https://doi.org/10.1007/978-981-15-9516-5_20

1 Introduction

The study of plant characteristics and identification of plants is very important to diverse areas like agriculture, forestry, nature conservation, etc. The plant leaf recognition, an object recognition method, is one of the low costs and the convenient method for identifying the plants from the images of leaves. This method is challenging to researchers due to varieties of plants and varieties of features which can be taken out from plant leaf images. Building a database of plant leaves and recognizing plants from that helps in conservation and preservation of plants.

In the last few decades, various feature extraction methods were discovered. These feature extraction methods extract numerical values of features from image objects. These numerical features are used for recognition and classification of objects. The features are mainly of color, texture and shape-based features. For plant leaf recognition, it has been observed shape of leaves is most useful distinguishing features. Shape features may be of contour-based features or region-based features. For extraction of shape features, the color image is converted to a binary image such that contour points, centroid or other shape features can be easily extracted. The combination of color and or texture features along with shape features yields better accuracy.

There are numerous shape feature extraction methods are available, which are successful for recognizing shapes in different application areas. Some of these are grid-based method [1], turning function-based method [2], centroid-radii-based method [3], etc. The grid-based method is a region-based method. In this method, the shape is normalized. The shape becomes rotation invariant and scaling invariant. The shape is represented as grid of cells of 0 and 1 and this grid is compared with the grid of other shapes. The turning function-based method actually considers boundary of the shape as shape descriptor. The turning angle is calculated from tangent in counter-clockwise direction to the contour of the shape. But turning function is sensitive to noise and does not give correct result in present of noise. Tan KL et al. proposed centroid-radii model-based shape recognition method, which was very efficient. The shape feature vector is represented by radii of the shape, i.e., distances from centroid of the shape to the contour of the shape. The method was invariant to scaling and translation. Modified centroid-radii model [4] made this method invariant to rotation and solves the problem of multiple major axes. But still the portion of the radius falls outside the shape boundary of specific shapes produces inaccurate result. The proposed method of resultant radial distances minimizes the effect by excluding the portion of radius falls outside the boundary.

The construction of the machine learning model from large number of training leaf images is required. There are different classification models which can be used for leaf classification. The nearest neighbor algorithm and its variants are simple algorithms classifies objects based on similarity measures. Devi VS and Murty MN [5] have applied variants of the nearest neighbor algorithms for classifications. The similarity of different feature vectors is computed using distance measures like Euclidean distances, etc. But the nearest neighbors-based algorithms become slow when volumes of data become large. The decision tree [6]-based classification is comparatively faster one. The classification trees are tree structures where leaves represent class labels with discrete set of target variable and nodes represent feature or combination of features. The decision trees classify an object in leaf according to the values of features in node. But the decision trees tend to overfit the training dataset for highly irregular features. The random forest solves this problem. The random forest [7] is an ensemble learning algorithm using bagging as the ensemble methods and decision tree is the individual model. It is trained on different parts of same dataset and randomly select predictor for decision splits of each dataset. It makes the final prediction of class with majority votes. The problem of overfitting is solved by reducing the variances. The support vector machine (SVM) [8] uses binary linear classifier which separates two categories of feature data with gap between them wide as much as possible. This linear classification model can be converted to nonlinear classification model using kernel trick, mapping input to higher dimensional feature space. The success of any object recognition method depends on both, i.e., choosing the right feature extraction method with the right classification model.

Though there are different leaf recognition methods available now, still the work is going on for improved accuracy rate for classification with any type of leaf shapes. Also, there are works for identifying the leaf shapes which are deformed. Our work has aimed to classify the leaf shape with higher accuracy rate which includes normal along with some deformed leaf shapes.

The paper organizes as follows. Section 2 describes recent related works of leaf recognition system. The proposed method is explained in Sect. 3, whereas Sect. 4 compares the result of this system. Section 5 concludes this work and Sect. 6 gives a brief idea about the future work.

2 Related Works

Most of the previous researches on plant leaf recognition consider shape of the leaf for classification of plants. Prasad et al. [9] proposed relative sub-image-based coefficient features from leaf images and SVM-based classification to implement the automatic leaf recognition system. This method achieved 95% accuracy. Munisami et al. [10] proposed plant leaf recognition using different shape features like leaf length, leaf width, leaf area, leaf perimeter, leaf hull area, leaf hull perimeter, centroid-based radial distances, etc., and obtained accuracy 83.5%. The use of the color histogram method increases the accuracy to 87.3%. The k-nearest neighbor algorithm is used for classification.

Novotny et al. [11] extracted features from leaf boundary and leaf texture. Image moments, Fourier descriptor and leaf size are the features used in this system. Neural network classifier classifies the leaves accurately. Chaki et al. [12] characterized and recognized plant leaves using a feature combination from leaf shape and leaf texture. A set of curvelet transform coefficients and moments used as shape features. Gabor filter with gray level co-occurrence matrix (GLCM) are used as texture features. Hu et al. [13] proposed the multi-scale distance matrix. The multi-scale distance matrix is shape descriptor, which is invariant to translation, invariant to rotation, invariant to scaling and invariant to bilateral symmetry. Dimensionality of the descriptor was reduced to improve the discriminative power of the system.

Wu et al. [14] employed probabilistic neural network (PNN) to implement the leaf recognition model. They used morphological features like aspect ratio, rectangularity, etc. The features are derived from basic geometric features like leaf diameter, leaf length, leaf width, leaf area and leaf perimeter. The principal component analysis (PCA) is for feature selection. The average accuracy of the method is more than 90%. Valliamal et al. [15] also extracted combined features from leaf images. The extracted features are optimized through GA and KPCA separately. Support vector machine (SVM) is used for classification. This method achieved 92% accuracy. Herbal medicine plant leaves are identified using a method by Luna et al. [16]. Shape feature length, width, leaf vein area and leaf vein density are extracted and used as input to the artificial neural network. The system provides more than 98% accuracy.

Tsloakidis et al. [17] computed Zernike moments for shape features and histogram-oriented gradients (HOG) for texture features of leaves. They used support vector machine for plant leaf image classification and recognition and yielded high accuracy. Caglayan et al. [18] extracted morphological features as shape features for plant leaf recognition. They used two sets of color features. The mean and standard deviation of three color components and their average are one sets of color features. Other sets of color feature are derived using histogram. The use of random forest algorithm achieves more than 96% accuracy.

Afifi and Ashour [19] extracted moments of color distribution as color features color matching and retrieval. RGB color moments are used as color features. Wang et al. [20] proposed Chord Bunch Walk (CBW) descriptor for shape-based leaf recognition which can distinguish deformable leaf shapes. The chord bunch groups multiple pairs of chords to reflect the contour features. The method achieved higher accuracy and lower computational cost.

Wallelign et al. [21] has used convolution neural network (CNN) leaf image classification problem. They have used it plant disease identification in soybean plant. The model was designed based on the LeNet architecture. The classification accuracy is more than the existing man-made model. The main advantage of this representation learning is it can learn features automatically from large number of leaf images in natural environment. Hall et al. [22] combined hand-crafted features with ConvNet features and achieved more than 97% accuracy. The method is tested in changing conditions of translation, rotation, scale, illumination and occlusion. Random forest classifier is used for classification.

3 The Proposed Method

The proposed method is divided into two stages (i) training and (ii) testing. In the training stage, the color leaf image is converted to binary image. The perimeter-based features are used to find differences of contours of the leaf images. The proposed system uses resultant radial distances for obtaining the leaf shapes. RGB color moments are used as the color feature. At the time of testing, the shape and color feature vector of query leaf is extracted and classified by already created random forest. Figure 1 shows the data flow diagram of the proposed system.

3.1 Training Stage

After segmentation and conversion to binary image, the shape features can be calculated from the leaf image. The feature extraction methods are given as following:

Perimeter-Based Shape Features Computation. The smallest convex polygon containing the leaf shape is found. It is called the convex hull of the leaf shape. At first, leaf perimeter is calculated. Then, the convex hull perimeter of leaf shape is calculated. The ratio of these perimeters is called perimeter convexity. The ratio takes low values for leaves with smooth edges and high values for leaves with jagged edges. Figure 2 shows a sample leaf image, its binary image and convex hull. Similarly, the ratio between leaf perimeter and the square root of the leaf area is also calculated and it is called perimeter ratio. The perimeter ratio of leaf increases as the

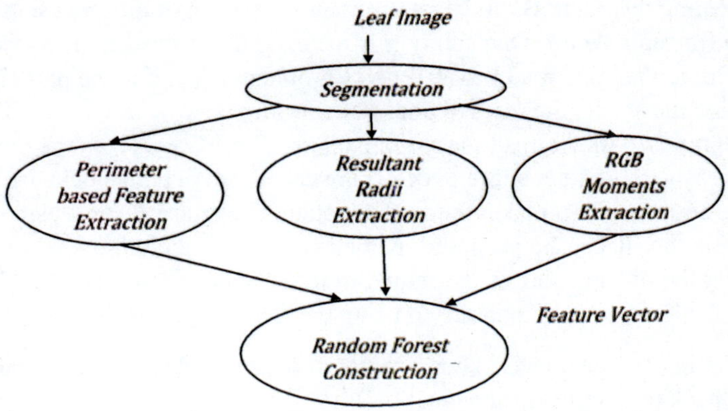

Fig. 1 Data flow diagram of the training stage of the proposed method

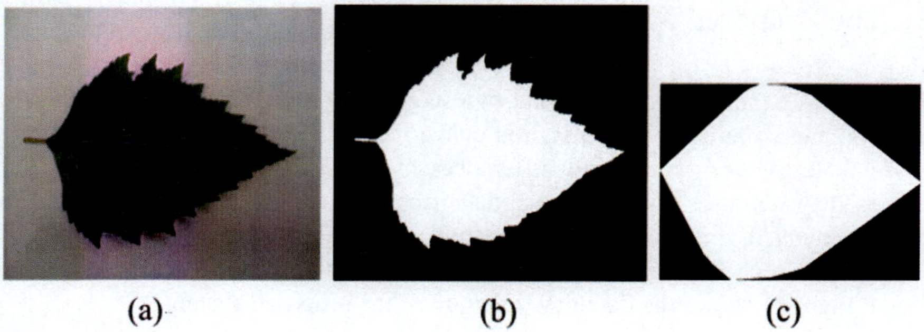

Fig. 2 **a** Sample leaf image, **b** binary image of the leaf, **c** convex hull of the leaf image

leaf edges become more complex. The perimeter convexity and perimeter ratio are represented by the following equations:

$$\text{Perimeter Convexity} = \frac{\text{Perimeter}_{\text{leaf}}}{\text{Perimeter}_{\text{convexhull}}} \tag{1}$$

$$\text{Perimeter Ratio} = \frac{\text{Perimeter}_{\text{leaf}}}{\sqrt{\text{Area}_{\text{leaf}}}} \tag{2}$$

Resultant Radial Distances Computation. In centroid-radii method, lengths from centroid to boundary of the shape are called radii and these radii represent the shape. Only distances are calculated, coordinates are not considered. Hence, the method was invariant to translation. The radii are divided by major axis length to make the method scaling invariant. But it is not invariant to rotation, if different starting point is chosen for radii vector. One solution is to rotate the major axis to x-axis. But if there are more than one major axis, it gives problem. Also, if some part of a radius falls outside the shape, the method does not consider it.

In the proposed method, all the major axes are considered. From the intersection of major axis with perimeter, radii vectors are calculated in both clockwise and anti-clockwise direction. This makes the method rotation invariant. If some part of a radius falls outside the shape, the proposed method calculates resultant radial distance by subtracting the outside part of the radius from the total radial distance. The shape vector of the leaf shape is constructed using the following steps.

Step1: Initialize image matrix L, with values of foreground pixels are equal to 1 and values of background pixels are equal to 0.

Step2: Assign the size of the image to $[m\ n]$.

Step3: Calculate the area of shape A as following:

$$A = \sum_{i=1}^{m} \sum_{j=1}^{n} L \tag{3}$$

Step4: Create two matrices X and Y of size $[m\ n]$.
Step5: Set all elements of X to x-coordinates of the elements.
Step6: Set all elements of Y to y-coordinates of the elements.
Step7: Calculate centroid (x, y) using the following formula:

$$x = \left(\sum_{i=1}^{m}\sum_{j=1}^{n} L.*X\right)\Big/ A \quad (4)$$

$$y = \left(\sum_{i=1}^{m}\sum_{j=1}^{n} L.*Y\right)\Big/ A \quad (5)$$

Step8: Calculate a point s on the contour of the shape.

Step9: Starting from point s, searching the neighborhood of the current pixel in anti-clockwise direction using 8-connectivity, contour matrix C is formed.

Step10: Radius r is calculated using following formula:

$$r = \sqrt{(C_x - x)^2 + (C_y - y)^2} \quad (6)$$

here (C_x, C_y) is the coordinate of a pixel on the contour C.

Step11: R is the shape vector, which consists of radii equal distant from each other.

$$R = [r_1\ r_2\ r_3 \ldots r_n] \quad (7)$$

where n is the length of the shape vector.

Step12: R is normalized, dividing by maximum length radius or major axis.

$$R = [r_1/r \text{ major } r_2/r \text{ major } r_3/r \text{ major} \ldots r_n/r \text{ major}] \quad (8)$$

where r major is the length of major axis.

Step13: The intersection of major axis with perimeter of the shape is the starting point of calculation. The first element of shape vector is 1.

Figure 3 shows 9 radii of a leaf shape. The major axis is drawn with bold line.

Fig. 3 Radial distances of a leaf shape

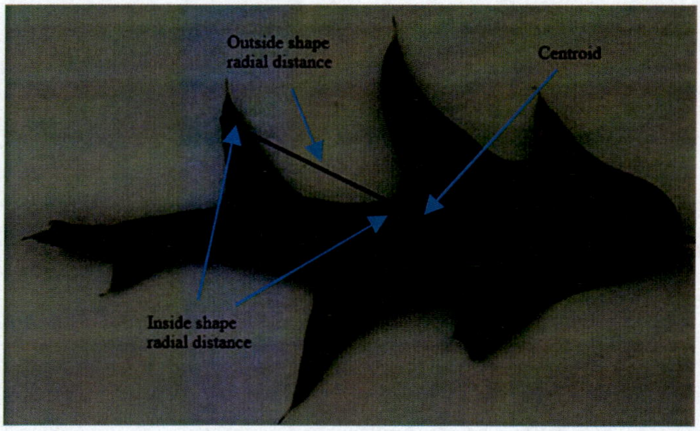

Fig. 4 One radius of a leaf shape where some part of radius falls outside shape

Step14: For some leaves, some part of radius may be outside the actual leaf shape. Assume this is denoted by Oi for radius ri. If full radius is inside the shape then $O_i = 0$.

Step15: Resultant radial distances considering radial distance outside the shape will be,

$$\begin{aligned} RR &= [r_1/r_{\text{major}} - O_1/r_{\text{major}} r_2/r_{\text{major}} - O_2/r_{\text{major}} \ldots r_n/r_{\text{major}} - O_n/r_{\text{major}}] \\ \text{or } RR &= [l_1 l_2 l_3 \ldots l_n] \end{aligned} \tag{9}$$

Figure 4 shows one radius of a leaf shape where some part of radius falls outside shape.

RGB Color Moments. Color moments are important color features for image recognition. Three color moments are mean, standard deviation and skewness. The color moments will be computed for each color component or color channel. So, for RGB color model, there will be nine color moments. The average color value of all pixels of the image is called mean. The square root of variance of pixels of the image is called as standard variation. Skewness is the measure of asymmetry of the image pixels.

Mean E_i of a color image for the ith color will be as following:

$$E_i = \sum_{j=1}^{N} (1/N) * p_{ij} \tag{10}$$

where N is the number of pixels in the image and for ith color, the value of the jth pixel of the image is denoted as pij. Standard deviation of a color image of the ith color is denoted by σi and calculated as following:

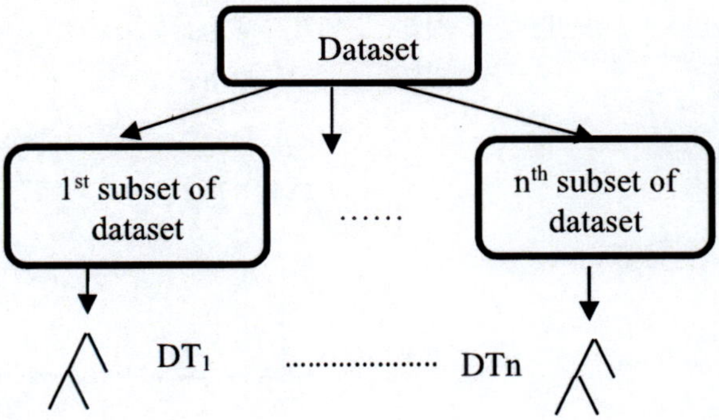

Fig. 5 Construction of random forest

$$\sigma_i = \sqrt{\left(\frac{1}{N} * \sum_{j=1}^{N} (p_{ij} - E_i)^2\right)} \tag{11}$$

Skewness S_i of the image is computed as following:

$$S_i = \sqrt[3]{\frac{1}{N} * \sum_{j=1}^{N} (p_{ij} - E_i)^3} \tag{12}$$

Random Forest Construction. A random forest is constructed from the extracted leaf features for training images. The leaf names are used as class label attributes. Figure 5 shows the pictorial representation of random forest. DT_1, DT_2,…,DTn are n decision trees formed from n subset of original feature dataset. The non-terminal nodes of the decision trees are splitting condition based on the different shape and color features of the leaf. The terminal nodes are the class label attributes or leaf names.

3.2 Testing Stage

In the testing stage, query leaf image is classified by the random forest created in the training stage. The resultant radii features, perimeter-based features and RGB color moments-based features of the query image are extracted by same procedure of training stage. The query leaf image is classified by each of the decision tree. The class of the leaf image is the class that is specified by majority voting of the decision trees. One vote is counted for each of the decision tree class name. The class of the query leaf is the class that gets maximum votes or maximum count of the class. Figure 6 shows the pictorial representation of classification of leaf shape by random forest.

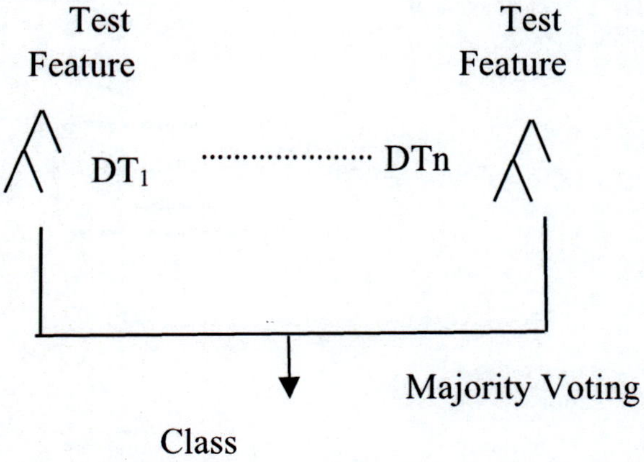

Fig. 6 Classification by the random forest

4 Experimental Result

The experiment is implemented using MATLAB. Plant leaf images are used in this experiment downloaded from leafsnap.com [23] and of collected images. 250 plant images are trained to create the random forest model, and 80 plant images are tested to find the accuracy of the proposed method. Figure 7 shows the sample leaf images used in this experiment. The recognition rate for the combination of the resultant radial distances, perimeter-based features and RGB color moments are given in Table 1. The recognition rate of the proposed method is 98%. The recognition rate is higher than the recognition rate of most of the other leaf recognition systems.

5 Conclusion

In the proposed method of leaf recognition, the resultant radial distances, perimeter-based features, RGB color moments-based feature extraction methods and random forest classifier are used. The proposed method achieves a recognition rate higher than most of other recent methods of leaf recognition. This is due to the accurate calculation of shape features by the resultant radial distances-based method. The perimeter-based feature increases the accuracy by distinguishing smooth and jagged contour of the shapes. The RGB color moments are useful for differentiating the leaf with different colors. The random forest classifies the leaves accurately.

Fig. 7 Sample leaf images used in the experiments

6 Future Work

The proposed method will be tested against the other frequently used leaf databases. Texture-based features can be combined for classifying the leaves based on leaf venation. The other classifiers will be used to check the accuracy of the leaf recognition method. Convolution neural network (CNN) can be applied for leaf recognition also.

Table 1 Recognition rate of different methods for leaf recognition

Author	Feature extraction method	Classifier	Recognition rate (%)
Hu R et al	Multi-scale distance matrix-based method [13]	Nearest neighbor	98
Luna RG et al	Geometric and morphological features-based method [16]	ANN	98
Tsolakidis DG et al	Zernike moments and HOG-based method [17]	SVM	98
Wallelign S et al	CNN-based method on soybean plant [21]	CNN	99
Hazra D et al	Proposed method	Random forest	98

References

1. A. Sajjanhar, G. Lu, A grid based shape indexing and retrieval method. Special Issue Austr. Comput. J. Multimedia Storage Arch. Syst. **29**(4), 131–140 (1997)
2. E.M. Arkin, L.P. Chew, D.P. Huttenlocher, K. Kedem, J.S.B. Mitchell, An efficiently computable metric for comparing polygonal shapes. IEEE Trans. Pattern Anal. Mach. Intell. (PAMI'91) **13**(3), 209–216 (1991)
3. K.L. Tan, B.C. Ooi, L.F. Thiang, Retrieving similar shapes effectively and efficiently, in *Multimedia Tools and Applications* (Kluwer Academic Publishers, 2003), pp. 111–134
4. N. Baik, D. Hazra, D. Bhattacharyya, Shape recognition based on mapreduce and in-memory processing on distributed file system. Int. J. Grid Distributed Comput. **11**(2), 21–30 (2018)
5. V.S. Devi, M.N. Murty, *Pattern Recognition an Introduction* (University Press, Hyderabad, India, 2011).
6. L. Breiman, J. Friedman, R. Olshen, C. Stone, *Classification and Regression Trees* (Routledge, New York, 1984).
7. L. Breiman, Random forests. Mach. Learn. **45**, 5–32 (Kluwer Academic Publishers, 2001)
8. C. Cortes, V. Vapnik, Support vector networks. Mach. Learn. **20**, 273–297 (Kluwer Academic Publishers, 1995)
9. S. Prasad, K.M. Kundiri, R.C. Tripathi, Relative sub-image based features for leaf recognition using support vector machine, in *Proceedings of The International Conference on Communication, Computing &Security*, Rourkela, Odisha, India (2011), pp. 343–346
10. T. Munisami, M. Ramsurn, S. Krishnah, S. Pudaruth, Plant leaf recognition using shape features and colour histogram using k-nearest neighbourclassifiers. Proc. Comput. Sci. **58**, 740–747 (2015)
11. P. Novotny, T. Suk, Leaf recognition of woody species in central Europe. Biosyst. Eng. **115**(4), 444–452 (Elsevier Ltd, 2013)
12. J. Chaki, R. Parekh, S. Bhattacharya, Plant leaf recognition using texture and shape features with neural classifiers. Pattern Recogn. Lett. **58**, 61–68 (Elsevier Ltd, 2015)
13. R. Hu, W. Jia, H. Lin, D. Huang, Multi-scale distance matrix for fast leaf recognition. IEEE Trans. Image Process. **21**(11), 4667–4672 (2012)
14. S.G. Wu, F.S. Bao, E.Y. Xu, Y.X. Wang, Y.F. Chang, Q.L. Xiang, A leaf recognition algorithm for plant classification using probabilistic neural network, in *IEEE International Symposium on Signal Processing and Information Technology*, Giza, Egypt (2007), pp. 11–16
15. N. Valliammal, S.N. Geethalakshmi, A optimal feature subset selection for leaf analysis. Int. J. Comput. Electrical Autom. Control Information Eng. **6**(2), 191–196 (2012)
16. R.G. Luna, R.G. Baldovino, E.A. Cotoco, A.L.P. Ocampo, I.C. Valenzula, A.B. Culaba, E.P. Dadios, Identification of Philippine herbal medicine plant leaf using artificial neural network, in *Proceedings of the IEEE 9th International Conference on Humanoid, Nanotechnology,*

Information Technology, Communication and Control, Environment and Management. Manila. Philippines (2017), pp. 1–8

17. D.G. Tsolakidis, D.I. Kosmopoulos, G. Papadourakis, Plant leaf recognition using Zernike moments and histogram of oriented gradients, in *Artificial Intelligence: Methods and Applications, LNCS*, vol. 8445, ed. by A. Likas, K. Blekas, D. Kellas (Springer, Cham, 2014), pp. 406–417
18. A. Caglayan, O. Guclu, A.B. Can, A plant recognition approach using shape and colour features in leaf images, in *Image Analysis and Processing, ICIAP 2013, Part-II, LNCS*, vol. 8157, ed. by A. Petrosino (Springer, Heidelberg, 2013), pp. 161–170
19. A.J. Afifi, W.M. Ashour, Image retrieval on content using color features. ISRN Comput. Graphics. Int. Scholarly Res. Netw. **2012**, 1–11 (2011)
20. B. Wang, Y. Gao, C. Sun, M. Blumenstein, J.L. Salle, Can walking and measuring along chord bunched better describe leaf shape? in *Proceedings of The IEEE Conference on Computer Vision and Pattern Recognition (CVPR)*. Honolulu, HI (2017), pp. 2047–2056
21. S. Wallelign, M. Polceanu, C. Buche, Soyabean plant disease identification using convolution neural network, in *Proceedings of The Thirty-First International Florida Artificial Intelligence Research Society Conference* (2018), pp. 146–151
22. D. Hall, C. McCool, F. Dayoub, N. Suenderhauf, B. Upcroft, Evaluation of features for leaf classification in challenging conditions, in *Proceedings of the IEEE Winter Conference on Applications of Computer Vision*, ed. by S. Das, S. Sarkar, B. Parvin, F. Proikili (IEEE, United States of America, 2015), pp. 797–804
23. LeafSnap Dataset. https://leafsnap.com/dataset/. Last accessed 2019/11/02

A Two-Level Hybrid Intrusion Detection Learning Method

K. Gayatri, B. Premamayudu, and M. Srikanth Yadav

Abstract The goal of this analysis is to identify intrusions on CSE-CIC-IDS2018 datasets. The methods used are split into two sections as a one-tier and two-level hybrid methods. In this analysis, we treated the data collection using machine learning techniques: convolutional neural network, random forest, light gradient boost system (RNN + Random Forest), (LGBM + Random Forest). The best intrusion detection system (RNN + Random Forest) was found to be a 98% accuracy rate and 0.86 macro-F-score. Additionally, hyperparameter optimization was performed with grid search, and the effect of synthetic minority over-sampling technique and high correlated features on detection was investigated. The study is unique because of that is the first time used the two-level hybrid multiclassifying on CSE-CIC-IDS2018 dataset.

Keywords Intrusion detection system · Convolutional neural network · Recurrent neural network

1 Introduction

Intrusion detection systems (IDS) are designed to detect and prevent network attacks from outside or inside. Intrusion detection systems signature- and anomaly-based intrusion detection programs can mostly be separated into two. Signature-based systems are generated by preserving already observed and established attack forms in a database while anomaly-based systems test real-time device inconsistencies with standard and usual packages. Machine learning techniques usually found these anomalies.

When analyzing the latest attack identification datasets, there are a few new threat complexity datasets. The research observed attacks against the 2018 CSE-CIC-IDS [1] dataset, which was prepared in 2018 and had a wide variety of attacks. There is

K. Gayatri (✉) · B. Premamayudu · M. S. Yadav
Department of IT, VFSTR (Deemed to be University), Guntur, AP 522213, India
e-mail: kg_itp@vignan.ac.in

D. Bhattacharyya and N. Thirupathi Rao (eds.), *Machine Intelligence and Soft Computing*, Advances in Intelligent Systems and Computing 1280,
https://doi.org/10.1007/978-981-15-9516-5_21

no consistent study classifying all types of attacks on this dataset when studies in the literature are analyzed, and the study has a unique value.

In this analysis, a single-level approach and two-level hybrid system, the effective identification of the CSE-CIC-IDS2018 data collection, were evaluated. The random forest, LGBM, and RNN classification algorithms and attack types on the dataset are classified on a single-level method. However, in order to correctly classify the observations classified by classifiers used by the single-level method, it was investigated whether the two-level hybrid method (a type of first-level attack and second-level multiple classifications) would influence classification success. The proposed model has two levels, Level 1 and Level 2 in the two-phase hybrid (Level 1 + Level 2) solution. At Level 1, random forest, LGBM, and RNN algorithms were tested separately to find out if a dataset attack and binary classification were performed. At Level 2, several observations were made using all observations identified as Level 1 attacks as test data for the random forest algorithm. After examining the experimental findings, it was found that the use of the two-level hybrid method increased the success of attack detection on the dataset. The most high-performing model is the two-level approach, which classifies the random forest method after binary classification with the RNN algorithm. The method used to detect attacks with an accuracy rate of 98 percent and an average F-score of 0.86.

Throughout the second section of the review, the literature includes experiments utilizing artificial learning and profound learning techniques for attack detection. Detailed data on CSE-CIC-IDS2018 is given in the third section, and the two-stage hybrid model proposed in the study is explained in detail, and the results analyzed in the fourth section. The findings obtained in the last segment are analyzed, and future studies listed.

2 Related Works

In their research at Wankhede and Kshirsagar in 2018 [2], they only observed attacks on DoS attacks on a specific day by using the random neural network (ANN) methods in the CICIDS2017 data collection. Further, the efficiency of the random forest and multi-layer perceptron—MLP algorithms was compared with the impact of the number of observations to predict success, using 20–80% observation of training results. It was noticed that the most efficient random forest method with 99.95 percent accuracy rate and 50 percent random forest method and 30% MLP method is optimal.

Sharafaldin et al., in their 2018 report [1], due to insufficient diversity of attacks in datasets created for network traffic rest and intrusion detection, the CICIDS2017 dataset was created. On this dataset, k-nearest neighbor (K-nearest neighbor—k-NN), random woodland, repeated binary tree, AdaBoost, MLP, naive Bayes, and quadratic discriminant analysis, using intrusion detection methods. The best performance of the ID3 algorithm was 0.98 F-score.

In its 2018 report, Aksu and Aydın detected port scanning attacks on the CICIDS2017 dataset using their MLP and SVM methods [3]. Attack detection for the data collection with port scanning attacks 0.65 with MLP, 0.95 F with SVM, and efficient seven-layer MLP threat detection.

In their 2019 report, Kanimozhi and Jacob [4] described botnet attacks using the MLP approach in the CSE-CIC-IDS2018 (CIC-AWS-2018) dataset. They also optimized the hyperparameter with the grid search method of the model, which was overfitted with the hyperparameters supposedly. The suggested approach classifies botnet attacks with a 99.97% accuracy score.

In their research in 2019, [5] Zhou and Pezaros analyzed the effectiveness of a model trained on the form of attacks (Zero-Day) CSE-CIC-IDS2018 dataset they never saw. Six different machine learning methods were tried for a tenfold cross-validation attack detection on this dataset (random forest, Gaussian Naive Bayes, decision tree, MLP, KNN, QDA). The tests were conducted by comparing binary traffic to regular traffic for each attack type, and the decision tree method was determined to be the most successful. The model was then tagged and qualified as usual and attack data collection, One-week regular test data traffic, and 6 new attacks (Zero Entry, DDoS Bot Obscurity, Google Docs, macdocs, Bitcoin miner, Drowor Worm, Nuclear Ransomware, Fake code invasion, Ponmocup Trojan). In this test set, attack detection with a 96% precision was carried out using the decision tree method.

Yulianto et al. studied in 2019 [6] better AdaBoost algorithm output on the CICIDS2017 dataset. Output enhancement has been proved by the usage of CICIDS2017 methods of primary component analysis (PCA), SMOTE, and ensemble function selection (EFS), unbalanced in attack forms. Based on comparative results, the method that best improved AdaBoost's performance with 0.90 F-score was the use of SMOTE and EFS algorithms.

When analyzing the latest attack identification datasets, there are a few new threat complexity datasets. The research observed attacks against the 2018 CSE-CIC-IDS [1] dataset, which was prepared in 2018 and had a wide variety of attacks. There is no consistent study classifying all types of attacks on this dataset when studies in the literature are analyzed, and the study has a unique value.

In this analysis, a single-level approach and two-level hybrid system, the effective identification of the CSE-CIC-IDS2018 data collection, were evaluated. The random forest, LGBM, and RNN classification algorithms and attack types on the dataset are classified on a single-level method. However, to correctly classify the observations classified by classifiers used by the single-level method, it was investigated whether the two-level hybrid method (a type of first-level attack and second-level multiple classifications) would influence classification success. In the hybrid two-level approach (Level 1 + Level 2), the suggested two-stage layout consists of IDS Level 1 and Level 2. In stage 1, random wood, LGBM, and RNN algorithms were evaluated independently to decide whether a dataset attack was carried out. At Level 2, several observations were made using all observations identified as level 1 attacks as test data for the random forest algorithm. After examining the experimental findings, it was found that the use of the two-level hybrid method increased the success of attack detection on the dataset. The most high-performing model is the two-level approach, which classifies the random forest method after binary classification with the RNN algorithm. The method used to detect attacks with an accuracy rate of 98% and an average F-score of 0.86.

3 Dataset Properties

The dataset used in CSE-CIC-IDS2018 is the updated version and the most recent known attack traffic dataset for CICIDS2017. The dataset was collected on the Amazon AWS LAN network by the Canadian Cybersecurity Institute (CIC) and the Communications Security Establishment (CSE) [7]. In this data collection, the Brute Force (Internet, XSS, FTP, SSH) Botnet, DoS (Hulk, SlowHTTPTest, GoldenEye, Slowloris), [7, 8] DDoS (HOIC, LOIC UDP, LOIC-HTTP), Database Injection (SQL), and Network intrusion (14 specific attack styles) including six intrusion forms are included. CICFlowMeter-V3 [9] packs were converted into network traffic flows, and 80 attributes were presented. The type of attack and distribution of observation numbers of these attacks are provided in Table 1.

The distribution of the observation numbers of the attack forms in the dataset is unbalanced, as can be seen from the table. The intrusion in the dataset is somewhat close to regular traffic since; unlike other threats, it fits the penetration direction in the network. This infiltration is usually done by email with malicious software sent to the victim or by creating a backdoor that allows you to use tools like Nmap and ports on the victim's computer, taking advantage of sensitive software weaknesses

Table 1 Observation numbers in the dataset

Attack type	Number of observations
Normal	13,390,249
Bot	286,191
Brute Force Web	611
Brute Force XSS	230
DDOS HOIC	686,012
DDOS LOIC UDP	1730
DDoS LOIC HTTP	576,191
DoS GoldenEye	41,508
DoS Hulk	461,912
DoS SlowHTTPTest	139,890
DoS Slowloris	10,990
FTP Brute Force	193,354
Infiltration	160,639
SQL Injection	87
SSH Brute Force	187,589

such as Adobe Acrobat Reader. Intrusion assaults were not observed effectively in this study, as in literature reviews. The difficulty with machine learning and deep learning methods was shown in identifying intrusion attacks [10].

4 Proposed Method and Experimental Results

Complete detail on the models used to identify intrusions is given below. The proposed approach after the data preprocessing phase is addressed as one and two levels within the framework of the analysis. The goal of the two rates is to sequentially inject the data in the test set into two separate models. In the first stage, it is determined whether the observation is an attack (binary classification). In the second level, the type of attack of observations described as a first-level attack is determined. The hybrid model is, therefore, expected to increase overall performance.

4.1 Data Preprocessing

In the shared dataset, CICFlowMeter-V3 extracted attributes, and the dataset deleted Flow Name, Source IP, Source Path, and Destination IP attributes. The timestamp attribute was also omitted from the dataset in an attack. Since the time of the attack was insignificant and unrelated, the time of the attack was unrelated [11]. The flow bytes/s and flow pkts/s attribute values are filled with their medians. The attack detection attributes are not needed for PSH flags Bwd, FwdByts/b Avg, Fwd Pkts/b Avg, BwdByts/b Avg, Bwd Pkts/b Avg, or Bwd Blk Rate Avg, but they are ineffectual and have been deleted from the dataset [12].

Moreover, infinite-value measurement values were omitted from the dataset. The test dataset is allocated to approximately 25% (4,034,296 observations) and the remaining 85% (12,102,887 observations) to reduce the time of training, make the training set more stable, and reduce the number of normal traffic observations to 500,000. Just from the training sample were 9.5 million common traffic measurements taken randomly. As can be seen in Table 2, the training package consists of 2,560,176, and the examination package consists of 4,034,296 observations.

The explanation for the number of regular traffic observations in training to 500,000 is the famous DDOS HOIC assault (514,590 observations) on the dataset.

If attributes with a zero standard deviation are discarded, characteristics that have a high correlation with one another have been identified so that the training period with the remaining 70 features is reduced. The correlation threshold value that determines the attributes to be thrown was determined to be between 0.9 and 1.00 using the elbow method. Figure 1 lists the correlation threshold value number of attributes to be deleted for this method. In Fig. 2, the significance spectrum graph defined in Fig. 1

Table 2 Observation numbers in training and test cluster

Attack_Type	Training_Set observations	Test_Set observations
Normal	500,000	3,347,538
Bot	214,555	71,636
Brute Force Web	454	157
Brute Force XSS	180	50
DDOS HOIC	514,590	171,422
DDOS LOIC UDP	1308	422
DDoS LOIC HTTP	431,871	144,320
DoS GoldenEye	31,042	10,466
DoS Hulk	346,618	115,294
DoS SlowHTTPTest	104,927	34,963
DoS Slowloris	8264	2726
FTP Brute Force	144,998	48,356
Infilteration	120,659	39,980
SQL Injection	68	19
SSH Brute Force	140,642	46,947
Total	2,560,176	4,034,296

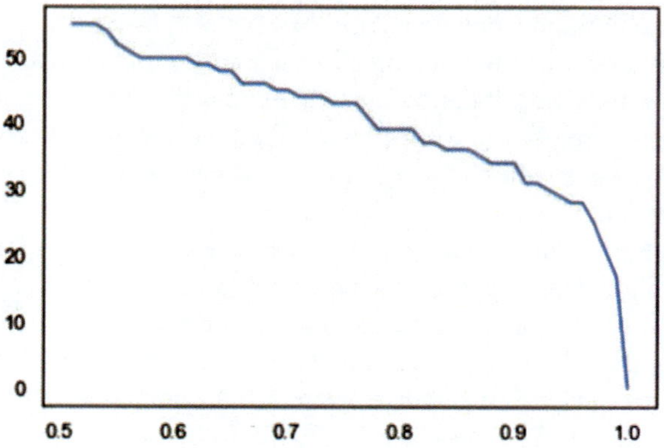

Fig. 1 Correlation threshold value number of attributes to be discarded

is shown, and the associated threshold value of 0.96 is calculated. The data collection, of which 0.96 threshold meaning high correlation attributes were removed, was only included in this experiment because it improves the efficiency for the single-level LGBM experiment alone.

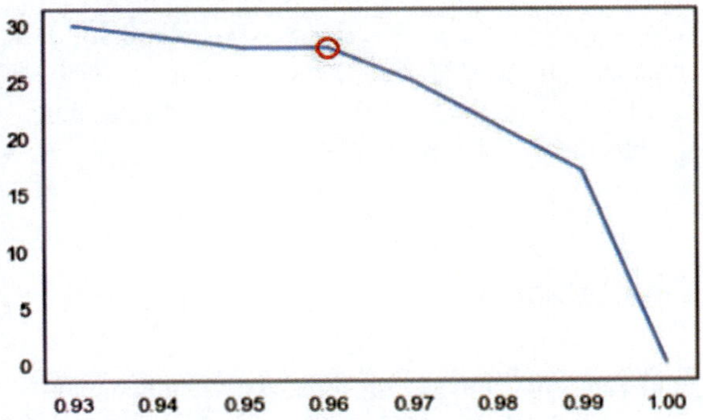

Fig. 2 Correlation threshold value determined according to the elbow method

4.2 *Smote*

SMOTE approach was attempted to resolve the dataset disparity. Table 3 provides the results of the F-score obtained with and without the SMOTE of random forest and LGBM methods [12]. By considering the above graph, it was observed that, in the case of random forest, no improvements were found in the efficiency of Brute Force Network, Brute Force XSS, attacks of only 0.03, and 0.06, respectively. In LGBM [13], the results of the assaults of Brute Force Network, Brute Force XSS, DDOS LOIC UDP, DDoS LOIC HTTP, DoS GoldenEye, DoS SlowHTTPTest, and

Table 3 SMOTE effect on random forest and LGBM

Attack Type/Method (F-score)	Random Forest	Random Forest SMOTE	LGBM	LGBM_SMOTE
Normal	0.96	0.96	0.98	0.84
Bot	1	1	1	1
Brute Force Web	0.43	0.46	0.03	0.06
Brute Force XSS	0.80	0.86	0.09	0.26
DDoS HOIC	1	1	1	1
DDoS LOIC UDP	0.90	0.90	0.72	0.87
DDoS LOIC HTTP	0.99	0.99	0.98	0.99
DoS GoldenEye	0.99	0.99	0.96	1
DoS Hulk	0.99	0.98	0.99	0.98
DoS SlowHTTPTest	0.60	0.60	0.60	0.61
DoS Slowloris	1	0.9	0.81	0.95
FTP Brute Force	0.78	0.78	0.78	0.78
Infilteration	0.15	0.15	0.24	0.05
SQL Injection	0.66	0.66	0	0.11
SSH Brute Force	0.9	0.99	0.9	0.99

DoS Slowloris have improved, respectively, to 0.03, 0.17, 0.15, 0.01, 0.03, 0.01 and to 0.14 and decreased to 0.19. Because of the weak effect of the SMOTE method on the results, no effective method for the detection of intrusions was found. The SMOTE approach has also not been seen [14].

4.3 Single-Level Method

LGBM, RNN, and random forest algorithms were independently evaluated in a single-level approach for the different classification of attack forms. The architecture of the method is detailed in Fig. 3. The model is trained and checked with test data after the data preprocessing phase. For the LGBM [15]process, the hyperparameter "boosting type" was chosen as "dart," and training was performed by extracting the attributes at a higher correlation value of 0.96. There are two one-dynamic convolution layers, 2 × 2 filtering (pooling) layer for the RNN process, a flatten layer, two related layers with a "ReLu" feature, and 512 neurons with a 10% dropout and 15 neurons, softmax activation mechanism [16]. A network was used with the output layer. Figure 4 shows the iteration accuracy graph of the model with 100 iterations.

Hyperparameter selection and optimization of the random forest method was done with the grid search method with threefold cross-validation. The tested hyperparameters are given below.

- max_features = "sqrt," 02, 0.3, 0.4, 0.5, 0.6
- min_samples_split = 2:5 (2, 3, 4, 5), 6:10, 15, 17, 30, 45, 70, 100, 200.

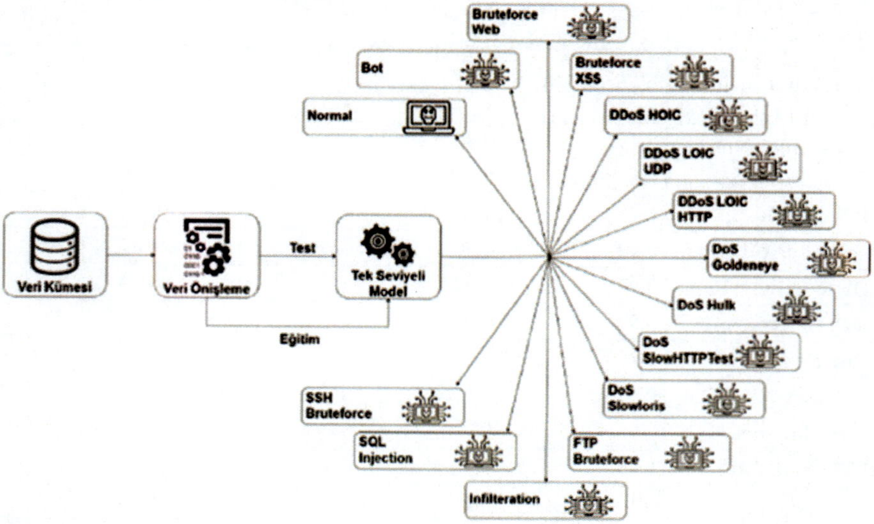

Fig. 3 Architecture of the single-level method

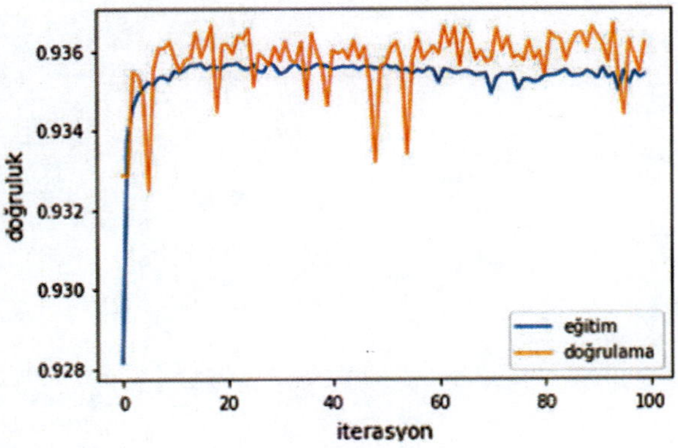

Fig. 4 Iteration accuracy graph for single-level RNN

Table 4 Cross-validation results

Evaluation criterion	F-score Macro-average
Fold-1	0.8831
Fold-2	0.8793
Fold-3	0.883
Average	0.8820

- min_samples_leaf = 1, 2, 3, 4, 8
- n_estimators = 20, 40, 80.

The values 0.4 for max features, 5 for min samples split, 1 for min samples leaf, and 80 for n estimators were determined among the tested hyperparameters, given the best results. The results of the cross-validation are given in Table 4.

4.4 Two-Level Method

Figure 5 shows the general structure of the two-level method. As shown in the figure, the model consists of two stages in this method. Information at first level to assess if a cluster attack exists, binary classification has been conducted separately with random forest, LGBM, and RNN approaches. In the second step, findings observed as a first-level attack have been used as test results for the second level of the random forest model. Level 1 is trained for training following binary labeling, while Level 2 is trained immediately after data preprocessing. The following subtitles give comprehensive details on rates 1 and 2—the following subtitles.

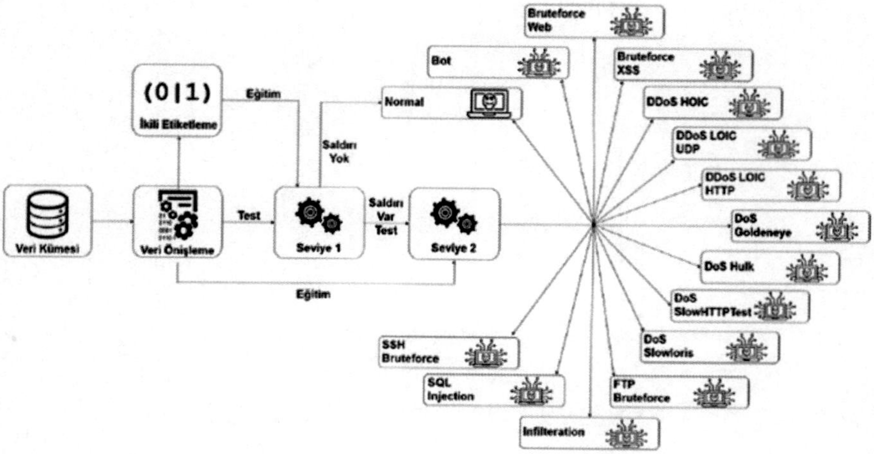

Fig. 5 Architecture of two-level hybrid method

Level 1

The dataset is marked as binary, attack, and usual traffic at the first level. The tagged data have been tested for random forest, LGBM, and RNN methods. The aim is to filter the regular traffic, which is heavily observed in the dataset. The network structure of the RNN model is similar to the network structure of the single-level RNN model. As such, attributes whose correlation is above the 0.96 threshold value were not eliminated. Differently from the LGBM model level. For max features, the random forest model used 0.2, for min samples split, for min samples leaf, for min samples and n estimators, 75.

Table 5 displays the strategies and their accomplishments. When the results obtained in the table are evaluated, as the F-score values are close to one another, the results of all models are used as Level 2 input.

Level 2

Observations identified as attacks at Level 1 are classified by the random forest method at Level 2. For the random forest model at Level 2, the same hyperparameters used in the single-level model were used. Random forest model (Level 2) is tested with the estimates of the models used in Level 1. When the results obtained for the two-level model were examined, it was seen that using the RNN model for Level

Table 5 Level 1 achievements

Evaluation Criterion	LGBM	Random Forest	RNN
F-Score	0.95	0.94	0.95
Accuracy rate	0.98	0.98	0.98

1 and then classifying with the random forest algorithm (RNN + Random Forest) increased the performance.

When education approaches for Level 2 are considered, in the two-level model, while training with Level 1 in the whole dataset, two approaches can be applied for the training in the Level 2 model. The first of these approaches is education with the whole dataset, and the other is education with the normal traffic-free (only attacks) dataset. In the study, both approaches were tried at Level 2. Both approaches have their advantages and disadvantages. The advantage of the entire dataset and education approach is that Level 1 is false (True Negative) for observations that False Positive is misleading as an attack. The disadvantage is the possibility of False Negative observing that Level 1 is correct (True Positive) observations as regular traffic. The advantage of training approach only with attack set is that since it does not see normal traffic during training, it does not confuse observations that Level 1 describes as accurate (True Positive) with normal traffic. The disadvantage is that the observations that Level 1 misrepresents (False Positive) as an attack are not corrected.

Both approaches (Random Forest + Random Forest, LGBM + Random Forest, and RNN + Random Forest) have been tried for each model at Level 2 to decide between the two approaches. When evaluating the outcomes for the two-stage hybrid model suggested for the analysis, it was observed that the usage of the learning method for all Level 2 knowledge was best with all three versions. In Table 6, the

Table 6 Training approach selection results for two level

Attack_Type/approach (F-score)	Education with the whole dataset	Training with the attack cluster only
Normal	0.98	0.98
Bot	0.99	0.99
Brute Force Web	0.46	0.40
Brute Force XSS	0.90	0.46
DDOS HOIC	0.99	0.99
DDOS LOIC UDP	0.91	0.91
DDoS LOIC HTTP	0.99	0.99
DoS GoldenEye	0.99	0.99
DoS Hulk	0.9	0.9
DoS SlowHTTPTest	0.60	0.60
DoS Slowloris	1	1
FTP Brute Force	0.78	0.78
Infilteration	0.28	0.27
SQL Injection	0.80	0.13
SSH Brute Force	0.9	0.98
Macro-Average	0.84	0.76
Accuracy rate	0.96	0.96

F-score results for the Random Forest + Random Forest model for both approaches are presented separately. This model is only given because the macro-F-score gap is the largest. The effectiveness of the education method for the entire dataset becomes apparent when the table becomes analyzed, in particular for the Brute Force XSS and the SQL attack forms. As the use of all of the training data gives better results, training for the proposed hybrid model was performed with the entire Level 2 dataset.

When all of the experimental results obtained were analyzed, the first approach to the hybrid models to be used to detect an attack achieved the highest performance using (RNN + Random Forest) method. Intrusion attacks were found to have reduced the identification effectiveness and to be the assault most closely resembled regular network traffic.

5 Conclusion and Future Work

In this study, it was suggested that the performance of the CSE-CIC-IDS2018 dataset could be increased by using a two-tier hybrid model as a result of misclassifications for attack detection. For assault detection purposes, LGBM, RNN, and random forest methods were attempted, and a two-level hybrid model was developed with the single-tier random forest approach used according to the binary classification of these methods. The two-step approach, which utilizes RNN and random forest approaches combined for Level 1 and Level 2, had the best performance in evaluating the experimental results with an overall macro-score of 0.86.

For future research, a platform with different machine learning and deep learning methods on the hybrid model should be created, which will boost the efficiency of attack detection and simultaneous attack detection.

References

1. I. Sharafaldin, A.H. Lashkari, A.A. Ghorbani, Toward generating a new intrusion detection dataset and intrusion traffic characterization, in *ICISSP*, Prague, Czech Republic (2018), pp. 108–116
2. S. Wankhede, D. Kshirsagar, DoS attack detection using machine learning and neural network, in *2018 Fourth International Conference on Computing Communication Control and Automation (ICCUBEA)*, Pune, India (2018), pp. 1–5. Conference on Information Systems Security and Privacy (ICISSP), Portugal, (2018)
3. D. Aksu, M. Ali Aydin, Detecting port scan attempt with comparative analysis of deep learning and support vector machine algorithms, in *2018 International Congress on Big Data, Deep Learning and Fighting Cyber Terrorism (IBIGDELFT)*, Ankara, Turkey (2018), pp. 77–80
4. V. Kanimozhi, T.P. Jacob, Artificial intelligence-based network intrusion detection with hyper-parameter optimization tuning on the realistic cyber dataset CSE-CIC-IDS2018 using cloud computing, in *2019 International Conference on Communication and Signal Processing (ICCSP)*, Chennai, India (2019), pp. 33–36
5. Q. Zhou, D. Pezaros,*Evaluation of Machine Learning Classifiers for Zero-Day Intrusion Detection—An Analysis on CIC-AWS-2018 Dataset*. ArXiv abs/1905.03685v1 (2019)

6. A. Yulianto, P. Sukarno, N. AnggisSuwastika, Improving AdaBoost-based Intrusion Detection System (IDS) performance on CIC IDS 2017 dataset. J. Phys. Conf. Ser. 1192 (2017)
7. I. Ullah, Q.H. Mahmoud, A two-level hybrid model for anomalous activity detection in IoT networks, in *2019 16th IEEE Annual Consumer Communications & Networking Conference (CCNC)*, Las Vegas, NV, USA (2019), pp. 1–6
8. A.R. Wani, Q.P. Rana, U. Saxena, N. Pandey, Analysis and detection of DDoS attacks on cloud computing environment using machine learning techniques, in *2019 Amity International Conference on Artificial Intelligence (AICAI)*, Dubai, United Arab Emirates (2019), pp. 870–875
9. CICFlowMeter: Network Traffic Flow Analyzer. https://netflowmeter.ca/netflowmeter.html. Accessed 28 July (2018)
10. Y. Yang, K. Zheng, C. Wu, X. Niu, Y. Yang, Building an effective intrusion detection system using the modified density peak clustering algorithm and deep belief networks. Appl. Sci. **9**(2), 238 (2019). https://doi.org/10.3390/app9020238
11. S. Yılmaz, S. Sen, Early detection of botnet activities using grammatical evolution.Theor. Appl. Models Comput. 395–404 (2019)
12. R. McKay, B. Pendleton, J. Britt, B. Nakhavanit, Machine learning algorithms on botnet traffic: ensemble and simple algorithms, in *The International Conference on Compute and Data Analysis 2019 (ICCDA)* (2019)
13. M.A. Ferrag, L. Maglaras, Delivery coin: an IDS and blockchain-based delivery framework for drone-delivered services. Computers **8**(58) (2019)
14. P. Lin, K. Ye, C.Z. Xu, Dynamic network anomaly detection system by using deep learning techniques, in *Cloud Computing—CLOUD 2019*. CLOUD 2019. Lecture Notes in Computer Science, ed. by D. Da Silva, Q. Wang, L.J. Zhang. vol. 11513 (Springer, Cham, 2019)
15. F.S. de Lima Filho, F.A.F. Silveira, A. de Medeiros Brito Jr., G. Vargas-Solar, L.F. Silveira, Smart detection: an online approach for DoS/DDoS attack detection using machine learning. Security Commun. Netw. vol. 2019, Article ID 1574749, 15 (2019)
16. V. Kanimozhi, T. Prem Jacob, Calibration of various optimized machine learning classifiers in network intrusion detection system on the realistic cyber dataset CSE-CIC-IDS2018 using cloud computing. Int. J. Eng. Appl. Sci. Technol. **4**(6), 209–213 (2019)

Supervised Learning Breast Cancer Data Set Analysis in MATLAB Using Novel SVM Classifier

Prasanna Priya Golagani, Tummala Sita Mahalakshmi, and Shaik Khasim Beebi

Abstract Chest carcinoma is the increasing problem, particularly among women. In this paper, we developed a breast cancer data set containing the biomolecules concentration in breast cancer cells measured using mass spectrometry. Based on the concentrations of these biomolecules, the tumor in the breast can be classified into benign and malignant which is specified in the data set. The biomolecules in different signal transduction pathways like PKB (protein kinase B), MAPK (mitogen-activated protein kinase), MTOR (mammalian target of rapamycin), Fas ligand (Type-II transmembrane protein), notch (single-pass transmembrane receptor), SHH (Sonic Hedgehog), Tnf (tumor necrosis factor), Wnt (wingless/integrated) are taken into account. In this work, a framework research program for the categorization of cancer and tumor is built using supervised machine learning. When selecting a function, we extract the bases set throughout the kernel capacity and then expand the edge-construct function discretion program. We are starting to traverse a variety of characteristic collection and filtration approaches as well as incorporate the optimum feature subcategories with multiple learning categorization methods, such as KNN (K-nearest neighbor), PNN (probalistic neural network), and SVM (support vector machine) classifiers. The highest categorization efficiency for diagnosis of chest carcinoma is achieved at 99.17% connecting diameter and durable attributes using the SVM sorter.

Keywords Chest carcinoma · Attribute draft · SVM categorization · Supervised learning

P. P. Golagani (✉) · T. S. Mahalakshmi
Department of Computer Science and Engineering, GITAM University, Visakhapatnam, AP, India
e-mail: prasannapriya18@gmail.com

S. K. Beebi
Department of Biotechnology, GITAM University, Visakhapatnam, AP, India

D. Bhattacharyya and N. Thirupathi Rao (eds.), *Machine Intelligence and Soft Computing*, Advances in Intelligent Systems and Computing 1280,
https://doi.org/10.1007/978-981-15-9516-5_22

1 Introduction

Chest carcinoma (cancerous organ excrescency) is a carcinoma of the chest flesh. There is also a requirement for such a dependable and unbiased screening method to identify and diagnose instances of breast cancer, benign, or infectious. Artificial intelligence and semantic network support us by offering a stronger and accurate recognition method.

The semantic network constitutes of an integrated category of false axons, [1] and process information using only a relation methodology of calculation. Present-day semantic matrix is random methods which are used for modeling statistical data. Machine learning is an evolving technique to identify and compute outcomes that help professionals make decisions in a world of ambiguity and deception. The necessity for a semantic network is that, dissimilar conventional dense enumerating, it is designed to handle the ubiquitous incoherence of the modern world. As a consequence, the leading concept of the semantic matrix is to leverage liberality for deception, ambiguity, and part of the veracity in order to attain traceability. The semantic matrix mock-up can be categorized and claimed to different parameters, including their teaching methods, design, deployment, procedure, and soon.

The aim of the task is to design the challenge with the desired load and product data file, so that the evolving matrix can possess customizable specifications that have been modified by the superintended education law. Receiving the class of directed education, experience is one type of neural network conducting a category in a two-dimensional space. The support vector machine is such a procedure that has a straight impact on apparatus education. The support vector apparatus should be used as the strongest categorization method for identifying any form of data file, even a random 1. In this paper, the data file on cancer and tumor instances of chest carcinoma is used as training load and is categorized in the hyper plane area, and its classified output is also measured.

2 Literature Review

There are several new techniques that have evolved with the growth of breast cancer prediction technology. The research pertaining to this area shall be outlined as heed shortly.

Azar et al. [2] have suggested a new strategy for chest carcinoma detection. The attitude employed three categorization programs called spiral base purpose, hand-picked semantic matrix, and multiple-coating perception. The art also applied the features of the chest carcinoma data file and test design. Device output is measured according. System performance is weighed in terms of certain guide of measurement of machine learning efficiency such as precision, specificity, and sensitivity. For training and research, multi-coating perception got achievability of 96.50% and 96.55%, respectively. The authors demonstrated a breast cancer recognition method that is implemented with GA features for the two separate Wisconsin chest carcinoma data files which removes the data that is not required and gives the correct information that can make the system fast. Several apparatus education arts are given for categorization requirement. The greater correctness 98.59% is observed by Spiral Garden program with GA constructed feature choice.

Ahmad et al. [3] suggested a breast cancer screening art. The hereditary program is applied for pin pointing the correct features from all the features. The method deviates the data file into three parts called training (49%), testing (24%), and validation (24%). The number of connections, the size of hidden nodes, and the number of selected functions are also considered as a measure of success. The programs with the union of goals achieved the better and average cases, 98.85% and 98.10% accuracy, respectively. The paper also compares with current techniques which have already been applied in this area.

Islam et al. [4] created a categorizer for breast cancer treatment with the multi-gene genetic programming symbolic regression. The model uses ninefold verification technique. The scheme provides a genetic multi-gene programming. The model employs the methodology of ninefold verification. The scheme gives a mathematical representation of data set attributes. The result presented in the paper is not clear whether it is training or research. The model's accuracy is 99.28% along with 0.1303 RMSE.

The authors in [5] demonstrated a groundbreaking technique for chest carcinoma prediction by categorizing the breast cancer data set attributes using a hybrid neuro genetic system consisting of hereditary program and training load straight back propagation. Therefore, the system is conditioned to keep one out of fashion, causing over fitting. The framework's overall performance is 97 percent. In this paper, also a comparison analysis is seen. For breast cancer detection, the most commonly used machine learning techniques called RF (random forest), BN (Bayesian networks), and SVM (support vector machine) are introduced in [1, 6]. To prevent over fitting, the device used ninefold cross-validation. The authors implemented the program in WEKA and measured the measure of success.

3 Materials and Methods

Throughout the past, a filtering approach [7] was used to select a function. It does not take into profile the deviation of the initiation program. KNN and PNN were included in the earlier grouping. The proposed method uses a novel approach to the selection of features, namely the right round initiative, and the categorization approach is SVM. Right round initiation [4] uses an initiation program and manages extensive data files and LOOCV suspicions.

Feature Selection in Kernel Space

Step1: Building a deviation set by either kernel GP (or) kernel PCA.

Step2: Scheming mass by kernel RELIEF.

Step3: Rating attributes by weight.

Step4: Select features based on the rank.

Step: 1 Including subsets:

Algorithm 1
FSKGP (FEATURE SELECTION IN KERNEL GRAM SCHMIDT PROCESS)
Input: data $x\ (i)\ (i = 1 \ldots N)$
Output: an orthogonal set of basis vectors for $i = 1$ to N do
$v\ (i) = -(x\ (i))s$
for $j = 1$ to $i - 1$ do
$v\ (i)$ □$v\ (i) - h-(x\ (i)),\ v\ (j)\ iv$
(j) end for
Normalize: $v\ (i)$ □$v\ (i)/\|v\ (i)\|$ end for
Output: basis set $\{v^{(i)}\}$

Algorithm 2
FSKSPCA (FEATURE SELECTION IN KERNEL SPACE PRINCIPAL COMPONENT ANALYSIS):
Input: training data *xi*, label *yi*
Output: Chosen attributes in the kernel area
1: Building a deviation set by either kernel GP or kernel PCA
2: Calculating *wi* by kernel relief
3: Categorize implicit attributes by *wi*, take attributes on the class
4: Showing the information into the educated area (Fig. 1)

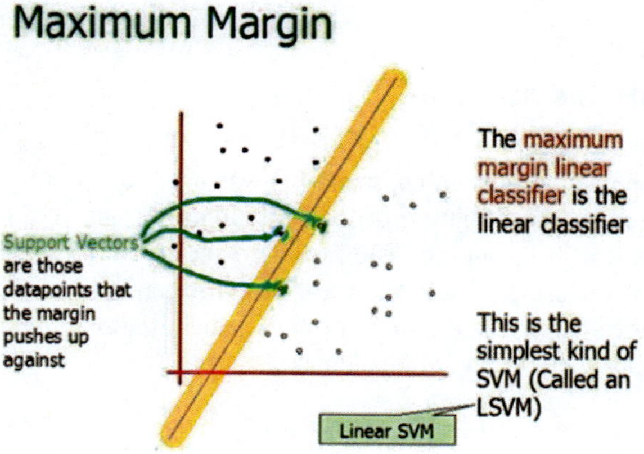

Fig. 1 Linear SVM

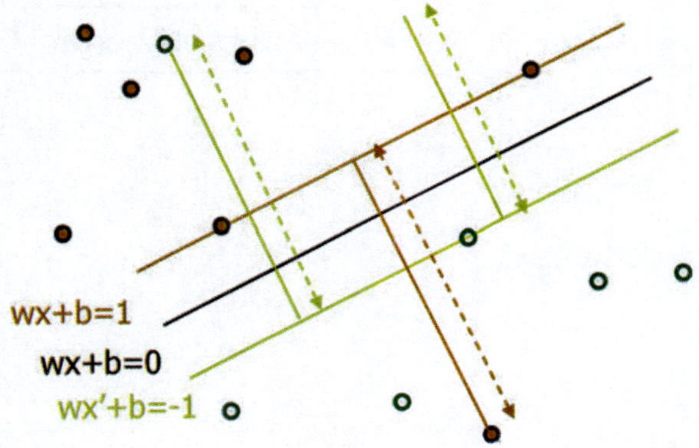

Fig. 2 Representation of hyper plane

Categorization Models

3.1 Linear SVM

Expression for maximum margin is given as

$$\textbf{margin} = \underset{\mathbf{x}\in D}{\arg\min}\, d(\mathbf{x}) = \underset{\mathbf{x}\in D}{\operatorname{argmin}} \frac{\lceil \mathbf{x}\cdot\mathbf{w} + b \rceil}{\sqrt{\sum_{i=1}^{d} w_i^2}}$$

Representation of Hyper Plane

It is in contrast method which breaks down the question of multiclass identification into equivalences and each chotomizer must distinguish a single class from all of the others [6] (Fig. 2).

1. If $Yi = +1$; $wxi + b \geq 1$
2. If $Yi = -1$; $wxi + b \leq 1$
3. For all i; yi

   $+b) \geq 1$ X □vector point
   W □weight maximum margin
   $M = 2/||w||$

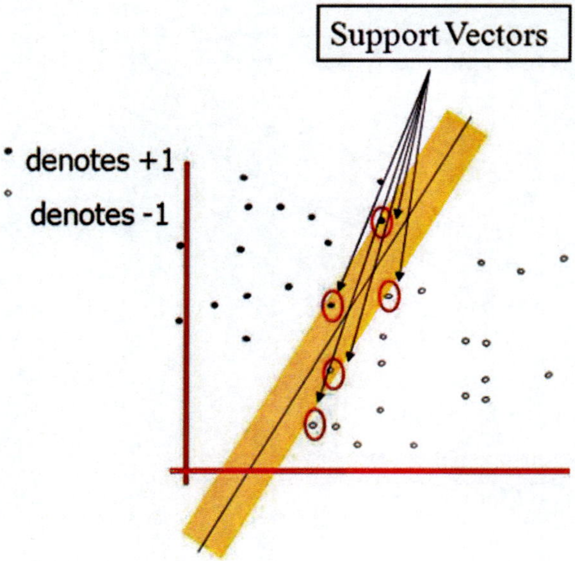

Fig. 3 Representation of support vectors

4 Representation of Support Vectors

The solution involves constructing a dual problem and where a Langlier's multiplier αi is associated. We need to find w and b such that Φ (w) = ½ |w'||w| is minimized and for all {(xi, yi)}: yi (w * xi + b) ≥ 1

Now solving: we get that

$w = \Sigma \alpha i * xi$

$b = yk - w * xk$ for any xk such that $\alpha k \neq 0$

Now the classifying function will have the following form:

$F(x) = \Sigma \alpha i\ yi\ xi * x + b$ (Fig. 3).

5 Results and Discussions

During the recruitment, all health information is split into two sections for training and research. It was performed dynamically by the SVM classifier. Since the analysis of the SVM, we can obtain the product of the classification of the percentages in Figs. 4, 5, 6, and 7.

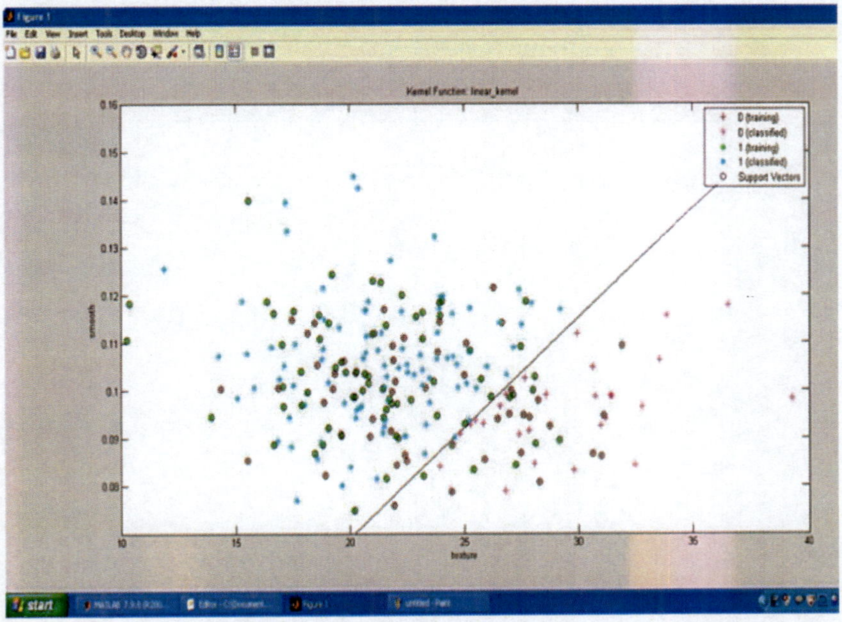

Fig. 4 Texture versus smooth classification performance (CP) = 52.89%

6 Conclusions and Recommendations

We have also used SVM to classify many of the features and also to obtain a classification output percentage. Try comparing all the characteristics of the compact and radius to the limit. Throughout this study, feature selection throughout SVM conducts practice and examine of information categorization, visualization for the diagnosis of chest carcinoma. Experimental findings, together with the proportion of classification, indicate that when comparing all features, the characteristics with the highest percentage of success categorization, such as radius and compactness, are classified as the best features used in the diagnosis of breast cancer. Approach of discrimination by apparatus education program states that the support vector machine program is a robust and effective technique that it can be used for any kind of categorization given that sufficient data sets are usable. Not only in the medical sector, this form of classification by means of support vector machines can be carried out in all emerging and unknowing fields such as stock market exchange, weather, forecasting natural calamities, automobile MPG predictions, and so on. The best categorization performance for breast cancer diagnosis is 99.17% between the radius and the compact features using the SVM method.

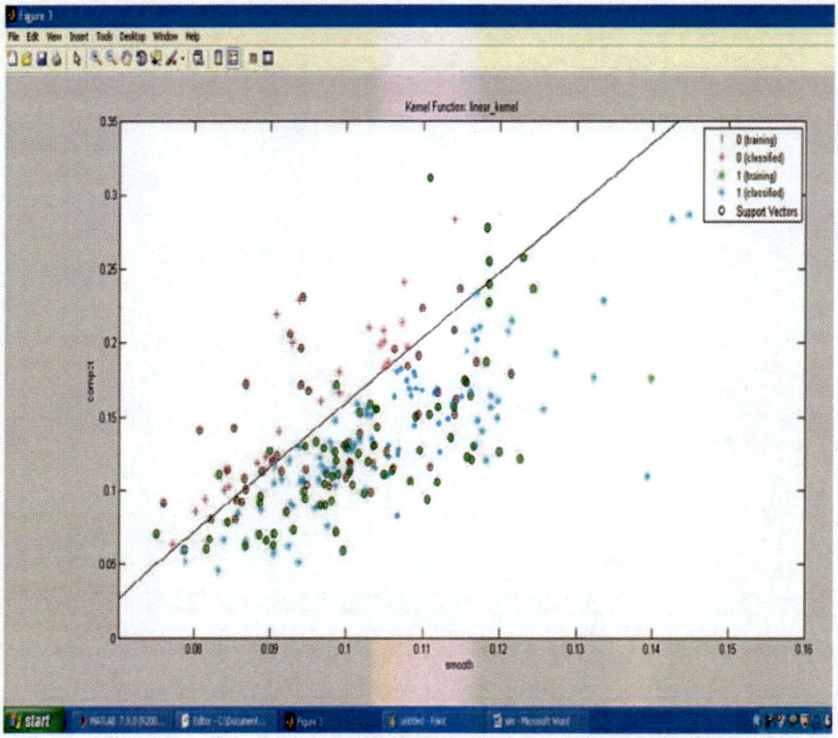

Fig. 5 Smooth versus compact classification performance (CP) = 59.50%

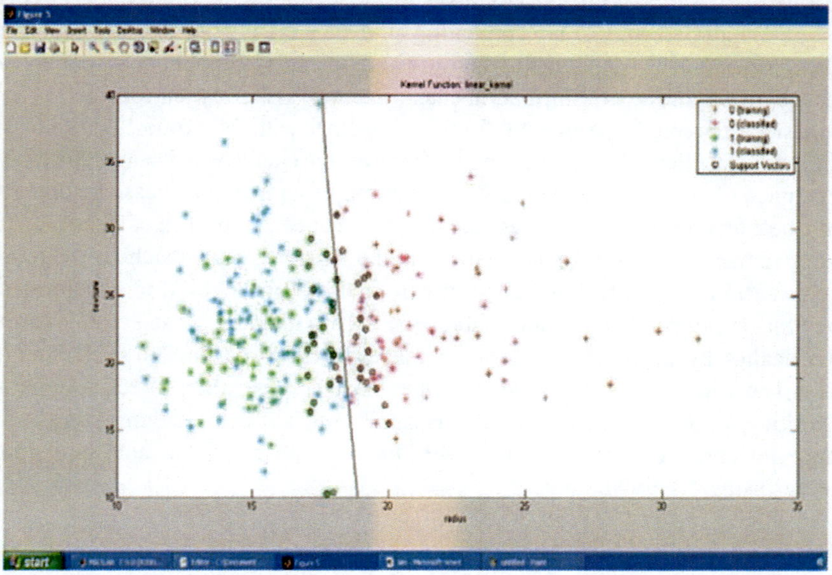

Fig. 6 Texture versus radius CP = 96.69%

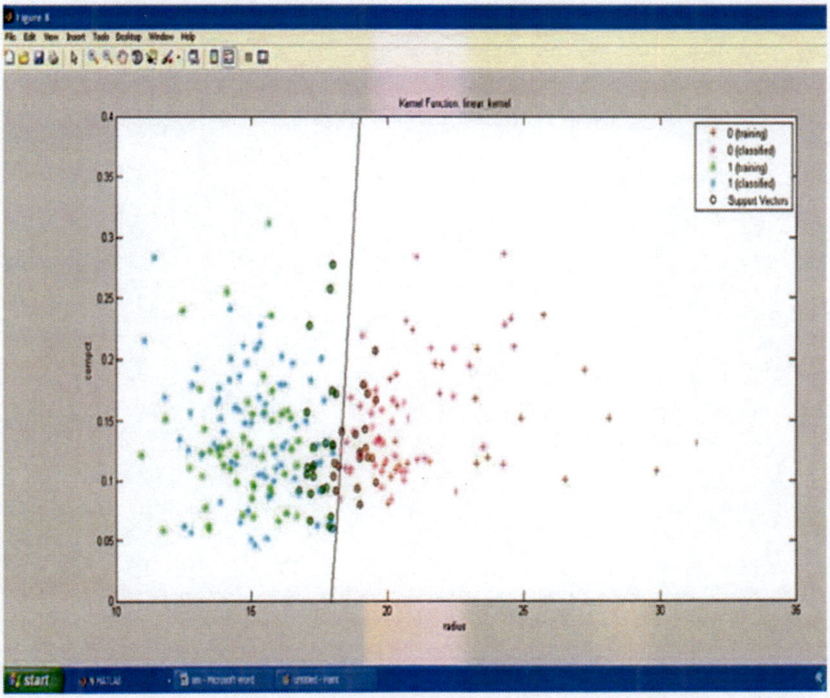

Fig. 7 Radius versus compact CP = 99.1%

References

1. D. Bazazeh, R. Shubair, Comparative study of machine learning algorithms for breast cancer detection and diagnosis, in *2016 5th International Conference on Electronic Devices, Systems and Applications (ICEDSA)*, Ras Al Khaimah, 1–4 (2016)
2. A.T. Azar, S.A. El-Said, Probabilistic neural network for breast cancer classification. Neural Comput. Appl. **23**, 1737–1751 (2013)
3. F. Ahmad, N.A.M. Isa, Z. Hussain, S.N. Sulaiman, A genetic algorithm- based multi-objective optimization of an artificial neural network classifier for breast cancer diagnosis. Neural Comput. Appl. **23**(5), 1427–1435 (2013)
4. M.K. Hasan, M.M. Islam, M.M.A. Hashem, Mathematical model development to detect breast cancer using multigene genetic programming, in *2016 5th International Conference on Informatics, Electronics and Vision (ICIEV)*, Dhaka (2016), pp. 574–579
5. G. James, D. Witten, T. Hastie, R. Tibshirani, *An Introduction to Statistical Learning*, 1st edn (2013)
6. A.T. Azar, S.A. El-Said, Performance analysis of support vector machines classifiers in breast cancer mammography recognition. Neural Comput. Appl. **24**(5), 1163–1177 (2014)
7. H. AttyaLafta, N. KdhimAyoob, A.A. Hussein, Breast cancer diagnosis using genetic algorithm for training feed forward back propagation, in *2017 Annual Conference on New Trends in Information & Communications Technology Applications (NTICT)*, Baghdad (2017), pp. 144–149

Retrieving TOR Browser Digital Artifacts for Forensic Evidence

Valli Kumari Vatsavayi and Kalidindi Sandeep Varma

Abstract The TOR browser is the most popular browser for surfing the Internet while being anonymous. This paper studies the digital artifacts left behind by TOR browser over the network and within the host. These artifacts give the most crucial forensic evidence for digital investigators to prove any unauthorized or unlawful activities. The paper also presents methods for retrieving more useful artifacts when compared to previous works and also investigates on Firefox, Chrome Incognito, and Internet Explorer. The results show that even the much-acclaimed TOR browser also leaves evidence and traces behind.

Keywords Private browsing · Memory forensics · TOR · Internet explorer · Privacy

1 Introduction

The most popular application used for surfing the World Wide Web is the browser. Like with any other software application, the Web browser has vulnerabilities. A few well-known threats on regularly used browsers like Chrome, Firefox, and Internet Explorer are session hijacking, buffer overflow, etc. The data generated by the Web browser application has the potential for financial gains or targeting a specific user. Hence, volumes of data are collected every day. Incidents of the data breach are also increasing year by year. To avoid the data breach, many users and organizations are now concerned about preserving privacy and anonymity. On the other hand, anonymity and privacy attract malignant users to perform malicious activities.

V. K. Vatsavayi (✉)
Department of CSSE, Andhra University, Visakhapatnam, India
e-mail: vallikumari@gmail.com

K. S. Varma
Department of CSE, GIT, GITAM (Deemed to Be University), Visakhapatnam, India
e-mail: skalidin@gitam.edu

D. Bhattacharyya and N. Thirupathi Rao (eds.), *Machine Intelligence and Soft Computing*, Advances in Intelligent Systems and Computing 1280,
https://doi.org/10.1007/978-981-15-9516-5_23

A popular browser used for user anonymity and privacy is the TOR browser, which uses the onion protocol to conceal user's identities. This feature attracts attackers to perform malicious activities.

This paper presents methods to collect digital artifacts and traces of evidence in the usage of TOR browser for digital forensics. The paper is organized as follows. Section 2 discusses the background and related work. Section 3 introduces tools and explains the methodology to collect digital artifacts. It also discusses the experimental setup, evidence collection process, and memory dump analysis. Section 4 discusses the results. Section 5 concludes the paper.

2 Background and Related Work

Ever since, Snowden has revealed information about the global surveillance programs and warned the world about individual privacy, since then, the necessity for anonymity has grown, and people became aware of it. TOR (The Onion Router) [1] browser became popular among the net surfers who wanted to remain anonymous. TOR browsers have a peculiar feature that prevents third-party trackers and advertisers from following a user's Web site visit. Other advantages of TOR are that cookies are automatically deleted, and browsing history cannot be traced.

The silk road market in the dark Web started using TOR to buy and sell illegal drugs online. This helped them remain anonymous. But later on, it was shutdown.

Anonymity is different from confidentiality. Anonymity is related to individual user privacy. The data about the individual can be made public, but the relation between the data and the individual must not be known. That is, the person remains anonymous, but the person's data can be utilized for different purposes.

With the increased usage of Internet-based communication of data, users started realizing the privacy and anonymity of their data. Hence, several methods for anonymous mail sending, anonymous storage, and retrieval of data are proposed.

One of the first known anonymous emailing was done through anonymous remailer [2], which used a mapping service wherein a user-id is replaced with anonymous id. However, it was shut down later due to legal problems. This was followed by cryptography-based techniques like Cyberpunk remailer, mixminion remailer, torguard, anonymousemail.me, guerilla mail, and many more without registration too.

Many works on anonymous storage services [3] were proposed for storing the data anonymously. Freenet was proposed to be a distributed anonymous information storage and retrieval system, an adaptive peer-to-peer network-based system. The model discussed claims to protect the origin and destination of the file. Eternity [4] is a service that helps store information anonymously and permanently. Freehaven [5] is another interesting method to publish documents anonymously. It is a distributed anonymous storage system. Publius [6] is another protocol designed to publish systems anonymously.

Many works have published different techniques for extracting security-sensitive information in Windows operating system [7].

Private browsing mode was introduced in many popular browsers. The main objectives of private browsing mode in the browsers are that:

i. No traces must be left after a user browses in private mode. This should prevent the local attacker or Web attacker from learning the Web user usage behavior.
ii. Identity of the Web user must be hidden from the Web sites that they visit.

However, it is seen in [8] that this is not the case, and many traces were found even in private browsing mode.

Extensions and plugins [9] when used in private browsing mode lead to a lot of problems. There is no security verification mechanism to prove that there is violation of privacy in private mode when the browsers use extensions and plugins.

It can be seen from [10] that for forensics, evidence collection becomes difficult in private browsing mode. From artifacts, it is not just the session data but the relation to a specific user's identity can also to be found.

Internet Explorer, Firefox, Chrome, Safari suffer from vulnerabilities in private browsing mode too [11].

The TOR browser in recent times has become very popular with common users.

While many papers were published about TOR's features and functionalities, very few works have focused on the credibility of TOR anonymity.

In this paper, we investigate different ways of revealing the traces and artifacts left in the system after using the TOR.

3 Finding the Artifacts After Browsing with TOR

This section explains the how the artifacts were retrieved after browsing with TOR.

3.1 Experimental Setup

Virtual machine was used for the experimentation. VMware 10 workstations (VM) was installed on all four suspect machines. These virtual machines were running on Windows 10. The fundamental purpose behind utilizing this virtual machine is to have similar conditions for all programs utilized in this experiment.

To make data extracting easy, we uninstalled all currently installed Web browsers from the suspect machines, cleared all cookies, cache, history, bookmarks, etc. On each suspect machine VM, one browser was installed. The Web browsers was installed were TOR, Firefox, Microsoft Internet Explorer, and Google Chrome Incognito (the term used by Google for the private browsing mode).

3.2 Collection of Data

Three tools have been used in this experimentation, and they are Autopsy [16], Dumpit [15], and Bulk extractor viewer [14]. Five laptops were used running Windows 10. Four laptops were used as suspect machines, and the fifth one was used as a forensics workstation. These virtual machines were also running on Windows 10 to maintain a similar environment for all browsers used in the experiment. To make the data extraction easy, except for the specific browser used in that virtual machine, the other remaining browsers have been uninstalled after clearing all browsing history, cache, bookmarks, cookies, and downloads, etc.

TOR claims to be the most secure and private browser mode. As this work primarily focuses on privacy in TOR, the virtual machine with the TOR browser installed is taken. A few browsing activities are done like searching for information from Google, log in into email, searching images, videos, etc,. are done on the virtual machine installed with the TOR browser.

The autopsy tool is used on the private browsing mode (PBM) of Google Chrome, IE, and Firefox. Then autopsy tool was installed on forensics workstation machine and analyzed the browsing sessions in PBM of Google Chrome, IE, and Firefox. The autopsy results of Web bookmarks, Web history, Web downloads, emails, and Web cache in the PBM of chrome IE have been captured. Finally, the reports are generated for chrome and IE, where private mode search results are found by using this tool.

Dumpit tool was installed in the suspect machine, and memory dump files are taken. And different memory dumps can be taken in the suspect machines, and those memory dumps can be analyzed in the forensic workstation. The bulk extractor tool has been installed on the forensic workstation machine to analyze the memory dump files.

To carry out the forensic investigation, the TOR browser was installed first. Then, browsing was done by opening Gmail account. Next, search was done for certain images and videos in Google search engine. Once the search is finished, the browser is closed and is uninstalled.

Later memory dump files were collected and analyzed.

3.3 Analyzing the Memory Dumps

The steps for analyzing the memory dumps using bulk extractor viewer is as follows,

a) Start the Bulk Extractor from the start menu of the Windows system from the following path: Menu → Bulk Extractor (Version No) → Bulk Extractor Viewer (32/64 bit).
b) Next, select the tools options from the menu bar and click on the "Run bulk_extractor" from the dropdown menu or press "Ctrl + R." as shown in Fig. 1.

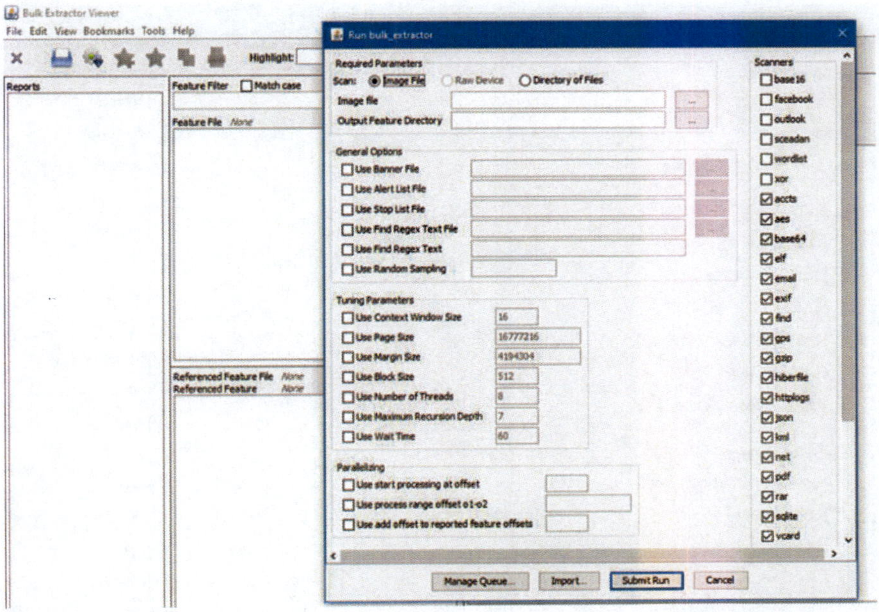

Fig. 1 Bulk_extractor window is opened

c) Select scan in the Bulk_extractor window and click on the image file radio button as shown in Fig. 2.
d) Now, upload the image file and select the directory where the output file has to be stored. Then, select the different types of scanners to find the respective information from the image as in Fig. 3.
e) The bulk extractor viewer scans the uploaded image file, and the output file will be created and stored in the destination folder.
f) After the scanning is completed, the bulk extractor viewer generates the reports based on the selected scanners. See Fig. 4. These reports are analyzed to find the artifacts that are stored in the memory dump.
g The investigator can see histograms of features, referenced feature files, and specific features in context. Figure 5 shows the reports of the image files. The highlighted area shows the email information of the suspect. Similarly, the investigator can search the AES keys, email histogram, URLs, URL services, URL searches, IP addresses, Ethernet addresses, etc.

4 Results

The reports of the bulk extractor tool captured from a virtual machine containing TOR browser contains Web search content like email data, Web data, Web downloads

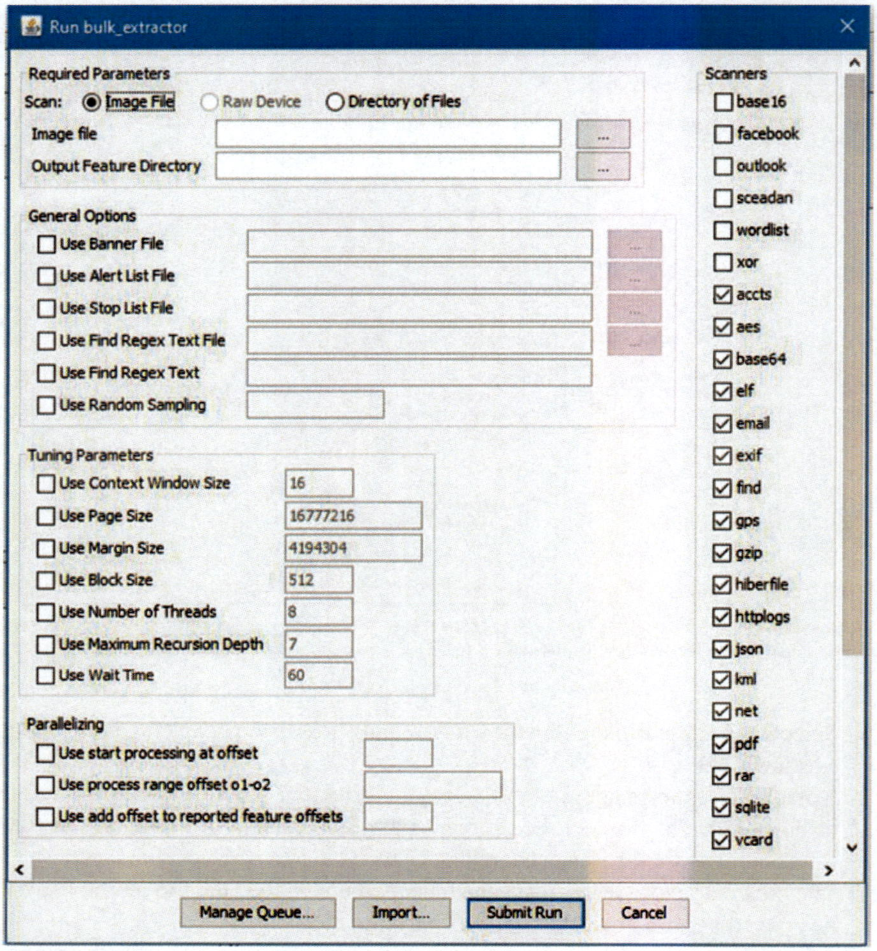

Fig. 2 Bulk extractor image file selection

even after uninstalling the TOR browser. As can be seen in Fig. 6, the traces of URLs visited were still found after the uninstallation of TOR.

Similarly, the evidence when further examined revealed email data (Fig. 7).

Finally, it was also seen TOR browser left traces about itself even after uninstallation. See Fig. 8.

5 Conclusions

This research proposed a new approach for detecting artifacts left by TOR Browser. The approach consists of memory forensics methods on TOR. The memory capture

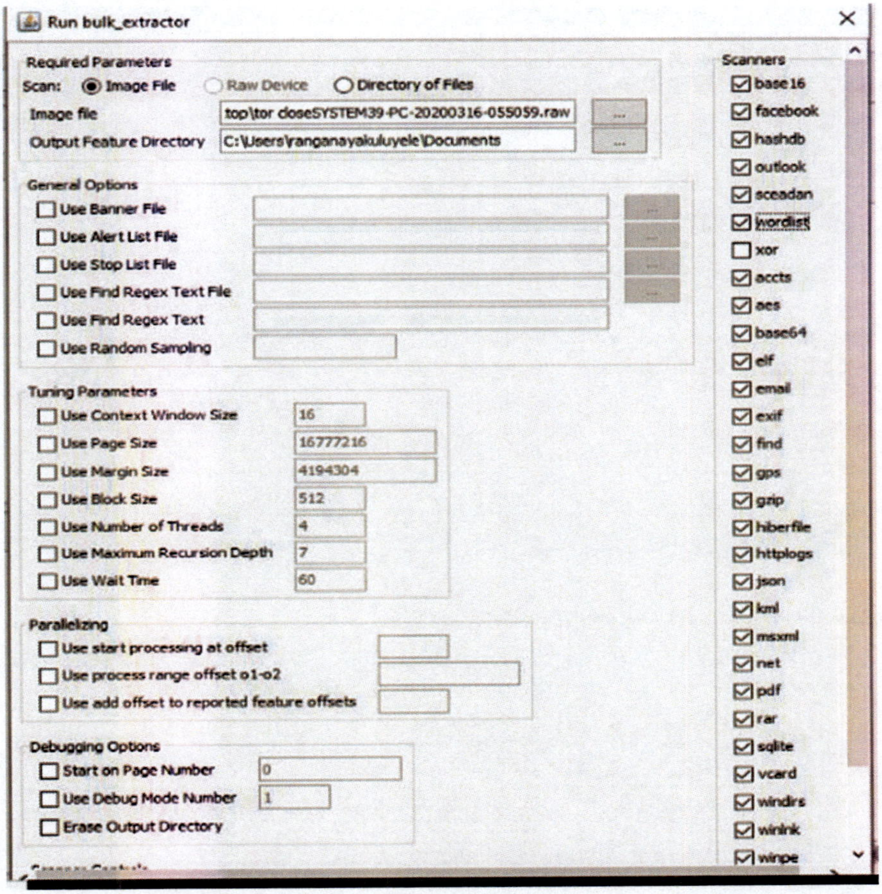

Fig. 3 Image file upload and select an output directory

was performed, and the TOR traces were found. It was found that through memory forensics, it is possible to retrieve forensically valuable information about suspect's activity, such as sites visited, Internet searches, and traces of email communication even after the browser were closed. These artifacts are enough to constitute a link between the data and the suspect. The experimental result analysis shows that the TOR's claim of anonymity is invalidated through memory forensics. The set of steps are simple and take minimum effort for performing evidence analysis when compared to other works published in literature.

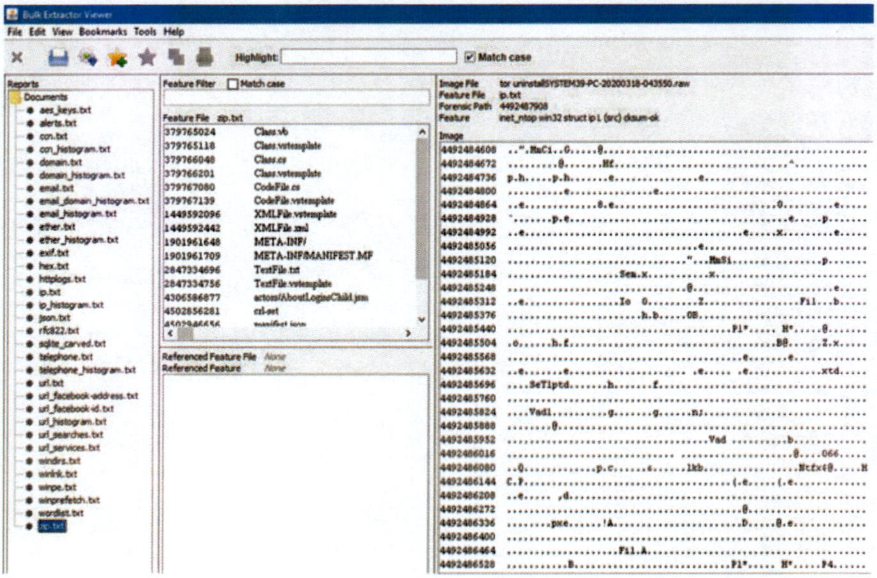

Fig. 4 Reports generated by the bulk extractor viewer

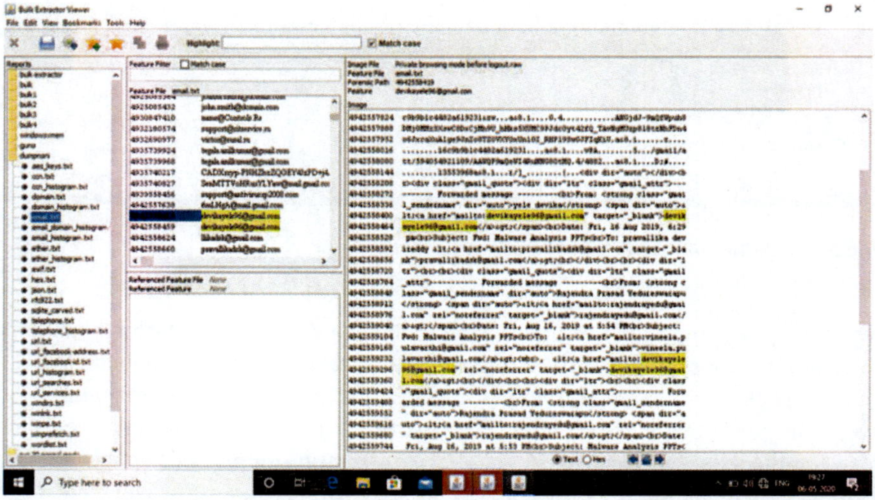

Fig. 5 Highlighted areas showing the email addresses of the suspect

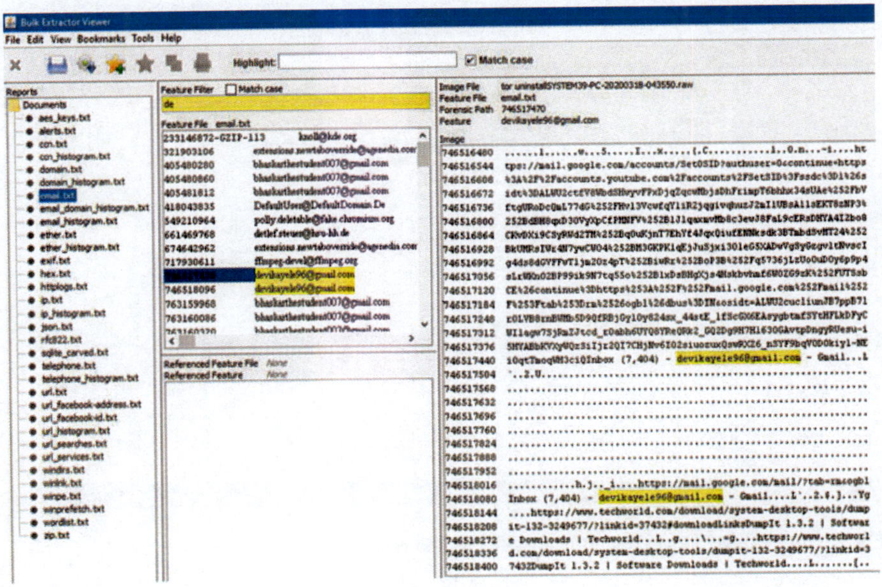

Fig. 6 Finding email data

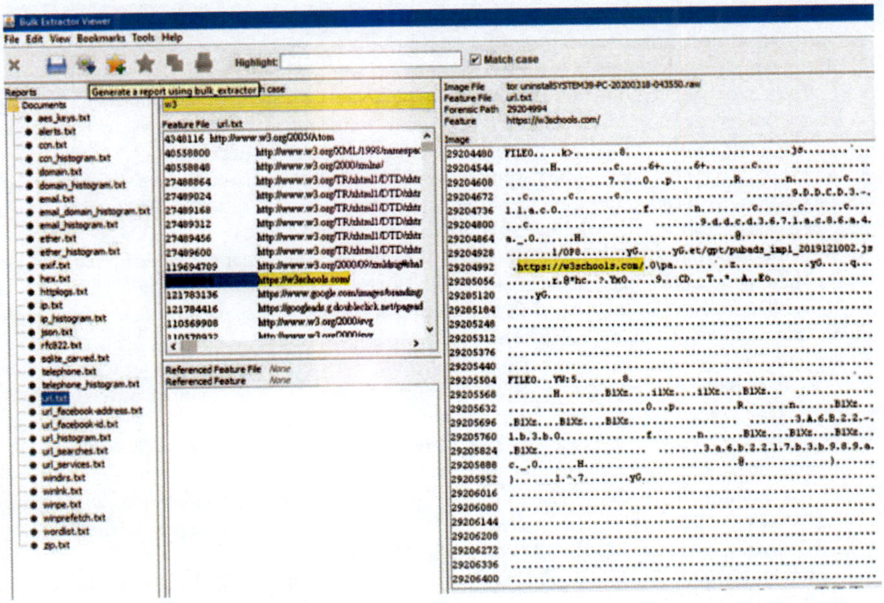

Fig. 7 Finding Web (URL) content

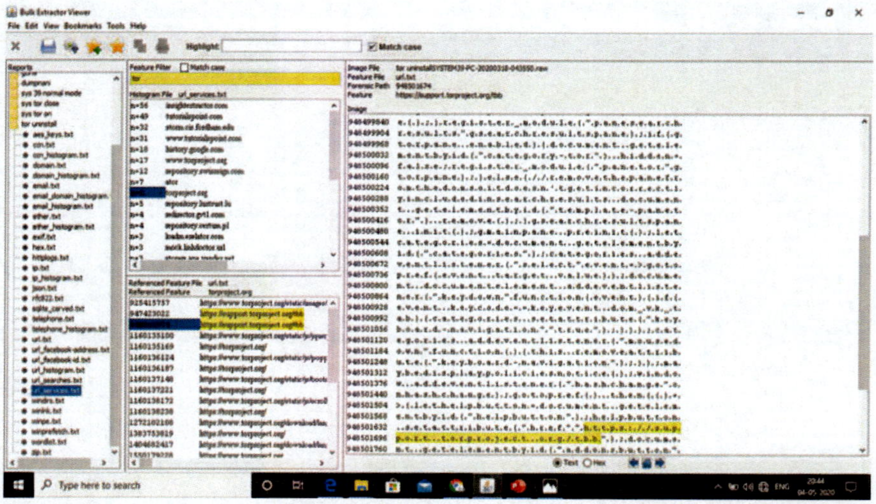

Fig. 8 Highlighted areas show TOR installed in the suspect machine

References

1. The TOR Project. https://www.torproject.org/
2. Anonymous remailer. https://en.wikipedia.org/wiki/Anonymous_remailer
3. Clarke I, Sandberg O, Wiley B, Hong TW (2001) Freenet: a distributed anonymous information storage and retrieval system. In: Federrath H (eds) Designing privacy enhancing technologies. Lecture Notes in Computer Science, vol 2009. Springer, Berlin, Heidelberg
4. Anderson R (1996) The Eternity Service. In: First international conference on theory and applications of cryptography, Prague. https://www.cl.cam.ac.uk/~rja14/Papers/eternity.pdf
5. Freehaven. https://www.freehaven.net. Last seen on 15 June (2020)
6. Waldman M, Rubin DA, Cranor LF (2000) Publius: a robust, tamper-evident, censorship-resistant, web publishing system. In: 9th USENIX security symposium
7. Hejazi SM, Talhi C, Debbabi M (2009) Extraction of Forensically sensitive data from windows physical memory. Comput Investig 6:121–131
8. Aggarwal G, Bursztien E, Jackson C, Boneh D (2010) An analysis of private browsing modes in modern browsers. In: Conference: 19th USENIX security symposium, Washington, DC, USA, 11–13 Aug (2010)
9. Mahendrakar A, Irving J, Patel S (2010) Measurable analysis of private browsing mode in popular browsers. http://mocktest.net/paper.pdf
10. Ohana DJ furthermore, Shashidhar N (2013) Do private and versatile internet browsers leave implicating evidence? A measurable examination of leftover relics from private and compact web browsing meetings. EURASIP J on Inf S 201, 6:1–13
11. Satvat K, Forshaw M, Hao F (2014) What's more, Toreini E., On the privacy of private browsing—a forensic methodology. Diary Inf Secur Appl 19:88–100
12. 20 sites to send anonymous emails. https://www.hongkiat.com/blog/anonymous-email-providers/. Last seen on 15 June (2020)
13. Unpredictability Foundation: accessible online at: http://www.volatilityfoundation.org/
14. Bulk Extractor, accessible online at: https://github.com/simsong/bulk_extractor/wiki
15. Dump it, accessible online at: https://github.com/thimbleweed/All-In-USB/tree/master/utilities/DumpIt
16. Autopsy, accessible online at: http://www.autopsy.com

Post-COVID-19 Emerging Challenges and Predictions on People, Process, and Product by Metaheuristic Deep Learning Algorithm

Vithya Ganesan, Pothuraju Rajarajeswari, V. Govindaraj, Kolla Bhanu Prakash, and J. Naren

Abstract COVID-19 has been posing unprecedented challenges to people, process, and product. Deadly COVID-19 is randomly depleting human emotions and leads stark to low mental health in daily routines, financial traits, jobs, and business. A wide zoom process is required to ensure a proper ecosystem for the pandemic disease. COVID-19 researches compare the functional and nonfunctional sectors to support quality assured and accurate products with supportive technologies to avoid further losses. The current research work proposes a deep learning mapping model for finding functional sector with different age group of people (p), and it reflects in the development of process (p) and product (p). Metaheuristic deep learning algorithm (MHDL) develops a model between functional and nonfunctional sectors by comparing the usage of information and communication technology to support process and product. MHDL model proves information communication technology (ICT) redeems communication between sectors and leads to less economic losses.

Keywords Post-COVID-19 · COVID-19 human emotion · Metaheuristic deep learning for COVID-19 · COVID-19 in process and product sector

V. Ganesan · P. Rajarajeswari · K. B. Prakash
Department of CSE, Koneru Lakshmaiah Education Foundation (KLEF), Vaddeswaram, AP, India
e-mail: vithyamtech@gmail.com

P. Rajarajeswari
e-mail: rajilikitha@kluniversity.in

K. B. Prakash
e-mail: drkbp1981@gmail.com

V. Govindaraj
IT, St. Joseph's Institute of Technology, OMR, Chennai 600119, India
e-mail: govicse15@gmail.com

J. Naren (✉)
Research Scholar, Department of CSE School of Computing, SASTRA University, Thanjavur, Chennai, India
e-mail: naren.jeeva3@gmail.com

D. Bhattacharyya and N. Thirupathi Rao (eds.), *Machine Intelligence and Soft Computing*, Advances in Intelligent Systems and Computing 1280,
https://doi.org/10.1007/978-981-15-9516-5_24

1 Introduction

COVID-19 is creating a big challenge to people, process, and products. To find a solution for pandemic COVID-19 is by screening, tracking, and forecasting to control the outbreak. In this herculean scenario, statistical machine learning, prediction, forecasting, inferences, and deep learning models are crucial to derive a solution for improving the ecosystem by technologies such as AI, ICT devices, and digital infrastructure for improving process and products. Among all, communication web intelligence tools such as WebEx, Zoom, Skype, Bitly, Blue Button networks take major role in sectors like IT and education. But some sectors like environment and agricultural, manufacturing, and production are most affected due to non-automation/improper communication technologies. The paper also proposes the tradeoff between the functioning of sectors with ICT and nonfunctioning of sectors due to improper usages of technologies.

2 Problem Formulation

An engineering approach is required for analyzing COVID-19 data from different journals, magazine, and health center data [1–3]. There are some predictions, lesson, and precautions for people/population, process, and product sectors for upskilling the word economy [4, 5]. It is by

- Rehabilitation and improving mental health in people
- Maintaining social distancing (contactless interfaces and interactions)
- Strengthened digital infrastructure by ICT devices [6]
- Better monitoring, tracking by IoT and Big Data [7]
- AI-enabled drug development, telemedicine [8]
- More secured online shopping [9, 10]
- Increased reliance on robots inpatient care [11, 12]
- More digital events for cash transaction and communication [13, 14]
- Rise in COVID-19 e-sports games [15].

A framework has been proposed for a MHDL system that uses Big Data and deep learning to give accurate resources for the functional and nonfunctional sectors [16, 17]. The approach involves the following steps:

(1) Creating the usage of ICT by functional sector during COVID-19. (2) Constructing a Framework that involves nonfunctional sectors. (3) Implementing a model for pending work. The system will alleviate all the disadvantages and makes use of the latest available technology to recommend the most accurate information [18].

Knowledge of emerging technologies leads to a solution for COVID-19 to protect the people, improve product and process by the following

1. Interfacing of process and product sector
2. Metaheuristic deep learning methodology (MHDL).

3 Problem Solution

3.1 Process and Product Support for COVID-19

A mapping formula and model is required for people, process, and product for identifying nonfunctioning sectors during COVID-19. A model is developed to identify their interdependency, individual growth, gap, and sustainability. In the current COVID scenario, process is split into two parts

1. Functional sector by using information and communication technology during COVID-19.
2. Nonfunctional sector due to improper inclusion of advanced technologies during COVID-19.

Table 1 shows the quantitative analysis of functional sectors by using web intelligence/ICT is.

Nonfunctional sectors during COVID-19 are

1. Manufacturing industries
2. Automobile and other production industries
3. Civil, aviation, and maritime engineering
4. Warehousing and lean management.

The materialized challenges of current and post COVID-19 on process and product of nonfunctional sector of COVID-19 are:

1. Innovation of product, process to sustain in any disaster like COVID-19
2. Up scaling the existing technologies to bridge the robustness for COVID-19 scenario
3. Focus on more healthcare devices to live with any worst-case decease to avoid complete lock down.

Table 1 Sector and its ICT usages

Sectors	Web intelligence	Video conferencing tool (Zoom, Web Ex, Skype)
Information technology	Yes	Yes
Education	Yes	Yes
E-commerce	Yes	Yes
Health sectors	Yes	Yes

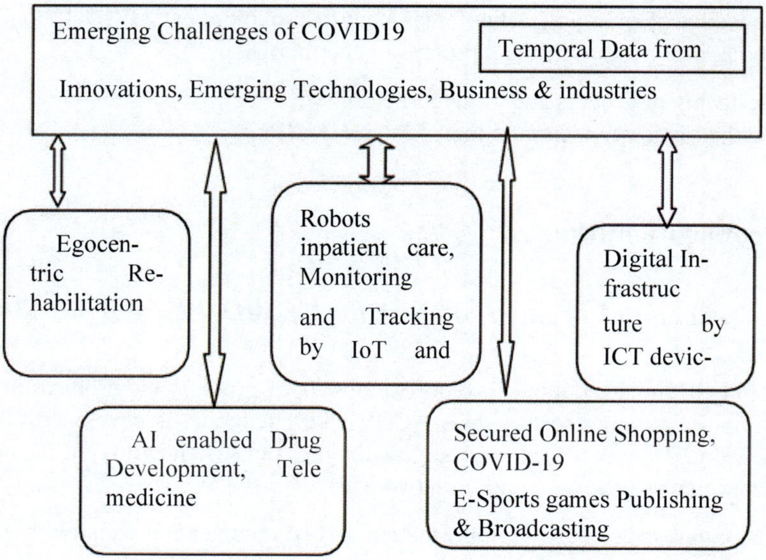

Fig. 1 Sector transformations to support COVID-19

Figure 1 shows the emerging sectors and the need of transformations to support COVID-19.

3.1.1 Innovation of Product, Process to Sustain in Any Disaster

The people, process, and products (PPP) are interlinked by major sectors such as automation, innovation, acquisition, and quality and its dependencies are identified and shown in Fig. 2.

3.1.2 Up Scaling the Existing Technologies to Bridge the Robustness for COVID-19 Scenario

The nonfunctioning existing technologies related with pandemic COVID-19 are health, environment, and agricultural engineering. The sectors under each technology are as follows

The sectors in health engineering (HE):

- Drinking water engineering
- Sanitization
- ICT for health product
- Public healthcare engineering.

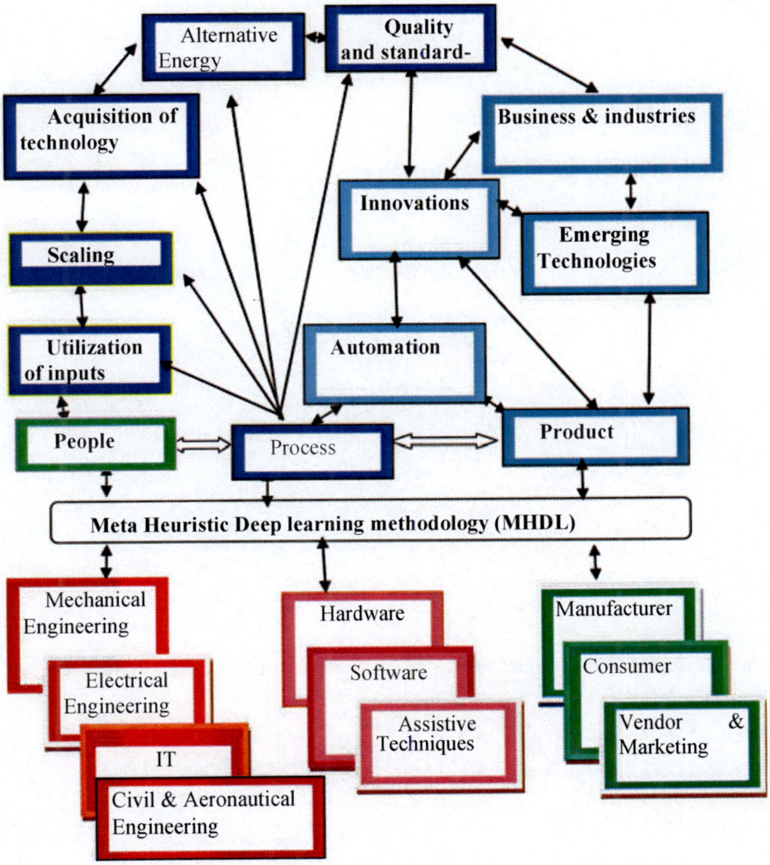

Fig. 2 Interlinking of sectors on current and post-COVID-19

E-HRD (high resource development) The sectors in Environment Engineering (EE):

Pollution control
Risk engineering
Climatic change estimation
Green technology
Civic engineering.

The sectors in agricultural engineering (AE):

- Plant, harvest engineering
- Irrigation and soil engineering
- Food technology and products.

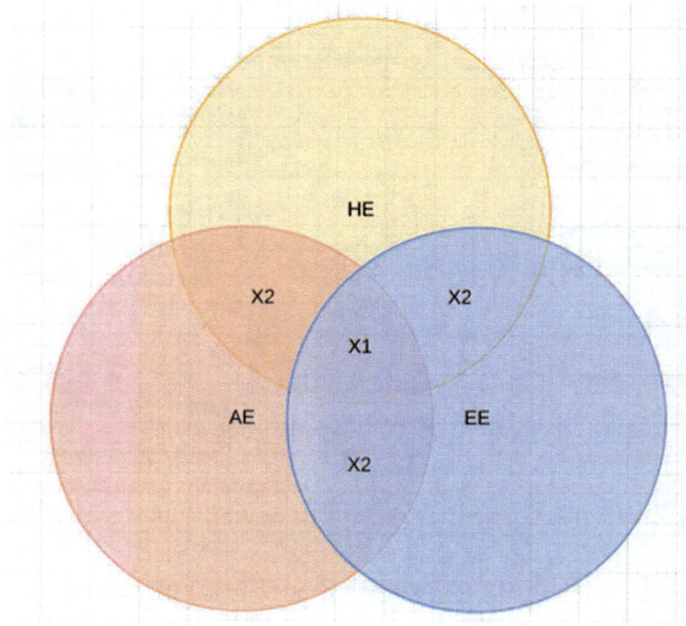

Fig. 3 Interdependencies between COVID-19 sectors and its domain $X1 = (\text{HE}, X2, \text{EE}) \cap (\text{EE}, X2, \text{AE}) \cap (\text{AE}, X2, \text{HE})$

The interdependencies between sectors are identified by the following domain

- Emerging Technologies ($X1$): Hardware, software, Assistive technique
- Business & Industries ($X2$): Manufacturer, Vendor, Marketing, and Consumer

Figure 3 shows the interdependencies between $X1$ and $X2$ by the sectors health engineering (HE), environment engineering (EE), and agricultural engineering (AE)

$X1$ is the intersection of all sectors and its domain. Sector transformations to COVID-19 are a digital infrastructure done by cognizant ICT devices, AI-enabled drug and telemedicine, secured online shopping and egocentric emotional analysis, COVID-19 e-sports games, online publishing and rehabilitation support for COVID-19.

Feasibility of bridging the sectors and the domains, which are analyzed by the following factors

- Process design and knowledge enhancement
- Profitability analysis
- Sensitivity analysis.

3.1.3 Focus on More Healthcare Devices to Live with Any Worst-Case Decease to Avoid Complete Lock Down

The healthcare industry consists of hospitals, medical devices, clinical trials, outsourcing, telemedicine, medical tourism, health insurance, and medical equipment to support any pandemic situations.

Metaheuristic deep learning algorithm invokes stochastic process to find the different optimum solution. It is characterized by encoding the hypothesis, fixing the fitness ranges, selecting the maximum or minimum optimization of data by its training model and validations.

3.2 Metaheuristic Deep Learning Methodology (MHDL)

MHDL is interfaced with Training model and validation to deploy a model for COVID-19. Figure 4 explains the model of MHDL algorithm and explanation.

To generate a population (p) for the chromosome people (pe), process (pr), and product (pt) and its length L is initialized. MHDL algorithm follows as

1. Initialize the hypothesis values of people (pe), process (pr), and product (pt) to truss toward COVID-19 for finding fitness function

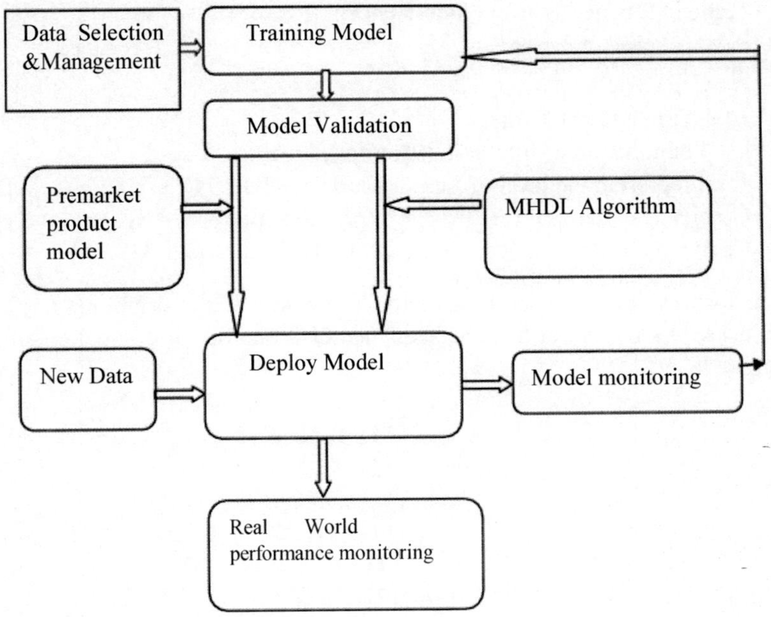

Fig. 4 Model of MHDL algorithm

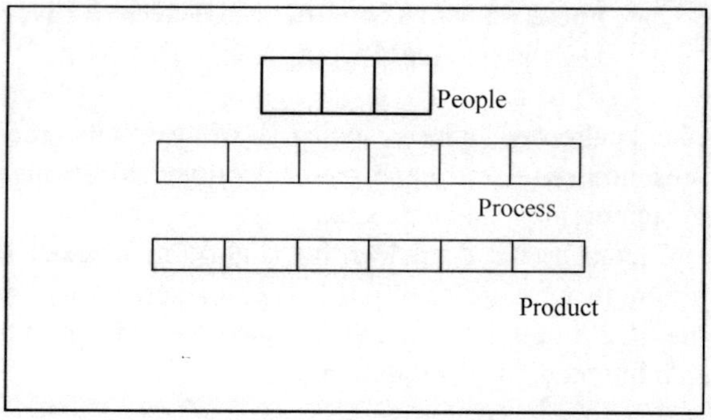

Fig. 5 COVID-19 population

People: Group of people with different ages process (pr): All sectors and its e-commerce by utilization (u), acquisition (a), QoS (q), alternate energy (AE), and scaling (S) toward COVID-19 as (pr (u, a, q, AE, S) ∈ L).
Product (pt): Sectors and sub-sectors with its domains as automation (A), innovation (I), emerging technologies (ET), and business industries (BI) to support COVID-19 as (pt (A, I, ET, BI) ∈ L)
Figure 5 explains the COVID-19 population by its hypothesis initializations. COVID-19 fitness function (fρ) is calculated by crossover for people, process, product and mentioned by ρ, C-chromosome, n-optimized fitness function.

While (max, fitness (ρ) <= fitness threshold
{Fitness function (fρn) = fitness (ρi)/fitness (sρ)}

where

ρ = (C(pe), C(pr), C(pt))
sρ = summation of fitness function for ρ
1…n represent the level of truss toward COVID-19 C(pe) = (Pe(n, s, Dp)
C(pr) = (pr1, u1…n) (pr2, a1…n), (pr3, q1…n), (pr4, AE1…n) (pr S1…n)
C(pt) = (pt1, A1…n) (pt2, I1…n), (pt3, ET1…n), (pt4, BI1…n)

2. Threshold value (wth) and time factor (t) are added as a weight matrix (hidden layer) to the input and find the sequence of n number of optimal solutions as output layer.

$$W\text{th} = \begin{matrix} W11 & W12 & W13 \\ W21 & W22 & W23 \\ W31 & W32 & W33 \end{matrix}$$

$$t_n = \begin{matrix} t11 & t12 & t13 \\ t21 & t22 & t23 \\ t31 & t32 & t33 \end{matrix}$$

Optimal function (Of) = fρn + wth + t_n.

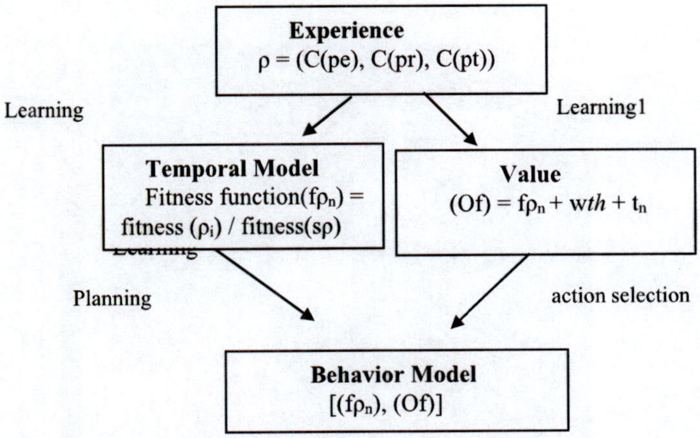

Fig. 6 Behavioral model for post-COVID-19

Table 2 A sample data set for functional sectors: information technology

Sample data	Production engineering
Age group	20–80
Balance of work	10,000 units
Duration	30 days
Campaign	0–60
Previous work	20%
Work on pandemic days	More than 75%

3. As a part of MHDL, a behavior model to trust toward COVID-19 is generated to find the number of true positive and false negative conditions. Figure 6 shows the behavioral model for process, product invoked in post COVID-19.
4. The goal of metaheuristic deep learning is for identifying functioning and nonfunctioning sectors during COVID-19.

Metaheuristic method is for identifying functional sectors, and its process by balance of work, duration, campaign, previous work, work on pandemic days are calculated by correlation matrix.

Table 2 shows the nonfunctional sectors during COVID-19.

3.3 Experimental Result

The proposed metaheuristic deep learning methodology (MHDL) algorithm is demonstrated by using the table data 2 data. Figure 7 explains the correlation matrix, and Fig. 8 shows the parameters of functional sectors during COVID-19.

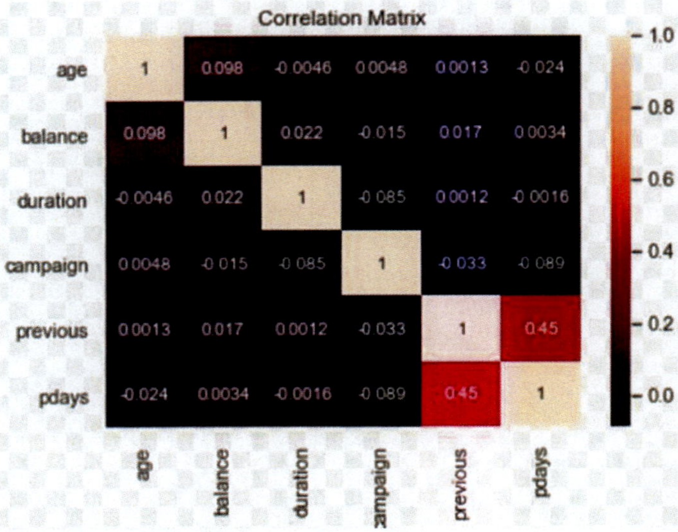

Fig. 7 Functional sector of COVID-19

Figure 9 shows the nonfunctional sector on COVID-19 due to improper usage of technologies.

4 Conclusion

MHDL is acting as an interface between people, process, and product. It identifies the tradeoff between the process and products in the functional and nonfunctional unit of COVID-19 lockdown scenario. The MHDL optimization algorithm is used to achieve the best solution depends on selecting the parameter, objective, threshold, and correct fitness function. In future, it is a very active area of research in the field of computational intelligence, time series mapping, micro- and macro-service analysis to identify the gap, need for upskilling people, process, and product.

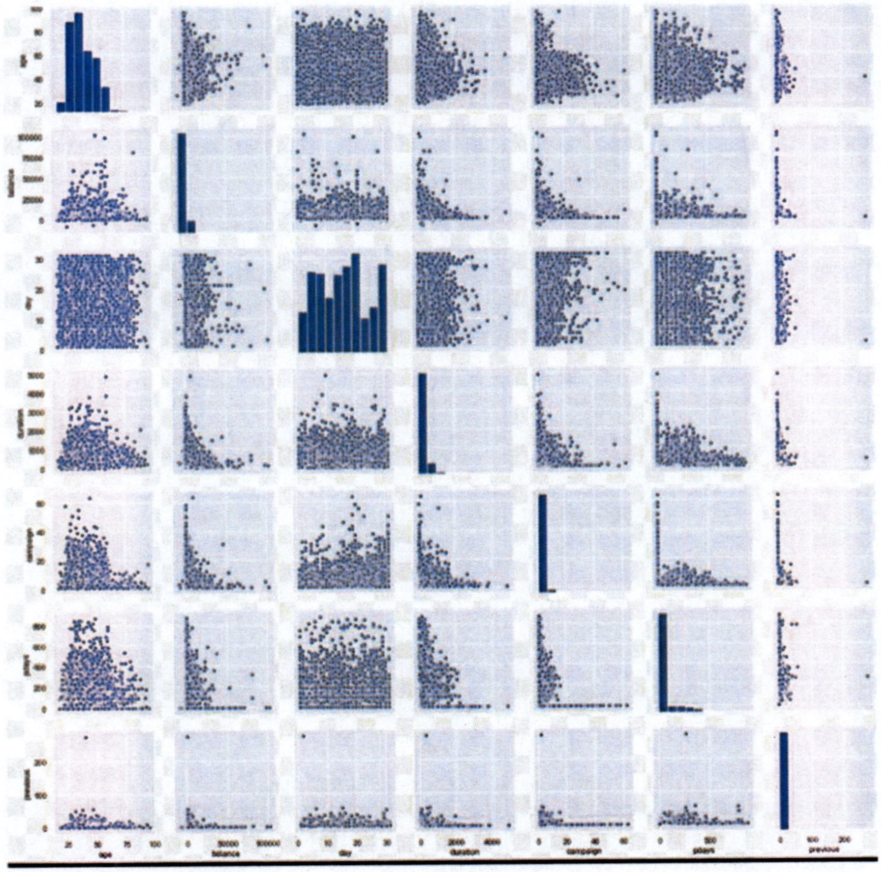

Fig. 8 Scatter plot for functional unit parameters on COVID-19

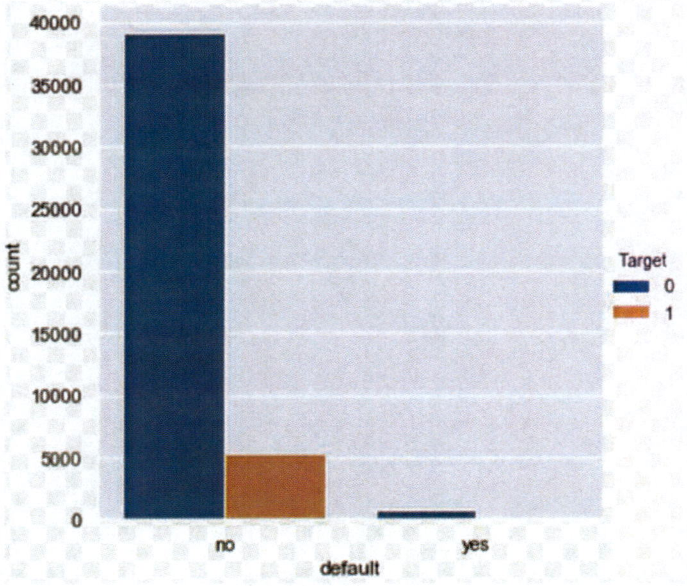

Fig. 9 Nonfunctional sectors of COVID-19

References

1. Attacks BT (2017) Preparedness and Response to. Distribution, 2 (November)
2. M.R. Mehra, S.S. Desai, F. Ruschitzka, A.N. Patel, Articles Hydroxychloroquine or chloroquine with or without a macrolide for treatment of COVID-19: a multinational registry analysis. The Lancet **6736**(20), 1–10 (2020). https://doi.org/10.1016/S0140-6736(20)31180-6
3. Koven S (2020) Engla, Journal—2010—New England Journal. New Engl J Med 1–2. https://doi.org/10.1056/NEJMp2009027
4. Guerrieri V, Lorenzoni G, Straub L, Werning I (2020) Macroeconomic implications of COVID-19: can negative supply shocks cause demand shortages? SSRN Electron J 1–36. https://doi.org/10.2139/ssrn.3570096
5. Wu Z, McGoogan JM (2020) Characteristics of and important lessons from the coronavirus disease 2019 (COVID-19) outbreak in China. Jama 2019. https://doi.org/10.1001/jama.2020.2648
6. Vithya (2017) Inpatient critical stage monitoring in smart hospitals by contextual Fuzzy based QoS routing for WBMS network nurse call system. Wirel Pers Commun: An Int J. https://doi.org/10.1007/s11277-016-3361-2
7. Z. Wang, K. Tang, Combating COVID-19: health equity matters. Nat. Med. **26**(4), 458 (2020). https://doi.org/10.1038/s41591-020-0823-6
8. Ge Q (2020) A Noel intervention recurrent autoencoder for real time forecasting and non-pharmaceutical intervention selection to curb the spread of Covid-19 in the world. Medrxiv. https://doi.org/10.1101/2020.05.05.20091827
9. SAFETY TIPS : AGAINST COVID -19 Disclaimer: (2020). April, 1–19
10. U.S. Department of Labor: Occupational Safety and Health Administration (2020) Guidance on preparing workplaces for COVID-19. Osha, 35
11. R. Kimmig, R.H.M. Verheijen, M. Rudnicki, Robot assisted surgery during the COVID-19 pandemic, especially for gynecological cancer: a statement of the society of european robotic

gynaecological surgery (SERGS). J Gynecol Oncol **31**(3), 1–7 (2020). https://doi.org/10.3802/jgo.2020.31.e59
12. G.Z. Yang, B.J. Nelson, R.R. Murphy, H. Choset, H. Christensen, S.H. Collins, P. Dario, K. Goldberg, K. Ikuta, N. Jacobstein, D. Kragic, R.H. Taylor, M. McNutt, Combating COVID-19-the role of robotics in managing public health and infectious diseases. Sci Robot **5**(40), 1–3 (2020). https://doi.org/10.1126/scirobotics.abb5589
13. Sirina Keesara MD (2020) Covid-19 and health care's digital revolution. New Engl J Med 1–2. https://doi.org/10.1056/NEJMp2009027
14. https://www.ncr.com/coronavirus/banking/how-financial-institutions-are- using- digital-to-keep-customers-connected-in-times-of-crisis
15. Evans AB, Blackwell J, Dolan P, Fahlén J, Hoekman R, Lenneis V, McNarry G, Smith M, Wilcock L (2020) Sport in the face of the COVID-19 pandemic: towards an agenda for research in the sociology of sport. Eur J Sports Soc (0):1–11. https://doi.org/10.1080/16138171.2020.1765100
16. Javaid M, Haleem A, Vaishya R, Bahl S, Suman R, Vaish A (2020) Industry 4.0 technologies and their applications in fighting COVID-19 pandemic. Diabetes Metab Syndr: Clin Res Rev 14(4):419–422. https://doi.org/10.1016/j.dsx.2020.04.032
17. Vithya G (2019) A study on hybrid recommender system with deep learning and deployment in Big Data. TEST Eng Manag 81:1869–1875
18. https://economictimes.indiatimes.com/news/company/corporate-trends/manufacturing-sector-anticipates-severe-workforce-crunch-post-covid-19

An Analytics Overview & LSTM-Based Predictive Modeling of Covid-19: A Hardheaded Look Across India

Ahan Chatterjee and Swagatam Roy

Abstract As it is said history repeats itself, in 2020, the world witnessed another epidemic with the name of novel corona virus (*2019-nCoV*). The dataset has been prepared through the information from John Hopkins University, WHO source, along with the data of Indian Government. In this paper, we will take a deeper view through the epidemiological data on corona virus affected patients. We have assessed various trends and discovered patterns through the data. We have studied the outbreak progression of the *2019-nCoV* in India through exponential growth model. Various patterns are analyzed like gender affected, age group of people being affected the most and percentage people survived based on assessing delays between symptom onset and initialization of treatment by the professionals. Through modeling of maximum likelihood function, we have drawn an approximate conclusion on number of unreported cases of *2019-nCoV* patients (initially as per data). Serial interval taken for other known corona-viruses (nCoV), Severe Acute Respiratory Syndrome (SARS), and Middle East Respiratory Syndrome (MERS) are used as an approximate measure to predict the number of unreported cases. The counts of unreported cases are more vital than its counterpart as it can cause severe outbreak if not checked. Along with that, we have proposed a LSTM-based architecture to predict the number of cases in India.

Keywords LSTM · MLE modeling · Covid-19 · Deep learning · SARS-2

A. Chatterjee (✉) · S. Roy
Department of Computer Science and Engineering, The Neotia University, Diamond Harbour, South 24-Pargana 743368, India
e-mail: chatterjeeahan02@gmail.com

S. Roy
e-mail: swagatamr577@gmail.com

D. Bhattacharyya and N. Thirupathi Rao (eds.), *Machine Intelligence and Soft Computing*, Advances in Intelligent Systems and Computing 1280,
https://doi.org/10.1007/978-981-15-9516-5_25

1 Introduction

The end of 2019 marked a severe outbreak of an acute illness that became one of the most discussed topics. The respiratory illness was caused due to novel corona virus (*2019-nCoV*) it is one of the major outbreak of the century. In India, first case 30th January and later the number of cases increases exponentially. It created a worldwide impact due to its high morbidity and mortality rate. As per latest statistics, number of affected patients stands as 7,774,263 (as on 06/14/2020) of which death toll stands as 428,927 (6%) around 53,964 people are in serious condition, 3,304,424 people are in mild condition (as on 06/14/2020). The outbreak of corona virus in Wuhan, China is rapid and it is being counted as a worldwide pandemic.

The coronaviruses are being enveloped in the non-separated positive RNA virus belongs to the mother family of Coronaviridae and Nidovirales. Initially, it was assumed that *2019-nCoV* is only transmitted from animal-human contact, but later found that human-human transmission is also occurring. Although most of the cases of the corona virus are mild in nature, only 2 of its class became an epidemic namely, Severe Acute Respiratory Syndrome Coronavirus (SARS-CoV) and Middle East Respiratory Syndrome Coronavirus (MERS-CoV). The mortality rate for SARS-CoV and MERS-CoV was around 10 and 37%.

In December 2019, an unknown series of pneumonia cases are being registered in the Wuhan Province, China. Professional medical checkup and deep sequencing analysis it seemed, there is a pneumonia tract in the lower respiratory tract and samples indicated it as novel corona virus [1, 2].

We aim to draw some conclusive statements using statistical operation on the dataset available from Indian Government on outbreak of Covid-19 across the country. Section 2, contains various trend analysis regarding the corona virus status, gender affected probability has been carried out, i.e., count of males/females died per count of being affected, age class analysis also been carried out in this section. Another critical analysis that has been carried out is the survival rate of percentage people based on assessment of delay for symptom onset and treatment days. Section 3 contains approximated number of unreported cases of corona virus. This has been modeled through maximum likelihood function. Section 4 contains estimation of number of outbreak of coronavirus affected individuals in India through time series analysis and autoregressive modeling. Section 5 contains predictive modeling of Covid-19 across India.

2 Trend Analysis for Different Parameters of COVID-19

The most basic analysis for a conclusive decision is analyzing the trend. In this section, we have analyzed various trends to draw a pattern in how the Covid-19 virus is affecting the human race.

2.1 Gender Trend Analysis

We have taken a dataset of John Hopkins University. The count has been taken as how many gave positive test for Covid-19. Table 1 shows infected people, i.e., the gender analysis of the people giving positive test to Covid-19.

We can see, there is lift in the gender section. From Table 1, It is evident that probability of getting affected by the virus is more in case of males and in terms of percentage, it is around 2.5% more than that of females. Table 2 shows the mortality rate for each gender class.

From Tables 1 and 2, it is being evident that Covid-19 virus is attacking the male gender more. It can be justified as dissimilar genetic development and biological lifestyle factor. Scientists suggest that cause behind such statistics can be weaker immune system as in general men do lot more smoking than women.

In a broader picture, female gender class holds more probability to be infected in coming time in such pandemic condition. In this pandemic condition, there are a lot of female health workers, going and giving treatments to those quarantined infected patients can affect a number of females. According to WHO, in the healthcare and social care sector, there are 70% females across 104 countries. Figure 1 shows statistics from WHO.

Initially, the affection and mortality rate for female are low but as number of female worker is huge in the healthcare sectors, it is expected that in near time percentage affected for female section can increase rapidly. But the mortality rate may not increase as it is generally affected by immune system power.

Table 1 Percentage infected gender from Covid-19

Males infected (percentage)	Females infected (percentage)
69%	31%

Table 2 Mortality rate of gender from Covid-19

Mortality rate for male	Mortality rate for females
4.7%	2.8%

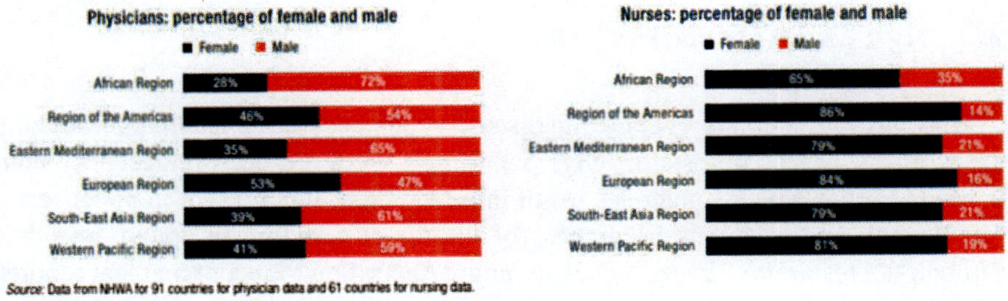

Fig. 1 Data representation of gender classification in health care sector. *Source* WHO Report

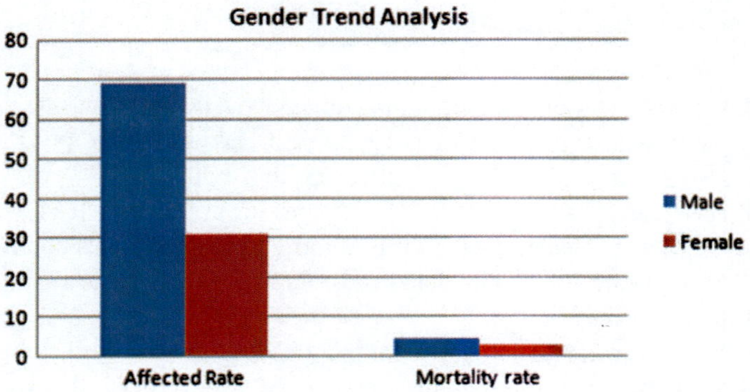

Fig. 2 Gender trend analysis of Covid-19. *Source* Created by Author, based on dataset

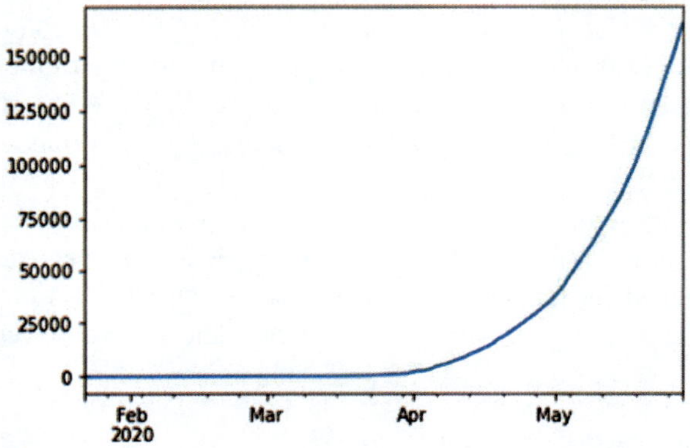

Fig. 3 Infected rate of Covid-19 in India. *Source* Created by Author, based on Dataset

Figure 2 shows that in time period how the affected, mortality, and survived figure increased. This graph is irrespective of the gender bias.

From Fig. 3, it is very much evident that the total cases in India are growing significantly and this is a big reason to worry.

2.2 *Age Class Trend Analysis*

Corona virus infection is an epidemic disease which becomes pandemic throughout the world. It causes fibrosis of lungs; i.e., lung tissues are been replaced by non-working tissues which sometimes result in death. First, it is present in upper respiratory track and it causes the dryness of the threat resulting in cough, and then through the respiratory track, it goes to lungs. According to recent medical reports, the mutation rate of the virus is very high. The current number of cases, i.e., registered is more than 774,260. The morbidity rate as well as fatality rate is increasing day-by-day throughout the world. The fatality rate is calculated as [3, 4]:

$$\mathrm{Fr} = \frac{\text{no of deaths occured}}{\text{Total number of cases registered}}$$

It is increasing rapidly creating the pandemic situation. The exponential rise in death rate has become a big reason to worry (Fig. 4).

The cases are increasing day-by-day, and so, for our analysis purpose, we have taken sample out the population (Fig. 5).

Corona virus affects the respiratory tracks of human body. Severe cough and cold, fever is some of the major symptoms sometimes leading to pneumonia. It mainly

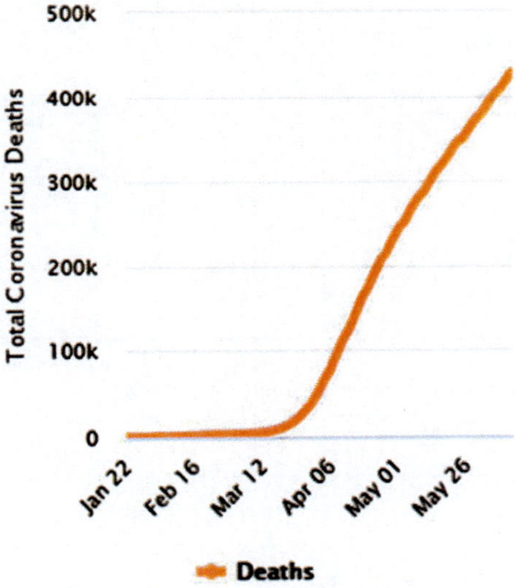

Fig. 4 Represents the increasing fatality rate. *Source* worldometers.info

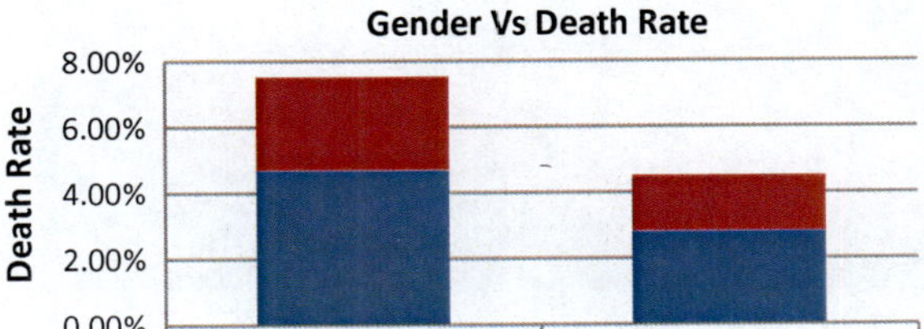

	confirmed cases	all cases
Female	2.80%	1.70%
Male	4.70%	2.80%

Fig. 5 Represents the gender versus mortality rate

follows human-human transmission. So, it is contagious and so spreading as well as transmission is quite easy through interaction and crowd gathering. So, most of the middle-aged are affected by this virus due to high interaction with different people at work place. Some patients affected by the virus are getting recovered also and this depends on many factors one of which is the immunity system. Immunity system contains t cells and immune cells. With the growing age, the number of *T* cells produced becomes less as a result get easily affected by the foreign bodies like virus and also take long time to respond to them and thus delaying to produce antibodies. Thus, the old-aged people get easily affected (Fig. 6).

In Covid-19, the most affected is the old-aged people mainly who are above 80 years. Also, the mortality or the death rate is maximum for them which are around 15% and it increasing day-by-day. From the sample, the trend has been analyzed that with the increasing age the death rate rapidly increases from age 40 to 80 (Fig. 7).

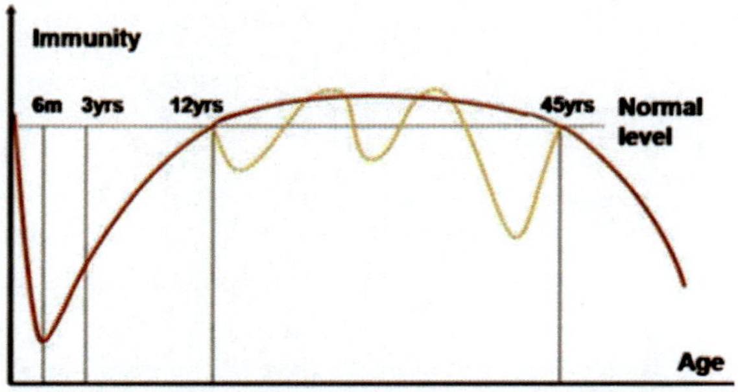

Fig. 6 Represents the immunity versus age. *Source* Biolifenz

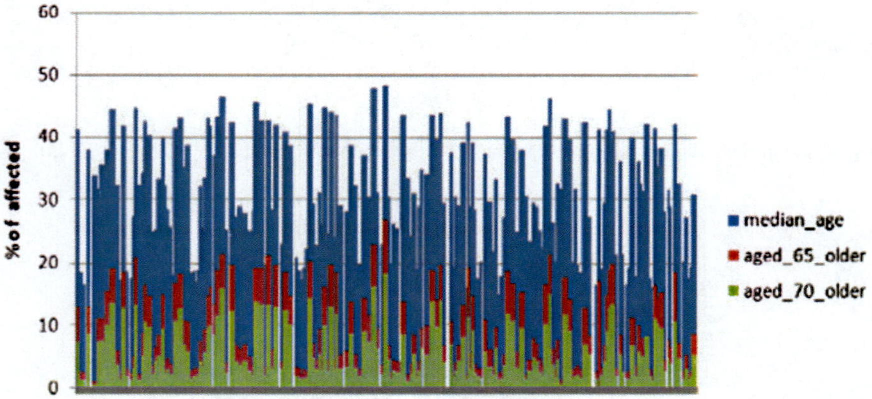

Fig. 7 Represents the graph of age-wise affected rate. *Source* Created by author

2.3 *Survival Rate Analysis Based on Delay Time Between Symptom Onset and Treatment*

One of the most important factors in detection of survival rate analysis is the reporting time of patients to the hospitals after detection of symptoms. Survival analysis is mainly a statistical approach to analyze and summarize the medical data which may include many invents like death rate, survival rate, symptoms-onset, illness onset, recovery time, etc. It may work on panel data. It helps in establishing a medical threshold to fight against a disease and to provide a better service to the nation. There are different approaches like *time to event, censoring, Zero time, Stage migration,* etc.

Time to event approach gives the comparison study between the groups using *T*-test and linear regression neglecting *censoring,* and censoring takes the help of some machine learning classification algorithms like logistic regression analyzing the different classes like recovered or death, etc. By implementing the fuzzy logic, we can measure the degree of uncertainty of the different parameters creating a comparative study of different factors.

The suspected patients are isolated, and the mean incubation period is around 3–4 days. The affected patients are going under treatments but due to high mutation rate of the virus and the patients suffers a lot and the disease proceed to most threatening stage leading to death. Approximately, the death rate is around 4.2% and it is increasing rapidly. Analysis shows that if the time of reporting is more, it can be threat to life. Analysis shows that the delay time is decreasing over the period of time (Fig. 8) [5–7].

The awareness of media and other factors decreases the delay time of patients. Decrease in delay time increases the probability of recovery from the disease. With the help of statistical analysis, it is evident that reporting to the clinic and symptoms-onset has a relation. If the delay is more than 4 days, then the probability of death is higher than the survival, except some of the exception (Table 3).

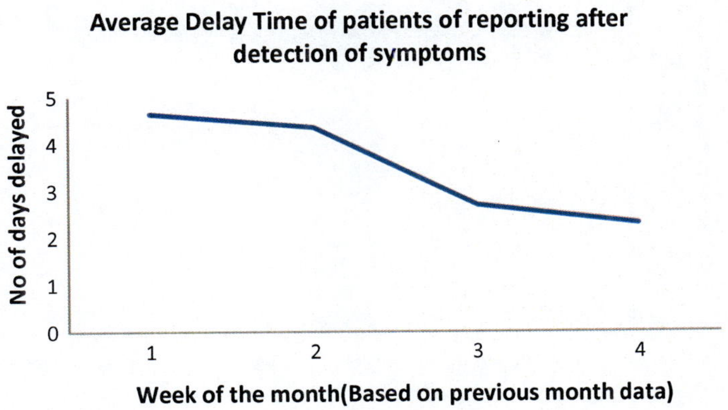

Fig. 8 Delay time curve. *Source* Created by author

Table 3 Mortality rate on delay time of Covid-19

Average number of days	Probability of death
4	High (more than 0.6)

3 Approximation of Number of Unreported Covid-19 Case

The question is not how to get cured, but how to live
—Joseph Conrad.

The count of the people victimized under corona virus increasing day-by-day and it is spreading like a wild fire all over the world. But, the count is done only on the basis of reported case, and so, the count of the unregistered is still unknown to us. The reports and analyst predicts that the actual total count may be 7 or 8 times higher. Once a patient is suspected, the person is isolated and not released until the test has been done and person is declared safe. According to study published in journals, it is evident that the unregistered cases are one of the crucial driver of spreading of the Covid-19. According to the reports, in the mid of January, the unreported cases are around 80% (Fig. 9).

It is advised by medical experts to report on the clinic or hospital as soon as possible or else the number of victims may get increased.

$$\text{No of unreported} \propto \text{Rate of contigency}$$

The cumulative cases during the period of 1–15 January had a saturated value of around and off 41. Similarly, in India during current run, the increase in count is very small, whereas the outbreak curve is exponential in nature. Thus, we have run a maximum likelihood function to get some idea of unreported cases.

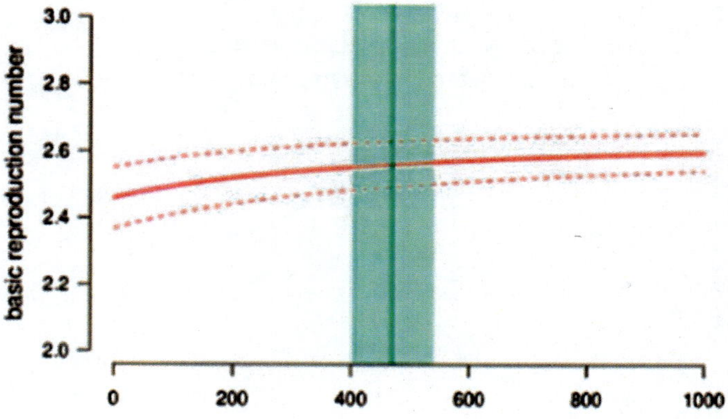

Fig. 9 Represents the graph between reproduction numbers of cases versus number of unreported. *Source* MDPI

An idea of unreported case is very important as the reported persons are majorly kept at quarantine but those who did not report government officials cannot put them into quarantine and from them huge number of infections can spread.

Let R_0 be the reported numbers of unreported cases
The time series data is taken from authentic source of John Hopkins University
Let δ be the number of unreported cases in covid-19 pandemic
The cumulative number of reported case for ith day be C_i
The cumulative number of un-reported case for ith day be c_i
Thus total number of cases is represented by Eq. 1 [8, 9].

$$C = C_i + c_i \tag{1}$$

As the outbreak of Covid-19 follows the exponential model, we have modeled our Poisson distribution model in exponential terms.

For MLE, the confidence interval C.I. taken was 95%.

The P_{Value} in the Poison distribution shows as 0.00; thus, we can accept the results of MLE modeling (Fig. 10).

The D.O.F. taken in the modeling is 1 and $\chi^2_{\text{pr}} = 0.95$.

Thus, the formula for reproduction number is represented in Eq. 2.

$$R_0 = \frac{1}{M(-\gamma)} \tag{2}$$

where γ is the susceptibility and it is taken as 100%.

And M is laplace transformation, used for series transformation.

The trend of covid-19 is still unpredictable; thus, we have used SI to predict the trend; the estimation parameters are disease of corona family namely SARS and MERS.

```
Poisson regression                              Number of obs   =        307
                                                LR chi2(1)      =      23.75
                                                Prob > chi2     =     0.0000
Log likelihood = -16601.028                     Pseudo R2       =     0.0007
```

id	Coef.	Std. Err.	z	P>|z|	[95% Conf.	Interval]
age	.0010814	.0002219	4.87	0.000	.0006465	.0015164
_cons	5.384797	.0115555	465.99	0.000	5.362149	5.407445

Fig. 10 Poisson test result of Covid-19. *Source* Created by Author, based on Dataset

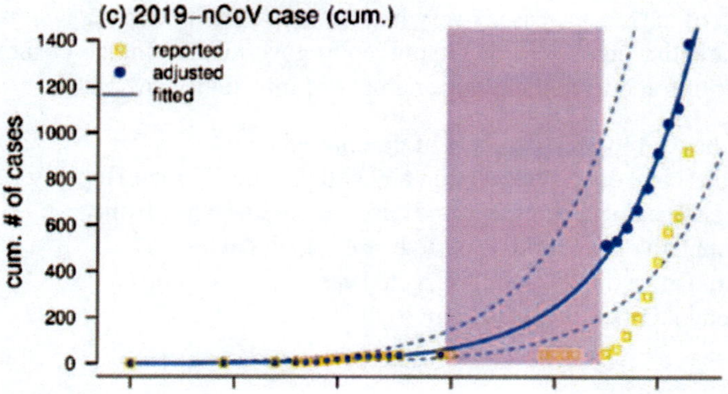

Fig. 11 MLE outbreak curve of Covid-19

We have model $r(t)$ as mean and SD of SARS as 8 and 3.6, respectively, and 7.6 and 3.4 for MERS, respectively [10, 11].

$$r(t) = f(\gamma) \tag{3}$$

The basic function ability model is designed on Poisson modeling. The Poisson model follows the probability mass function; it belongs to the generalized family of regression model, and regression analysis model is being used. It is mainly used to model count data and in contingency tables analysis. Sometimes, Poisson model is considered as log-linear model for its nature and spread.

When the input in Poisson model is θ and x as input vector, the mean predicted associated with Poisson distribution is given in Eq. 2.

$$E\left(\frac{Y}{X}\right) = e^{\theta x} \tag{4}$$

The results of this model gives us around 3,52,000 cases can be there by following the exponential outbreak curve and our MLE, and Poisson modeling.

It shows that there could be around 3094 unreported cases as per as our data taken (Fig. 11).

4 A Time Series Approach for COVID-19

The epidemic broke at a rapid pace and creating an exponential growth. Thus, in this situation, we frame a time series approach with AR model to find out how the situation is going and how it can go forward. The disease is spreading at human contact and spreading at a rapid pace as visualized in Fig. 6 (Fig. 12).

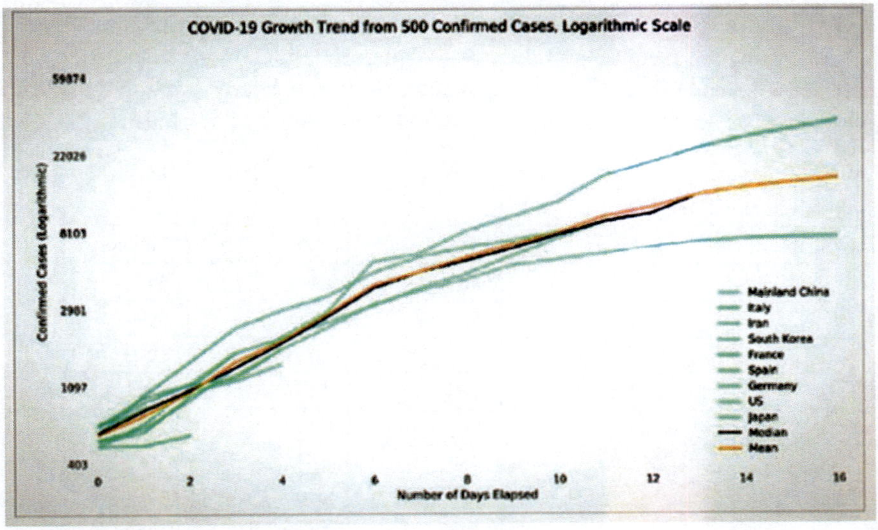

Fig. 12 Time series plot of Covid-19

To compute the relation how Covid-19 is spreading around the globe we have proposed a Pearson correlation matrix of all cases.

It is a descriptive statistical model which measures the linear correlativity between 2 variables X and Y. It measures the strength of association between the two variables. The range of Pearson's correlation coefficient is from [+1, −1].

- Here, 0 denotes there is no correlation between the variables.
- Any value greater than 0 denotes there is positive correlation between them (Fig. 13).

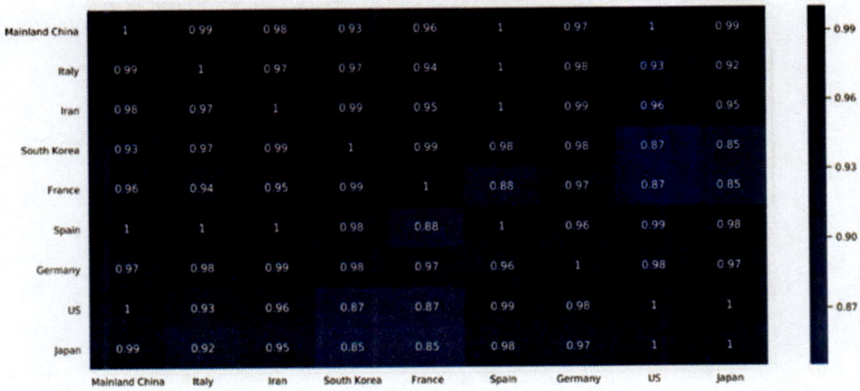

Fig. 13 Correlation matrix plot of Covid-19

Any value lesser than 0 denotes there is negative correlation between them

The correlation factor shows there is a high correlation among countries where the disease spread. This shows that spread of disease is nearly same in the world, following the exponential growth at its peak. Table 4 shows IQ range (Fig. 14) [12].

Table 4 Initial IQR result

Days elapsed	Mean	25%	75%	Median
0	584	540	629	593
1	764	653	874	799
2	1033	871	1196	1040
3	1539	1260	1817	1399
4	2004	1667	2342	1910
5	2804	2452	3155	2682
6	4120	3314	4929	3948
7	4872	4031	5712	4840
8	5915	4719	7111	5601
9	7003	5637	8370	6522
10	8227	6450	10,005	7708

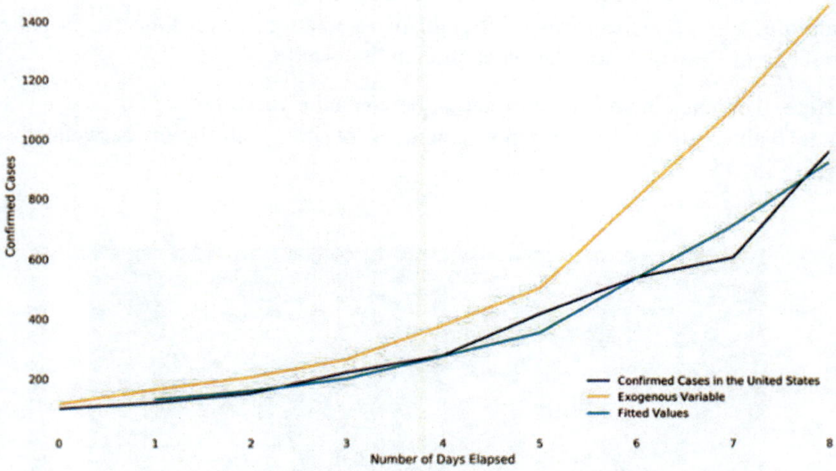

Fig. 14 Covid-19 AR plot

5 ANN Based Predictive Modeling of Covid-19 in India

The conceptualization of artificial neural network comes from biology which is inspired from the neurons of the nervous system where there are a large amount of interconnected systems, and the execution is going on which is done simultaneously and parallel. ANN is a data-driven model based on mathematics and the concept of large interconnected architecture. Generally, when we have only the features and the large amount of execution has to be done with high accuracy, then ANN is used. ANN used to establish and model the complex relationship between the input features and the output node and forms the chain of input and output. It mainly consisted of 3 layers: Input, hidden, and output layers.

They are activated by the activated functions. Some of the common activation functions are (Fig. 15).

ANN has 2 important components which is feed forward and back propagation. By the feed forward, there is frontward movement that is the weight along with biases are entering the node which is activated by suitable activation functions, and the prediction is done in the output node, and the error in the prediction is monitored and to make the prediction better, it is back propagated through the network and it continues and comes out with the best results.

The recurrent networks has the capability of process a successive recursion by a transition function to internal hidden vector state of input.

$$H(t) = \begin{cases} 0 \text{ when } t = 0 \\ f(H_{t-1}, ip_t)\text{when } t > 0 \end{cases}$$

It is a state tom state transition, and by establishing a cyclic interconnection and keeping time, a factor it makes a better model for the prediction of sequence of data with comparison to feed forward network. It nourishes activations of the network from the previous duration as a step and impacts the current condition and with that does the prediction (Fig. 16).

But in RNN calculation of vanishing gradient and understanding the correlation of remote factor and the contribution to the main event cannot be understood properly.

LSTM is the modified version of RNN which is basically done to overcome the drawbacks of recurrent network. It has some special blocks known as memory block

Name	Equation	First derivative	Output range
sigmoid	$\sigma(x) = \frac{1}{1+e^{-x}}$	$\sigma'(x) = \sigma(x)(1-\sigma(x))$	$\in 0, 1$
tanh	$\sigma(x) = \frac{e^{x}-e^{-x}}{e^{x}+e^{-x}}$	$\sigma'(x) = 1-\sigma(x)^2$	$\in -1, 1$
relu	$\sigma(x) = \begin{matrix} 0, & x<0 \\ x, & x \geq 0 \end{matrix}$	$\sigma(x) = \begin{matrix} 0, & x<0 \\ 1, & x \geq 0 \end{matrix}$	$\in >0, \infty$

Fig. 15 Representing the activation functions. *Source* Brian S. Freeman, Graham Taylor, Bahram Gharabaghi & Jesse Thé (2018)

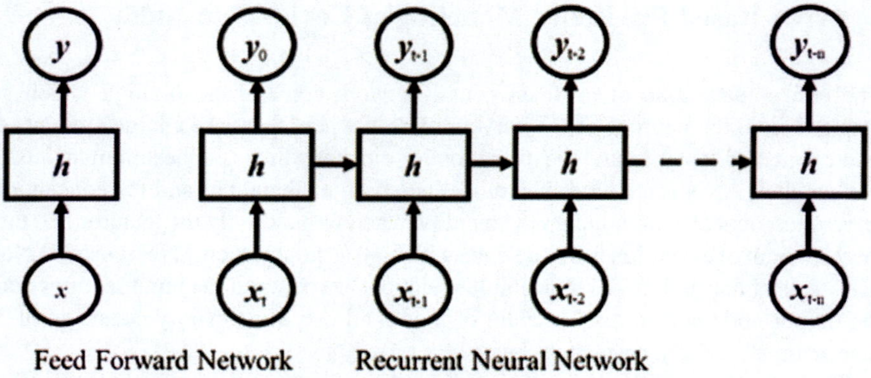

Fig. 16 Schematic representation of the architecture of RNN. *Source* Brian S. Freeman, Graham Taylor, Bahram Gharabaghi & Jesse Thé (2018)

in the hidden layer which are having a good self-connection network which stores the information of the temporary, state and the flow is operated by the input and output gates (Fig. 17).

LSTM uses concentration of gates and feedback loops for training the network, and thus, itself acts a feed forward network. The input goes through the gates, and the processed and trained weights gets to the next network through the output gate (Fig. 18).

LSTM is widely for predication of univariate time series models where a single series of observation which varies with the time is taken into the account. In our

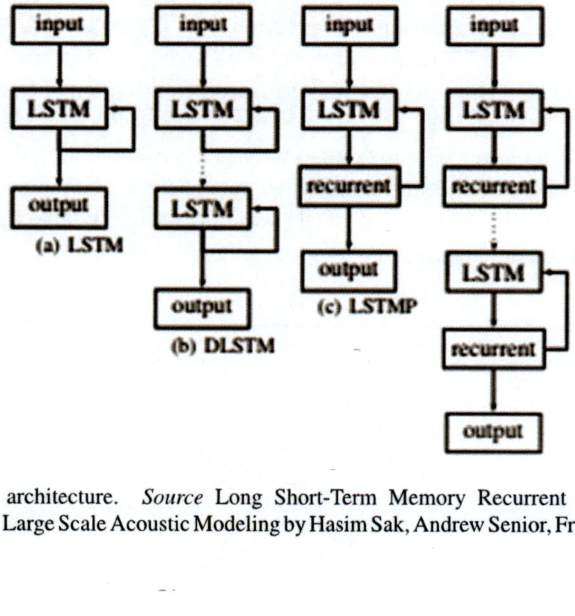

Fig. 17 LSTM architecture. *Source* Long Short-Term Memory Recurrent Neural Network Architectures for Large Scale Acoustic Modeling by Hasim Sak, Andrew Senior, Francoise Beaufays

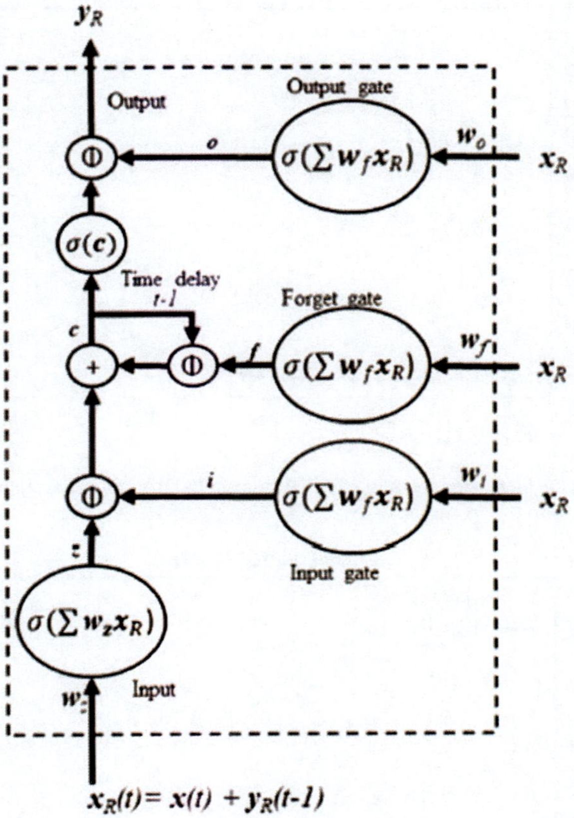

Fig. 18 Working of LSTM. *Source* Brian S. Freeman, Graham Taylor, Bahram Gharabaghi & Jesse Thé (2018)

paper, the observation of total cases of COVID19 is considered and it is modeled (Fig. 19) [13].

We can see that it is having an increasing trend. Using LSTM, the predication can be done better. In Figs. 20 and 21, a predicted graph is presented. Here, the number of cases is scaled in terms of 0–1.

From the graph, it is evident that it is growing exponentially and the rate of contamination is very high. Also, we can visualize the week-wise prediction in Fig. 21.

As we know that health is the major wealth of human and from the nature of the predicted graph that worst is coming very soon. The crisis will increase leading to the destruction of human lives as well as pillars of the economy and we all know that the lion's share of GDP is in healthcare department, and so, this department has a huge contributing factor. From the different reports of UNICEF, it is evident that there is a huge estimated shortfall in the health care centers as and the healthcare infrastructure should be improved. In West Bengal, the pillars of the health care department are not

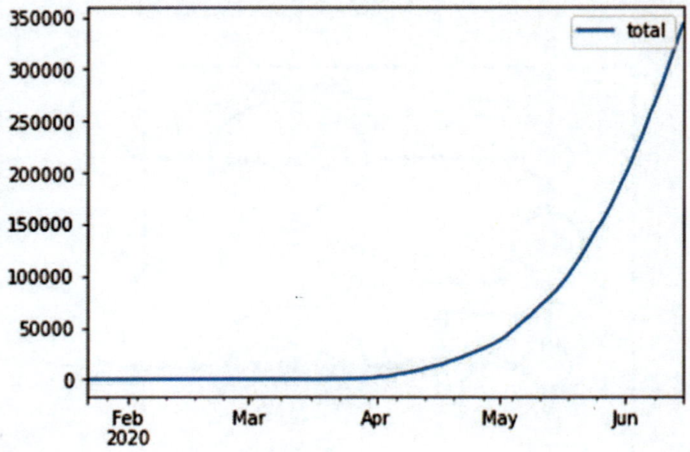

Fig. 19 Showing the increasing trend of covid19 cases in INDIA. *Source* Created by author

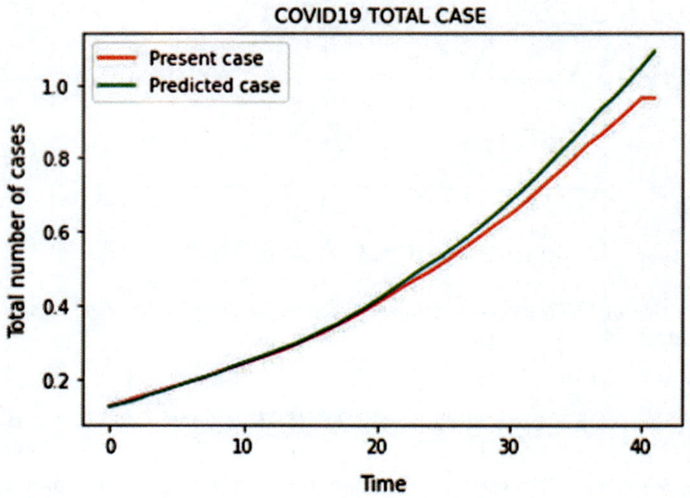

Fig. 20 Prediction using LSTM. *Source* Created by author

too strong to gear up the infrastructure and the service to serve the nation. Let's dive deeper into risk analysis which is performed using Bayesian Hierarchical modeling. Here, the data of last few years have been taken and the probability of death or death risk is calculated. We will look at the health infrastructure and the cause and effect in normal circumstances (Fig. 22).

From the figure, significant risk factors can be seen and analyzed due to the mainly lack of infrastructure, lack of center, lack of doctors, and health care workers. One of the major factors is the lack of centers

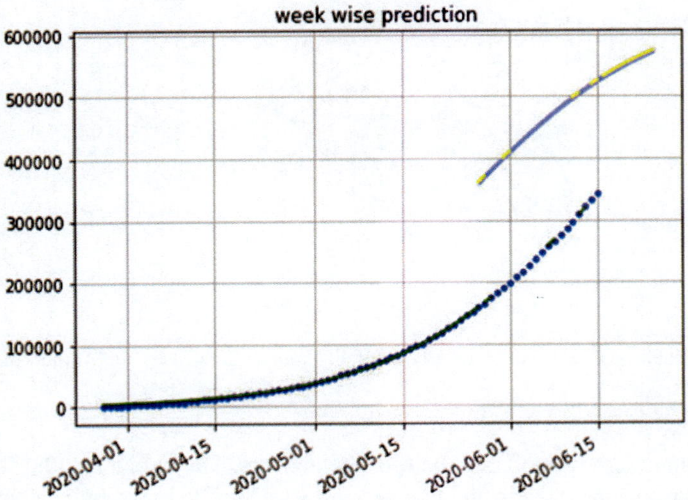

Fig. 21 Week-wise prediction of total case. *Source* Created by author

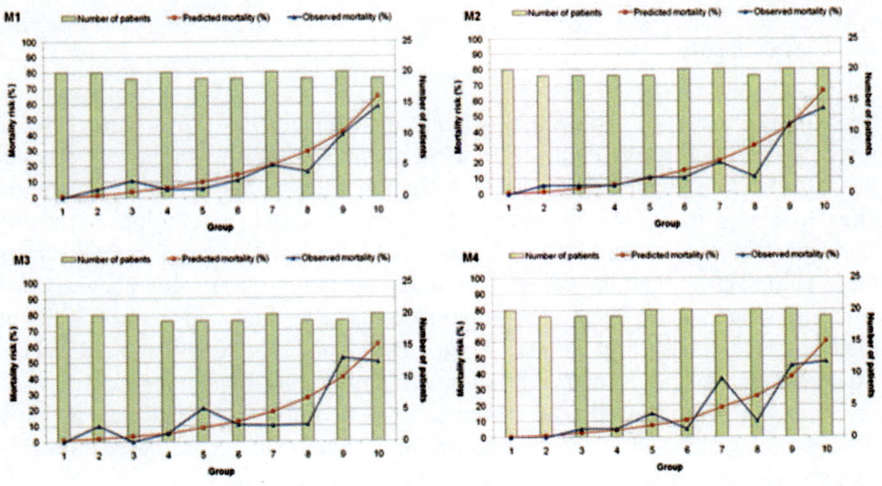

Fig. 22 Mortality rate in West Bengal in normal circumstance. *Source* Created by the Author, based on the dataset

- Primary Health Centers (PHCs)
- Sub Health Centers (SCs)
- Community Health Centers (CHCs).

In normal circumstance, there would be huge shortfall which can be visualized by Fig. 23.

Current situation is pandemic and in West Bengal, the total affected as per 06/14/2020 is more than 9000 and if the rate goes like this, there would be high

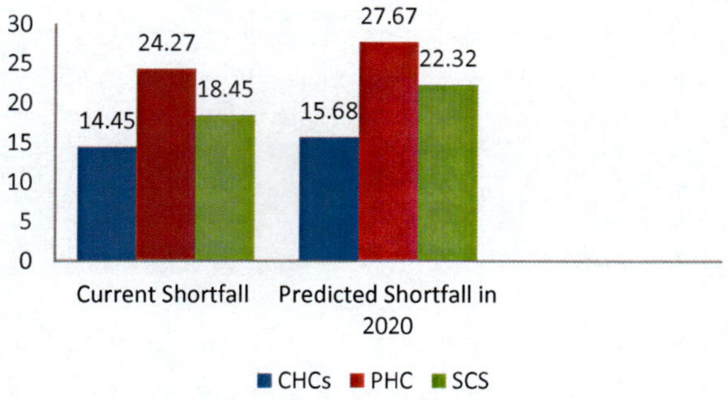

Fig. 23 Visualized health infrastructure shortfall. *Source* Created by the Author, based on the dataset

shortfall and there would be no bed for the patients. The public health sectors now needs a serious observation and planned infrastructure to make India a better place to live.

6 Conclusion

Through our analytics approach, we found that females have lower affected and mortality rate but on the contrary as most of the healthcare workers are female, they are standing at more severe breakthrough, but the mortality rate might not get affected as it is generally constituent due to immune system and biological factors. The age trend analysis shows that lower and higher aged group of people shows more vulnerability than the others. The delay symptom onset study suggests that admitting a patient after 4 days of symptom onset can be fatal, and the probability of their death could be as high as 60%. Through MLE and time series, we modeled the exponential growth of the outbreak. There could be around 3049 unreported cases as per statistical data received. The economic slowdown due to the outbreak is pretty obvious and we showed a study how its getting affected. The LSTM-based modeling also showed there is still a lot to go before the relief from this virus.

Now, in a situation when unlock 1, 2 is on the cards, planning for an optimal exit strategy keeping in mind, the tradeoff between life versus livelihood is the need of the hour. It is clear that though the absolute number of spread (based on our model from July) is increasing the rate of spread except the month of May and June, has slowed down. Certainly, implementing expert advice effectively is not easy in a country of 1.3 billion, where widespread inequality, poverty, and malnutrition are deeply rooted in the society. There are also biological, technological, economic, and logistic reservations that require utmost caution in putting together an optimal exit strategy.

References

1. Zhao S, Musa SS, Lin Q, Ran J, Yang G, Wang W, Lou Y, Yang L, Gao D, He D, Wang MH (2020) Estimating the unreported number of novel coronavirus (2019-nCoV) cases in China in the first half of January 2020: a data-driven modelling analysis of the early outbreak
2. Knobler S, Mahmoud A, Lemon S, Mack A, Sivitz L, Berholtzer K Learning from SARS: preparing for the next disease outbreak
3. Guan W, Ni Z, Hu Y, Liang W, Ou C, He J, Liu L, Shan H, Lei C, Hui DSC, Du B, Li L, Zeng G, Yuen K-Y, Chen R, Tang C, Wang T, Chen P, Xiang J, Li S, Wang J, Liang Z, Peng Y, Wei L, Liu Y, Hu Y, Peng P, Wang J, Liu J, Chen Z, Li G, Zheng Z, Qiu S, Luo J, Ye C, Zhu S, Zhong N (2019) Clinical characteristics of coronavirus disease 2019 in China
4. El Deeba O, Jalloulc M (2020) The dynamics of COVID-19 spread in Lebanon. arXiv, arXiv-2005
5. Gopal R, Chandrasekar VK, Lakshmanan M (2020) Dynamical modelling and analysis of COVID-19 in India. arXiv preprint arXiv:2005.08255
6. Roy S, Roy Bhattacharya K (2020) Spread of COVID-19 in India: a mathematical model. Available at SSRN 3587212
7. Sharma VK, Nigam U (2020) Modelling of Covid-19 cases in India using regression and time series models. medRxiv
8. Gupta S, Raghuwanshi GS, Chanda A (2020) Effect of weather on COVID-19 spread in the US: a prediction model for India in 2020. Science of the Total Environment 138860
9. Singh BP, Singh G (2020) Modeling Tempo of COVID-19 pandemic in India and significance of lockdown. medRxiv
10. Sharma VK, Nigam U (2020) Modeling and Forecasting for Covid-19 growth curve in India. medRxiv
11. Gupta S (2020) Epidemic parameters for COVID-19 in several regions of India. arXiv preprint arXiv:2005.08499
12. Shekhar H (2020) Prediction of spreads of COVID-19 in India from current trend. medRxiv
13. Wu Z, Jennifer M (2019) Characteristics of and important lessons from the coronavirus disease 2019 (COVID-19) outbreak in China: summary of a report of 72,314 cases From the Chinese Center for Disease Control and Prevention. Huang C, Wang Y, Li X, Ren L, Zhao J, Hu Y, Zhang L, Fan G, Xu J, Gu X, Cheng Z, Yu T, Xia J, Wei Y, Wu W, Xie X, Yin W, Li H, Liu M, Xiao Y, Gao H, Guo L, Xie J, Wang G, Jiang R, Gao Z, Jin Q, Wang J, Cao B (2019) Clinical features of patients infected with 2019 novel coronavirus in Wuhan, China

Studies on Optimal Traffic Flow in Two-Node Tandem Communication Networks

N. Thirupathi Rao, K. Srinivas Rao, and P. Srinivasa Rao

Abstract This paper tends to the original thought of using compound Poisson binomial process for creating and investigating a two hub couple communication coordinate with two-phase landings and dynamic data transfer capacity portion (DBA). Here it is accepted that two hubs are associated pair and messages touch base to the first and second supports are associated with an irregular number of bundles and put away in cradles for forwarding transmission. Entries are portrayed by compound Poisson binomial procedures in the two cradles which coordinate close with the reasonable circumstance. The transmission forms in both the transmitters are expected to take after unique transfer speed allotment which is portrayed by stack reliant on time. Utilising distinction differential conditions and joint likelihood producing capacity the transient conduct of the framework is examined. With reasonable cost contemplations, the ideal working strategies of the communication networks are determined and broke down. It had watched that the compound Poisson binomial mass landings dissemination parameters have a noteworthy impact on framework execution measures. Dissecting the two-phase coordinate landings enhances the system execution in cradles and mean postponements.

Keywords Tandem networks · Binomial bulk arrivals · DBA · Optimal analysis

---

N. Thirupathi Rao (✉)
Department of Computer Science & Engineering, Vignan's Institute of Information Technology (A), Visakhapatnam, AP, India
e-mail: nakkathiru@gmail.com

K. Srinivas Rao
Department of Statistics, Andhra University, Visakhapatnam, AP, India
e-mail: ksraoau@yahoo.com

P. Srinivasa Rao
Department of Computer Science & Systems Engineering, AUCE (A), Andhra University, Visakhapatnam, AP, India
e-mail: peri.srinivasarao@yahoo.com

D. Bhattacharyya and N. Thirupathi Rao (eds.), *Machine Intelligence and Soft Computing*, Advances in Intelligent Systems and Computing 1280,
https://doi.org/10.1007/978-981-15-9516-5_26

1 Introduction

Communication networks displaying is essential for plan and investigation of numerous communication frameworks. It is hard to lead research centre examinations under factor stack conditions; the communication organises models created with different suspicions on landing forms, transmission forms, designation, steering and stream control systems [1–3]. For better use of assets and to enhance nature of administration parcel exchanging utilised over circuit or message exchange. Much work had accounted for in writing in regards to communication systems with blockage control procedures. Bit dropping is one of the typical strategies received for blockage control. In this technique, the thought is disposing of the particular segment of benefit, for example, substantial minimum bits with a specific end goal to diminish the heap. In any case, the bit dropping makes changes in voice quality because of a progressively fluctuating piece rate amid a cell transmission [4, 5]. To keep up nature of administration and to lessen the clog in cradles another transmission system dynamic data transfer capacity designation technique is used as an option and effective control procedure.

In every one of the papers alluded above, they expected that the landings are single and take after Poisson process. In any case, in packetised exchanging, the message that touches base to the source are changed over into an arbitrary number of parcels and land to the cushions in mass [6–8]. In any case, in these papers likewise, the creators considered that the entries to the system are to be the first cradle as it were. In some communication frameworks like satellite and remote interchanges, there is a two-phase landing; i.e. the entries of bundles are to the first cradle and furthermore to the second support specifically. For instance, in media communications, there are some neighbourhood calls, and some STD calls where the STD calls may straightforwardly touch base to the second cushion [9, 10]. To break down this kind of frameworks, a two hub pair communication networks coordinate with dynamic data transfer capacity allotment having two-phase coordinate compound binomial Poisson entries is produced and broke down.

The various models considered earlier are as follows (Fig. 1).

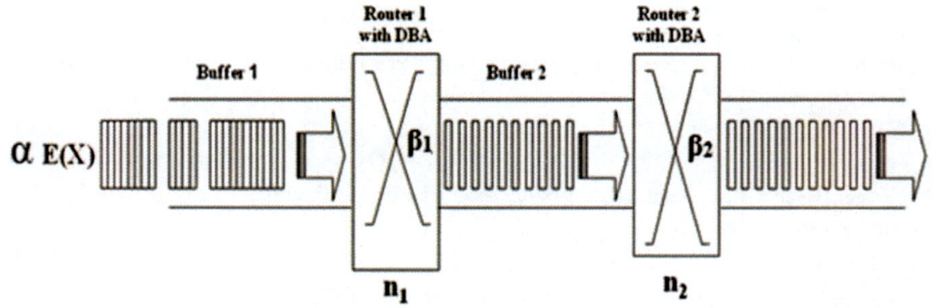

Fig. 1 Communication network model

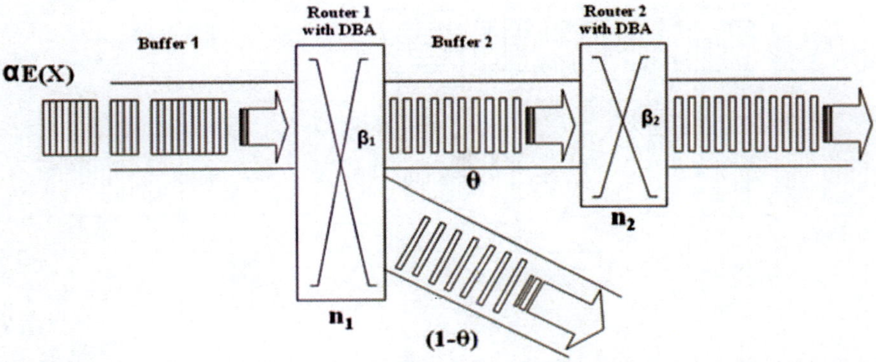

Fig. 2 Communication network with dynamic bandwidth allocation strategy

The diagram represents the normal arrival of packets to the first node and the leaving of packets from the second node (Fig. 2).

In the above model of the network, the arrival of packets to the network was single and the leaving of the packets from node 2 and a phase type communication was introduced at the node 1 such that to identify the number of packets being leaving from node 1 without reaching node 2 in the communication network.

2 Queuing Model

Consider two-transmitters pair correspondence organises in which the messages touch base to the system are changed over into a random number of bundles. The landing procedure of the messages is arbitrary, and various parcels (X) that a message can be changed and overtakes after binomial dissemination with parameters m and p; i.e., the entry modules takes after a compound Poisson binomial process with composite entry rate $\alpha_1 . E(X), \alpha_2$. Here, it accepts that the landing of bundles takes after compound Poisson process with parameters α_1 and α_2 and the number of transmissions at hubs 1 and two additionally take after compound Poisson forms with parameters β_1 and β_2 (Fig. 3).

3 Optimal Policies of the Model

In this segment, we infer the ideal working arrangements of the correspondence organises under investigation. Here, it is expected that the specialist co-op of the correspondence arrange is keen on augmentation of the benefits work at a given time t. Let the specialist co-op gets a measure of R_i units per each unit of time of the framework occupied at the ith transmitter ($i = 1, 2$). As it were, he gets an income

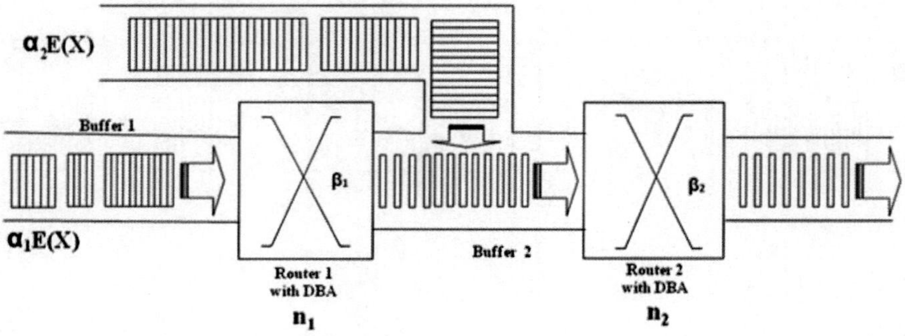

Fig. 3 Communication network with two-stage bulk arrivals

of R_i units per each unit of throughput of the ith transmitter. In this manner, the aggregate income of the correspondence arrange at time t is,

$$\begin{aligned} R(t) &= R_1.(\text{Number of packets transmitting through transmitter 1}) \\ &\quad + R_2.(\text{Number of packets transmitting through transmitter 2}) \end{aligned} \tag{1}$$

$$\begin{aligned} R(t) = R_1.\beta_1\Bigg[1 - \exp\Bigg[\alpha_1 \sum_{k_1=1}^{m_1}\sum_{r=1}^{k_1} \frac{{}^{m_1}C_{k_1} p_1^{k_1}(1-p_1)^{m_1-k_1}}{1-(1-p_1)^{m_1}}\,{}^{k_1}C_r \\ (-1)^{3r}\frac{\left(1-\mathrm{e}^{-r\beta_1 t}\right)}{r\beta_1}\Bigg]\Bigg] \\ + R_2.\beta_2.\Bigg\{1 - \exp\Bigg\{\Bigg[\alpha_1 \sum_{k_1=1}^{m_1}\sum_{r=1}^{k_1}\sum_{J=0}^{r}(-1)^{3r-J}\frac{{}^{m_1}C_{k_1} p_1^{k_1}(1-p_1)^{m_1-k_1}}{1-(1-p_1)^{m_1}} \\ ({}^{k_1}C_r)({}^{r}C_J)\left(\frac{\beta_1}{\beta_2-\beta_1}\right)^r\left(\frac{1-\mathrm{e}^{-[J\beta_2+(r-J)\beta_1]t}}{J\beta_2+(r-J)\beta_1}\right)\Bigg] \\ +\alpha_2\Bigg[\sum_{k_2=1}^{m_2}\sum_{s=1}^{k_2}\frac{{}^{m_2}C_{k_2} p_2^{k_2}(1-p_2)^{m_2-k_2}}{1-(1-p_2)^{m_2}}({}^{k_2}C_s)(-1)^S\left(\frac{1-\mathrm{e}^{-\mu_2 k_2 t}}{\mu_2 k_2}\right)\Bigg]\Bigg\}\Bigg\} \end{aligned} \tag{2}$$

$$\begin{aligned} C(t) &= A - C_1 * (\text{standard waiting time of a client in transmitter 1}) \\ &\quad - C_2 * (\text{ standard waiting time of a client in transmitter 2}) \end{aligned} \tag{3}$$

$$C(t) = A - C_1 . \frac{\frac{\alpha_1}{\beta_1}\left[\sum_{k_1=1}^{m_1} \frac{{}^{m_1}C_{k_1} p_1^{k_1} (1-p_1)^{m_1-k_1}}{1-(1-p_1)^{m_1}} . k_1 \left(1 - e^{-\beta_1 t}\right)\right]}{\beta_1 \left[1 - \exp\left[\alpha_1 \sum_{k_1=1}^{m_1} \sum_{r=1}^{k_1} \frac{{}^{m_1}C_{k_1} p_1^{k_1} (1-p_1)^{m_1-k_1}}{1-(1-p_1)^{m_1}} {}^{k_1}C_r (-1)^{3r} \frac{\left(1-e^{-r\beta_1 t}\right)}{r\beta_1}\right]\right]}$$

$$- C_2 . \frac{\frac{\alpha_1}{\beta_2}\left(\sum_{k_1=1}^{m_1} \frac{{}^{m_1}C_{k_1} p_1^{k_1} (1-p_1)^{m_1-k_1}}{1-(1-p_1)^{m_1}} k_1\right)\left[\left(1-e^{-\beta_2 t}\right) + \frac{\beta_2}{\beta_2-\beta_1}\left(e^{-\beta_2 t} - e^{-\beta_1 t}\right)\right] + \frac{\alpha_2}{\beta_2}\left(\sum_{k_2=1}^{\infty} \frac{{}^{m_2}C_{k_2} p_2^{k_2} (1-p_2)^{m_2-k_2}}{1-(1-p_2)^{m_2}} k_2\right)\left(1-e^{-\beta_2 t}\right)}{\beta_2 . \left\{1 - \exp\left\{\left[\alpha_1 \sum_{k_1=1}^{m_1} \sum_{r=1}^{k_1} \sum_{J=0}^{r} (-1)^{3r-J} \frac{{}^{m_1}C_{k_1} p_1^{k_1} (1-p_1)^{m_1-k_1}}{1-(1-p_1)^{m_1}} ({}^{k_1}C_r)({}^{r}C_J)\left(\frac{\beta_1}{\beta_2-\beta_1}\right)^r \left(\frac{1-e^{-[J\beta_2+(r-J)\beta_1]t}}{J\beta_2+(r-J)\beta_1}\right)\right] + \alpha_2\left[\sum_{k_2=1}^{m_2}\sum_{s=1}^{k_2} \frac{{}^{m_2}C_{k_2} p_2^{k_2} (1-p_2)^{m_2-k_2}}{1-(1-p_2)^{m_2}} ({}^{k_2}C_s)(-1)^s \left(\frac{1-e^{-\beta_2 k_2 t}}{\beta_2 k_2}\right)\right]\right\}\right\}} \tag{4}$$

Substituting the values of $R(t)$ and $C(t)$ from Eqs. (3) and (4), respectively, we get total cost function as

$$P(t) = R_1 . \beta_1 \left[1 - \exp\left[\alpha_1 \sum_{k_1=1}^{m_1} \sum_{r=1}^{k_1} \frac{{}^{m_1}C_{k_1} p_1^{k_1} (1-p_1)^{m_1-k_1}}{1-(1-p_1)^{m_1}} {}^{k_1}C_r (-1)^{3r} \frac{\left(1-e^{-r\beta_1 t}\right)}{r\beta_1}\right]\right]$$

$$+ R_2 . \beta_2 . \left\{1 - \exp\left\{\left[\alpha_1 \sum_{k_1=1}^{m_1} \sum_{r=1}^{k_1} \sum_{J=0}^{r} (-1)^{3r-J} \frac{{}^{m_1}C_{k_1} p_1^{k_1} (1-p_1)^{m_1-k_1}}{1-(1-p_1)^{m_1}} ({}^{k_1}C_r)({}^{r}C_J)\left(\frac{\beta_1}{\beta_2-\beta_1}\right)^r \left(\frac{1-e^{-[J\beta_2+(r-J)\beta_1]t}}{J\beta_2+(r-J)\beta_1}\right)\right]\right.\right.$$

$$\left.\left. + \alpha_2\left[\sum_{k_2=1}^{m_2}\sum_{s=1}^{k_2} \frac{{}^{m_2}C_{k_2} p_2^{k_2} (1-p_2)^{m_2-k_2}}{1-(1-p_2)^{m_2}} ({}^{k_2}C_s)(-1)^s \left(\frac{1-e^{-\mu_2 k_2 t}}{\mu_2 k_2}\right)\right]\right\}\right\}$$

$$- C_2 . \frac{\frac{\alpha_1}{\beta_2}\left(\sum_{k_1=1}^{m_1} \frac{{}^{m_1}C_{k_1} p_1^{k_1} (1-p_1)^{m_1-k_1}}{1-(1-p_1)^{m_1}} k_1\right)\left[\left(1-e^{-\beta_2 t}\right) + \frac{\beta_2}{\beta_2-\beta_1}\left(e^{-\beta_2 t} - e^{-\beta_1 t}\right)\right] + \frac{\alpha_2}{\beta_2}\left(\sum_{k_2=1}^{\infty} \frac{{}^{m_2}C_{k_2} p_2^{k_2} (1-p_2)^{m_2-k_2}}{1-(1-p_2)^{m_2}} k_2\right)\left(1-e^{-\beta_2 t}\right)}{\beta_2 . \left\{1 - \exp\left\{\left[\alpha_1 \sum_{k_1=1}^{m_1} \sum_{r=1}^{k_1} \sum_{J=0}^{r} (-1)^{3r-J} \frac{{}^{m_1}C_{k_1} p_1^{k_1} (1-p_1)^{m_1-k_1}}{1-(1-p_1)^{m_1}} ({}^{k_1}C_r)({}^{r}C_J)\left(\frac{\beta_1}{\beta_2-\beta_1}\right)^r \left(\frac{1-e^{-[J\beta_2+(r-J)\beta_1]t}}{J\beta_2+(r-J)\beta_1}\right)\right] + \alpha_2\left[\sum_{k_2=1}^{m_2}\sum_{s=1}^{k_2} \frac{{}^{m_2}C_{k_2} p_2^{k_2} (1-p_2)^{m_2-k_2}}{1-(1-p_2)^{m_2}} ({}^{k_2}C_s)(-1)^s \left(\frac{1-e^{-\beta_2 k_2 t}}{\beta_2 k_2}\right)\right]\right\}\right\}} \tag{5}$$

To obtain the optimal values of β_1 and β_2, maximising $P(t)$, concerning β_1 and β_2 and verify the hessian matrix $\frac{\partial P(t)}{\partial \beta_1} = 0$ implies

$$\frac{\partial P(t)}{\partial \beta_1} = R_1.\beta_1\left[1-\exp\left[\alpha_1\sum_{k_1=1}^{m_1}\sum_{r=1}^{k_1}\frac{{}^{m_1}C_{k_1}p_1^{k_1}(1-p_1)^{m_1-k_1}}{1-(1-p_1)^{m_1}}\,{}^{k_1}C_r(-1)^{3r}\frac{\left(1-e^{-r\beta_1 t}\right)}{r\beta_1}\right]\right]$$
$$+R_2.\beta_2.\left\{1-\exp\left\{\left[\alpha_1\sum_{k_1=1}^{m_1}\sum_{r=1}^{k_1}\sum_{J=0}^{r}(-1)^{3r-J}\frac{{}^{m_1}C_{k_1}p_1^{k_1}(1-p_1)^{m_1-k_1}}{1-(1-p_1)^{m_1}}({}^{k_1}C_r)({}^{r}C_J)\left(\frac{\beta_1}{\beta_2-\beta_1}\right)^r\left(\frac{1-e^{-[J\beta_2+(r-J)\beta_1]t}}{J\beta_2+(r-J)\beta_1}\right)\right]\right.\right.$$
$$\left.\left.+\alpha_2\left[\sum_{k_2=1}^{m_2}\sum_{s=1}^{k_2}\frac{{}^{m_2}C_{k_2}p_2^{k_2}(1-p_2)^{m_2-k_2}}{1-(1-p_2)^{m_2}}({}^{k_2}C_s)(-1)^s\left(\frac{1-e^{-\mu_2 k_2 t}}{\mu_2 k_2}\right)\right]\right\}\right\}$$
$$-C_1.\frac{\frac{\alpha_1}{\beta_1}\left[\sum_{k_1=1}^{m}\frac{{}^{m_1}C_{k_1}p_1^{k_1}(1-p_1)^{m_1-k_1}}{1-(1-p_1)^{m_1}}.k_1\left(1-e^{-\beta_1 t}\right)\right]}{=\beta_1\left[1-\exp\left[\alpha_1\sum_{k_1=1}^{m_1}\sum_{r=1}^{k_1}\frac{{}^{m_1}C_{k_1}p_1^{k_1}(1-p_1)^{m_1-k_1}}{1-(1-p_1)^{m_1}}\,{}^{k_1}C_r(-1)^{3r}\frac{\left(1-e^{-r\beta_1 t}\right)}{r\beta_1}\right]\right]}$$
$$-C_2.\frac{\frac{\alpha_1}{\beta_2}\left(\sum_{k_1=1}^{m_1}\frac{{}^{m_1}C_{k_1}p_1^{k_1}(1-p_1)^{m_1-k_1}}{1-(1-p_1)^{m_1}}k_1\right)\left[\left(1-e^{-\beta_2 t}\right)+\frac{\beta_2}{\beta_2-\beta_1}\left(e^{-\beta_2 t}-e^{-\beta_1 t}\right)\right]+\frac{\alpha_2}{\beta_2}\left(\sum_{k_2=1}^{\infty}\frac{{}^{m_2}C_{k_2}p_2^{k_2}(1-p_2)^{m_2-k_2}}{1-(1-p_2)^{m_2}}k_2\right)\left(1-e^{-\beta_2 t}\right)}{\beta_2.\left\{1-\exp\left\{\left[\alpha_1\sum_{k_1=1}^{m_1}\sum_{r=1}^{k_1}\sum_{J=0}^{r}(-1)^{3r-J}\frac{{}^{m_1}C_{k_1}p_1^{k_1}(1-p_1)^{m_1-k_1}}{1-(1-p_1)^{m_1}}({}^{k_1}C_r)({}^{r}C_J)\left(\frac{\beta_1}{\beta_2-\beta_1}\right)^r\left(\frac{1-e^{-[J\beta_2+(r-J)\beta_1]t}}{J\beta_2+(r-J)\beta_1}\right)\right]+\alpha_2\left[\sum_{k_2=1}^{m_2}\sum_{s=1}^{k_2}\frac{{}^{m_2}C_{k_2}p_2^{k_2}(1-p_2)^{m_2-k_2}}{1-(1-p_2)^{m_2}}({}^{k_2}C_s)(-1)^s\left(\frac{1-e^{-\beta_2 k_2 t}}{\beta_2 k_2}\right)\right]\right\}\right\}}=0 \quad (6)$$

$\frac{\partial P(t)}{\partial \beta_2}=0$ Implies

$$\frac{\partial P(t)}{\partial \beta_2} = R_1.\beta_1\left[1-\exp\left[\alpha_1\sum_{k_1=1}^{m_1}\sum_{r=1}^{k_1}\frac{{}^{m_1}C_{k_1}p_1^{k_1}(1-p_1)^{m_1-k_1}}{1-(1-p_1)^{m_1}}\,{}^{k_1}C_r(-1)^{3r}\frac{\left(1-e^{-r\beta_1 t}\right)}{r\beta_1}\right]\right]$$
$$+R_2.\beta_2.\left\{1-\exp\left\{\left[\alpha_1\sum_{k_1=1}^{m_1}\sum_{r=1}^{k_1}\sum_{J=0}^{r}(-1)^{3r-J}\frac{{}^{m_1}C_{k_1}p_1^{k_1}(1-p_1)^{m_1-k_1}}{1-(1-p_1)^{m_1}}({}^{k_1}C_r)({}^{r}C_J)\left(\frac{\beta_1}{\beta_2-\beta_1}\right)^r\left(\frac{1-e^{-[J\beta_2+(r-J)\beta_1]t}}{J\beta_2+(r-J)\beta_1}\right)\right]\right.\right.$$
$$\left.\left.+\alpha_2\left[\sum_{k_2=1}^{m_2}\sum_{s=1}^{k_2}\frac{{}^{m_2}C_{k_2}p_2^{k_2}(1-p_2)^{m_2-k_2}}{1-(1-p_2)^{m_2}}({}^{k_2}C_s)(-1)^s\left(\frac{1-e^{-\mu_2 k_2 t}}{\mu_2 k_2}\right)\right]\right\}\right\}$$
$$-C_1.\frac{\frac{\alpha_1}{\beta_1}\left[\sum_{k_1=1}^{m}\frac{{}^{m_1}C_{k_1}p_1^{k_1}(1-p_1)^{m_1-k_1}}{1-(1-p_1)^{m_1}}.k_1\left(1-e^{-\beta_1 t}\right)\right]}{=\beta_1\left[1-\exp\left[\alpha_1\sum_{k_1=1}^{m_1}\sum_{r=1}^{k_1}\frac{{}^{m_1}C_{k_1}p_1^{k_1}(1-p_1)^{m_1-k_1}}{1-(1-p_1)^{m_1}}\,{}^{k_1}C_r(-1)^{3r}\frac{\left(1-e^{-r\beta_1 t}\right)}{r\beta_1}\right]\right]}$$
$$\frac{-C_2.\frac{\alpha_1}{\beta_2}\left(\sum_{k_1=1}^{m_1}\frac{{}^{m_1}C_{k_1}p_1^{k_1}(1-p_1)^{m_1-k_1}}{1-(1-p_1)^{m_1}}k_1\right)\left[\left(1-e^{-\beta_2 t}\right)+\frac{\beta_2}{\beta_2-\beta_1}\left(e^{-\beta_2 t}-e^{-\beta_1 t}\right)\right]+\frac{\alpha_2}{\beta_2}\left(\sum_{k_2=1}^{\infty}\frac{{}^{m_2}C_{k_2}p_2^{k_2}(1-p_2)^{m_2-k_2}}{1-(1-p_2)^{m_2}}k_2\right)\left(1-e^{-\beta_2 t}\right)}{\beta_2.\left\{1-\exp\left\{\left[\alpha_1\sum_{k_1=1}^{m_1}\sum_{r=1}^{k_1}\sum_{J=0}^{r}(-1)^{3r-J}\frac{{}^{m_1}C_{k_1}p_1^{k_1}(1-p_1)^{m_1-k_1}}{1-(1-p_1)^{m_1}}({}^{k_1}C_r)({}^{r}C_J)\left(\frac{\beta_1}{\beta_2-\beta_1}\right)^r\left(\frac{1-e^{-[J\beta_2+(r-J)\beta_1]t}}{J\beta_2+(r-J)\beta_1}\right)\right]+\alpha_2\left[\sum_{k_2=1}^{m_2}\sum_{s=1}^{k_2}\frac{{}^{m_2}C_{k_2}p_2^{k_2}(1-p_2)^{m_2-k_2}}{1-(1-p_2)^{m_2}}({}^{k_2}C_s)(-1)^s\left(\frac{1-e^{-\beta_2 k_2 t}}{\beta_2 k_2}\right)\right]\right\}\right\}}=0 \quad (7)$$

The determinant of the Hessian matrix is,

$$|D| = \begin{vmatrix} \frac{\partial^2 P(t)}{\partial \beta_1^2} & \frac{\partial^2 P(t)}{\partial \beta_1 \partial \beta_2} \\ \frac{\partial^2 P(t)}{\partial \beta_1 \partial \beta_2} & \frac{\partial^2 P(t)}{\partial \beta_2^2} \end{vmatrix} < 0 \quad (8)$$

Substituting the values of β_1^* and β_2^* in Eq. (6), we get the optimal value of the profit at given time t as,

$$\begin{aligned}
P*(t) &= R_1.\beta_1^*\left[1-\exp\left[\alpha_1\sum_{k_1=1}^{m_1}\sum_{r=1}^{k_1}\frac{{}^{m_1}C_{k_1}p_1^{k_1}(1-p_1)^{m_1-k_1}}{1-(1-p_1)^{m_1}}\,{}^{k_1}C_r(-1)^{3r}\frac{\left(1-e^{-r\beta_1^* t}\right)}{r\beta_1^*}\right]\right]\\
&+R_2.\beta_2^*.\left\{1-\exp\left\{\left[\alpha_1\sum_{k_1=1}^{m_1}\sum_{r=1}^{k_1}\sum_{J=0}^{r}(-1)^{3r-J}\frac{{}^{m_1}C_{k_1}p_1^{k_1}(1-p_1)^{m_1-k_1}}{1-(1-p_1)^{m_1}}({}^{k_1}C_r)({}^{r}C_J)\left(\frac{\beta_1*}{\beta_2*-\beta_1*}\right)^r\left(\frac{1-e^{-[J\beta_2*+(r-J)\beta_1*]t}}{J\beta_2*+(r-J)\beta_1*}\right)\right]\right.\right.\\
&\left.\left.+\alpha_2\left[\sum_{k_2=1}^{m_2}\sum_{s=1}^{k_2}\frac{{}^{m_2}C_{k_2}p_2^{k_2}(1-p_2)^{m_2-k_2}}{1-(1-p_2)^{m_2}}({}^{k_2}C_s)(-1)^s\left(\frac{1-e^{-\beta_2*k_2t}}{\beta_2*k_2}\right)\right]\right\}\right\}\\
&-C_1.\frac{\frac{\alpha_1}{\beta_1^*}\left[\sum_{k_1=1}^{m_1}\frac{{}^{m_1}C_{k_1}p_1^{k_1}(1-p_1)^{m_1-k_1}}{1-(1-p_1)^{m_1}}.k_1\left(1-e^{-\beta_1*t}\right)\right]}{\beta_1*\left[1-\exp\left[\alpha_1\sum_{k_1=1}^{m_1}\sum_{r=1}^{k_1}\frac{{}^{m_1}C_{k_1}p_1^{k_1}(1-p_1)^{m_1-k_1}}{1-(1-p_1)^{m_1}}\,{}^{k_1}C_r(-1)^{3r}\frac{\left(1-e^{-r\beta_1*t}\right)}{r\beta_1*}\right]\right]}\\
&-C_2.\frac{\frac{\alpha_1}{\beta_2^*}\left(\sum_{k_1=1}^{m_1}\frac{{}^{m_1}C_{k_1}p_1^{k_1}(1-p_1)^{m_1-k_1}}{1-(1-p_1)^{m_1}}k_1\right)\left[\left(1-e^{-\beta_2*t}\right)+\frac{\beta_2*}{\beta_2*-\beta_1*}\left(e^{-\beta_2*t}-e^{-\beta_1*t}\right)\right]+\frac{\alpha_2}{\beta_2*}\left(\sum_{k_2=1}^{\infty}\frac{{}^{m_2}C_{k_2}p_2^{k_2}(1-p_2)^{m_2-k_2}}{1-(1-p_2)^{m_2}}k_2\right)\left(1-e^{-\beta_2*t}\right)}{\beta_2*.\left\{1-\exp\left\{\left[\alpha_1\sum_{k_1=1}^{m_1}\sum_{r=1}^{k_1}\sum_{J=0}^{r}(-1)^{3r-J}\frac{{}^{m_1}C_{k_1}p_1^{k_1}(1-p_1)^{m_1-k_1}}{1-(1-p_1)^{m_1}}({}^{k_1}C_r)({}^{r}C_J)\left(\frac{\beta_1*}{\beta_2*-\beta_1*}\right)^r\left(\frac{1-e^{-[J\beta_2*+(r-J)\beta_1*]t}}{J\beta_2*+(r-J)\beta_1*}\right)\right]+\alpha_2\left[\sum_{k_2=1}^{m_2}\sum_{s=1}^{k_2}\frac{{}^{m_2}C_{k_2}p_2^{k_2}(1-p_2)^{m_2-k_2}}{1-(1-p_2)^{m_2}}({}^{k_2}C_s)(-1)^s\left(\frac{1-e^{-\beta_2*k_2t}}{\beta_2*k_2}\right)\right]\right\}\right\}}
\end{aligned} \tag{9}$$

4 Arithmetical Analysis

In the current section, we exhibit the answer process during an arithmetical illustration. Think the examination supplier to a communication network is functioning a two-transmitter tandem communication network. Let the approximate price of income per unit throughput of transmitter 1 is R_1 varies from 0.3 to 0.6, and the revenue per unit of throughput of transmitter 2 R_2 varies from 0.3 to 0.6 (Fig. 4; Table 1).

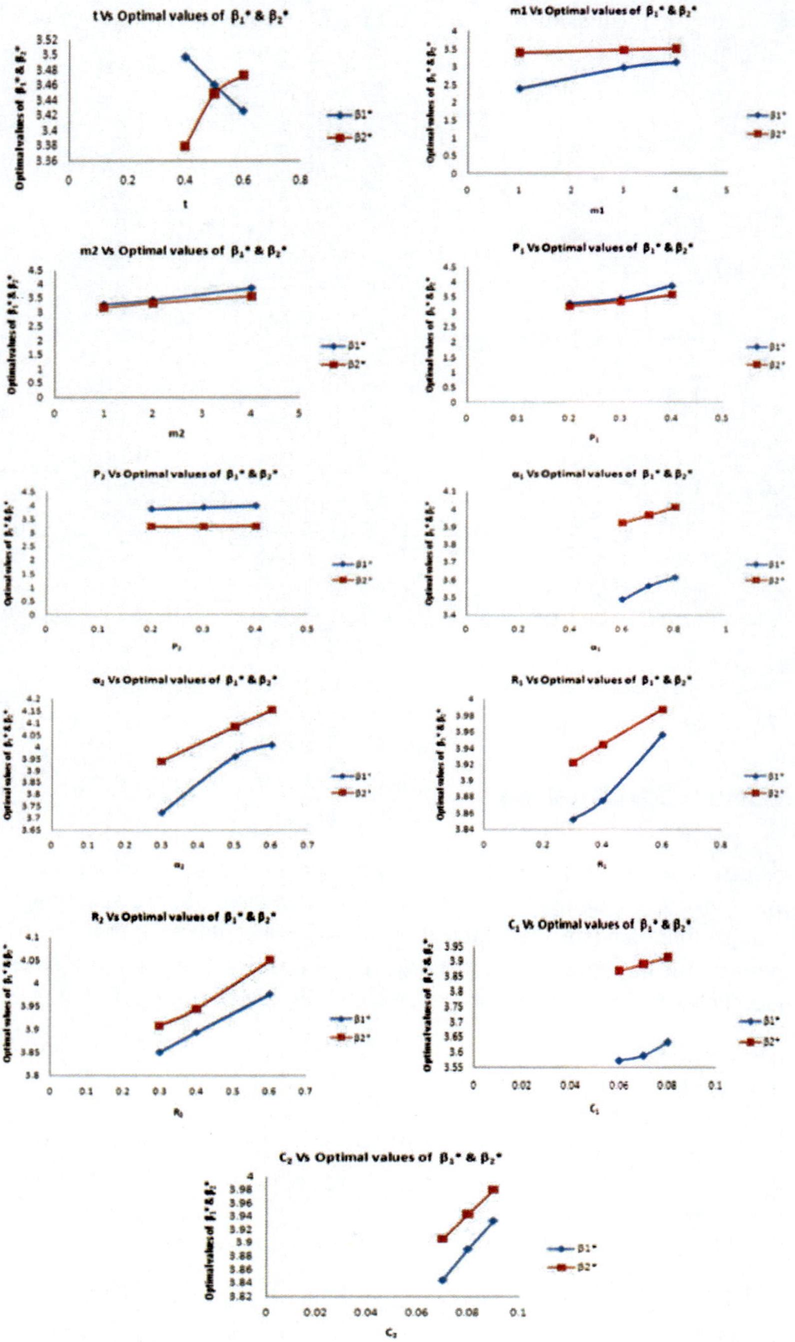

Fig. 4 Optimal performance of the current model for various input considerations

Table 1 Arithmetical symbol of best value of β_1*and β_2*

t^*	m_1	m_2	p_1	p_2	α_1#	α_2#	R_1	R_2	C_1	C_2	β_1*$	β_2*$	R	D_1
0.6	6	5	0.3	0.4	0.4	0.3	0.6	0.7	0.2	0.3	3.498	3.380	3.015	−0.564
0.7	6	5	0.3	0.4	0.4	0.3	0.6	0.7	0.2	0.3	3.461	3.449	3.038	−0.564
0.8	6	5	0.3	0.4	0.4	0.3	0.6	0.7	0.2	0.3	3.426	3.473	3.045	−0.564
0.4	**3**	5	0.3	0.4	0.4	0.3	0.6	0.7	0.2	0.3	2.387	3.392	2.466	−0.927
0.4	**4**	5	0.3	0.4	0.4	0.3	0.6	0.7	0.2	0.3	2.973	3.475	2.829	−0.927
0.4	**5**	5	0.3	0.4	0.4	0.3	0.6	0.7	0.2	0.3	3.125	3.511	2.920	−0.927
0.4	6	**2**	0.3	0.4	0.4	0.3	0.6	0.7	0.2	0.3	3.293	3.180	2.863	−0.656
0.4	6	**3**	0.3	0.4	0.4	0.3	0.6	0.7	0.2	0.3	3.451	3.335	2.999	−0.656
0.4	6	**4**	0.3	0.4	0.4	0.3	0.6	0.7	0.2	0.3	3.876	3.568	3.271	−0.656
0.4	6	5	**0.4**	0.4	0.4	0.3	0.6	0.7	0.2	0.3	3.874	3.220	3.159	−0.189
0.4	6	5	**0.5**	0.4	0.4	0.3	0.6	0.7	0.2	0.3	3.942	3.224	3.190	−0.189
0.4	6	5	**0.6**	0.4	0.4	0.3	0.6	0.7	0.2	0.3	3.998	3.229	3.216	−0.189
0.4	6	5	0.3	**0.5**	0.4	0.3	0.6	0.7	0.2	0.3	3.013	3.442	2.827	−0.496
0.4	6	5	0.3	**0.6**	0.4	0.3	0.6	0.7	0.2	0.3	3.214	3.442	2.936	−0.496
0.4	6	5	0.3	**0.7**	0.4	0.3	0.6	0.7	0.2	0.3	3.458	3.442	3.060	−0.496
0.4	6	5	0.3	0.4	**0.5**	0.3	0.6	0.7	0.2	0.3	3.487	3.922	3.270	−0.823
0.4	6	5	0.3	0.4	**0.6**	0.3	0.6	0.7	0.2	0.3	3.564	3.965	3.326	−0.823
0.4	6	5	0.3	0.4	**0.7**	0.3	0.6	0.7	0.2	0.3	3.612	4.008	3.367	−0.823
0.4	6	5	0.3	0.4	0.4	**0.3**	0.6	0.7	0.2	0.3	3.721	3.938	3.391	−0.270
0.4	6	5	0.3	0.4	0.4	**0.5**	0.6	0.7	0.2	0.3	3.961	4.083	3.561	−0.270
0.4	6	5	0.3	0.4	0.4	**0.6**	0.6	0.7	0.2	0.3	4.010	4.153	3.611	−0.270
0.4	6	5	0.3	0.4	0.4	0.3	**0.3**	0.7	0.2	0.3	3.852	3.922	3.445	−0.319
0.4	6	5	0.3	0.4	0.4	0.3	**0.4**	0.7	0.2	0.3	3.875	3.944	3.465	−0.319
0.4	6	5	0.3	0.4	0.4	0.3	**0.6**	0.7	0.2	0.3	3.956	3.987	3.519	−0.319
0.4	6	5	0.3	0.4	0.4	0.3	0.6	**0.3**	0.2	0.3	3.851	3.907	3.438	−0.645
0.4	6	5	0.3	0.4	0.4	0.3	0.6	**0.4**	0.2	0.3	3.894	3.944	3.473	−0.645
0.4	6	5	0.3	0.4	0.4	0.3	0.6	**0.6**	0.2	0.3	3.976	4.050	3.554	−0.645
0.4	6	5	0.3	0.4	0.4	0.3	0.6	0.7	**0.06**	0.3	3.575	3.870	3.293	−0.702
0.4	6	5	0.3	0.4	0.4	0.3	0.6	0.7	**0.07**	0.3	3.591	3.893	3.310	−0.702
0.4	6	5	0.3	0.4	0.4	0.3	0.6	0.7	**0.08**	0.3	3.635	3.915	3.340	−0.702
0.4	6	5	0.3	0.4	0.4	0.3	0.6	0.7	0.2	**0.07**	3.845	3.907	3.436	−0.139
0.4	6	5	0.3	0.4	0.4	0.3	0.6	0.7	0.2	**0.08**	3.891	3.944	3.472	−0.139
0.4	6	5	0.3	0.4	0.4	0.3	0.6	0.7	0.2	**0.09**	3.934	3.980	3.506	−0.139

5 Sensitivity Investigation

The sensitivity examination of the transmission rate parameters β_1* and β_2* and the aggregate cost work $p*(t)$ are contemplated regarding the parameters t, m_1, m_2, p_1, p_2, α_1, α_2, R_1, R_2, C_1 and C_2. The following data has been considered for the sensitivity analysis.

$t = 0.7$ s,
$m_1 = 1$,
$m_2 = 3$,
$p_1 = 0.3$,
$p_2 = 0.2$,
$\alpha_1 = 0.5 \times 10^4$ packets/s,
$\alpha_2 = 0.4 \times 10^4$ packets/s,
$R_1 = 0.7$,
$R_2 = 0.6$,
$C_1 = 0.4$ (Table 2).

The execution measures influenced with the variety in time (t) and the bunch estimate dissemination parameters of entries. As time (t) increments by 15%, the average number of bundles broadcast through the two cushions increments alongside the two transmitters and the lending rate of the parcels increments. As the group measure appropriation parameter p increments to 15%, the average number of parcels broadcast through the two cradles increments alongside the two transmitters and the entry rate of the bundle's increments. Overall investigation of the parameters mirrors that effective data transfer capacity distribution methodology for clog control immensely diminishes the mean postponement in correspondence and enhance influence excellence by lessening business in cushions.

6 Conclusion

The current paper dealing with a good and fresh correspondence arranges to demonstrate with mass landings having two-phase coordinate entries. Here it is expected that the messages arrive straightforwardly to the first cushion and second cradle which are associated pair. Advance expected that the communication is changed over into a random number of parcels and put away in cushions for forwarding transmission. The landing forms in both the cradles portrayed with compound Poisson binomial procedures. With appropriate cost contemplations, the ideal working approaches of the correspondence organise are additionally determined. Through mathematical representations, the affectability of the adjustments in input parameters and expenses on the ideal working arrangements likewise examined. This correspondence organises much helpful for execution assessment, control and observing of correspondence systems at information/voice transmissions, satellite interchanges, LLAN, WAN planning

Table 2 Sensitivity analysis

Parameter	Performance measure	% change in parameters						
		−15%	−10%	−5%	0	+5%	+10%	+15%
$t = 0.7$	β_1*	4.013	4.052	4.087	4.105	4.126	4.152	4.178
	β_2*	3.950	3.970	3.991	4.011	4.031	4.051	4.071
	R	3.581	3.605	3.628	3.643	3.660	3.683	3.697
$P_1 = 0.3$	β_1*	3.724	3.789	3.818	3.851	3.885	3.910	3.942
	β_2*	3.897	3.920	3.942	3.964	3.986	4.007	4.029
	R	3.442	3.479	3.501	3.524	3.547	3.566	3.589
$P_2 = 0.2$	β_1*	3.896	3.924	3.952	3.976	3.998	4.018	4.036
	β_2*	3.888	3.923	3.957	3.991	4.024	4.057	4.090
	R	3.508	3.535	3.561	3.585	3.608	3.630	3.651
$\alpha_1 = 0.5$	β_1*	3.862	3.889	3.922	3.942	3.958	3.982	4.003
	β_2*	3.856	3.880	3.905	3.929	3.953	3.977	4.001
	R	3.481	3.502	3.526	3.545	3.561	3.581	3.600
$\alpha_2 = 0.4$	β_1*	3.796	3.823	3.849	3.878	3.892	3.918	3.947
	β_2*	3.763	3.794	3.825	3.855	3.885	3.915	3.945
	R	3.413	3.438	3.462	3.487	3.505	3.529	3.554
$R_1 = 0.7$	β_1*	3.781	3.804	3.831	3.862	3.886	3.911	3.937
	β_2*	3.803	3.827	3.851	3.875	3.899	3.922	3.946
	R	3.425	3.445	3.466	3.489	3.509	3.529	3.550
$R_2 = 0.6$	β_1*	3.751	3.787	3.812	3.841	3.874	3.898	3.921
	β_2*	3.876	3.910	3.943	3.976	4.009	4.042	4.074
	R	3.445	3.474	3.499	3.525	3.552	3.576	3.600
$C_1 = 0.4$	β_1*	3.851	3.898	3.925	3.968	3.987	4.019	4.041
	β_2*	3.854	3.879	3.903	3.927	3.951	3.974	3.998
	R	3.476	3.505	3.526	3.554	3.572	3.592	3.613
$C_2 = 0.2$	β_1*	3.471	3.492	3.524	3.552	3.581	3.603	3.634
	β_2*	3.967	3.396	3.426	3.455	3.483	3.512	3.540
	R	3.127	3.160	3.193	3.214	3.240	3.263	3.285
		−75%	**−50%**	**−25%**	**0**	**+25%**	**+50%**	**+75%**
$m_1 = 1$	β_1*	3.487	3.503	3.521	3.548	3.561	3.578	3.591
	β_2*	3.415	3.427	3.440	3.452	3.464	3.477	3.489
	R	3.146	3.159	3.174	3.193	3.204	3.219	3.230
$m_2 = 3$	β_1*	3.202	3.239	3.261	3.286	3.307	3.331	3.364
	β_2*	3.253	3.299	3.344	3.389	3.433	3.476	3.518
	R	2.927	2.966	2.997	3.030	3.060	3.091	3.127
	R	3.425	3.445	3.466	3.489	3.509	3.529	3.550

and web suppliers. This correspondence organises model can likewise be stretched out to non-Markovian change forms which require assist examinations.

References

1. Sriram K (1993) Methodologies for bandwidth allocation, transmission scheduling and congestion avoidance in broadband ATM networks. Comput Netw, ISDN Syst J 26:43–59
2. Ramasundari MV, Srinivasa Rao K, Srinivasa Rao P, Suresh Varma PS (2011) Three node communication network model with dynamic bandwidth allocation and Non homogeneous Poisson arrivals. Int J Comput Appl 31(1):19–27
3. Trinatha Rao P, Srinivasa Rao K, Reddy KVVS (2012) Performance of non-homogeneous communication with Poisson arrivals and dynamic bandwidth allocation. Int J Syst, Control Commun 4(3):164–182
4. Srinivasa Rao K, Varma PS, Reddy PVGDP (2007) A communication network with random phases of transmission. Indian J Math Math Sci 3(1):39–46
5. Srinivasa Rao K, Ramasundari MV, Srinivasa Rao P, Suresh Varma PS (2011) Three node communication network model with modified phase type transmission under DBA having NHP arrivals. Int J Comput Eng 4(1):17–29
6. K.K. Leung, Load dependent service queues with application to congestion control in broadband networks. Perform Eval **50**(1–4), 27–40 (2002)
7. L. Kleinrock, R.R. Muntz, E. Rodemich, The processor-sharing queuing model for time-shared systems with bulk arrivals. Int J Netw **1**(1), 1–13 (2006)
8. Suresh Varma P, Srinivasa Rao P, Prasad Reddy PVGD (2007) A communication Network with random phases of transmission. Indian J Math Math Sci 3(1):39–46
9. Ramasundari MV, Srinivasa Rao K, Srinivasa Rao P, Suresh Varma PS (2013) Tandem communication network model with DBA and modified phase type transmission having NHP arrivals for first node and Poisson arrivals for second node. Int J Comput Sci Issues 8(5):51–58
10. Thirupathi Rao N, Srinivasa Rao K, Srinivasa Rao P (2019) Optimal traffic flow in tandem communication networks. Int J Control Autom 11(12):35–46

Design and Implementation of a Modified H-Bridge Multilevel Inverter with Reduced Component Count

Madisa V. G. Varaprasad, B. Arundhati, Hema Chander Allamsetty, and Phani Teja Bankupalli

Abstract In the recent era, applications of the multilevel inverter (MLI) technology have expanded in the various areas of power conversion technology. MLI facilitates the efficient use of renewable energy sources, and power from renewable resources is DC in nature and is to be converted to AC for domestic/industrial loads. The efficient conversion of power from DC to AC is accomplished by these MLIs. There are various topologies of multilevel inverter, but the major drawbacks are related to their complexity and they require high count of power devices. In this regard, this paper proposes a topology with six unidirectional switches and two voltage sources to obtain five-level and seven-level voltage. This configuration produces improved quality of output waveform with reduced harmonic content and device count when compared to classical topologies. The proposed topology has been validated in MATLAB/SIMULINK environment.

Keywords Multilevel inverter · Modified H-bridge inverter · Total harmonic distortion (THD)

1 Introduction

Multilevel inverter technology in recent times has emerged as a significant substitute to conventional inverter topologies for obtaining improved output voltage and current waveforms. It is capable of providing a desired output AC voltage with less

M. V. G. Varaprasad (✉) · B. Arundhati
Vignan's Institute of Information Technology, Visakhapatnam, India
e-mail: mvgvaraprasad@gmail.com

H. C. Allamsetty
Gayatri Vidya Parishad College of Engineering for Women, Visakhapatnam, India

P. T. Bankupalli
Aditya Institute of Technology and Management, Tekkali, India

D. Bhattacharyya and N. Thirupathi Rao (eds.), *Machine Intelligence and Soft Computing*, Advances in Intelligent Systems and Computing 1280,
https://doi.org/10.1007/978-981-15-9516-5_27

than/equal to lower level of DC voltage input. It can generate sinusoidal output voltages, as in the shape of a stepped voltage waveform using DC voltage sources. In response to the emergent demand for medium and low-power applications, multilevel inverter technology has been drawing consideration in variable speed wind turbine and photovoltaic systems recently. In recent years, renewable energy power supply connected to the utility grid has drawn attention due to an increase in environmental pollution, energy demand boom, and most importantly, the drastic increase in fossil fuel prices [1–4]. The energy obtained from renewable sources has a lower voltage level DC and relatively low conversion efficiency. To get better efficiency and to convert the generated voltage into usable AC, multilevel inverter acts as a link between renewable power generation at one side and low-power load on the other end [5].

The conventional MLIs include diode clamped, flying capacitor, and cascaded H-bridge multilevel inverter. Among them, cascaded H-bridge multilevel is simpler than other inverter topologies because of the reduced component count; low harmonic distortion for over switching frequencies, circuit complexity is reduced, low cost, and size. The proposed topology uses cascaded H-bridge and produces five- and seven-level output voltages with less number of power elements and with less complex circuit. With increase in the number of output levels, the harmonic content decreases [5, 6].

2 Proposed Topology

The proposed system can be generalized as depicted in Fig. 1 using block diagram. The proposed MLI topology generates five-level and seven-level voltage across the load. The proposed topology has many advantages compared to other topologies. Due to the structure and working of the proposed topology, it can be considered as a modified version of the H-bridge inverter topology.

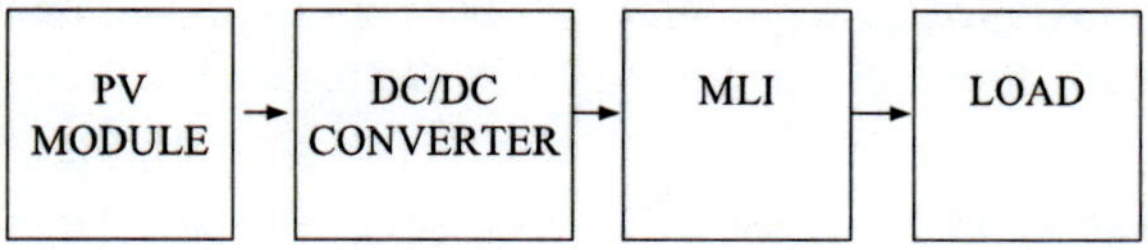

Fig. 1 Block diagram

Table 1 Switching states of proposed five-level MLI

$Q1$	$Q2$	$Q3$	$Q4$	$Q5$	$Q6$	V_0
Off	Off	Off	Off	**On**	Off	**0**
On	Off	Off	Off	**On**	**On**	**V**
Off	Off	**On**	**On**	**On**	Off	**2 V**
On	Off	Off	Off	**On**	**On**	**V**
Off	Off	Off	Off	**On**	Off	**0**
Off	**On**	**On**	**On**	Off	Off	**−V**
On	Off	Off	Off	**On**	**On**	**−2 V**
Off	**On**	**On**	**On**	Off	Off	**−V**

2.1 *Working of Modified H-Bridge Inverter for Five Level*

In this modified H-bridge, we use six switches and two voltage sources for placing an *R*-load. Normally, it operates in four modes and one short-circuit mode adding to a total of five modes of operation. Table 1 illustrates different switching states.

2.2 *Operation*

As previously said, it will be operated in four modes that will give complete five-level output. It consists of six switches, and it was named in the circuit diagram and two voltage sources named as V_1 and V_2, and output (V_0) is collected across the resistor (R). Here, for the triggering of MOSFETs, we are using sinusoidal pulse with modulation technique

Mode 1 Three switches are triggered at any instant of time out of six, in Mode-1, only switches-1, 6, and 5 are triggered with the help of a pulse generator. The current flows from V_1 positive to resistor with the path of switch-1, resistor, switch-6, switch-5, and V_1 negative. The source voltage V_1 will appear across the resistor which is equal to V_1 (Fig. 2).

Mode 2 In Mode-2, three switches, 5, 4, and 3, are triggered only with the help of a pulse generator. The current takes the path of V_2 positive, switch-5, switch-4, resistor, and switch-3, and return to V_2 negative. The source voltage V_2 will appear across resistor R, so the output is equal to V_2 (Fig. 3).

Mode 3 In Mode-3, only the switches-2, 3, and 4 are triggered with the help of a pulse generator. The current flows from V_1 positive to resistor with the path of switch-2, switch-3, resistor, switch-4, and V_1 negative. The source voltage across resistor R in the reverse direction is equal to $-V_1$ (Fig. 4).

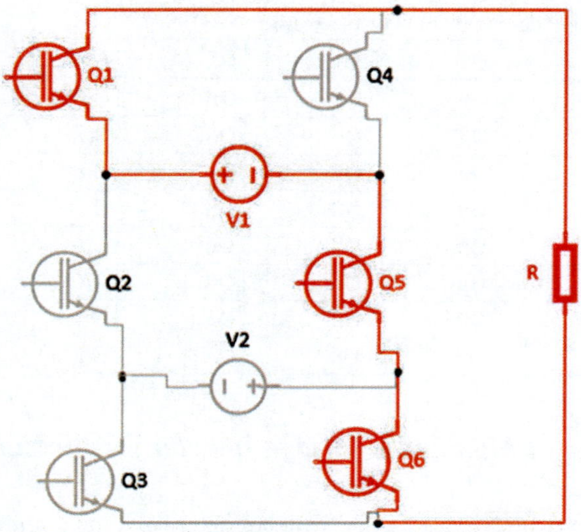

Fig. 2 Mode-1 where V_o is positive V_1

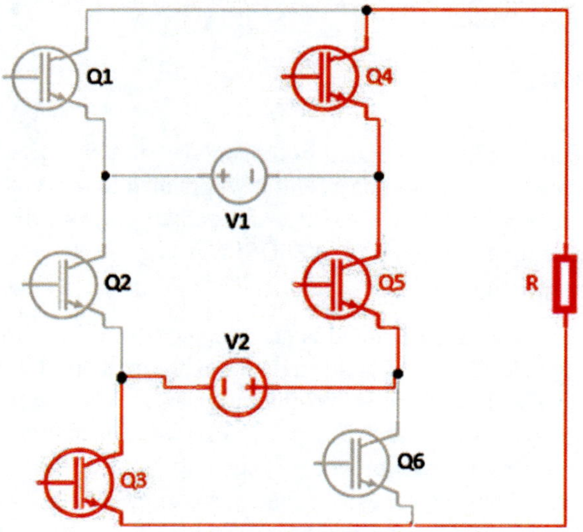

Fig. 3 Mode-2 where V_o is positive V_2

Mode 4 In Mode-4, the switches-6, 1, and 2 are triggered with the help of a pulse generator. The current takes the path of V_2 positive, switch-6, resistor, switch-1, switch-2, and return to V_2 negative. The source voltage V_2 will appear across resistor R in the reverse direction, so the output is equal to $-V_2$ (Fig. 5).

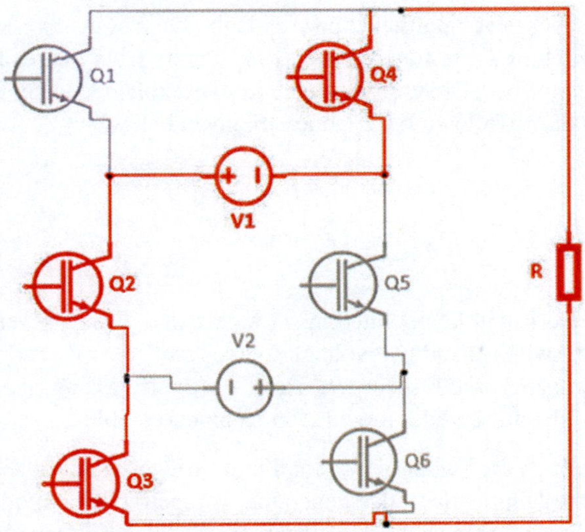

Fig. 4 Mode-3 where V_o is negative V_1

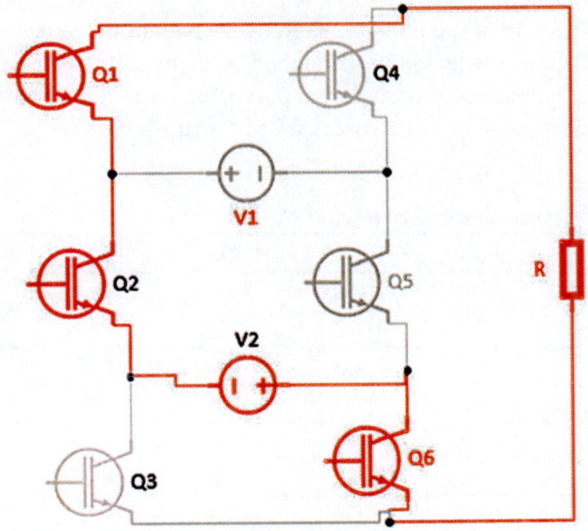

Fig. 5 Mode-4 where V_o is negative V_2

2.3 *Working of Modified H-Bridge Inverter for Seven Level*

The operation of modified H-bridge seven-level multilevel inverter is explained from the modes and switching logic that are shown in the below figures. The modified H-bridge for seven-level multilevel inverter is being done by using a resistive load (*R*-load), and the circuit also includes two DC voltage sources and six switches. The diagram of the seven-level multilevel inverter is the same as the five-level multilevel inverter, but switching logic for the seven level differs from that of five level. The operation of the multilevel inverter for seven level is explained using six modes, and the explanation of the following six modes are given below.

2.4 Operation

As previously said it will be operated in six modes and gives a seven-level output. It consists of six switches and two voltage sources named as V_1 and V_2, the output (V_0) is collected across the resistor (R). Here, for the triggering of MOSFETs, we are using sinusoidal pulse width modulation technique (Table 2).

Mode 1 In Mode-1, the switches-1, 5, and 6 are triggered out of the six switches with the help of the pulse generator. The path of this mode is explained as follows: The current flows from the V_1 positive terminal through the switch-1 to resistor, switch-6, switch-5, and back to V_1 negative. The source voltage V_1 will appear across the resistor which is equal to V_1 (Fig. 6).

Mode 2 In Mode-2, the switches-5, 4, and 3 are triggered out of the six switches with the help of the pulse generator. The path of this mode is explained as follows: The current flows from the V_2 positive terminal through switch-5, switch-4, resistor, switch-3, and pave back to the V_2 negative. The source voltage V_2 will appear across resistor R, so the output is equal to V_2 (Fig. 7).

Table 2 Switching states for proposed seven level

$Q1$	$Q2$	$Q3$	$Q4$	$Q5$	$Q6$	V_0
Off	Off	Off	On	On	On	0
On	Off	Off	Off	On	On	V
Off	Off	On	On	On	Off	2 V
On	Off	On	Off	On	Off	3 V
Off	Off	On	On	On	Off	2 V
Off	On	On	On	Off	Off	V
Off	Off	Off	On	On	On	0
Off	On	On	On	Off	Off	−V
On	On	Off	Off	Off	On	−2 V
Off	On	Off	On	Off	On	−3 V
On	On	Off	Off	Off	On	−2 V
Off	On	On	On	Off	Off	−V
Off	Off	Off	On	On	On	0 V

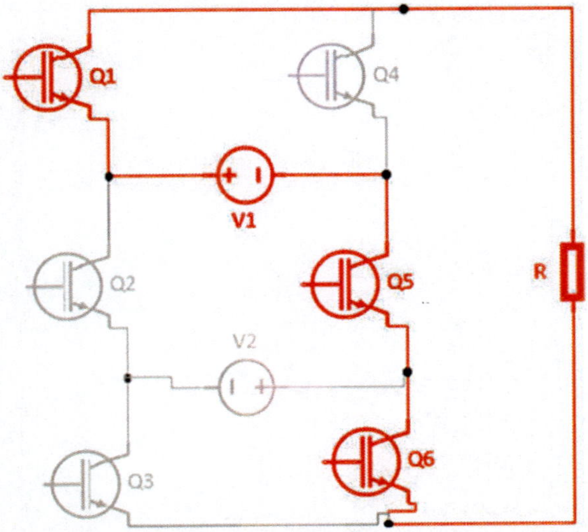

Fig. 6 Mode-1 where V_o is positive V_1

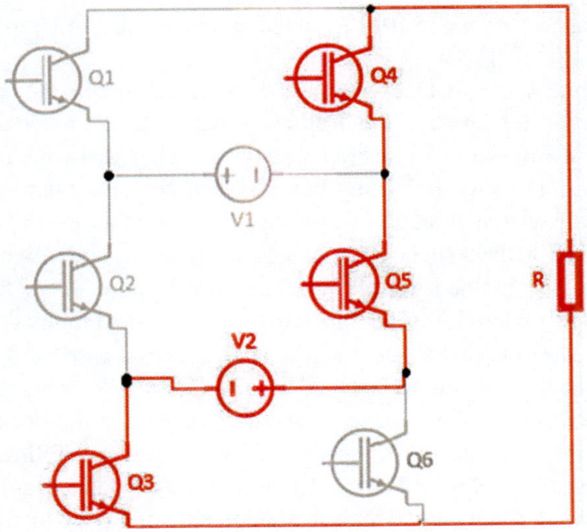

Fig. 7 Mode-2 where V_o is positive V_2

Mode 3 In Mode-3, the switches-1, 3, and 5 are triggered with the help of the pulse generator. In this method, both the voltage sources, V_1 and V_2, are included. The path of the following method is explained as follows: The current flows from the V_1 positive through switch-1, resistor, switch-3, and passes through the negative terminal of the V_2 and then passes through the switch-5 and returns to V_1 negative. The source voltages, V_1 and V_2, both

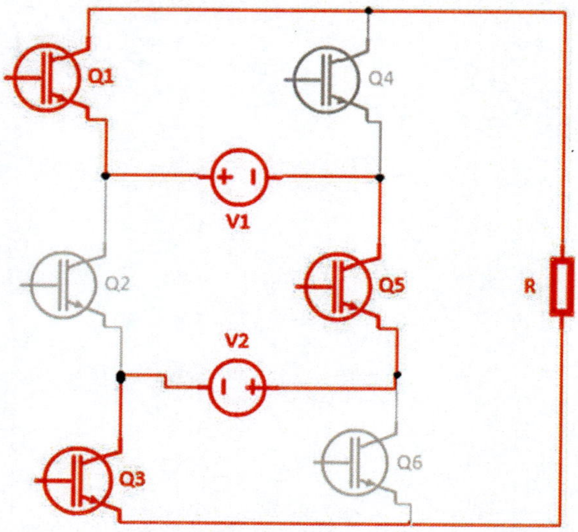

Fig. 8 Mode- 3 where V_o is positive $(V_1 + V_2)$

appears across the resistor R, so the output is equal to $V_1 + V_2$, i.e., $(V_0 = V_1 + V_2)$ (Fig. 8).

Mode 4 In Mode-4, the switches-2, 3, and 4 are triggered with the help of the pulse generator. The path of this following method is explained as follows: The current flows from V_1 positive through switch-2, switch-3, resistor, switch-4, and returns to the V_1 negative. The source voltage across the resistor R is in the reverse direction that is equal to $-V_1(V_0 = -V_1)$ (Fig. 9).

Mode 5 In Mode-5, the switches 6, 1, and 2 are triggered. The path of the following mode is explained as follows: The current flows from the V_2 positive through switch-6, resistor, switch-1, switch-2, and returns to the V_2 negative. The source voltage V_2 will appear across resistor R in the reverse direction, so the output is equal to $-V_2$ $(V_0 = -V_2)$ (Fig. 10).

Mode 6 In Mode-6, the switches-2, 6, and 4 are triggered out of the six switches with the help of a pulse generator. The path of the following method is explained as follows: The current flows from V_2 positive and flows through the switch-6, resistor, switch-4, and negative of V_1 and then passes through switch-2 and returns to the V_2 negative. The source voltage, V_1 and V_2, both appear across the resistor R in the reverse direction, so the output is equal to $(-V_1 - V_2)$ (Fig. 11).

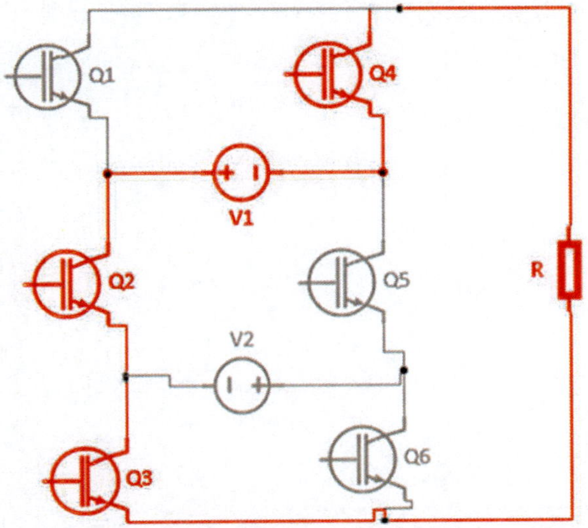

Fig. 9 Mode-4 where V_o is negative V_1

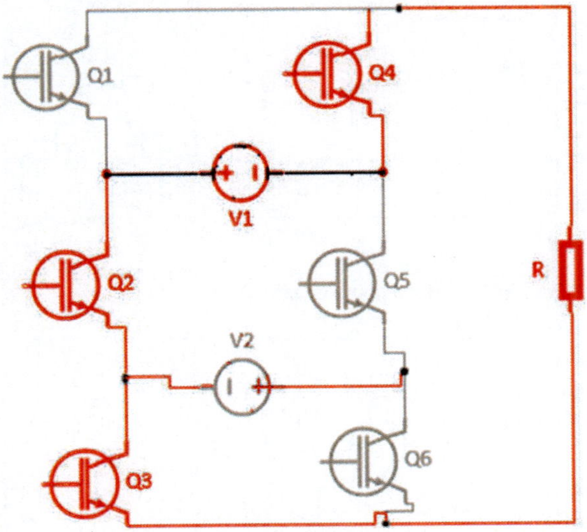

Fig. 10 Mode-5 where V_o is negative V_2

3 Simulation Results

The proposed modified H-bridge MLI has been validated in MATLAB/Simulink platform, and the results are shown in Figs. 4 and 6. The five-level and seven-level inverter output voltage is as shown as (Figs. 12, 13, 14, 15, 16 and 17).

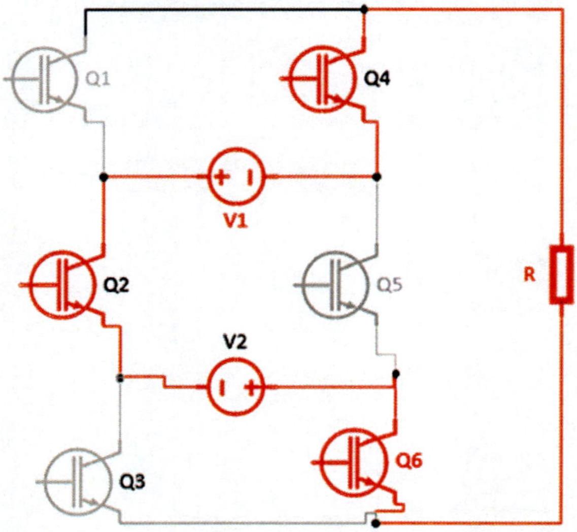

Fig. 11 Mode-6 where V_o is negative $(V_1 + V_2)$

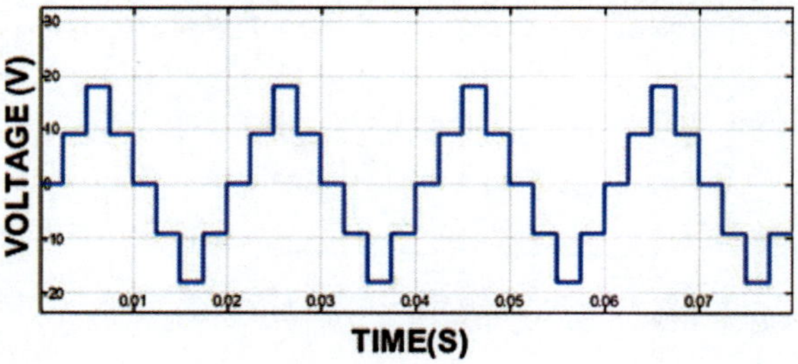

Fig. 12 Proposed five-level inverter output voltage

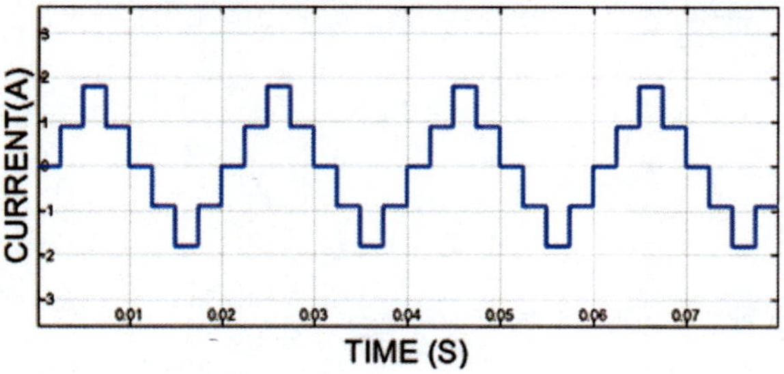

Fig. 13 Proposed five-level inverter output current

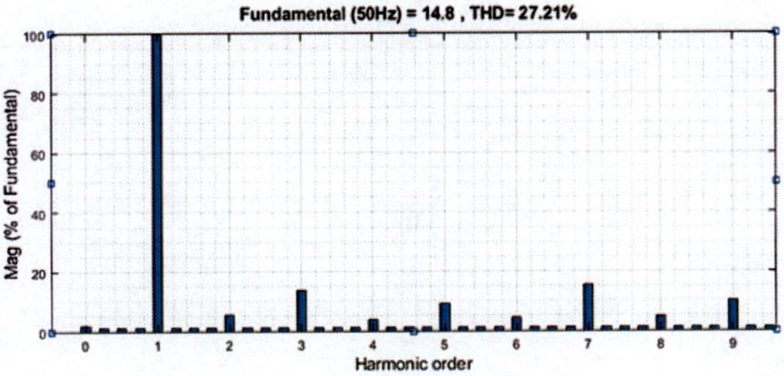

Fig. 14 THD of five-level inverter

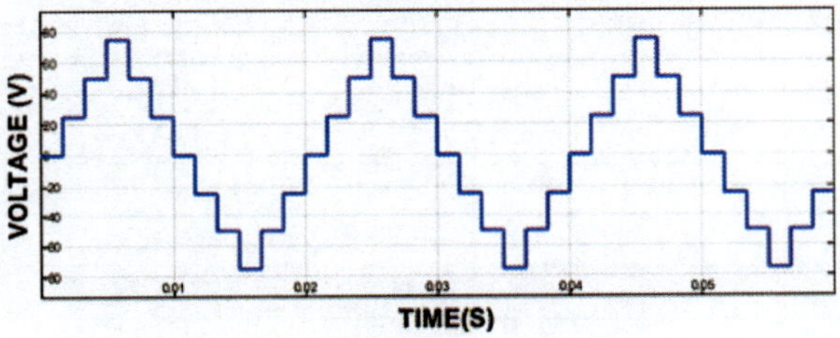

Fig. 15 Proposed seven-level inverter output voltage

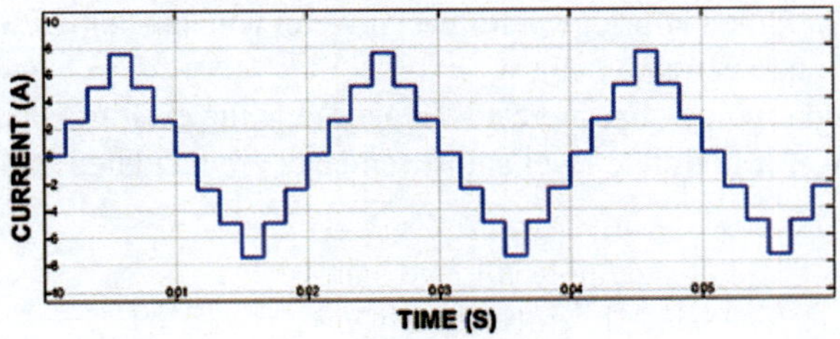

Fig. 16 Proposed seven-level inverter output current

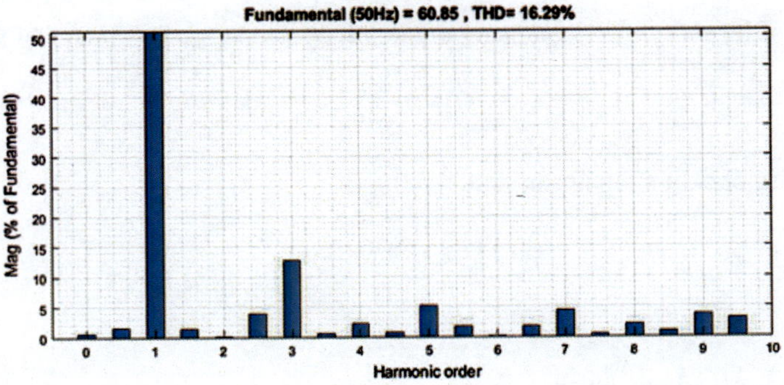

Fig. 17 THD of seven-level inverter

Table 3 Comparison of the five-level inverter with existing topologies

Topology/parameter	Diode clamped [1]	Flying capacitor [2]	Cascaded H-bridge [3]	Proposed topology
Power switches	4	4	8	6
DC sources	1	1	2	2
Capacitors	4	10	–	–
Clamping diode	12	–	–	–
THD%	42.69	40.98	39.41	27.21

Table 4 Comparison of the seven-level inverter with existing topologies

Topology/parameter	Diode clamped [1]	Flying capacitor [2]	Cascaded H-bridge [3]	Proposed topology
Power switches	12	12	12	6
DC sources	6	6	3	2
Capacitors	6	15	–	–
Clamping diode	30	6	–	–
THD%	24.65	19.45	16.31	16.29

This paper aims to generate seven-level and seven-level outputs of single-phase voltages using modified H-bridge multilevel inverter with the help of sinusoidal pulse width modulation technique.

The proposed topology has more advantages than the existing topologies; hence, the comparisons for both five level and seven level are tabulated in Tables 3 and 4, respectively.

4 Mathematical Modeling

Number of levels synthesized using the topology can be generalized and given by:

$$L = 2^{(n+1)} - 1 \tag{1}$$

The numbers of switches used are:

$$Q = 2 * (n + 1) \tag{2}$$

where

L number of voltage levels,
n number of voltage sources,
Q number of switches.

The losses in the MOSFET consist of conduction loss and switching loss.

4.1 Conduction Loss

The conduction losses are given by

$$\mathbf{P_L = V_{ce(sat)} * D * I_c} \tag{3}$$

where

$V_{ce(sat)}$ collector to emitter saturation voltage,
D duty cycle,
I_c collector current.

4.2 Switching Loss

The switching loss in the MOSFET is the product of switching energies and the switching frequency.

The energy loss during the turn-on process is given by

$$W_{ON} = \frac{(V_{ce} * I_c * t_{on})}{6} \tag{4}$$

The power loss during turn-on time is given by

$$P_{ON} = \frac{(V_{ce} * I_c * t_{on}) f_s}{6} \tag{5}$$

The power loss during turn-off time is

$$P_{OFF} = \frac{(V_{ce} * I_c * t_{off}) f_s}{6} \tag{6}$$

The losses in the MOSFET consists of conduction and switching losses. The conduction loss is the power loss that occurs during the conduction period, and the switching losses during the transition from off state to on state.

The total switching power loss can be calculated as

$$P_{sw} = \frac{(V_{ce} * I_c * (t_{on} + t_{off}) f_s}{6} \tag{7}$$

where

V_{ce} collector to emitter voltage,
I_c collector current,
t_{on} turn-on time,
t_{off} turn-off time, and
f_s switching frequency.

5 Conclusion

This paper proposed a modified H-bridge multilevel inverter topology with reduced component count for renewable and industrial drive applications. The structure of the proposed MLI along with this operation has been discussed. The detailed procedure for evaluating the THD of five and seven levels in each stage has been detailed. Further, it has been shown that the proposed inverter employs only six switches which is less compared to the number of switches used in conventional five-level and seven-level inverter topologies. This results in reduced switching losses and circuit complexity. The proposed topology also effectively reduces the harmonics of lower order. Hence, an effective reduction of THD is achieved.

References

1. Pan Z, Peng FZ, Stefanovic V, Leuthen M (2004) A diode-clamped multilevel converter with reduced number of clamping diodes. In: Proceedings of 19th annual applied power electronics conference and exposition, IEEE, pp 820–824. APEC-04
2. Abdul Kadir MN (2006) Novel control strategy for three stage 18-level hybrid multilevel. In: The proceedings of International multi conference on signals and system devices
3. Prajapati P, Jayaraman M, Sreedevi VT (2016) Harmonic elimination in a five-level multi-level inverter. In: The proceedings of international conference on computer communication and informatics
4. B. Arundhati, K. Alice Mary, M. Suryakalavati, Implementation of thrust ripples reduction for a permanent magnet linear synchronous motor using adaptive feed forward controller. J Power Electron **14**(4), 687–694 (2014)
5. Siddique MD, Saad M, Shah NM, Adil S, Atif I, Mudasir A (2019) A new multilevel inverter topology with reduced switch count. IEEE Access
6. Varaprasad, MVG, Ravi Sankar RS, Alice Mary K (2013) Reduction of torque &flux ripple of 3-Φ asynchronous machine using multilevel inverter. Int J Adv Res Electr, Electron Instru Eng 2(2):4781–4793

Private Cloud for Data Storing and Maintain Integrity Using Raspberry Pi

Harsha Vardhan Reddy Padala, Naresh Vurukonda, Venkata Naresh Mandhala, Deepshika Valluru, Naga Sai Reddy Tangirala, and J. Lakshmi Manisha

Abstract A cloud is a virtual platform that lets you work efficiently. When it comes to deep digging into the cloud, it contains a physical data center which is attached at its backend where no one looks about it. As per the scenario, the concept of storage class can be built from any physical storage device when it combined with powerful microprocessors to form a cloud storage device. Cloud storage is of different levels, like standard level, business level, and enterprise level. We plan to build a standard-level storage class from a microprocessor, where these components are not suitable for higher interaction rates. A storage class depends upon speed, agility, performance which is high in enterprise level. As per the planning of concept, we tend to provide a standard storage class with a combination of a microprocessor to handle all types of storage acquisition from a prototype level. A microprocessor like Raspberry Pi or any other can perform the same type of complex operations, and Raspberry Pi lets you work in a feasible manner where it can be used as a computer with low cost. When it comes to the cloud maintenance level, it handles with insane capabilities and best performance with low latency. Own cloud runs efficiently when it bound with Raspberry Pi and a physical storage device. When these are confined properly, a standard storage class with a good performance level will be the output, where some couple of users can easily work with it. All these operations will smoothly work in local host. When it comes to accessing from a different location, it needs to be Port Forwarded. This lets you access from anywhere around the Globe.

Keywords Raspberry Pi · Cloud storage · Integrity · Own cloud

---

H. V. R. Padala (✉) · N. Vurukonda · V. N. Mandhala · D. Valluru · N. S. R. Tangirala · J. Lakshmi Manisha
Department of Computer Science and Engineering, Koneru Lakshmaiah Education Foundation, Vaddeswaram, Guntur, Andhra Pradesh 522502, India
e-mail: harshareddy196@gmail.com

N. Vurukonda
e-mail: naresh.vurukonda@gmail.com

V. N. Mandhala
e-mail: mvnaresh.mca@gmail.com

D. Bhattacharyya and N. Thirupathi Rao (eds.), *Machine Intelligence and Soft Computing*, Advances in Intelligent Systems and Computing 1280,
https://doi.org/10.1007/978-981-15-9516-5_28

1 Introduction

This project mainly comprises a module which deals with cloud [1–3]-related topics. In general, we know all the topics that are related to the cloud such as service models and deployment models [1–3]. When it comes to practical, we know that the cloud [1] is nothing but a group of data centers located in remote places. When it comes to cloud computing [3] it can be determined as processing, accessing, and retrieving of data. When it comes to this project, we try to intervene the cloud in a single microprocessor named Raspberry Pi [4, 5]. In general, Raspberry Pi is a credit card-sized microprocessor that can handle any sort of operations that a real computer can handle. The sort of cloud that we have used is an OwnCloud platform [4] which provides us an interface to combine both the storage device and Raspberry Pi with a single medium. It creates a friendly environment which can be understandable to everyone who is going to use it. As it is of free platform, everyone can easily configure and the flexibility provided by the OwnCloud platform enhances them to do better ideas that can be matched with the present tech industry as it is an IoT [6] device that can be accessed from any location. In this, we tried to integrate a cloud provider named Owncloud with a microprocessor named Raspberry Pi to have better enhancements towards the technology. More Innovations can be implemented with Raspberry Pi as it's an IoT Device.

1.1 Theoretical Analysis

The collection or group of data centers is situated on different locations and can be accessed from any remote location [1–3]. Processing, accessing, and retrieving data over the cloud resources are termed as cloud computing [1–3]. It is on-demand availability. It is of scalable, affordable, secure [7, 8]. For your changing needs, the cloud service [1] can quickly scale up and down. It provides better services for a cheaper price [1]. It does not charge you for unnecessary things. It is fully protected [7, 8] and also consists of high encryption standards. These are of three types [1] "SAAS, PAAS, IAAS".

There are various types of clouds [1]—"public, private, hybrid, community clouds". The public cloud is mainly for public usage [1]. Public cloud application frameworks are perfect for growing and steadily rising organizations [1]. The private cloud is mainly for an organizational purpose. It is more secure than the public [7, 8]. The hybrid cloud integrates both private and public clouds [1]. It is ideal for scalability, flexibility, and security [7, 8]. The community cloud offers a group of organizations that tries to incorporate with a single cloud from their community purpose [1]. Only those have accessibility rights. Each deployment model consists of their functionalities and their cloud architectures [1]. Each model is identical as follows (Fig. 1).

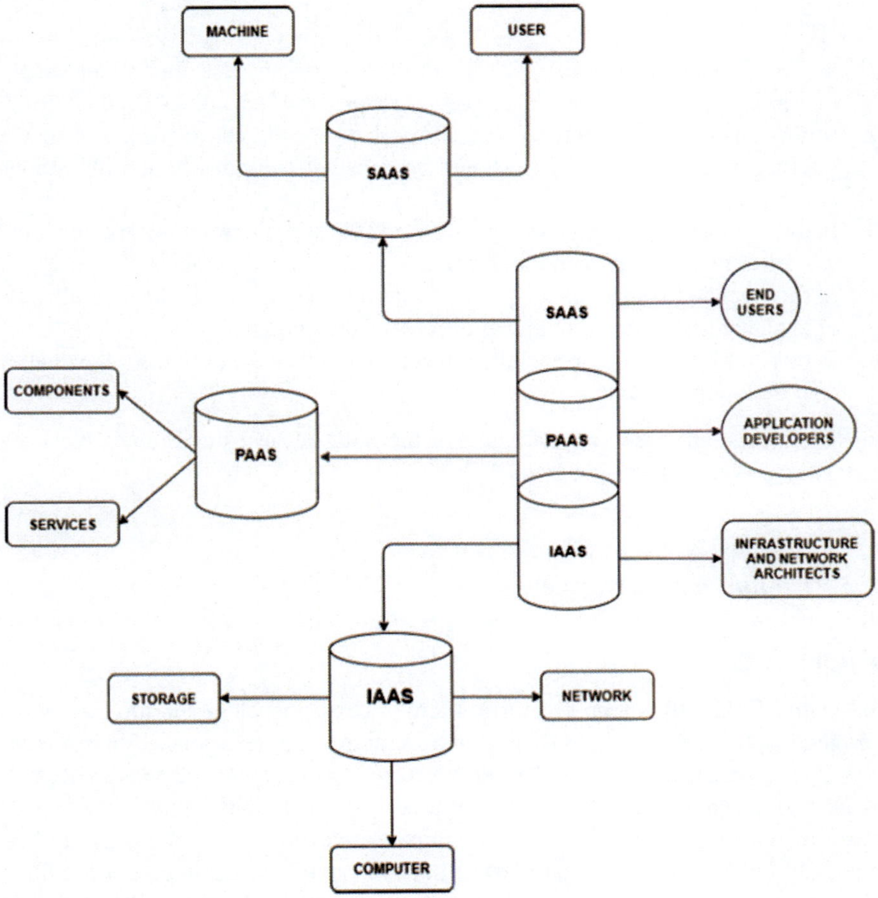

Fig. 1 Blueprint of cloud services

1.2 Cloud Computing Risks and Rewards

Rewards

1. It is of cost efficient [1] because it follows the pay as you grow mechanism.
2. It is of an on promise technology nowadays.
3. It is of high speed [1] because we can easily deploy the required service in fewer clicks.
4. It consists of "excellent accessibility [1, 9]" because it can be accessed from anytime, anywhere, whenever it is required.
5. It consists of a "back up and restores data [3, 8, 9]" facility where it can easily be retrieved from anywhere where it is a promising technology.
6. It also consisting of a "manageability [1]" factor where you can easily manage data whenever it is required.

Risks

1. In this, “vulnerable to attacks [7, 8]” is the key issue because altering the data in the cloud may lead to severe challenges perhaps it is of an online-based platform.
2. In this, “network connectivity” is the issue where the entire work is over the Internet. Internet connectivity should be constant and consistent with various payloads.
3. In this, “down time” is the key issue where if the system gets down the company will be in severe losses and the reputation gets down in any means.
4. In this “vendor lock-in” is the concept where the company should face all sorts of challenges where one vendor is different from another.
5. In this “limited control” is the major factor where the customer faces the control over their deployments.

Till now no bank has its infrastructure on the cloud because there comes a security issue that is data tampering may take place.

1.3 Components Overview

Raspberry Pi

Raspberry Pi [5, 10] is a tiny onboard microprocessor developed in the UK by the former Raspberry Pi foundation in which to promote computer science by demonstration [5, 10] in various countries. The introductory model is not much popular because it does not reach market tolerance at that time. It failed because it not consisting of basic computer [5] components such as mouse and keyboard. According to that company by the end of February 2015, they are capable of dealing with 5 million Pis’ which had been made as to the top-selling computer of British. By the end of March 2018, company sales had reached 19 million. There are different models of the Raspberry Pi by which those are made for various uses. The present models tend to use a micro-SD card for file storage and operating system. It can be turned on by using micro-USB port, and it consists of a single HDMI post one audio and video jack socket along with a GPIO connector of 40 pins. Below is the list of the present versions of Raspberry Pi and its features as represented in Fig. 2.

1.3.1 Assets of Raspberry Pi Module

Anyone can install an operating system [11] in a full stretch like Ubuntu, windows, and many others. It is also used as a daily driver. The presence of GPIO is to produce the output to the user by user-specific formats that can be useful for the real world. Newer models of the Pi’s are coming with built-in Wi-Fi, also the Bluetooth model. This is used to take the projects in a wireless architecture [5, 10]. It is of open source,

Fig. 2 Specifications of Raspberry Pi

and it can also be used as a supercomputer [5]. It is capable of multitasking and also it is of energy efficient and acts as a small Web server. It also can handle Web traffic.

1.3.2 Limitations of Raspberry Pi Module

1. It does not consist of fuse protection which leads to the damage if the pins regularity is incorrectly matching.
2. It is of not as fast as the general present-day computers but it does its better performance of 35$ bucks.
3. It does not have any built-in analog to virtually convert the GPIO pins like Arduino. Here comes the play of ADC chip to come up and to work where you need higher accuracy. Continuous power is supplied for the redundant usage of this processor.
4. It cannot run X86 operating systems. As it is a low processor, it is unable to do much complex multitasking. Big businesses, it is of not suitable.

1.4 OwnCloud

1. It is a sharing type service where you can sync your files in an open-source manner where it is of a self-hosted manner.
2. It is of like Google Drive [12] and Dropbox in which you are having the capability of accessing your files [1], contacts, etc.

3. You can sync your files with others and along with your own devices. Own cloud can handle many more tasks that you tend to perform.
4. OwnCloud.org is a solution of the cloud [1, 8, 9] for tiny and medium-size businesses in which they tend to not to compromise about the security [8, 9] of their particular data.
5. It mainly comprises of a little effort and no high IT costs.

1.4.1 The Own Cloud Enterprise Features Include the Following

1. Professional support
2. Collaborative editing
3. Unlimited use free of charge, etc.

2 Methodology

The main methodology under this project is to inhibit the cloud [1]-related software into the microprocessor [5] and to process the data with cloud-related peculiarities and to access it from anywhere around the Globe. As the used cloud is OwnCloud which is just an interface provided by that company. As it is a freeware, it can be used for the development to exhibit new ideas for future use. It provides a flexible environment for everyone to implement new ideas. In this, we try to install this software into a Debian-based Linux known as Raspbian OS [11] which provides a command-line execution process where a cloud can be installed on this basis. It is somehow related to the Linux [11] that is available in the present day. We try to install it by using a lamp server which comprises MySQL, apache, etc. By this, some packages get installed that supports OwnCloud-related factors. This makes us install in a much clearer way to avoid flaws in the implementing process. It also can be used as a weather forecasting device [4] where it takes the data from the other microchip such as Arduino which collects the data directly and it tries to figure out the things that are obtained to a storable format that can be stored directly to the OwnCloud [1, 4]. It can use as a video server [13] where the live video forecast can be done with no delays and also can be stored for further usage. When it comes to gaming it can be used as a retro gaming machine that can handle those smoothly. The concept of Raspberry Pi is also used in robotics [14] to create prototypes for the industrial requirements. For the present-day Web hosting techniques, it can also be used for the serial Web hosting where anyone has a chance to develop a Web server mechanism from a trusted domain [7, 8] for absolutely free for the lifetime from any place of globalwise.

3 Implementation of OwnCloud on Raspberry Pi

Advanced IP scanner is of user-friendly along with it is of free, fast, and powerful network IP scanner. It can locate all the available computers that are connected to the ports.

PUTTY is a program that executes over the terminal emulation. It supports Telnet, SSH, and serial ports. The predefined port is "22". Used to login to a specified IP (Figs. 3, 4 and 5).

VNC termed as virtual network computing. It always allows you to remotely control the other computer. It is of platform-dependent. VNC server operates different types of operating systems (Fig. 6).

4 Integrity in OwnCloud

As we know that OwnCloud provides the best security [7, 8] than some of the clouds because it mainly depends upon the registered IP at the time of registering with the specified IPs. At the time of installing, the OwnCloud, it automatically tends to create a specific file called config.php. It consists of an array where we need to keep the specific array value with the IP that you are going to alter at the time you need. The array index value is represented as:

0 => "192.168.XXX.XXX"

Along with that client, URL is also represented as,

"overwrite.cli.url" => "https://192.168.XXX.XXX/owncloud"

These are to be changed in all perspectives when you change the network or the systems that you are going to be used or you are going to work. This is a "security barrier" by OwnCloud where it tends to stop the users or attackers if they try to type the unregistered IP. This helps in maintaining the privacy of every user who tries to use the OwnCloud Service.

4.1 Overview of Array Index Validation

When it comes to the term "integrity [15–17]", let us consider an Admin who has the right to create another "Admin" along with the "Group". Here, he can add multiple numbers of Admins and Groups with separate login credentials for them. There are provided to specific required ones as per the need. In this, if two Admins try to log in at the same time, there will be no misconception of interchanging the accounts because the OwnCloud tries to maintain all the necessary needs to the users in all the ways. There will be no loss of data that is of case sensitive. There is also a concept that cloud treats the case-sensitive data [9] and normal data in the same manner. But

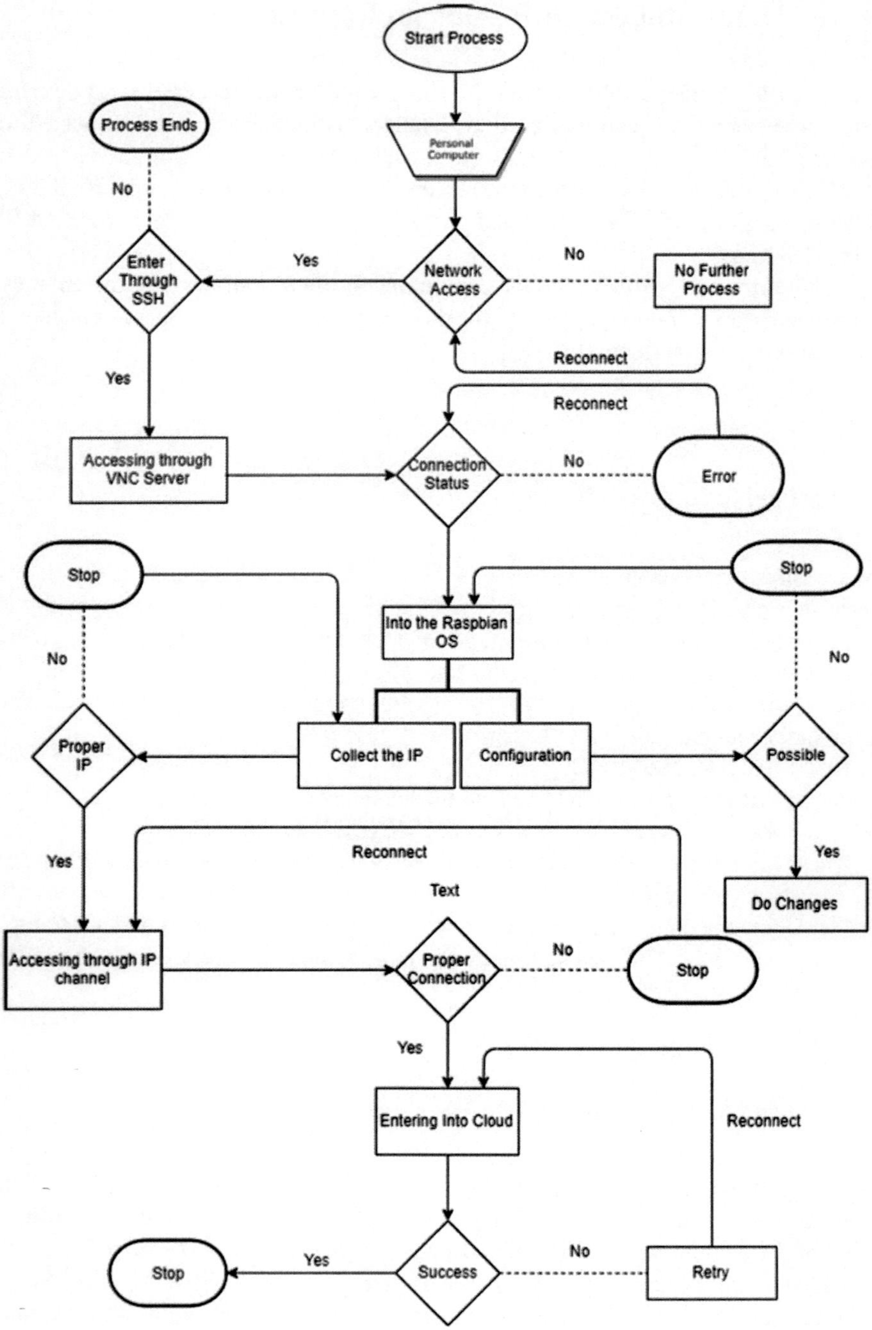

Fig. 3 Path of implementation

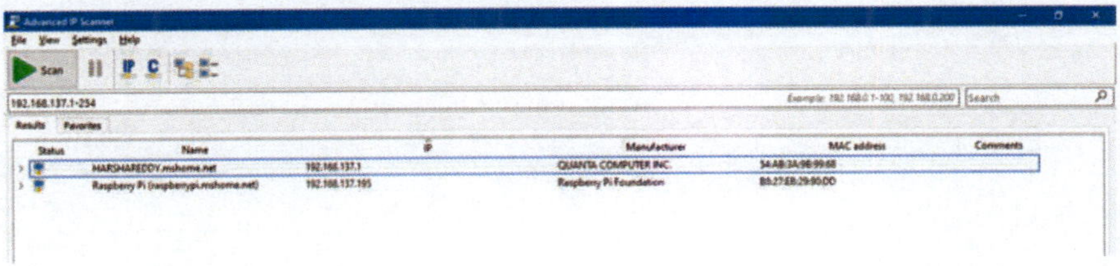

Fig. 4 Locating IP address

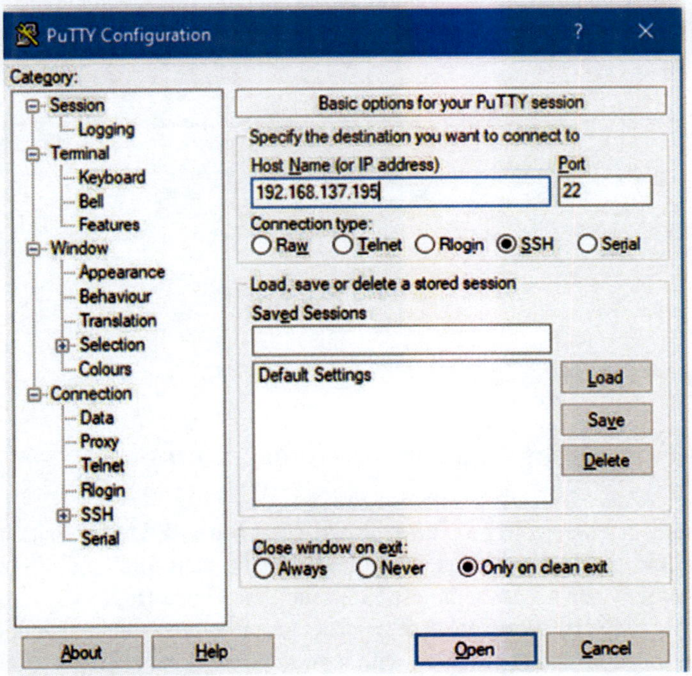

Fig. 5 Accessing via PUTTY

OwnCloud treats the data as the most sensitive one and provides all sorts of security [7, 8] concerns to that. OwnCloud has a feature to share the things via links where we can create the life span of the link and can be shared in all types of social media formats. You can also share the data among the limited "Admins" or some sort of limited "Groups" (Figs. 7 and 8).

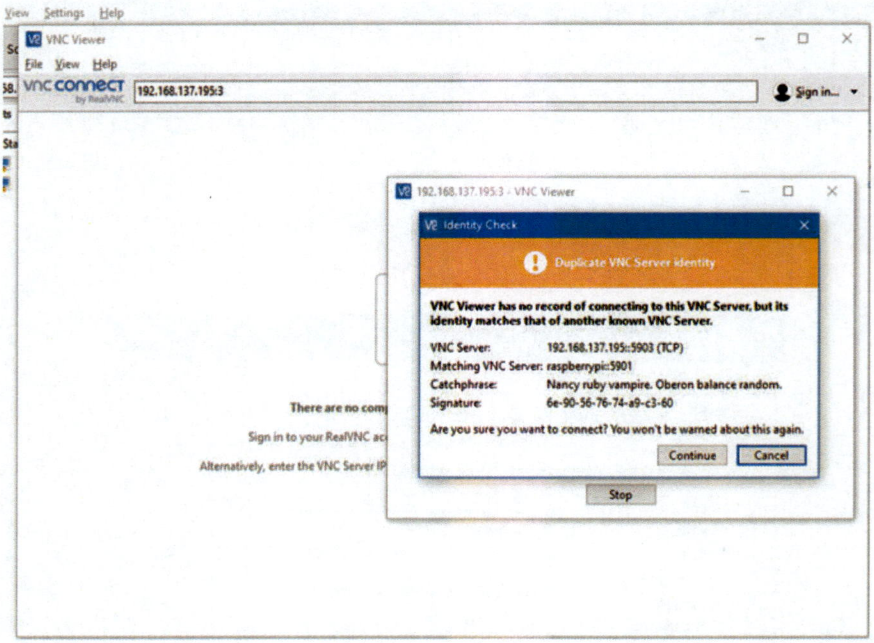

Fig. 6 GUI using VNC viewer

5 Port Forwarding Concept for Global Access

This is a concept where you can access the particular site with fixed IP that is allocated for some domestic purposes. That particular IP can be shared among "N" number of users so that they can access remotely [6] around the Globe (Fig. 9).

In this basically, the IP or WAN of a particular system is collected and it is made bound to a router which is provided with a good Internet connection. After making bound to the router that particular IP is shared between the users who can access it globally.

5.1 Port Forwarding for OwnCloud

To make OwnCloud of particular IP for global access, the IP of the system through which the network is shared should get registered in the "Routers Port Forwarding [6]" option. Then after providing the required settings, the WAN IP should be get registered in the array index in "config.php". There, the IP should be replaced for the global access and save it. Finally, the OwnCloud of the specific user by registered IP can be accessed around the Globe [6] (Figs. 10, 11 and 12).

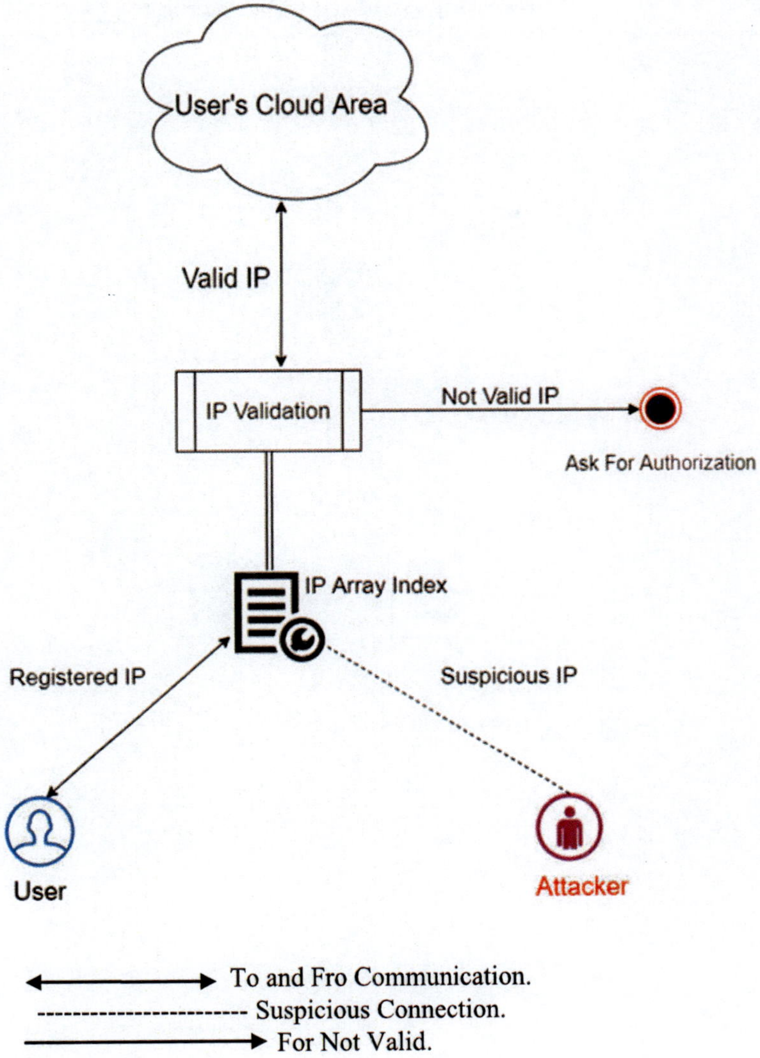

Fig. 7 Array index validation

6 Conclusion

By the end of this project, we can understand better about the present importance of cloud in the fields that requires the state of enhancement in all sort of areas to lead a technological world and along with those microprocessors are the basic computers of present modern technology that fruitfully makes the usage of technology. Finally, we are going to prove that incorporating the open-source cloud for data storing and

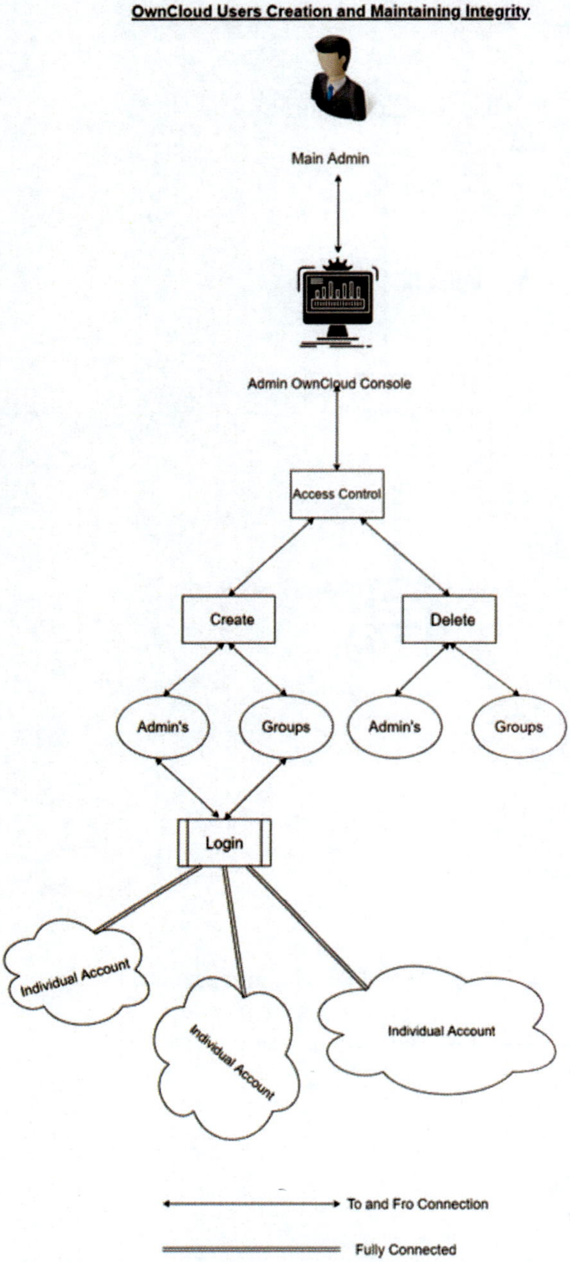

Fig. 8 Integrity and creating of specified users like admins and groups

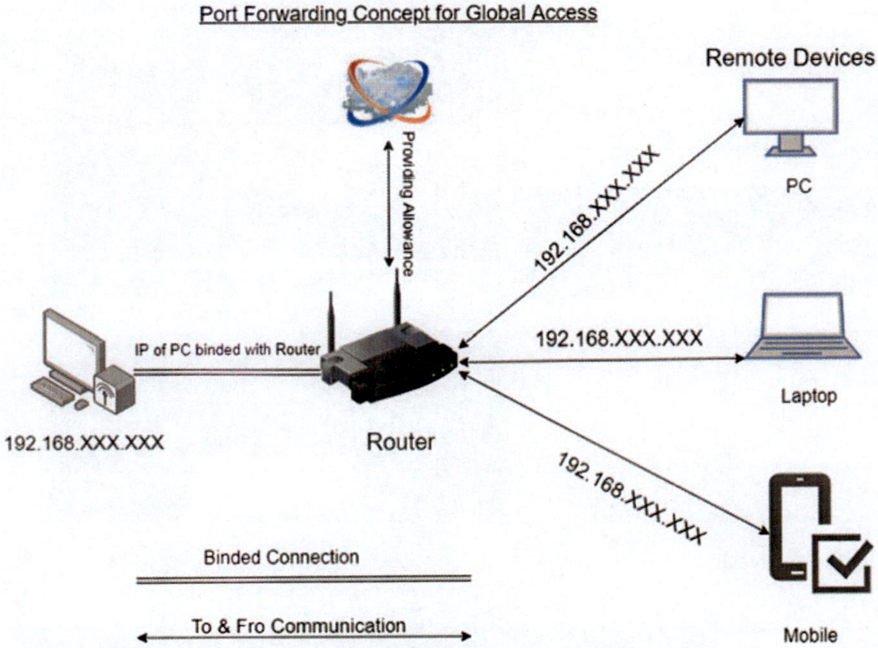

Fig. 9 Concept of port forwarding

Fig. 10 Raspberry Pi setup

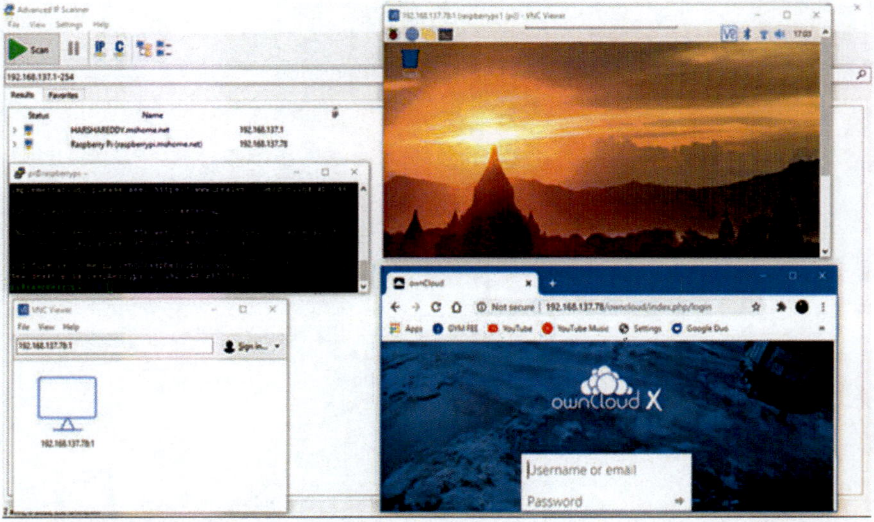

Fig. 11 End result

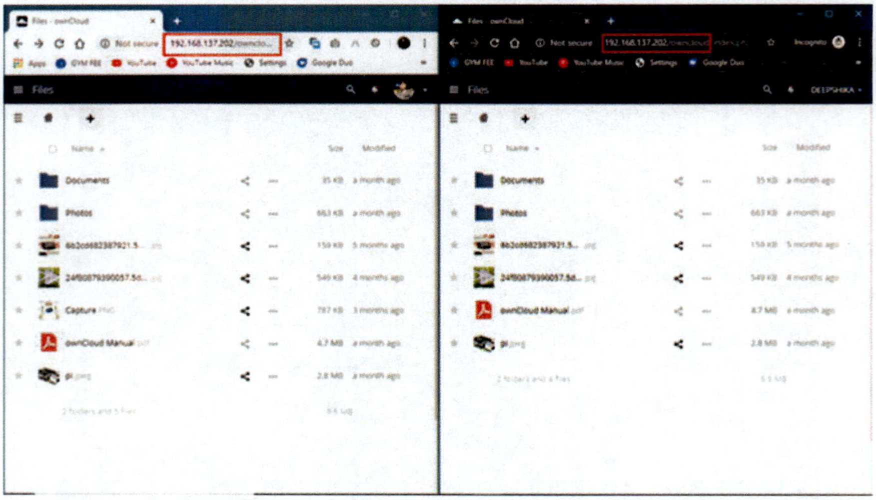

Fig. 12 Integrity b/w users

maintaining integrity on a Raspberry Pi is possible and the microprocessor which we had used is cable of maintaining the tasks properly.

7 Future Scope

As we are in the Modern era we need to be up to date by learning new technologies. When it comes to children they can grasp the knowledge in a short time & they can produce new ideas in terms of Technology.

References

1. R.K. Ramesh, Understanding cloud computing and its architecture. J. Comput. Math. Sci. **10**(3), 519–523 (2019)
2. K. Kamakshaiah, K. Venkateswara Rao, M. Subrahmanyam, SABE: efficient and scalable-filtered access control in distributed cloud data storage, in *Smart Computing and Informatics* (Springer, Singapore, 2018), pp. 31–42
3. T. Sasidhar, P.K. Illa, S. Kodukula, A generalized cloud storage architecture with backup technology for any cloud storage providers. Int. J. Comput. Appl. **2**(2), 256–263 (2012)
4. S.E. Princy, K.G.J. Nigel, Implementation of cloud server for real time data storage using Raspberry Pi, in *2015 Online International Conference on Green Engineering and Technologies (IC-GET)* (IEEE, 2015), pp. 1–4
5. B. Balon, M. Simić, Using Raspberry Pi computers in education, in *2019 42nd International Convention on Information and Communication Technology, Electronics and Microelectronics (MIPRO)* (IEEE, 2019), pp. 671–676
6. N. Verma, A. Jha, Extending port forwarding concept to IoT, in *2018 International Conference on Advances in Computing, Communication Control and Networking (ICACCCN)* (IEEE, 2018), pp. 37–42
7. N. Vurukonda, B.T. Rao, B.T. Reddy, A secured cloud data storage with access privileges. Int. J. Electr. Comput. Eng. **6**(5), 2338–2344 (2016)
8. B.T. Rao, A study on data storage security issues in cloud computing. Procedia Comput. Sci. **92**, 128–135 (2016)
9. B.L. Dhote, G. Krishna Mohan, Trust and security to shared data in cloud computing: open issues, in *International Conference on Advanced Computing Networking and Informatics* (Springer, Singapore, 2019), pp. 117–126
10. N.S. Yamanoor, S. Yamanoor, High quality, low cost education with the Raspberry Pi, in *2017 IEEE Global Humanitarian Technology Conference (GHTC)* (IEEE, 2017), pp. 1–5
11. Z. Youssfi, Making operating systems more appetizing with the Raspberry Pi, in *2017 IEEE Frontiers in Education Conference (FIE)* (IEEE, 2017), pp. 1–4
12. S. Mischie, L. Mâţiu-Iovan, G. GăŠpăresc, Implementation of Google assistant on Rasberry Pi, in *2018 International Symposium on Electronics and Telecommunications (ISETC)* (IEEE, 2018), pp. 1–4
13. F. Salih, S.A. Mysoon Omer, Raspberry Pi as a video server, in *2018 International Conference on Computer, Control, Electrical, and Electronics Engineering (ICCCEEE)* (IEEE, 2018), pp. 1–4
14. B.V.S. Krishna, J. Oviya, S. Gowri, M. Varshini, Cloud robotics in industry using Raspberry Pi, in *2016 Second International Conference on Science Technology Engineering and Management (ICONSTEM)* (IEEE, 2016), pp. 543–547
15. Y. Chen, L. Li, Z. Chen, An approach to verifying data integrity for cloud storage, in *2017 13th International Conference on Computational Intelligence and Security (CIS)* (IEEE, 2017), pp. 582–585
16. P. Parida, S. Konhar, B. Mishra, D. Jena, Design and implementation of an efficient tool to verify integrity of files uploaded to cloud storage, in *2017 7th International Conference on Communication Systems and Network Technologies (CSNT)* (IEEE, 2017), pp. 62–66

17. W. Luo, G. Bai, Ensuring the data integrity in cloud data storage, in *2011 IEEE International Conference on Cloud Computing and Intelligence Systems* (IEEE, 2011), pp. 240–243

Prediction of Swine Flu (H1N1) Patient's Condition Based on the Symptoms and Chest Radiographic Outcomes

Pilla Srinivas, Debnath Bhattacharyya, and Divya Midhun Chakkaravarthy

Abstract H1N1 is one of the rarely found viruses which is spreading from one country to another with high increasing rate. Although the death rate of H1N1 is low when compared to other viruses, the severity of symptoms and the spreading rate is high. So in order to avoid the severity risks behind the H1N1, in this paper, we are going to identify the major symptoms of H1N1 patients and based upon the symptoms, the suspects will get the order for taking the chest X-rays. To evaluate whether the symptoms of swine flu H1N1 patients and the chest radiographs of the patients helps in predicting the clinical outcomes and the actual condition of the patients (Aviram et al. in Radiology 255:252–259, 2010 [1]; Al-Nakshabandi in Pol J Radiol 76:45–48, 2011 [2]). To predict the condition of patients by correlating the major symptoms and clinical outcome of the H1N1 patients, early cases of prediction leads to minimal the severity risks and helps to get rid of the death risks. If the virus is predicted early, then the suspects can be diagnosed early and can be recovered early.

Keywords Swine flu (H1N1) · Chest X-rays (CXR) · Symptoms · Influenza · Radiographs

P. Srinivas (✉) · D. M. Chakkaravarthy
Department of Computer Science and Multimedia, Lincoln University College, Kota Bharu, Malaysia
e-mail: Srinivasp3@gmail.com

D. M. Chakkaravarthy
e-mail: divya@lincoln.edu.my

D. Bhattacharyya
Department of Computer Science and Engineering, K L Deemed to be University, KLEF, Guntur 522502, India
e-mail: debnathb@gmail.com

D. Bhattacharyya and N. Thirupathi Rao (eds.), *Machine Intelligence and Soft Computing*, Advances in Intelligent Systems and Computing 1280,
https://doi.org/10.1007/978-981-15-9516-5_29

1 Introduction

In Mexico, March 2009, new respiratory disease was noticed by the health authorities. Within one week of new influenza swine origin, it was found in California and slowly later few weeks it spread among globally all over the world by June 11, and the World Health Organization (WHO) declared swine influenza as pandemic [3, 4]. The majority of cases included mild influenza and cause illness. According to the health report of the H1N1 patients in USA, the virus includes symptoms like severe illness, pneumonia, and finally acute respiratory syndrome [5, 6].

In USA, among 272 patients, 40% of patients who had gone through chest radiographies are suffering from pneumonia [6]. In California among 1088 hospitalization cases or death, 833 patients who have gone through chest radiography are having pneumonia and acute respiratory syndrome [7, 8] and some are suffering from severe illness have been taken to intensive care unit. Among 66 H1N1 influenza patients, 38% are diagnosed [9, 10].

The suspected patients have gone through chest radiography, but the real suspect need to be gone through chest radiographies, the symptoms when correlated with the initial chest radiographies are suggested to go through the radiography. The chest X-rays help in identifying the severity levels of H1N1 virus. The radiologic finding include the lower parts of the lung zone [11, 12]. Some radiologic characteristics viewed in chest radiographs have led to the poor clinical outcome. In this paper, in order to predict the best clinical outcome, the symptoms and the chest radiographs studies have been done [13–15].

2 Materials and Methods

The adult patients who are suffering from H1N1 influenza and check their front lower zonal radiographies and shift to emergency and observed his condition for a period of 24 hours. The records and the charts of the patient's chest radiographies have been included in the data and the image findings, blood urine, and sputum cultures have been excluded [1, 16, 17].

All the patients' initial chest radiographies have been in posterior–anterior projection. Generally, posterior–anterior projections are taken general X-ray unit. Chest radiographies are taken for people who are having fever 37.80 °C (1000 F) or sometimes higher, cough, and sore throat [18].

3 Study Design

The selected patient's background and past details health track records are observed which may include some health conditions like (obesity, asthma, ongoing pregnancy, diabetes, and smoking). It is also reviewed about how many days the medication has been done and whether they had gone through intensive care unit, mechanical ventilation and death records are recorded. Other minor details like whether the patients currently under medication and also what is the current going status are also recorded.

4 Results

Among the data of 179 patients, H1N1 influenza patients, 97 patients have gone through chest radiographs during the admission time. 39 out of 97 people are having abnormal radiologic findings representing similar symptoms of H1N1 virus and among them 5 are having adverse results. 58 among 97 are having normal radiographs and 2 are having adverse results. The image findings of people individuals who have adverse outcomes are having both opacities in common and also the involvement of multiple lung zones when compared with other patients with good outcomes.

Table 1 represents the background diseases report and the symptoms of patients who have recovered after the illness without any unusual happening when compared with people who are having adverse outcomes. Based on the symptoms of the H1N1, patients' data are gathered individually regarding all of the symptoms which occur during the influenza virus and based upon the symptoms reading graphs are designed to show the differentiation among the normal CXR and abnormal CXR (Fig. 1).

While comparing the symptoms with the H1N1 patient's X-rays, it is graphically represented that the abnormal CXR has the symptoms in abnormal range. Cough and sore throat in normal CXR are normal in range when compared to abnormal CXR. Normal CXR has low cough and sore throat than abnormal CXR. Among the radiographs, they were classified as normal and abnormal radiographs based on their abnormalities. The abnormal radiographs are further classified for finding the ground glass opacity, nodular opacity, and pleural effusion. Among the data study, 55% has the normal chest X-ray and remaining 45% has the abnormal chest X-ray (Tables 2 and 3).

While taking the one of the major symptoms cough we are going to calculate positive and negative PCR values and also individual *P*-values based on the chi-squared test and also considering normal CXR and abnormal CXR (Figs. 2 and 3).

Based on the readings of PCR values, graphs are designed. Graph represents the positive PCR and negative PCR reading of normal CXR with and without cough and Abnormal CXR with and without cough values in percentage and also in the normal values. H1N1 radiologic findings are observed in order to identify the abnormal

Table 1 Conditions and symptoms of 91 H1N1 patients

Characteristic	No mechanical ventilation (n = 90)	Mechanical ventilation/death (n = 7)	*P*-value
Age(y)*	39 ± 15	54 ± 20	0.016
Background			
Male sex	50 (56)	3 (43)	0.698
Diabetes mellitus	9 (10)	0	0.62
Obesity	10 (11)	1 (14)	>0.999
Asthma	15 (17)	0	0.368
Chronic obstructive pulmonary disease	10 (11)	1 (14)	>0.999
Ischemic heart disease	5 (6)	1 (14)	0.37
Congestive heart failure	2 (2)	1 (14)	0.203
Smoker	28 (31)	2 (29)	>0.999
Pregnant	4 (4)	1 (14)	0.318
Symptoms			
Sore throat	26 (29)	1 (14)	0.699
Myalgia	38 (42)	3 (43)	>0.999
Rhinitis	31 (34)	1 (14)	0.420
Diarrhea	19 (21)	2 (29)	>0.999
Vomiting	19 (21)	1 (14)	>0.999
Cough	80 (89)	6 (86)	>0.999
Dyspnea	29 (32)	4 (57)	0.224
Fever > 37.80 °C	76 (84)	6 (86)	>0.999

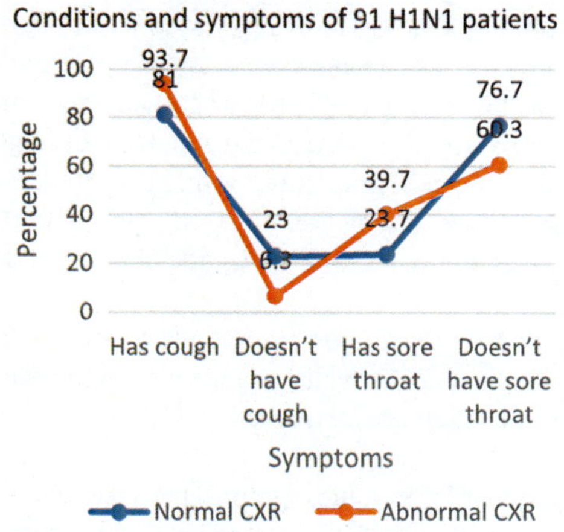

Fig. 1 Graph showing cough and sore throat symptoms in H1N1 patients versus chest X-rays (CXR) in H1N1 patients

Table 2 Statistical relation between cough and polymerase chain reaction (PCR) in percentage according to chi-squared test

	PCR		*P*-value
	+ve ($n = 123$)	−ve ($n = 44$)	
Cough	112 (91.1%)	32 (72.7%)	0.006*
No cough	11 (8.9%)	12 (27.3%)	

Table 3 Statistical relation between cough and polymerase chain reaction (PCR) in decimal values according to chi-squared test (normal CXR/abnormal CXR)

	PCR		*P*-value
	+ve ($n = 123$)	−ve ($n = 44$)	
Cough	112 (68/44)	32 (21/11)	0.006*
No cough	11 (11/0)	12 (8/4)	

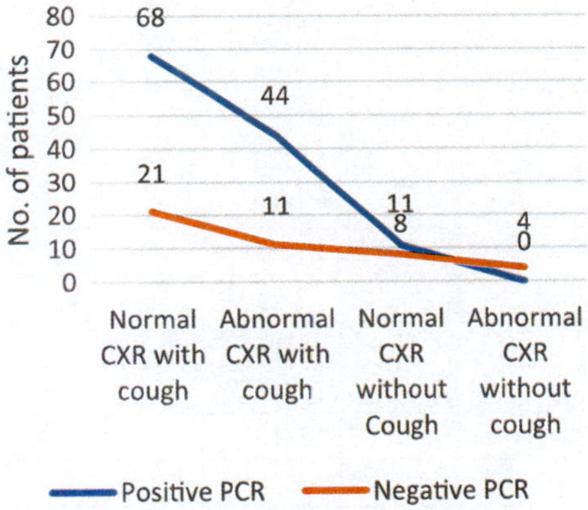

Fig. 2 Graph representing PCR versus CXR and cough in the number of H1N1 patients

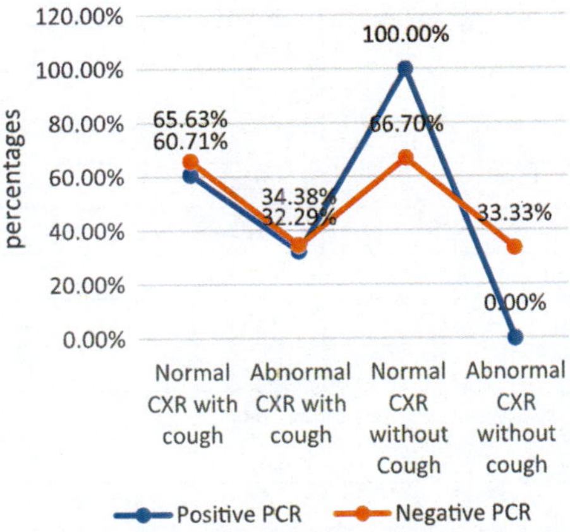

Fig. 3 Graph representing the percentage of PCR versus CXR and cough in H1N1 patients

values for abnormal chest radiographs. 39 patients' readings like opacity, lung zone, and distribution are taken and are tabulated (Table 4).

Data are raw numbers and the percentage is present in the parenthesis. Abnormal chest radiographs help us to identify the clear vision of all the parts and lower, upper, and middle zones in lungs by showing some patches and also irregularities which help to identify the severity based on the overall readings. Radiographs pictures give us a clear view of some of the irregularities. A 45-year-old woman chest radiograph was taken who is suffering from high fever, sore throat, cough, and illness which in that have shown some bilateral airspace opacification is a subset of the larger differential diagnosis for airspace opacification seen in the radiographs (Fig. 4).

Table 4 H1N1 patient's radiologic findings with abnormal chest radiograph

Characteristic	No. of patients ($n = 39$)
Opacity	
Ground glass	27 (69)
Consolidation	**23 (59)**
Patchy	16 (41)
Nodular	**11 (28)**
Confluent	2 (5)
Air bronchogram	**13 (33)**
Lung zone	
Right upper	**4 (10)**
Right middle	26 (66.7)
Right lower	**13 (33)**
Left upper	1 (3)
Left middle	**24 (62)**
Left lower	16 (41)
Distribution	
Central	24 (62)
Peripheral	**30 (77)**
No. of zones	
Single	11 (28)
Multiple	28 (72)
No. of sides involved	
Unilateral	15 (38)
Bilateral	24 (62)
Central	21 (54)
Peripheral	8 (21)
Pleural infusion	3 (8)

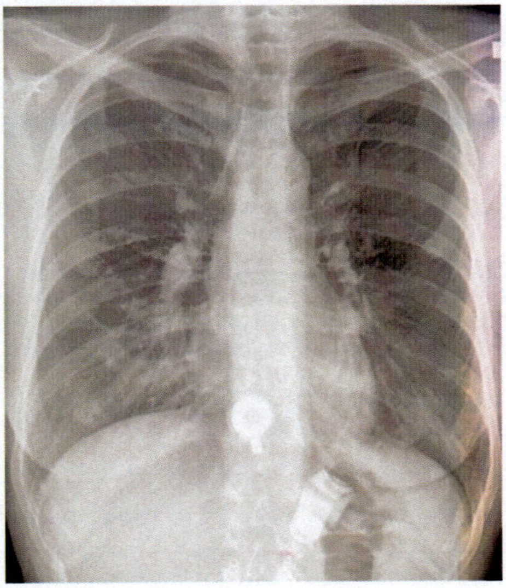

Fig. 4 Representing 45-year-old woman initial chest radiograph of a patient who is having fever and sore throat. Bilateral central ground opacities are seen in peripheral region. Patient has been discharged after three days

The influenza infection was distinguished among the patients based on the pattern of opacity. Among the patients, 82% are hospitalized and remaining 18% were discharge home from emergency ward. Together with illness, sore throat, and cough, dyspnea is also seen among the patients. Dyspnea is the common symptom among all the abnormal chest radiograph patients. Round nodular opacities were also seen in many radiographs. Our results suggest that many radiographs help in identifying the bilateral peripheral involvement and involving two or more lung zones found to predict the progression of respiratory system which in some cases leads to cause of ventilation and also in some severe cases leads to death. In this infection, it may include homogeneous and sometimes unilateral or bilateral patches. Serial radiographs show only poor patchy areas say 1–2 cm in diameter [13–16] (Table 5).

Data which are present are the raw numbers and the percentage is present inside the brackets in the above table. Observing the abnormal chest radiographs, we can predict the condition of patient like whether the patients need the ventilator or death [17]. A 21-year-old woman chest radiograph is taken and in that round patches were observed in lower zones of lung (Fig. 5).

Ground glass opacities were often observed in majority of patients who is having abnormal chest radiographs in the emergency department. These radiographs represent either severe viral infections acute respiratory distress syndrome or bacterial infection. The images of chest X-rays display the patches, opacities, and mild difference in the lower parts of lungs which help to identify the problem and diagnose [1, 18]. The symptoms among the swine flu (H1N1) patients and their radiographs help to find out the major similarities and dissimilarities and to predict the clinical

Table 5 Outcomes of H1N1 patients with abnormal chest radiographs

Characteristic	No mechanical ventilation (n = 34)	Mechanical ventilation/death (n = 5)	P-value
Opacity			
Ground glass	23 (67.6)	4 (80)	0.665
Consolidation	18 (52.9)	5 (100)	0.066
Air bronchogram	10 (29.4)	3 (60)	0.31
Patchy	13 (38.2)	3 (60)	0.631
Nodular	11 (32.4)	0	0.296
Confluent	1 (2.9)	1 (20)	0.243
Lung zone			
Right upper	3 (8.8)	1 (20)	>0.999
Right middle	21 (61.8)	5 (100)	0.149
Left upper	0	1 (20)	0.128
Left middle	21 (61.8)	3 (60)	>0.999
Left lower	13 (38.2)	3 (60)	0.631
Distribution			
Central	19 (55.9)	5 (100)	0.136
peripheral	23 (73.5)	5 (100)	0.318
Bilateral	21 (61.8)	3 (60)	>0.999
Central	18 (52.9)	3 (60)	>0.999
peripheral	5 (14.7)	3 (60)	0.049
Four or more zones	5 (2.9)	3 (60)	0.01

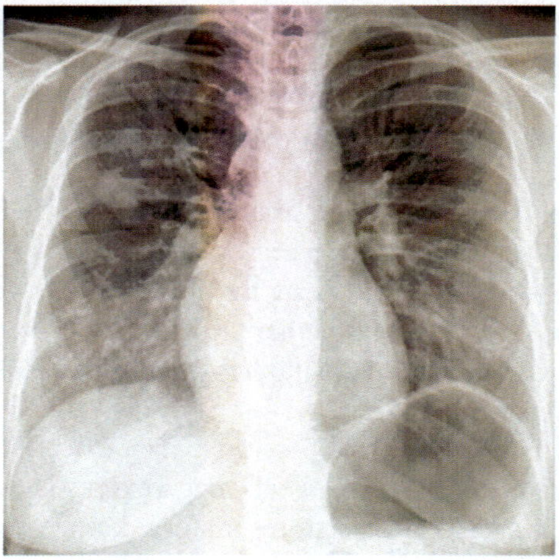

Fig. 5 Representing 21-year-old woman posterior–anterior chest radiograph who is having dyspnea. Patchy consolidation some with air bronchogram is distributed in multiple lung zones and central ground glass opacities

outcomes. The prediction may lead to early identification of the virus and the severity level of it and also helps for the doctors to identify and order for chest radiographs. Among the H1N1 patients data which is taken maximum number of cases were identified in the people who are having smoking habit have the abnormal chest. There was no major difference among the outcomes of the patients with normal and abnormal chest radiographs. The chest radiographs of various swine flu patients help to give the comparable indications among those which help the H1N1 suspects to treat them through beginning chest radiographs and furthermore help us to realize that there is a need to take the underlying chest radiographs dependent on the seriousness and the manifestations of the patients.

5 Conclusion

The suspected H1N1 patient with major symptoms who is having chest X-rays helps in predicting the clinical outcomes. The involvement of both lungs is evidenced by the presence of bilateral and multizonal peripheral opacities that are seen in adverse conditions. Minor symptoms associated with H1N1 do not require chest radiography until it is necessary. Initial chest X-rays are helpful in predicting the clinical outcomes and the condition of patient and it leads to better recovery, but the normal radiographies cannot eliminate adverse outcomes.

References

1. G. Aviram, A. Bar-Shai, J. Sosna et al., H1N1 influenza: initial chest radiographic findings in helping predict patient outcome. Radiology **255**(1), 252–259 (2010)
2. N.A. Al-Nakshabandi, Determining symptoms for chest radiographs in patients with swine flu (H1N1). Pol. J. Radiol. **76**(4), 45–48 (2011)
3. Centers for Disease Control and Prevention (CDC), Outbreak of swine origin influenza A (H1N1) virus infection—Mexico. MMWR Morb. Mortal. Wkly. Rep. **58**(17), 467–470 (2009)
4. Centers for Disease Control and Prevention (CDC), Swine influenza A (H1N1) infection in two children—Southern California. MMWR Morb. Mortal. Wkly. Rep. **58**(15), 400–402 (2009)
5. J.S. Peiris, L.L. Poon, Y. Guan, Emergence of a novel swine-origin influenza A virus (S-OIV) H1N1 virus in humans. J. Clin. Virol. **45**(3), 169–173 (2009)
6. J.K. Louie, M. Acosta, K. Winter et al., Factors associated with death or hospitalization due to pandemic 2009 influenza A(H1N1) infection in California. JAMA **302**(17), 1896–1902 (2009)
7. S. Jain, L. Kamimoto, A.M. Bramley et al., Hospitalized patients with 2009 H1N1 influenza in the United States, N. Engl. J. Med. **361**(20), 1935–1944 (2009)
8. G. Chowell, S.M. Bertozzi, M.A. Colchero et al., Severe respiratory disease concurrent with the circulation of H1N1 influenza. N. Engl. J. Med. **361**(7), 674–679 (2009)
9. J. Rello, A. Rodríguez, P. Ibanez et al., Intensive care adult patients with severe respiratory failure caused by influenza A (H1N1) v in Spain. Crit. Care **13**(5), R148 (2009)
10. D.J. Mollura, D.S. Asnis, R.S. Crupi et al., Imaging findings in a fatal case of pandemic swine-origin influenza A (H1N1). AJR Am. J. Roentgenol. **193**(6), 1500–1503 (2009)

11. P.P. Agarwal, S. Cinti, E.A. Kazerooni, Chest radiographic and CT findings in novel swine-origin influenza A (H1N1) virus (S-OIV) infection. AJR Am. J. Roentgenol. **193**(6), 1488–1493 (2009)
12. Interim guidance on specimen collection, processing, and testing for patients with suspected novel influenza A (H1N1) virus infection. U.S. Centers for Disease Control and Prevention. https://www.cdc.gov/h1n1flu/specimencollection.htm. Published May 13, 2009. Accessed Date 2009
13. E.A. Kim, K.S. Lee, S.L. Primack et al., Viral pneumonias in adults: radiologic and pathologic findings. Radio Graph. **22**, S137–S149 (2002)
14. E.Y. Lee, A.J. McAdam, G. Chaudry et al., Swine origin influenza A (H1N1) viral infection in children: initial chest radiographic findings. Radiology **254**(3), 934–941 (2010)
15. R. Fraser, J. Pare, N. Muller, N. Colman, *Fraser and Pare's Diagnosis of Diseases of the Chest* (Saunders, Philadelphia, PA, 1999).
16. W.J. Tuddenham, Glossary of terms for thoracic radiology: recommendations of the Nomenclature Committee of the Fleischner Society. Am. J. Roentgenol. **143**(3), 509–517 (1984)
17. J.T. Hagaman, G.W. Rouan, R.T. Shipley, R.J. Panos, Admission chest radiograph lacks sensitivity in the diagnosis of community-acquired pneumonia. Am. J. Med. Sci. **337**(4), 236–240 (2009)
18. J.K. Taubenberger, D.M. Morens, The pathology of influenza virus infections. Annu. Rev. Pathol. **3**, 499–522 (2008)

Low Energy Utilization with Dynamic Cluster Head (LEU-DCH)—For Reducing the Energy Consumption in Wireless Sensor Networks

S. NagaMallik Raj, Divya Midhunchakkaravarthy, and Debnath Bhattacharyya

Abstract The ability of wireless sensor network (WSN) can be observed in different fields like patient wearable monitoring, health applications, intruder detection, combat monitoring and battlefield surveillance in the last few years. In such applications, WSN contains a wide range of sensor nodes, which will monitor continuously, and passes the data to end-user through base station (BS) or sink node (SN). In such applications where users are unattended to replace the sensor node frequently, that indicates users can expect long-lasting of WSN. So in this paper, we concentrated on long-lasting of WSN, by minimizing the energy utilization of sensor node in WSN. So we are applying the clustering technique. So, clustering will minimize the energy usage of the sensor node, such that it will improve the lifetime of WSN. In this, we introduced our proposed protocol "Low Energy Utilization with Dynamic Cluster Head (LEU-DCH)" which reduces the energy consumption of sensor nodes. Besides, we also have shown the best way of selecting how many clusters we need.

Keywords Clustering · CH · Elbow · Silhouette · Dynamic CH · SN

S. NagaMallik Raj (✉) · D. Midhunchakkaravarthy
Department of Computer Science and Multimedia, Lincoln University College, Kuala Lumpur, Malaysia
e-mail: mallikblue@gmail.com

D. Midhunchakkaravarthy
e-mail: divya@lincoln.edu.my

D. Bhattacharyya
Department of Computer Science and Engineering, K L Deemed to be University, KLEF, Guntur 522502, India
e-mail: debnathb@kluniversity.in

D. Bhattacharyya and N. Thirupathi Rao (eds.), *Machine Intelligence and Soft Computing*, Advances in Intelligent Systems and Computing 1280,
https://doi.org/10.1007/978-981-15-9516-5_30

1 Introduction

WSN contains a wide range of sensor nodes, which will monitor continuously and passes the data to end-user through base station (BS) or sink node (SN). In the below Fig. 1, it was shown that some sensor nodes based on our requirement, will be distributed in a remote area to sense the data. If the nodes sense any data, it passes to the sink node, after that it passes to base station, at last to the end-user. WSN was used in a wide variety of applications, and some of the applications were patient wearable monitoring, health applications, intruder detection, combat monitoring, battlefield surveillance, etc. [1–3].

A sensor node in a sensor network was used to sense the given target, but the main drawback is its limited battery. In some applications where a human cannot access the target, in such situation, it is difficult to replace the dead sensor nodes. So to deploy the sensor nodes in the target area, we have to consider the battery size and cost, because the cost of sensor node will depend on the battery size. If sensor node cost is too high, it indicates we are using larger batteries, obviously the node size is large. If sensor node cost is low, it indicates we are using smaller size battery then node size is small.

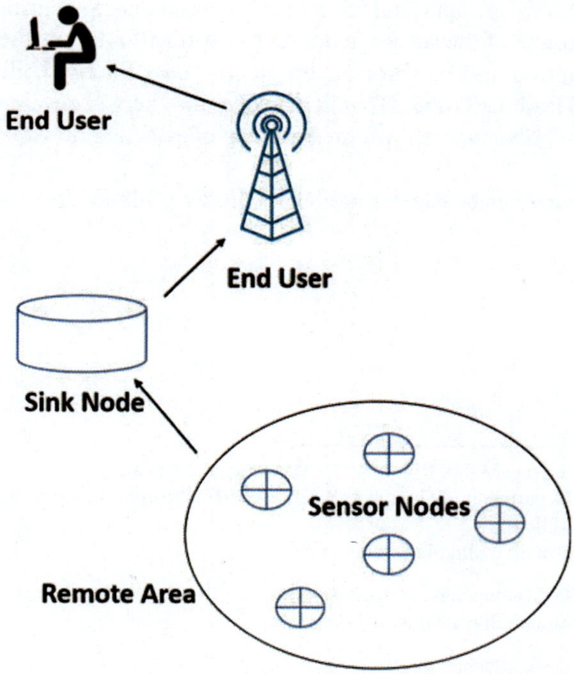

Fig. 1 Network model architecture [4]

1.1 Clustering

By applying the clustering technique, it minimizes the energy usage of the sensor node, such that it will improve the lifetime of WSN. An efficient method is needed to form some clusters in a network and how to elect a cluster head in the clusters. If we not using cluster technique, the majority of nodes will involve in the sensing process. Such energy consumption will not be wasted. So the lifetime of a network will become less. So to avoid that, we prefer clustering technique and by implementing our proposed protocol to identify the cluster head [1, 4].

Already some existing techniques were there to elect a node as a cluster head (CH). If a cluster head is static, then comparing to other nodes cluster head will do some additional things like receiving, processing and transmitting to sink node. So energy utilization will be more compared to another node. So with static cluster head network lifetime reduces fast. To overcome this problem, we implement dynamic CH by considering some parameters which will explain below sections [5, 6].

2 Proposed Protocol

This is proposed protocol "Low Energy Utilization with Dynamic Cluster Head (LEU-DCH)" that we implementing in clustering [7].

2.1 Algorithm Explanation

Step 1: Distribution of sensor nodes in *X* and *Y* coordinates.
Step 2: To implement clustering technique, we have to decide how many clusters will form in the given network.
Step 3: After that, we have to identify cluster centroid.
Step 4: To select the cluster head, we have to consider some parameters. Here, cluster head was dynamic. Every time CH will change.
Step 5: In this scenario, we consider mobile sink node.
Step 6: Depending on the energy levels of nodes in specified clusters and transmission energy, we are going to identify which node is acting as a cluster head.

After implementing all the above steps, energy consumption will be uniformly utilized among all nodes in a specified cluster, such that network life will be increased. This thing did not happen if the cluster head was static [8].

Let us see each step of a proposed algorithm LEU-DCH which was implemented in Python language.

3 Distribution of a Sensor Nodes in *X* and *Y* Coordinates

Here, we took ten nodes by names Node 1, Node 2 and so on. These ten nodes will be distributed among latitude and longitude. That was a show in below,

Name lat-long
Node 1, 28, 7
Node 2, 36, 5
Node 3, 32, 2
Node 4, 56, 8
Node 5, 47, 5
Node 6, 75, 9
Node 7, 34, 4
Node 8, 56, 9
Node 9, 28, 1
Node 10, 33, 6

In above Node 1 will be distributed in latitude 28 and longitude 7, Node 2 Lat 36 and long 5 and so on.

```
nodes = load_nodes_from_csv("data.csv")
plot_sensors(nodes,[−1 for x in range(len(nodes))])
```

To the above code, we have to provide node distribution details as a data set input. In the below figure, it is shown how ten nodes were distributed in latitudewise and longitudewise. Before clustering, the ten nodes will look like as shown in Fig. 2.

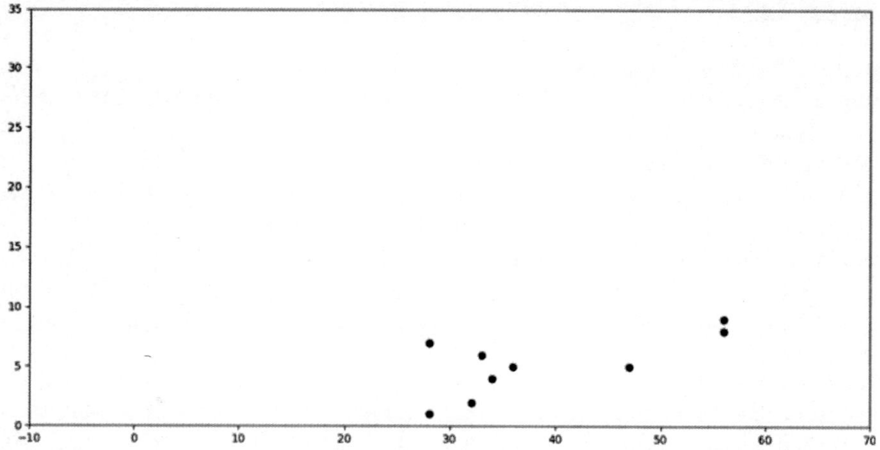

Fig. 2 Distribution of nodes in *X* and *Y* coordinates before clustering

4 To Select the Total Number of Clusters in Given WSN

To obtain the K value, i.e. number of clusters, so many existing techniques are there. Let us compare two existing techniques elbow method and average silhouette method.

4.1 Average Silhouette Method

To select the total number of clusters, i.e. K value, silhouette method can be used. The values were in the range of [−1, 1]. If the sample is far away to its neighbouring cluster, then its value is +1. If the sample is close to its neighbouring cluster, then its value is −1 [9, 10].

Thus, the silhouette $s(i)$ can be calculated as

```
s(i) = (b(i) - a(i))
--------------
max(b(i), a(i))
```

So $s(i)$ lies in the range of [−1, 1].

Sample code for average silhouette:

```
*** imported from the Scikit-Learn library***
def plot_silh(nodes):                                    *** fuction body***
  X = np.array([[node.lat,node.lng] for node in nodes])
  sil=[]
  for i in range(2,10):
    KM = KMeans(n_clusters=i,max_iter=500)
    KM.fit(X)
    lables= KM.labels_
    sil.append(silhouette_score(X,lables,metric='euclidean'))
  plt.clf()
  plt.plot(range(2,10),sil)
  plt.draw()
  print(sil.index(max(sil))+2)
  print(sil)
  return sil.index(max(sil))+2
K=plot_silh(nodes)                    *** Should write in main***
```

By using the above sample code, the following Fig. 3 will generate. In that below Fig. 3, a graph was generated by applying average silhouette. So, we can observe K value from that graph. Here, K value is 2.

4.2 Elbow Method

Now let us see another method to find out K value, by picking the elbow of the curve as the number of clusters to use [11, 12].

The below sample code is for finding K value by using the elbow method.

```
from clusters import plot_elbow
def plot_elbow(nodes):
    X = np.array([[node.lat,node.lng] for node in nodes])
    wss=[]
    for i in range(1,10):
        KM = KMeans(n_clusters=i,max_iter=500)
        KM.fit(X)
        wss.append(KM.inertia_)
    plt.clf()
    plt.plot(range(1,10),wss)
    plt.draw()
plot_elbow(nodes)          *** Should be written in main to plot nodes***
```

By using the above sample code, the following Fig. 4 will generate. In Fig. 4, a graph shows that the K value may be 2 or 3.

(a) Comparison Between Average Silhouette and Elbow Methods

By comparing two Figs. 3 and 4, the K value by using elbow method is 2 or 3. The K value by using the average silhouette method is 2.

So average silhouette method is the best method to find out K value. And also the accurate value can be obtained compared to the elbow method.

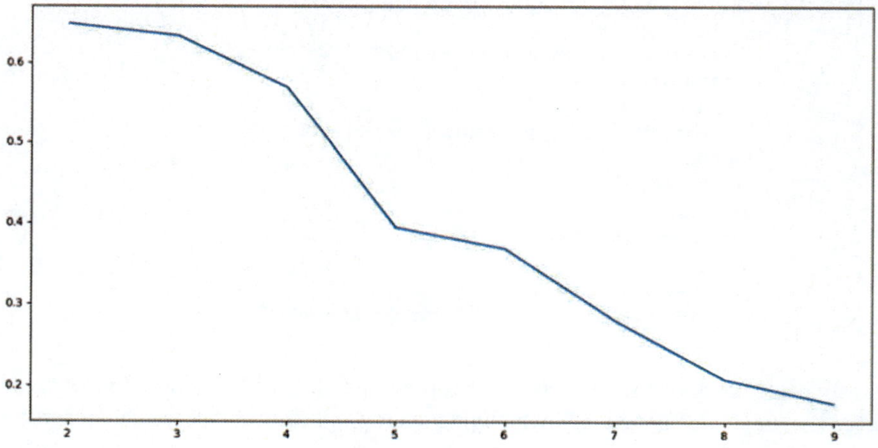

Fig. 3 K value is 2 by using the average silhouette method

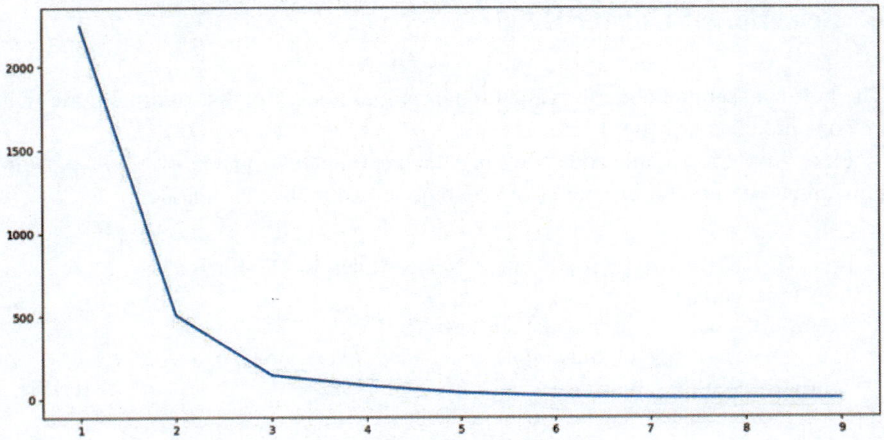

Fig. 4 To find out *K* value by using the elbow method *K* value is 2 or 3

5 Formation of Clusters ($K = 2$)

In Fig. 5, two clusters were formed for differentiation and the sensor nodes present in one cluster is in red, and the second cluster node was in green colour. So, centroid location was also pointed.

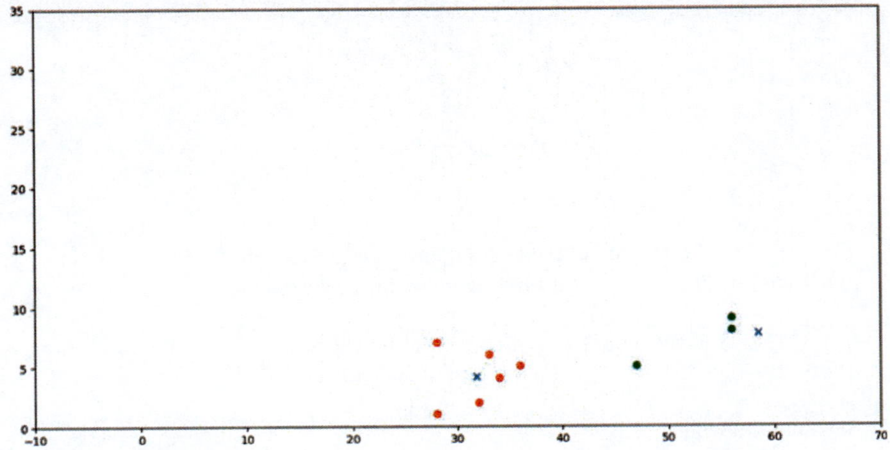

Fig. 5 Nodes after clustering along with centroid

6 Selection of Cluster Head

The below is sample code for calculating optimal node, i.e. that optimal node will be considered as a cluster head.

Here, from this sample code, we can calculate transmission energy (Tx), i.e. from how much energy will be spent from each node to a mobile sink node.

```
print("{:<16} {:<16} {:<16}".format("Name","Energy","Tx-Energy"))
for x in range(len(cluster)):
    network_node = [cluster[x].lat,cluster[x].lng]
    e_tx = get_energy_of_tramission(sink_node,network_node)
    print("{:<16}                                        {:<16.10f}
{:<16}".format(cluster[x].name,cluster[x].energy,e_tx))
    if cluster[x].energy >= optimal_node.energy:
        if e_tx<optimal_energy:
            optimal_energy = e_tx
            optimal_node = cluster[x]
print("optimal node for tramission is: {}".format(optimal_node.name))
return optimal_node
```

Output in Terminal:

2 **// No of Clusters//**

```
[[28  7]
 [36  5]
 [32  2]
 [56  8]
 [47  5]
 [75  9]
 [34  4]
 [56  9]
 [28  1]
 [33  6]]
```
//location of the nodes along x and y Co-ordinates//

[0 0 0 1 1 1 0 1 0 0] **//corresponding nodes in the clusters//**

```
[[31.83333333  4.16666667]
 [58.5        7.75     ]]
```
// Centroid of 1st cluster//
// Centroid of 2nd cluster//

Sink moment begins

```
cluster distance
  1     32.10
  2     59.01
nearest cluster is 1
```
//distance from mobile sink//

Name	Energy	Tx-Energy
NODE1	5.0000000000	0.00023335
NODE2	**5.0000000000**	**0.00020895000000000004**
NODE3	5.0000000000	0.00022360000000000001
NODE7	5.0000000000	0.00021640000000000003
NODE9	5.0000000000	0.00023575
NODE10	5.0000000000	0.00021875000000000003

optimal node for transmission is: **NODE2**

cluster distance // distance from sink//

1 31.93

2 58.83

the nearest cluster is 1

Name	Energy	Tx-Energy
NODE1	5.0000000000	0.00023417480468750004
NODE2	4.9997910500	0.0002096248046875
NODE3	5.0000000000	0.00022394980468750002
NODE7	**5.0000000000**	**0.00021696230468750002**
NODE9	5.0000000000	0.00023597480468750002
NODE10	5.0000000000	0.00021950605468750003

optimal node for transmission is: **NODE7**

cluster distance

1 31.66

2 58.53

the nearest cluster is 1

Name	Energy	Tx-Energy
NODE1	5.0000000000	0.00023524687500000002
NODE2	4.9997910500	0.00021064687500000002
NODE3	5.0000000000	0.000224596875
NODE7	4.9997830377	0.00021784687500000003
NODE9	5.0000000000	0.00023644687500000002
NODE10	**5.0000000000**	**0.000220571875**

Optimal node for transmission is: **NODE10**

In the above sample output, Node 2 contains less transmission energy (Tx) 0.00020895000000000004. So, Node 2 will be elected as a cluster head.

Name	Energy	Tx-Energy
NODE1	5.0000000000	0.00023417480468750004
NODE2	4.9997910500	0.0002096248046875
NODE3	5.0000000000	0.00022394980468750002
NODE7	5.0000000000	0.00021696230468750002
NODE9	5.0000000000	0.00023597480468750002
NODE10	5.0000000000	0.00021950605468750003

Here, Node 2 had less Tx-energy but the energy level is 4.9997910500. So it was not elected as a CH. To obtain uniform energy consumption, we go for next less Tx-energy is 0.00021696230468750002 for Node 7. So, Node 7 will be elected as a CH and so on. In every next stage, nodes with the highest energy and less Tx-energy will be elected as a CH.

Thus, we proved when to compare to static CH. Energy consumption of nodes will less in this proposed protocol. This process will repeat until all nodes in cluster energy levels will be uniform. So, wireless sensor network lifetime will increase [13–15].

7 Conclusion

Wireless sensor networks can be observed in different fields like patient wearable monitoring, health applications, intruder detection, combat monitoring and battlefield surveillance in the last few years. In such applications, WSN contains a wide range of sensor nodes, which will monitor continuously, and passes the data to end-user through base station (BS) or sink node (SN). In such applications where users are unattended to replace the sensor node frequently, that indicates users can expect long-lasting of WSN. So, in this paper, we concentrated on long-lasting of WSN, by minimizing the energy utilization of sensor node in WSN. So, we are applying the clustering technique. Clustering protocols divide the network into several clusters, which each perform their work independently and each cluster has ahead. Clustering can significantly affect the overall scalability of the system, the lifetime and energy efficiency. In this paper, with the proposed protocol, uniform distribution of energy levels for nodes and how CH was selecting every time to improve the WSN lifetime.

References

1. K.K. Kho, Y. Bissessur, D.S. Goolaub, An inland wireless sensor network system for monitoring seismic activity. Future Gener. Comput. Syst. **105**, 520–532 (2020)

2. S.A. Nikolidakis, D. Kandis, D.D. Vergados, C. Douligeris, Energy efficient automated control of irrigation in agriculture by using wireless sensor networks. Comput. Electron. Agric. **113**, 154–163 (2015)
3. M.E. Celebi, H. Kingravi, Deterministic initialization of the K-means algorithm using hierarchical clustering. Int. J. Pattern Recogn. Artif. Intell. **26**(7), 1250018 (2012)
4. Thakkar, K. Kotecha, CVLEACH: coverage based energy efficient LEACH algorithm. Int. J. Comput. Sci. Netw. (IJCSN) **1**(5), 10–21 (2012)
5. Thakkar, K. Kotecha, WALLACH: weight based energy efficient advanced LEACH algorithm. Comput. Sci. Inf. Technol. (CS & IT) **2**(4), 21–31 (2012)
6. R. Yadav, V. Kumar, R. Kumar, Energy-efficient hybrid clustering protocol for multilevel heterogeneous wireless sensor network, in *2014 International Conference on Computer and Communications Technologies (ICCCT)* (IEEE, 2014), pp. 1–5
7. S. Faisal, N. Javaid, A. Javaid, M. Khan, S.H. Bouk, Z. Khan, Z-SEP: zonal-stable election protocol for wireless sensor networks. arXiv preprint arXiv:1303.5364 (2013)
8. R. Pal, R. Sindhu, A.K. Sharma, SEP-E (RCH): enhanced stable election protocol based on redundant cluster head selection for HWSNs, in *International Conference on Heterogeneous Networking for Quality, Reliability, Security and Robustness* (Springer, Berlin, 2013), pp. 104–114
9. H. Jiang, S. Jin, C. Wang, Prediction or not an energy-efficient framework for clustering-based data collection in wireless sensor networks. IEEE Trans. Parallel Distrib. Syst. **22**(6), 1064–1071 (2011)
10. O. Younis, S. Fahmy, HEED: a hybrid, energy-efficient, distributed clustering approach for ad-hoc sensor networks. IEEE Trans. Mobile Comput. **3**, 366–379 (2004)
11. R. Soua, P. Minet, A survey on energy efficient techniques in wireless sensor networks. IEEE Technical Program at IFIP WMNC (2011)
12. T.N. Qureshi, T. Shah, N. Javaid, EESAA: energy efficient sleep awake aware intelligent sensor network routing protocol, in *Proceedings of 15th International Multi-Topic Conference (INMIC)*, Islamabad, PK (2012)
13. F. Akyildiz, W. Su, Y. Sankarasubramaniam, E. Cayirci, Wireless sensor networks: a survey. Comput. Netw. **38**(4), 25–32 (2011)
14. S. Lindsey, C.S. Raghvendra, PEGASIS: power efficient gathering in the sensor information system, Los Angeles, CA, 90009-2957
15. N. Thirupathi Rao, D. Bhattacharyya, T. Kim, Black hole attacks on WSNs using discrete simulator: an extensive review. Int. J. Grid Distrib. Comput. **11**(12), 33–42 (2018)

Efficient Cryptographic Technique to Secure Data in Cloud Environment

T. Lakshmi Siva Rama Krishna, Kalpana Kommana, Sai Sowmya Tella, Padmavathi Nandyala, and Venkata Naresh Mandhala

Abstract Cloud computing technology offers various services such as infrastructure, platform and application to its client on demand through its data center which consists of hardware and software components. Work with cloud environment needs to address several design challenges such as scalability, reliability, concurrency, transparency and security. In this paper, we address one of the design challenges of cloud computing which is security. The limitation of the existing cloud security technique with respect to data storage and access is that there is a possibility of third-party vendors to decrypt the data; therefore, the third-party vendors can access and misuse the data. We have proposed a new algorithm which is a combination of Paillier key encryption (PKC) with multiparty computation (MPC) to secure data from the third-party vendors. We name the proposed new algorithm as Paillier key with multiparty computation (PMPC). In addition to this, we have compared the proposed algorithm with the existing algorithms such as homomorphic encryption (HoE), multiparty computation (MPC), and the proposed PMPC algorithm results indicated that the proposed PMPC algorithm performs better than HoE and MPC algorithms.

Keywords Cryptography · Homomorphic encryption · Paillier key encryption · Multi-party computation · Cloud computing and security

1 Introduction

Cloud computing technology offers various services such as infrastructure, platform and application to its client on demand through its data center which consists of hardware and software components. Work with cloud environment needs to

T. Lakshmi Siva Rama Krishna (✉) · K. Kommana · S. S. Tella · P. Nandyala · V. N. Mandhala
Department of Computer Science and Engineering, Koneru Lakshmaiah Education Foundation, Vaddeswaram, Guntur, Andhra Pradesh, India
e-mail: sivamca.mtech@gmamil.com

V. N. Mandhala
e-mail: mvnaresh.mca@gmail.com

D. Bhattacharyya and N. Thirupathi Rao (eds.), *Machine Intelligence and Soft Computing*, Advances in Intelligent Systems and Computing 1280,
https://doi.org/10.1007/978-981-15-9516-5_31

address several design challenges such as scalability, reliability, concurrency, transparency and security. In this paper, we address one of the design challenges of cloud computing which is security. Providing security in cloud environment is important for business organizations as well as personal clients who are willing to take cloud services. Moreover, business organizations have legal obligations to keep their clients' data securely. Similarly, other business sectors have more stringent rules about data storage. Security is an essential part of cloud service, and clients should always check that whether the cloud service provider offers the desired levels of security according to client's requirement. We can provide security in the cloud environment by proposing encryption schemes [1–10].

Encryption is the method of converting the information or data into the ciphertext and is the product of encryption of the original data that can be interpreted, and even consumers may operate on it as if it is the original information that the consumers do not currently have. We can run the computations on the encrypted data without sacrificing the original data.

In this paper, we address one of the design challenges of cloud computing which is security. The limitation of the existing cloud security technique with respect to data storage and access is that there is a possibility of third-party vendors to decrypt the data; therefore, the third-party vendors can access and misuse the data. We have proposed a new algorithm which is a combination of Paillier key encryption (PKC) with multiparty computation (MPC) to secure data from the third-party vendors. We name the proposed new algorithm as MPC. In addition, the proposed algorithm was compared with the existing algorithms such as homomorphic encryption (HoE), multiparty computing (MPC) and the proposed PMPC algorithm.

The remainder of the paper is as follows; in Sect. 2, we explain the importance of confidentiality, integrity and interaction with respect to cloud security. We discuss the existing work on various encryption schemes available in the literature in Sect. 3. Later, we present the proposed algorithm and explain the working procedure with an example. In Sect. 5, we analyze the results of the existing and proposed algorithms. Finally, in Sect. 6, we conclude the paper along with future research perspectives.

2 Issues in Cloud Security

2.1 Confidentiality

Confidentiality is intended to prevent the attacker from reaching sensitive information, while ensuring that the authorized user has access to it. Services allow consumers to trust their data to the cloud. Yet data owners do not trust the software in the untrusted environment. Protection on the user side is thus important. Users encrypt their data with a public key before they save it in the cloud.

2.2 *Integrity*

Integrity of the data means data accuracy and trustworthiness. It ensures right estimation on critical data. Unauthorized user cannot alter the data.

2.3 *Availability*

Cloud service providers have multiple servers in this case. There is no question of unavailability as one system crashes, because another system is able to provide services.

3 Encryption Schemes

We have studied several encryption schemes proposed in the literature [1–10]. Among these schemes, homomorphic encryption (HoE) is one of the popular encryption schemes that modify the type of media or documents into the other while maintaining same connections between the data items from either of the sets of data. Confused numerical functions on the scrambled knowledge are conducted in homomorphic encryption. Homomorphic cryptography plays a significant role in distributed computing. We explain how HoE works with an example, and how it is used to perform operations in the cloud environment.

Let us consider an extremely busy business ABC, and we consider cloud provider (CP) to carry out activities. As we understand that only an authorized cloud provider can only have the entry to the encrypted range, consider the 30-20 difference and give the answer as 10.

A minor example of this is as follows

- An agency called "Creta" has important datasets details with numbers 10 and 20. To encrypt this, every element in it is multiplied by 3 which creates a new set as 30 and 60.
- Transfers encrypted information later on the cloud. Some other partner consults the "Creta" a few months later and asks for the details.
- While "Creta" is working with its specific job, it requests CP to conduct the procedure that has encrypted data access to find a total of 90 as well as return the response.
- Now, the Creta decrypts data and provides other party with 30 as output.
- Thus, the data is supplied with this protection.

The foundation for secure multiparty computation started in the late 1970s with the work on mental poker, cryptographic work that simulates game playing/computational tasks over distances without requiring a trusted third party.

Note that traditionally, cryptography was about concealing content, while this new type of computation and protocol is about concealing partial information about data while computing with the data from many sources and correctly producing outputs. This is known as multi-party computing; computation-preserving privacy also secures computing, which is in the field of cryptography that can also be used to measure a common joint function on all inputs. Secure MPC helps us to operate and evaluate details about people's health without disclosing very sensitive private information.

The protected multiparty computation guarantees that everybody knows about the right outcome of a shared computation, but nothing more about the inputs of someone else, except though any of the users who execute the computation might be aggressively or unconsciously malicious. For arbitrary computations and for any number of parties, secure MPC can be carried out. Thus, we can view secure MPC protocols as compilers that require a mapping specification as input, and output is a protocol that securely calculates the function. The secure MPC provides both confidentiality and integrity that is much better than fully homomorphic encryption and verifiable computing.

Another type of the homomorphic algorithms is the Paillier key cryptography (PKC). Users using this algorithm will be issued with a public and private keys. First, data is encrypted based on the public key and forwarded it to a specific receiver. Later, receiver is able to decrypt the information based on the available private key. For example, the sender is Kalpana, and the receiver is Sowmya. Kalpana has two messages which are encoded based on the "public key" of the recipient. Now, Kalpana will combine the information collected and forward them to the receiver. With his private key, Sowmya decodes the data and gets the answer as the sum of both messages.

As there are increasing numbers of customers involved in any of the calculations or figures in the arrangement, the number of customers in the framework must be limited as the number of customers out there is all random.

The effectiveness of different algorithms is compared on the basis of their safety features. These are algorithms of single homomorphic nature. In certain cases, these algorithms vary on the basis of the homomorphism. Various forms of five, totally, "homomorphic encryption algorithms" have been tested to assess the performance with respect to high data protection.

The GM algorithm is established using semantic-secure quadratic residue and which satisfies additional homomorphism along with low efficiency. In 1984, GM algorithm was implemented. The ELGAMAL algorithm was introduced in 1985, in order to provide data protection. This algorithm is based on the cryptosystem of the public key and the elliptic curve cryptosystem, satisfying any multiplicative homomorphism, and can be used for encryption. Additional homomorphism is fulfilled by the Hill cipher algorithm. The RSA algorithm is based on number theory and satisfies any encryption of homomorphism by multiplication. An algorithm called improved probabilistic encryption algorithm was implemented in 1994. It encrypts n bits of data at a time but can only perform a finite number of homomorphic addition operations. The security of the data will be guaranteed at low possible cost. The main problem with RSA is not semantically secure. This is a major disadvantage.

Another of the homomorphic algorithms is the Paillier key cryptography. Customers who use this formula will be given a free key and a private key. The information is scrambled first with the open key and sent to the receiver. The receiver will then decrypt the data using its private key. The main cryptography by Paillier introduces "additive homomorphism." For example, the sender is Padmavathi, and the receiver is Siva. Padmavathi has two messages which are encoded with the public key of the recipient. Padmavathi will add the messages together and send them to the receiver. With its private key, Siva decodes the data and gets the output as the sum of the two messages.

3.1 Homomorphic Encryption

With the growing cloud computing technology, users are storing ever more sensitive information and private data in the cloud. Cloud storage protection is one of the cloud computing major security problems. Cloud data should be stored in the form of ciphertext, in order to protect user data privacy. Homomorphic encryption technology allows ciphertext handling, under data protection. It can read, measure and count ciphertext directly in the cloud and return the results in ciphertext format to users. Homomorphic encryption technology is essential to ensuring data security in cloud environments.

3.2 Multiparty Computation

Confidentiality and integrity are established using MPC. Multiparty computing provides adequate security and therefore could be far more effective. After merging homomorphic encryption with multiparty computing, no single party knows about MPC data. Confidentiality of privacy and data is retained, and the latency is lower than homomorphic encryption but more than multiparty computing (Fig. 1).

3.3 Paillier Key Encryption

The Paillier cryptosystem, developed by Pascal Paillier and named after him in 1999, is a probabilistic asymmetric algorithm for public key cryptography. Paillier provides additive homomorphism. This means that messages can be added together while they

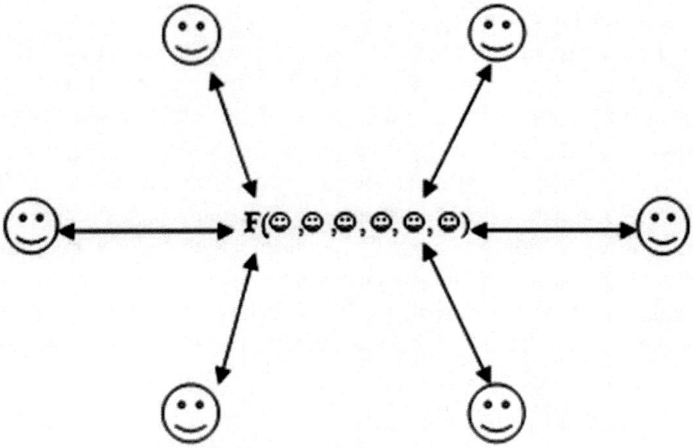

Fig. 1 Multiparty computation

are encrypted, and they will decrypt correctly. For example, if Bob has two messages (x and y) encrypted with the public key of Alice (creating cx and cy), then they can be added to Alice and sent. If Alice decrypts the message, the outcome is x + y (Fig. 2).

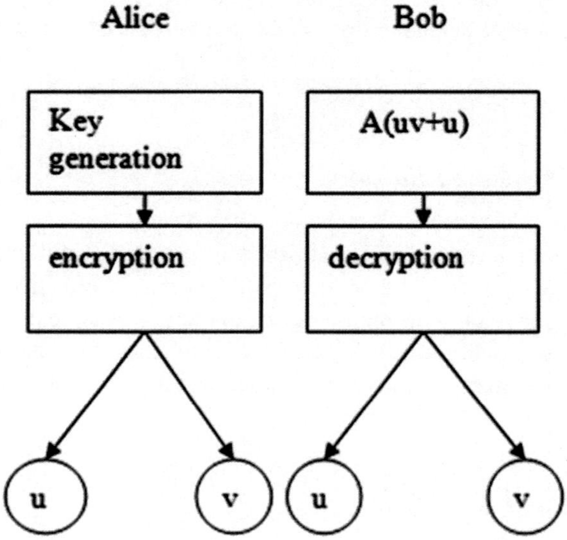

Fig. 2 Paillier key encryption

4 Proposed Scheme

In this section, we propose our scheme which is a combination of both paillier and multiparty computation, and we name the proposed scheme as Paillier key with multiparty computation (PMPC).

4.1 Paillier Key with Multiparty Computation (PMPC)

In 1999, paper by Pascal Paillier, Paillier cryptosystem is first delineated, which describes the relationship of uneven non-deterministic algorithmic rule of cryptography with homomorphic additive properties. Additionally, the cryptosystem offers disconnected signatures but not direct identities as given by RSA. One of the homomorphic algorithms is the Paillier key cryptography. Users using this algorithm will be issued with two keys: public and private keys. First, the data is encrypted with the public key and forwarded to the recipient. The receiver must then decrypt the data using its private key. Paillier key cryptography includes "homomorphic additives."

For example, the sender is Bob, and the receiver is Alice. Bob has two messages encoded with the public key of the recipient. Bob puts together the messages and sends them to the receiver. Alice decodes the data using his private key and obtains the result as the sum of the two messages.

Example: Voting System

We want to sign all the ballots with an election board, so third parties cannot check the votes. We want the election board to know who won the election. We do not wish to tell the election board that the elector has voted for that candidate. With ancient cryptography and coding, the election board has no approach to grasping who won the election while not decrypting every vote. Once again, the voting system can be constructed using additive homomorphism as follows: Voters encrypt their ballots with the public key of the election board. Voters send their encrypted ballots to a server under investigation. The investigative server removes all the ballots and sends them to the polls. The electoral board decrypts the entire vote and declares who won the election without ever reading the ballot of a candidate.

Algorithm: PMPC

1. Choose two random prime numbers say p and q, independent of one another that is gcd of pq, (p-1)(q-1) equal to 1
2. Compute n=p*q & compute lambda=lcm of p-1, q-1
3. Take a random integer 'g'
4. n divides the modulus of 'g' by checking the multiplicative inverse condition denoted by 'mu'

5. Public encryption key—(N, G)
6. Private encryption key—(lambda, mu)
7. Now, Encrypt the data and then decrypt it and finally produces the results.

4.2 Applications

1. *Electronic cash*: A new notion of self-blindness is a function. The versatility to convert one coded text into another when the decryption content is not dynamically. This program is useful when the vendors in a shop do not need to pay the sum through our credit card. Close to e-voting, the program. All the programs tend to be the same. By sharing our information and encouraging the notion of e-cash, we can purchase the items online
2. *Electronic voting*: The people had to cast their vote anonymously in electronic voting without disclosing on whom they voted to other people. Here, they encrypt the vote, thus casting their votes. However, the Election Commission collects certain votes to see how many people have cast their ballots.

Therefore, the election committee will have the results of the people's votes cast. You must decode the data and collect information to see how many of the total population have cast their ballots. Individuals will not know what others voted to.

5 Result Analysis

Below graph indicates protection standard for various algorithms. Multiparty computation (MPC) collectively calculates a function over its inputs while keeping those inputs private. The confidentiality of the data is maintained after combining homomorphic encryption and multiparty computing (HE + MPC). So, the data stored is more secure compared with only MPC. Combination of Paillier key encryption (PKC) and multiparty computation (MPC), i.e., (PKC + MPC), gives the good result compared with MPC and MPC + HE in terms of security. Homomorphic encryption has many disadvantages; in addition to its inefficiency, only numeric data are taken here, and we send several messages at a time with protection through this algorithm (Fig. 3).

In this section, we propose our scheme which is a combination of both Paillier and multiparty computation, and we name the proposed scheme as Paillier key with multiparty computation (PMPC).

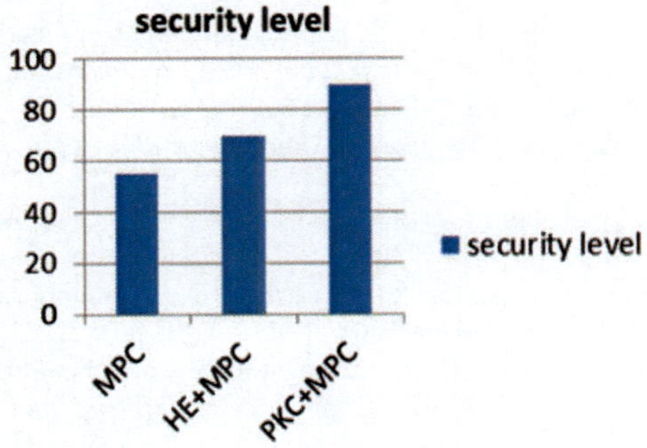

Fig. 3 Security-level comparison of MPC, HE + MPC and PKC + MPC schemes

6 Conclusion

In this paper, we have proposed a stabilized cloud computing model in which effective cryptographic technique based on homomorphic encryption (PKC) and multi-party computing (MPC) was used to encode user data, preceded by processes on their data while retaining privacy and confidentiality. The performance is the same as when the operations were carried out on raw data. A party may perform computations jointly without disclosing its data to the other party. Here, we built and configured stable homomorphic encryption and multiparty computing strategies uniquely suited to a protected, semi-trusted cloud environment. This environment helps developers to build the private cloud in accordance with the required cryptographic techniques (i.e., HE + MPC) to secure.

7 Future Scope

In future, we are planning to propose secure algorithms using Pallier key encryption and multiparty computation schemes for string and image types. Since the proposed approach is only used for digits, it can be expanded for strings and photos in the future.

References

1. P. Paillier, Public-key cryptosystems based on composite degree residuosity classes, in *International Conference on the Theory and Applications of Cryptographic Techniques* (Springer, Berlin, 1999), pp. 223–238
2. A.K.A. Hassan, Reliable implementation of Paillier cryptosystem. Iraqi J. Appl. Phys. **10**(4), 27–29 (2014)

3. Y. Yao, J. Wei, J. Liu, R. Zhang, Efficiently secure multiparty computation based on homomorphic encryption, in *2016 4th International Conference on Cloud Computing and Intelligence Systems (CCIS)* (IEEE, 2016), pp. 343–349
4. X. Yi, R. Paulet, E. Bertino, *Homomorphic Encryption and Applications*, vol. 3 (Springer, Heidelberg, 2014)
5. A. Chatterjee, K.M.M. Aung, *Fully Homomorphic Encryption in Real World Applications* (Springer, Berlin, 2019)
6. Z. Brakerski, C. Gentry, V. Vaikuntanathan, (Leveled) fully homomorphic encryption without bootstrapping. ACM Trans. Comput. Theor. (TOCT) **6**(3), 1–36 (2014)
7. K. Kogiso, T. Fujita, Cyber-security enhancement of networked control systems using homomorphic encryption, in *2015 54th IEEE Conference on Decision and Control (CDC)* (IEEE, 2015), pp. 6836–6843
8. P. Martins, L. Sousa, A. Mariano, A survey on fully homomorphic encryption: an engineering perspective. ACM Comput. Surv. (CSUR) **50**(6), 1–33 (2017)
9. G. Kalpana, P.V. Kumar, S. Aljawarneh, R.V. Krishnaiah, Shifted adaption homomorphism encryption for mobile and cloud learning. Comput. Electr. Eng. **65**, 178–195 (2018)
10. D. Boneh, G. Segev, B. Waters, Targeted malleability: homomorphic encryption for restricted computations, in *Proceedings of the 3rd Innovations in Theoretical Computer Science Conference* (2012), pp. 350–366

Secure Information Transmission in Bunch-Based WSN

S. NagaMallik Raj, B. Dinesh Reddy, N. Thirupathi Rao, and Debnath Bhattacharyya

Abstract Remote network incorporates a bigger favourable position in the present correspondence application like ecological, movement, military and well-being perception. To understand these applications, it is important to have a solid directing convention. The self-sorting out nature of MANETs makes them appropriate for some applications and henceforth, extensive exertion has been put into anchoring this kind of systems. Secure correspondence in a system is dictated by the unwavering quality of the key administration conspire, which is in charge of creating, circulating and looking after encryption/decoding keys among the hubs. In this paper, different key administration plans for MANETs are talked about. This examination work proposes a novel secure identity-based key management convention making utilization of cryptographic and information theoretic security.

Keywords MANET · Steering conventions · Key administration · Symmetric key

S. NagaMallik Raj (✉) · B. Dinesh Reddy · N. Thirupathi Rao
Department of Computer Science and Engineering, Vignan's Institute of Information Technology, Visakhapatnam 530049, India
e-mail: mallikblue@gmail.com

B. Dinesh Reddy
e-mail: dinesh4net@gmail.com

N. Thirupathi Rao
e-mail: nakkathiru@gmail.com

D. Bhattacharyya
Department of Computer Science and Engineering, K L Deemed to be University, KLEF, Guntur 522502, India
e-mail: debnathb@kluniversity.in

D. Bhattacharyya and N. Thirupathi Rao (eds.), *Machine Intelligence and Soft Computing*, Advances in Intelligent Systems and Computing 1280,
https://doi.org/10.1007/978-981-15-9516-5_32

1 Introduction

The ongoing advancement of specially appointed remote advances has permitted versatile impromptu systems (MANETs) to build unconstrained associations among cell phones with none foundation. Besides, with the development of sensor-empowered keen cell phones, MANETs turned into a fundamental part inside the foundation of shrewd city and Internet of things (IoT) circumstances because of people with savvy gadgets will openly and powerfully kind a self-arranging MANET to send, get and share information in an exceedingly confined zone [1].

In an exceedingly such a keen situation, MANETs, WSNs and WMNs speak to key innovations giving numerous IoT applications and administrations to clients. Besides MANETs have discovered a scope of uses in human services, front line correspondences, debacle recuperation, emergency administration administrations instruction associations, impromptu agreeable registering, social exercises and gathering lobbies [2, 3].

A MANET is an accumulation of self-governing hubs or terminals that speak with one another by shaping a multi-bounce radio system and keeping up a network in a decentralized way. Because of the constrained transmission scope of every portable hub, it might be essential for one versatile hub to enroll in the guide of different hubs in sending a bundle to its goal. Accordingly, in such condition, each hub in the system assumes the part of a switch by having the capacity to decide the ways of transmitting parcels to their goals [4]. Figure 1 outlines a case of a MANET which

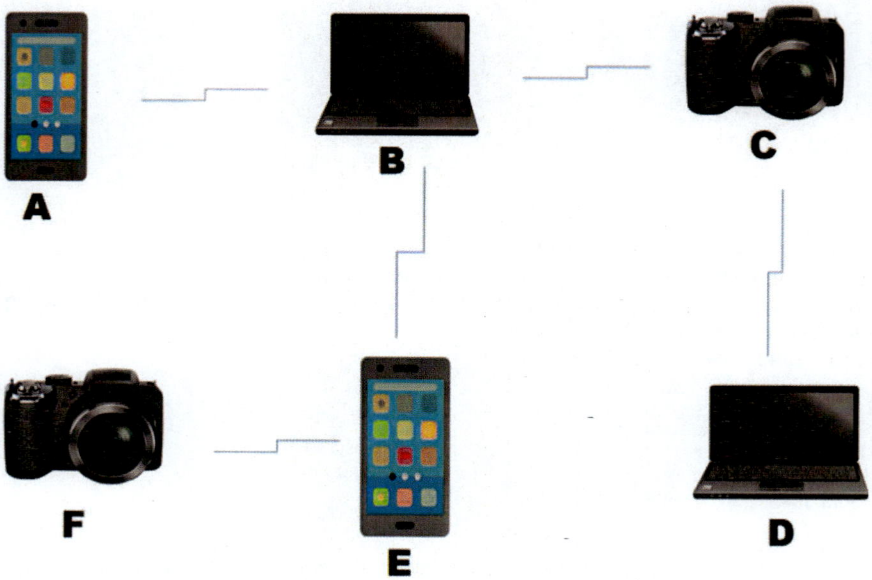

Fig. 1 Mobile ad hoc network [2]

contains two PCs, two PDAs and two computerized cameras. Since hub D is outside hub A's transmission extend, the information is retransmitted by hubs.

Despite the appealing utilization of MANETs, these frameworks still face a few difficulties and imperatives that need more examination before the across the board business arrangement of MANETs. The most requirements that may affect MANET configuration are as per the following: (1) the restricted vitality and lifetime of the battery, nature of administration (QoS), infrastructureless and self-sufficient set-up, dynamic system topologies, the versatility of hubs, remote connection unwavering quality, variety in hub abilities, multijump directing adaptability, multicast support and security dangers.

As it is hard to set up an ensured correspondence by key creation among flexible centres in a MANET, a malicious compact centre point can use a phoney character to make faked trust relations with various centre points, and a short time later strike the MANET. Such centre points would drop all of the data groups understood that they need to advance amid the whole reenactment. A solid steering convention for mobile ad hoc networks (MANETs) keeps the vitality utilization as low as conceivable. Despite what might be expected, in MANET with cell organize joining, it is practical to validate portable hubs previously any real key age. Keeping in mind the end goal to build up trust connection between any two portable hubs in the cell-based MANET, it is advantageous to exploit cell framework to empower a trustable and secure key age before correspondence [3, 5].

2 Directing Protocols

Directing is the demonstration of moving data from a source to a goal in an internetwork. Amid this procedure, not less than one middle of the road hub inside the internetwork is experienced.

The directing idea fundamentally includes two exercises: right off the bat, deciding ideal steering ways and besides, exchanging the data gatherings (called bundles) through webwork. The later idea is called bundle exchanging which is straight forward and the way assurance could be exceptionally perplexing.

Directing conventions are grouped into various classifications relying upon their properties [4].

- Centralized versus dispersed
- Static versus versatile
- Reactive versus proactive.

In incorporated estimations, all course choices are made at a central centre point, while in passed on computations, the count of the courses is shared among the framework centre points and another gathering of coordinating traditions relates to whether they change courses in light of the action input plans. In static computations, the course used by the source-objective sets is settled paying little personality to

Table 1 Investigation of major routing protocols [5, 6]

Protocols	Strength	Drawbacks
Leach protocol	Dynamic cluster head selection	Fixed number of clusters and random cluster head selection
Hybrid Energy Efficient Distributed Protocol (HEED)	Dynamic cluster head selection	Fixed number of clusters and inefficient processing time
Hierarchical, energy efficient routing protocol (HEERP)	It is an energy efficient protocol. It performs well in the dense network	Sensors are used for data aggregation process. Efficiency is low in case of less dense network
Hierarchical geographic clustering protocol (HGCP)	It uses virtual grids. Actor acts as a cluster head	It assumed that both the sensors and actors are static. Small grid area
LEACH-B	Effectively balance the network load and improve the utilization of energy, so as to extend the network life cycle	Data fusion issue on cluster head node. Only focuses on residual energy of cluster head

development conditions. It can simply change in light of a centre or association frustration. This kind of count cannot achieve high throughput under a wide assortment of movement input designs.

Most significant bundle frameworks use some sort of adaptable controlling where the courses used to course between source-objective sets may change in light of blockage. Proactive traditions reliably survey the courses inside the framework, with the objective that when a pack ought to be sent the course is starting at now known and can be in a flash used. Responsive traditions summon a course assurance methodology on request as it were. There is require another controlling unit for correspondence coordinate which join adaptable, flexible, and secure points of view in it. A portion of the group-based directing conventions is dissected beneath in Table 1.

3 Security Goals in MANET

When managing security in correspondence is organized, one is looked with the issue of accomplishing a few or the greater part of the accompanying objectives:

- **Availability**: This implies organized resources are accessible to approved gatherings when required and the system ought to guarantee the survivability of system administrations regardless of dissent-of-benefit (DOS) assaults, which could be propelled at any layer of correspondence arrange.
- **Genuineness**: In correspondence arrange, confirmation is fundamental for some authoritative errands (e.g. organize reinventing or controlling sensor hub obligation cycle). Information confirmation enables the collector to check that the

information was extremely sent by the asserted sender. More grounded levels of validness (e.g. express key validation) are given by some key foundation conventions.

- **Confidentiality**: A classified message is impervious to uncovering its importance to a meddler. Notwithstanding directing data in WAN needs to stay private, since it might be utilized to in a DOS assault. The standard answer for keeping touchy information mystery is to encode the information with a mystery key that exclusive the expected recipients have, thus accomplishing privacy.
- **Data uprightness**: Integrity measures guarantee that the information got is not modified in travel by an enemy. The uprightness administration can be furnished utilizing cryptographic hash works alongside some type of encryption. When managing system security, the honesty benefit is frequently given certainly by the validation benefit [7].
- **Scalability and self-association**: As opposed to general systems that do not place versatility in the principal need, WAN cannot use a keying plan that has poor scaling properties (either regarding vitality cost or dormancy) for building up and keeping up a key for the WAN all in all or for some vast subset of hubs. As an outcome, the WAN hubs must have the capacity to self-compose and select the proper keying instrument for the circumstance.

Remote correspondence systems are testing a result of the unusual conduct of the medium and the proactive impact of impedance. Contrasted with the wired systems, the level of the inconstancy of the condition of remote systems is very high. Additionally, the execution of the system, as far as postponement and throughput, is profoundly reliant upon the condition of the system. The layers must facilitate and adjust to the adjustment in arrange state. To manage the dynamic varieties in systems administration and processing assets nimbly, both the versatile registering condition and the applications likewise need to adjust their conduct contingent upon the accessible assets.

Late advances in the compactness, power, and abilities of remote gadgets and applications have brought about the expansion and expanded prominence of these gadgets. As the quantity of clients keeps on developing, remote steering conventions will be required to scale to progressively bigger populaces of hubs. Systems administration situations can require the arrangement of systems on the request of tens to several hubs, while numerous military applications can include thousands to a huge number of hubs. Moreover, as the arrangement of remote correspondence systems turns out to be more across the board, new applications may support the development of expansive correspondence systems. The primary target is to outline a safe steering convention for an extensive remote sensor arrange in which the hubs and besides the base station are versatile. The convention ought to be secure and vitality is proficient and versatile as for every single existing calculation.

4 Enter Management in MANET

Cryptography decreases the secrecy and respectability of a message to the privacy and uprightness of a key [8]. When utilizing symmetric cryptography, the gatherings included need to arrange a mystery key. A decent key foundation conspire gives element validation (all gatherings know the personality of alternate gatherings with whom they are setting up a key), key verification (all gatherings are guaranteed that no unapproved gatherings could have gotten the mystery key), and key affirmation (all gatherings are guaranteed that every single other gathering know about the mystery key).

Key foundation plans can be separated into three noteworthy classifications:

1. Key pre-conveyance plans,
2. Plans utilizing a confided in an outsider, and
3. Plans given open key cryptography.

Key pre-dispersion plans have gotten a considerable measure of consideration in the setting of advertisement hoc systems. They are extremely suited for advertisement hoc organizes as they do not require a confided in an outsider to be accessible consistently and are exceptionally effective. Plans utilizing confided in outsiders are not so much suited for promotion hoc organizes as they accept that the confided in an outsider is accessible to anyone. The detriment of open key-based plans is that they require authentications and those open key calculations are ineffective.

Symmetric key calculations are computationally exceptionally productive and are along these lines of high enthusiasm for MANETs.

The system is apportioned into bunches. The hub with the most extreme trust capacity is chosen as CH. Among bunch individuals in a bunch, k hubs with high trust esteem are chosen as PKG serving hubs. Hubs can move to start with one bunch then onto the next. System manager chooses CH. The disconnected CA appoints node_id for the hub that joins the system. Each new hub has a self doled out the open key and registers its data in CH. The private key of the hub is created by PKG serving hubs. The CH additionally goes about as one of the PKG serving hubs and assumes the job of the key combiner. Private key offers created by k PKG serving hubs are consolidated by CH to get the entire key. Beginning open key of CH is gotten by applying for one-way hash work on its id. CH open key fluctuates dependent on its trust level. New open key of CH is figured dependent on old pubic key and its new trust esteem. The private key of CH is at first appointed by the system head.

Later private key offers are processed by PKG hubs. The general population key of CH is dispersed to all bunch individuals in the comparing group. The portable specialist is a program section that gathers data about k trustable hubs in a group and data about hubs whose declaration is revoked [8]. The framework utilizes two dimensions of recurrence for correspondence: low dimension recurrence is utilized to set up a correspondence between bunch individuals and abnormal state recurrence is utilized to set up a correspondence between bunch heads.

The zone-based key administration conspire is a crossbreed key administration blueprint for MANETs, in light of crafted by [2, 3] and the Zone Routing Protocol

(ZRP) [4], where a zone is characterized for every hub and incorporates the hubs whose remove (e.g. in jumps) is at most some predefined number. This separation is alluded to here as the zone range, zone. Every hub utilizes symmetric key administration inside its zone and unbalanced key administration utilized for between zone security, without relying upon bunching. For effortlessness and consistency, the documentation described in [2] will be utilized in this work, except if generally determined.

In group key management schemes, assemble key is an extraordinary key that is doled out to a gathering of hubs. To build up a gathering key, the gathering needs to make and disseminate the way to all individuals.

5 Assaults on MANET

There are different sorts of assaults on a specially appointed system:

(A) **Location Disclosure**

Area exposure is an assault that objectives the protection necessities of a specially appointed system. By utilizing movement examination systems, more straightforward testing and observing methodologies, an assailant can identify the area of the hub.

(B) **Dark Hole**

In a dark gap, a vindictive hub infuses false course answers to course asks for, reporting it as having the briefest way to a goal. These phoney answers can be manufactured to redirect arrange activity through the vindictive hub for essentially to pull in all the movement towards it keeping in mind the end goal to play out dissent of administration assault by disposing of the got bundles.

(C) **Replay**

A replay assault is one of the assaults that debase extremely the execution of MANET. An answer assailant does this assault by block attempt and retransmission of substantially marked messages.

(D) **Wormhole**

In a wormhole assault, an assailant gets bundles at one point in the system, "burrows" them to another point in the system, and after that replays them into the system starting there.

(E) **Shakedown**

These assaults are pertinent against steering conventions that utilization instrument for the recognizable proof of noxious hubs and engenders messages and endeavour to confine honest to goodness hub from the system. The non-renouncement security criteria can end up being convenient.

(F) **Denial of Service**

Refusal of administration assaults goes for the total disturbance of the directing capacity and along these lines the whole activity of the specially appointed system. In a directing tangle flood assault, the vindictive hub surges the system with counterfeit course creation parcels to devour the assets of taking an interesting hub and upset the foundation of the lawful course.

(G) **Directing**

Table harming routing convention keeps up a table that holds data concerning the course of the system. In this kind of assault, the vindictive hub creates and sends manufactured flagging movement or adjust honest to goodness message from alternate hubs, with a specific end goal to include false sections in the tables of the taking an interesting hub.

(H) **Masquerading**

Amid the neighbour securing process, an outside interloper could disguise non-existence or existing IS by assaulting itself to correspondence interface and unlawfully participating in the steering conventions area by including verification framework.

(I) **Impersonation**

Pantomime assaults are a danger to the assailant can catch a few hubs in the system and influence them to look like cordial hubs. Subsequently, the bargained hubs can join the system as the ordinary hubs and start to lead the vindictive practices, for example, spread phoney steering data and increase improper need to get to some classified data.

(J) **Listening Stealthily**

It is another sort of assault that generally occurs in the versatile impromptu networks. Eavesdropping intends to get some classified data that ought to be kept mystery amid the correspondence. The classified data may incorporate general society key, private key, area and passwords of the hubs. Since such information is exceptionally private to the hubs, they ought to be kept mystery with the goal that unapproved cannot get to this.

6 Proposed Model

Following focuses are considered as pre-presumptions for outlining secure information transmission in bunch-based WSN:

1. The hubs in the system framework are versatile with accessible locally available memory, figuring capacity, correspondence data transfer capacity and accessible battery control.
2. The battery might be inexhaustible or not sustainable.
3. The sensor hubs are not solid and they should be confirmed.
4. The aggressors may listen in the radio transmissions. The aggressors can likewise send some indistinguishable hubs to misdirect genuine hubs.

For proposed calculation, three distinct kinds of keys are created, i.e. introductory mystery key (ISK), sensor-group entryway key (SCGK) and door base key (GBK) in the proposed display.

Each sensor hub has an initial key (at first stacked into the memory of every sensor hub) alongside their ID. Every sensor hub utilizes the ISK and ID at first to verify itself with the base station.

SCGK key is utilized for the correspondence inside a group among bunch head and sensor hubs. SCGK is produced by base station alongside bunch ID and send to all group heads. This key is then disseminated among all hubs utilizing ISK.

GBK is utilized to convey between door hub and base station.

Ventures of Proposed Calculation:

1. The confirmation stage comprises of following advances:
   Every hub sends validation demand to the base station. The ask for message contains ID encoded with ISK.
   The base station at that point decodes ID utilizing ISK of the hub.
   If ID coordinated at that point confirmed and base station doles out SBK to every hub.
2. After the organization of the sensor hubs in the field, the base station isolates the sensor field into a few bunches.
3. In each bunch relegate one cluster head nodes (CH). All sensor nodes in a bunch speak with the cluster head node lastly, the gateway nodes are in charge of transmitting information to the base station (as portrayed in Fig. 2).

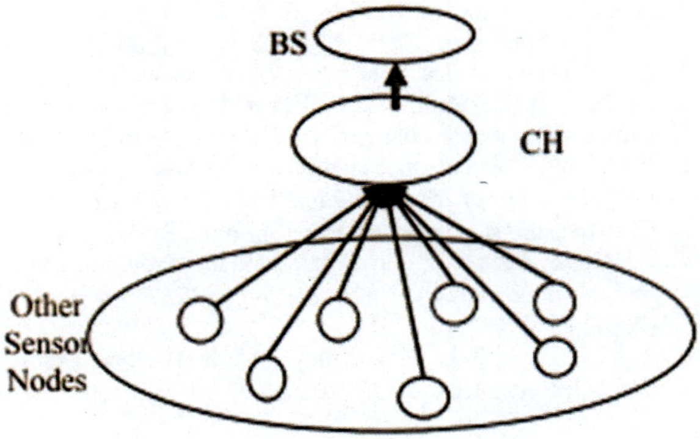

Fig. 2 Proposed model [7, 8]

On the off chance that any sensor nodes sense information from nature then it advances the information to its separate cluster head hub. After receiving the data (information) from sensor Nodes to Cluster Heads. Then Cluster Heads (CHs) will forward the sensed information to the Base Station (BS).

7 Conclusion

In this paper, a near investigation of remote sensor organizes (WSN) directing conventions. The conventions are partitioned into two classifications, i.e. bunch based and non-group based. It is watched that among all steering conventions group-based conventions are various levelled, dependable and vitality proficient directing convention that beats others. Thus, it can be gathered that group-based conventions are more mainstream among the examination network of WSN. One of the key issues in WSN is a security issue in which verification is a principal issue for a dependable system benefit. The hubs must have the capacity to recognize dependable from deceitful hubs in the neighbour revelation process, and they should have the capacity to check both steering message beginning and respectability. Any cryptographic validation conspire requires appropriate key administration. That is, common verification of the included gatherings is required amid the key set-up. This paper incorporates a thorough review of key administration techniques proposed for ad hoc systems. The appropriateness of personality-based open key plans for assurance of ad hoc steering data is likewise talked about.

References

1. National Institute of Standards and Technology (NIST), Wireless ad hoc network projects. Available at https://www.antd.nist.gov/wahn_home.shtml
2. A. Balasubramanian, S. Misha, R. Sridhar, A hybrid approach to key management for enhanced security in ad hoc networks. Technical report, University at Buffalo, NY, USA (2004)
3. A. Balasubramanian, S. Misha, R. Sridhar, Analysis of a hybrid key management solution for ad hoc networks, in *IEEE WCNC'05*, vol. 4 (2005), pp. 2082–2087
4. C.E. Perkins, *Ad Hoc Networking* (Addison-Wesley, Boston, MA, 2001)
5. M. Radi, B. Dezfouli, K. Abu Bakar, M. Lee, Multipath routing in wireless sensor networks: survey and research challenges. Sensors **12**, 650–685 (2012)
6. F. Stajano, R. Anderson, The resurrecting duckling: security issues for ad-hoc wireless networks, in *Proceedings of 7th International Workshop on Security Protocols*, ser. LNCS 1796, ed. by B. Christianson, B. Crispo, M. Roe (Springer, Berlin, 1999), pp. 172–194
7. J. Kong, P. Zervos, H. Luo, S. Lu, L. Zhang, Providing robust and ubiquitous security support for mobile adhoc networks, in *Proceeding of IEEE International Conference on Network Protocols* (2001)
8. Z. Yi, Z. Lina, F. Li, Key management and authentication in ad hoc network based on mobile agent. J. Netw. **4**(6), 20–29 (2016)

Student Performance Monitoring System Using Decision Tree Classifier

V. Ramakrishna Sajja, P. Jhansi Lakshmi, D. S. Bhupal Naik, and Hemantha Kumar Kalluri

Abstract The enormous issue of drop-out students or resigned is regarding scholarly accomplishment. The academic organization wants to catch up the prompting framework while counselors be supposed to direct the arranging educational programs to their advisors. The data mining methods are very much helpful to provide the qualitative education in the academic institutions and to analyze the student performance quickly. Different classification models which can be applied in academic data mining are focused in this paper. The student problems can be identified by applying different classification models. The goal of this paper is to enhance the performance of student and at that point anticipating the appropriated scholarly accomplishment in each major. To look at the investigation, we utilized 1200 students' data. Two measures like accuracy and error rate are assessed the framework. This method produces the 95.5% accuracy and 4.5% error rate.

Keywords Preprocessing · Classification · Decision tree · Training · Testing · Rule extraction · Student performance

V. Ramakrishna Sajja (✉) · P. Jhansi Lakshmi · D. S. Bhupal Naik · H. K. Kalluri
Department of CSE, VFSTR Deemed to be University, Guntur, India
e-mail: vramakrishnasajja@gmail.com

P. Jhansi Lakshmi
e-mail: laxmi.jhansi@gmail.com

D. S. Bhupal Naik
e-mail: dsbhupal@gmail.com

H. K. Kalluri
e-mail: hemanth_mtech2003@yahoo.com

D. Bhattacharyya and N. Thirupathi Rao (eds.), *Machine Intelligence and Soft Computing*, Advances in Intelligent Systems and Computing 1280,
https://doi.org/10.1007/978-981-15-9516-5_33

1 Introduction

Education is a major part for building up a nation. Management will look the academic achievement in part of evaluation. If the number of dropped students or resigned students is less then advisory system will be helpful to enhance the student performance. The number of resigned students can be decreased if curriculum planning in every semester has been done early. Information mining innovation can be applied in the training for enhancing an adapting, for example, the associating a learning holes, the anticipating the understudy achievement, and the finding the reasonable major. Numerous instruments are utilized in information digging for investigating an enormous dataset and discovering significant information known as learning. This study will be helpful to draw the useful conclusion. At first, the use of information mining was confined to business space yet realizes it is reached out to academics and is called as ADM.

Academic data mining (ADM) talks with data mining applications and innovative methods which examine the data at academic institutions to derive the knowledge [1]. The knowledge discovery database (KDD) derives the techniques in sight of statistics, machine learning, and database systems to evaluate the big data of academic information. The interesting patterns are extracted from enormous databases by KDD process. However, the competent algorithm is a key point of any application to discover the interesting patterns presented in the data considered.

In data mining [2] and machine learning, recognizing the problem which is related among different cases based on the given training is called as classification. For example, finding the tendency of customer at hardware shop whether he buys computer or not: "Yes" or "No". It will be decided based on the training given to the classifier across the attributes like age, student, income, credit rating, etc. Decision tree classifier [3] uses a decision tree. The item value which belongs to test of attribute is represented on the branches of tree. The class labels are presented on leaf nodes of a tree. The tree pruning is one type of decision tree which improves the efficiency of classifier.

The remaining paper is ordered as follows: literature survey is described in Sect. 2. Introduction of data mining techniques is presented in Sect. 3. Methods and procedures have been discussed in Sect. 4. Practical results of experimentation are discussed in Sect. 5. Section 6 describes the conclusion of paper.

2 Literature Survey

The process of discovering the knowledge from huge dataset is known as knowledge discovery in database (KDD). Academic data mining (ADM) is the piece of KDD that facilitates the student to enhance their academic percentage. ADM uses various other data mining techniques such as association rule discovery, classification, and clustering. Several researchers felt that decision tree algorithm is reasonably good for prediction of students' academic performance [4]. They executed different decision tree algorithms such as J48, multilayer perception, random forest, ID3, random tree, and rule based with the help of WEKA tool. Split technique or cross-validation technique has been used to assess the performance percentage. It is observed that student academic performance will be improved due to the usage of innovative data mining techniques like split method and cross-validation method. In the examination [5], they considered the abilities of data mining inside the setting of the following academic institute and attempts to obtain fresh explicit data by connecting information handling procedures to scholarly learning of technical scholastic Institute of Athens. Every department of Athens institute utilized the decision tree algorithm to examine the student's survey. Different hands, Ahmed and his group [6] discussed that last grade of students will be predicted based on the various classification models. They have chosen the decision tree (ID3) strategy for their examination. Counting to the examination [7] of Kaur and his team, they have discussed the various classification and prediction algorithms to forecast the slow learners in academic zones. Various classification algorithms used by Kaur and his team include Naive Bayes, J48, multilayer perception, J48, and REPTree in WEKA environment. The modern research was discovered with new outline, known as cloud-service decision tree classification for academic platform [8]. This examination considered the multitarget weight self-adjustment structure and proposed the cloud-administration classification with decision tree algorithm. At that point, their frame work is confirmed by different performance metrics.

3 Data Mining

Knowledge discovery in database (KDD) or data mining (DM) is a procedure to discover examples [9] and connection from enormous informational collections with the different techniques, for example, AI, database frameworks, and statistics [10, 11]. There are three procedures of information mining as the depicted after:

1. Preprocessing: sets up the reasonable dataset and will organize in a helpful arrangement. By applying the following techniques, noise has to be eradicated [12] from the original dataset. The noise-free dataset that has a connection and can be utilized to develop the framework.

   (i) Data cleaning: dispense mistake information or an immaterial dataset.
   (ii) Data integration: gather the dataset which derived by various bases.
   (iii) Data selection: pick the qualified characteristic which can be anticipated the great execution.
   (iv) Data transformation: modify the arrangement of properties that is anything but difficult to utilize or anticipate.

2. Modeling: utilize the chose information for making a model. Model has to be trained with various cases by selection of suitable technique. Once model is trained then it has to be tested with new cases.
3. Post-processing: Derive the knowledge from the results.

4 Methods and Procedure

In this examination, student's problems can be identified by applying different classification models and afterward assess the performance rate with the known metrics of accuracy and error rate. The test in this examination, we center on the data mining procedure, for example, warehouse construction, preprocessing, classification (modeling), and rules extraction (post-processing) which are portrayed underneath; Fig. 1 depicts the architecture of student performance prediction using decision tree.

4.1 Warehouse Construction

In this work, a dataset of 1200 students are collected for experimentation. The total information of 1200 students is collected across 20 attributes. Types of attributes are belonging to nominal, ordinal, binary, and numeric. Some of these details are collected from Vignan's information system portal and other details which cannot be retrieved from the portal are collected manually [13]. Table 1 describes about attributes which were used for experimentation.

4.2 Preprocessing

Preprocessing is an important technique for any classification problems. It reduces the computational complexity of algorithm and makes the data robust and noise free. Data preprocessing includes data cleaning, data reduction, data integration, and data transformation. These are explained in the following:

1. **Data cleaning**: It is the process of making the data as noise free [14]. During the process of data retrieval from university portal, values which were missed or wrongly loaded are filled by manual process. Some of the values can be predicted by smoothing techniques like bin means and bin boundary.
2. **Data integration**: It is the process of integrating the information which came from several sources. Tuple duplication is one of the integration techniques, and this method can be used to identify duplicate tuples [15]. Duplicate tuples are identified by computing the dissimilarity matrix of mixed type for the considered dataset of 1200 records using Eq. 1.

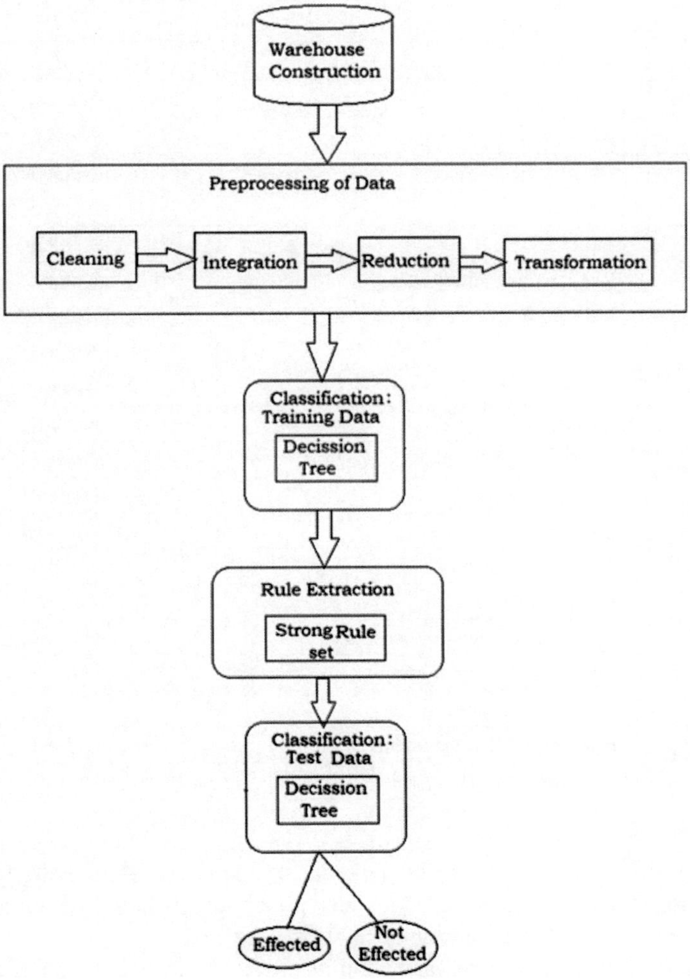

Fig. 1 Architecture of student performance prediction using decision tree

$$\text{dis}(i, j) = \frac{\sum_{a=1}^{n} \delta_{ij}^{(a)} \text{dis}_{ij}^{(a)}}{\sum_{a=1}^{n} \delta_{ij}^{(a)}} \tag{1}$$

where

- "n" indicates number of attributes participated in computation.
- $\delta_{ij}^{(a)}$ is 0 if either attribute value is missing or attribute value is 0. Otherwise 1.
- $\text{dis}_{ij}^{(f)}$ is computed depended on attribute's type.

Table 1 Attributes used to collect information of each student

ID	Attribute	Type
A1	Regd. no	Nominal
A2	Name	Nominal
A3	DOB	Numeric
A4	10th GP	Numeric
A5	Inter GP	Numeric
A6	Rank in competitive exams	Numeric
A7	UG CGPA	Numeric
A8	No. of backlogs	Numeric
A9	Medium up to SSC	Binary
A10	Annual income	Numeric
A11	Parent's educational background	Ordinal
A12	Parent's attention	Ordinal
A13	Social problems	Nominal
A14	Behavior of student	Ordinal
A15	Friend circle	Binary
A16	No. of hours spent for academics	Numeric
A17	No. of hours spent for non-productive work	Numeric
A18	Regularity of student	Binary
A19	Promptness on assignments	Ordinal
A20	Reaction toward counselor's action	Ordinal

3. **Data reduction**: It is the process of reducing the size of data. Attribute subset selection is one of the data reduction techniques, and this method can be used to remove the non-relevant or redundant attributes.
   Forward selection and backward elimination are the two important methods of attribute subset selection. Forward selection includes the attributes to the empty set which are more relevant. Backward elimination excludes the attributes from the wholes set which are more irrelevant. Hence, the attributes which would be considered for classification are $\{A4, A5, A6, \ldots, A19, A20\}$.
4. **Data transformation**: It is the process of consolidating the data. Normalization is one of the data transformation techniques, and this method can be used to normalize the non-data by using min–max normalization.

If the values of numeric attributes are scattered in multiple ranges, those values must be in normalized form during the classification process. Hence, min–max normalization is adopted to accomplish this goal. Equation 2 represents the process of min–max normalization.

$$v_i^{|} = \frac{v_i - \min_A}{\max_A - \min_A}(\text{new\_max}_A - \text{new\_min}_A) + \text{new\_min}_A \tag{2}$$

where

- v_i represents the value to be normalized.
- $\min_A$ represents the minimum value among the all values of attribute "*A*".
- $\max_A$ represents the maximum value among the all values of attribute "*A*".
- new_min_A represents the new minimum value considered for all values of attribute "*A*".
- new_max_A represents the new maximum value considered for all values of attribute "*A*".

4.3 Classification

Classification is a sort of data investigation that signifies models depicting significant data classes. Such models, called classifiers, predict class labels. A classification model can be constructed to classify bank advance applications as either safe or unsafe [16]. Such examination helps in superior comprehension of the data. Many such classification techniques have been introduced by researchers in AI, statistics, and pattern recognition.

Classification works in two phases, which includes training and testing phases. In training phase, the classifier model is built by training data and in testing phase, a tuple with unknown class label is given to classifier as an input and extracts output with a class label.

Various classifiers can be used for classifying the data. Classifiers are decision tree, Bayes, rule-based, SVM, backpropagation, ANN, etc. In this paper, decision tree [17] is adopted to classify the student data. Figure 2 shows a sample decision tree classification.

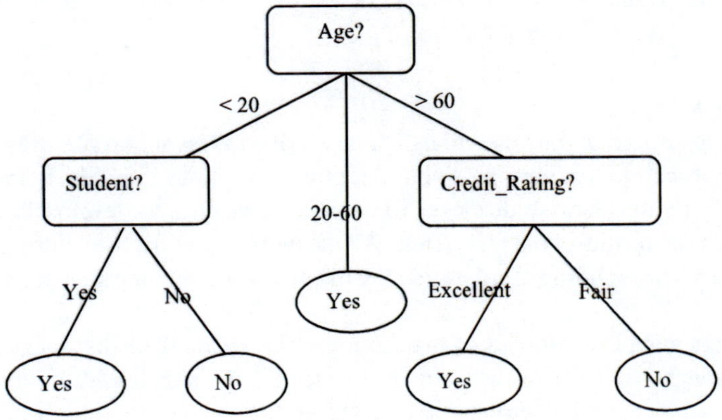

Fig. 2 Example decision tree

Decision Tree Induction

Decision tree induction is the learning of decision trees from class-labeled preparing instances [18]. A decision tree is a flowchart-like tree structure, where each inward node (non-leaf node) means a test on a characteristic, each branch speaks to a result of the test, and each leaf node holds a class name. The highest node in a tree is the root node. A common decision tree is appeared in Fig. 2. It speaks to the idea purchases PC. It predicts whether a client at All Electronics is probable to buy a PC. Inner nodes are depicted by square shapes and leaf nodes by ovals. Few decision tree algorithms generate binary trees, though others can create non-binary trees.

Decision Tree Algorithm

Step 1: Grant all training instances as a root node of the tree. Fix the current node as a root node.

Step 2: For each and every attribute.

(a) Split all the data instances with the help of attribute value at the node.
(b) Calculate the information gain (IG) ratio from the above partition.

Step 3: Identify the highest information gain (IG) ratio feature. Use this instance for partitioning process at the current node.

(a) If the highest IG ratio is zero, then tag the current node as a leaf node and return.

Step 4: Partition each and every instance based on the attribute value of the best feature.

Step 5: Tag each partition as a descendant node to the current node.

Step 6: For every descendant node:

(a) If the descendant node is "pure", denote it as a leaf node and return.
(b) If the descendant node is not "pure", then set the descendant node as a current node and go to step 2.

Attribute Selection

The splitting criterion discloses to us which ascribe to test at hub N by deciding the "best" approach to separate or divide the tuples in D into individual classes. The splitting criterion likewise discloses to us which branches to develop from hub N concerning the results of the picked test. All the more explicitly, the splitting criterion demonstrates the splitting ascribe and may likewise show either a split point or a split subset.

The part model is resolved so that, in a perfect world, the subsequent segments at each branch are as "pure" as would be prudent. A division is pure if all the tuples in it have a place with a similar class. At the end of the day, on the off chance that we split up the tuples in D as indicated by the fundamentally unrelated results of the part model, we trust in the subsequent segments to be as pure as would be prudent.

In this paper, a dataset with 20 attributes has to be transformed as non-binary tree which tells the paths from root node to leaf node that gives respective class labels. Attribute selection measure is the best splitting criterion to partition the data. Information gain, gain ratio, and Gini index are the attribute selection measures. One of the attribute selection measures can be adopted to split the data.

(a) **Information gain**: Information gain is one of the attribute selection measures. It will be computed for all the 17 attributes considered for construction of decision tree. The attribute which gets highest information gain will be the root node in a decision tree. The same process will be repeated till reach the class label. "*D*" denotes the whole dataset, and "*A*" denotes the respective attribute. Equations 3, 4, and 5 have been used to compute the information gain.

$$\text{Gain}(A) = \text{Info}(D) - \text{Info}_A(D) \tag{3}$$

$$\text{Info}(D) = -\sum_{i=1}^{m} p_i \log_2(p_i) \tag{4}$$

where "m" is the number of class labels. "p_i" is the probability for the ith class label.

$$\text{Info}_A(D) = \sum_{j=1}^{v} \frac{|D_j|}{D} X\ \text{Info}(D_j) \tag{5}$$

where "v" is the number of values presented for the attribute "A". "D_j" is the number of tuples for the jth value of attribute "A" in a whole dataset "D".

(b) **Gain ratio**: Gain ratio is one of the attribute selection measures. It can be taken as another measure and computed for all the 17 attributes considered for construction of decision tree. The attribute which gets highest gain ratio will be the root node in a decision tree. The same process will be repeated till reach the class label. "*D*" denotes the whole dataset and "*A*" denotes the respective attribute. Equations 6 and 7 have been used to compute the gain ratio.

$$\text{Gain ratio}(A) = \frac{\text{Gain}(A)}{\text{SplitInfo}_A(D)} \tag{6}$$

$$\text{SplitInfo}_A(D) = -\sum_{j=1}^{v} \frac{|D_j|}{D} X\log_2\left(\frac{|D_j|}{D}\right) \tag{7}$$

where "v" is the number of values presented for the attribute "A". "D_j" is the number of tuples for the jth value of attribute "A" in a whole dataset "D".

(c) **Gini index**: The Gini index splits attribute into two partitions. Assume that A is a discrete-valued attribute with "w" distinct values, $\{a_1, a_2, \ldots, a_w\}$, presenting in D. Best binary split will be determined at attribute "A" by cross-checking all the probable subsets that can be formed using known values of A.

For instance, if *performance* has three probable values, namely {*poor*, *average*, *good*} then the possible subsets are {*poor*, *average*, *good*}, {*poor*, *average*}, {*poor*, *good*}, {*average*, *good*}, {*poor*}, {*average*}, {*good*}, and { }. We exclude the power set, {*poor*, *average*, *good*}, and the empty set from consideration since, conceptually, they do not represent a split. Therefore, there are $2^v - 2$ possible ways to form two partitions of the data, *D*, based on a binary split on *A*.

"*D*" denotes the whole dataset, and "*A*" denotes the respective attribute. Equations 8, 9, and 10 have been used to compute the Gini index.

$$\Delta \text{Gini}(A) = \text{Gini}(D) - \text{Gini}_A(D) \tag{8}$$

$$\text{Gini}(D) = 1 - \sum_{i=1}^{m} p_i^2 \tag{9}$$

where "m" is the number of class labels. "p_i" is the probability for the ith class label.

$$\text{Gini}_A(D) = \frac{|D_1|}{D}\text{Gini}(D_1) + \frac{|D_2|}{D}\text{Gini}(D_2) \tag{10}$$

where "D_1" is the number of tuples considered from dataset "*D*" with respect to attribute "*A*" under first split. "D_2" is the number of tuples considered from dataset "*D*" with respect to attribute "*A*" under second split.

The resultant decision tree can be viewed in Fig. 3 with 69 number of leaves and 137 levels of the tree.

4.4 Rule Extraction Mining

Rule extraction mining is considered as post-processing technique. Rules can be extracted from the tree also called feature extraction. If "*n*" number of paths are generated from root node to each leaf node called as class label then "*n*" number of rules are derived. All those rules are called as strong rules. Rules include single-dimensional, multidimensional, and hybrid-dimensional association rules.

5 Experimental Results

WEKA is a popular data mining tool which solves the many data mining problems. Dataset with 1200 records have been taken for experimentation. 1000 records are utilized for training the classifier and 200 records are used for validation or testing. Table 2 demonstrates the dataset and its divisions.

```
UG_Perf<= 1
| Alcoholic_Parent<= 0.041401
| | Parents_Education<= 0.153846
| | | Social_Problems<= 0.000791
| | | | Medium_upto_SSC<= 0.015873
| | | | | SSC_Performance<=
0.014184
| | | | | | Communications_skills<=
0.01494: 0.012285012 (2.0/1.0)
| | | | | | Regularity_of_the_student>
0.025: 0.011056511 (3.0/2.0)
| | | | Medium_upto_SSC> 0.015873
| | | | | SSC_Performance<=
0.018913: 0.013513514 (7.0)
| | | | | SSC_Performance> 0.018913
| | | | | | Communications_skills<=
0.01494: 0.00982801 (2.0/1.0)
| | | | | | Communications_skills>
0.01494: 0.01965602 (2.0/1.0)
| | | Social_Problems> 0.000791
| | | | SSC_Performance<= 0.023641:
0.017199017 (3.0/1.0)
| | | | SSC_Performance> 0.023641
| | | | | UG_Perf<= 0.033333
| | | | | | Alcoholic_Parent<=
0.035032: 0.01965602 (3.0)
| | | | | | Alcoholic_Parent>
0.035032: 0.023341523 (3.0/1.0)
| | | | | UG_Perf> 0.033333:
0.022113022 (2.0/1.0)
| | Parents_Education> 0.153846
| | | Alcoholic_Parent<= 0.006369
| | | | Alcoholic_Parent<= 0.003185:
0.001228501 (2.0)
| | | | Alcoholic_Parent> 0.003185:
0.002457002 (4.0)
| | | Alcoholic_Parent> 0.006369
| | | | Alcoholic_Parent<= 0.012739
| | | | | Social_Conciousness<=
0.037115: 0.003685504 (2.0)
| | | | | Social_Conciousness>
0.037115: 0.004914005 (2.0)
| | | | Alcoholic_Parent> 0.012739:
0.011056511 (3.0/2.0)
| Alcoholic_Parent> 0.041401
| | Friend_circle<= 0.010287
| | | Parents_Education<= 0.061538
| | | |
Promptness_on_teacher_assignment<=
0.017699
| | | | | Medium_upto_SSC<=
| | | | | | Communications_skills>
0.01494: 0.025798526 (2.0/1.0)
| | | | | SSC_Performance> 0.014184
| | | | | |
Regularity_of_the_student<= 0.025
| | | | | | | Alcoholic_Parent<=
0.035032: 0.012285012 (8.0)
| | | | | | | Alcoholic_Parent>
0.035032: 0.018427518 (2.0/1.0)
| | | | | Medium_upto_SSC>
0.047619
| | | | | | SSC_Performance<=
0.073286: 0.084766585 (2.0/1.0)
| | | | | | SSC_Performance>
0.073286: 0.095823096 (2.0/1.0)
| | | |
Promptness_on_teacher_assignment>
0.017699
| | | | | Communications_skills<=
0.051566
| | | | | | Parents_Education<=
0.046154
| | | | | | | SSC_Performance<=
0.052009: 0.046683047 (2.0/1.0)
| | | | | | | SSC_Performance>
0.052009: 0.045454545 (3.0/1.0)
| | | | | | Parents_Education>
0.046154: 0.035626536 (2.0/1.0)
| | | | | Communications_skills>
0.051566
| | | | | | Behaviour_of_Student<=
0.047285
| | | | | | | IP_Performance<=
0.041667
| | | | | | | | UG_Perf<= 0.033333
| | | | | | | | |
Medium_upto_SSC<= 0.047619:
0.066339066 (2.0)
| | | | | | | | | Medium_upto_SSC>
0.047619: 0.046683047 (2.0/1.0)
| | | | | | | | UG_Perf> 0.033333:
0.061425061 (2.0)
| | | | | | | IP_Performance>
0.041667: 0.051597052 (5.0/1.0)
| | | | | | Behaviour_of_Student>
0.047285
| | | | | | | SSC_Performance<=
0.085106: 0.068796069 (2.0)
| | | | | | | SSC_Performance>
0.085106: 0.078624079 (2.0/1.0)
| | | Parents_Education> 0.061538
```

Fig. 3 Resultant decision tree

```
0.047619
| | | | | | UG_Perf<= 0.033333
| | | | | | |
Promptness_on_teacher_assignment<=
0.0059: 0.040540541 (3.0/1.0)
| | | | | | |
Promptness_on_teacher_assignment>
0.0059: 0.07002457 (4.0/2.0)
| | | | | | UG_Perf> 0.033333:
0.054054054 (2.0)
| | | | | Alcoholic_Parent> 0.079618:
0.041769042 (2.0/1.0)
| | Friend_circle> 0.010287
| | | Alcoholic_Parent<= 0.171975
| | | | UG_Perf<= 0.233333
| | | | | UG_Perf<= 0.133333
| | | | | | UG_Perf<= 0.033333
| | | | | | | SSC_Performance<=
0.106383: 0.119164619 (2.0/1.0)
| | | | | | | SSC_Performance>
0.106383: 0.092137592 (2.0/1.0)
| | | | | | UG_Perf> 0.033333:
0.171990172 (2.0/1.0)
| | | | | UG_Perf> 0.133333
| | | | | | Communications_skills<=
0.123373
| | | | | | | Financial_status<=
0.063814: 0.088452088 (2.0/1.0)
| | | | | | | Financial_status>
0.063814: 0.093366093 (2.0)
| | | | | | Communications_skills>
0.123373: 0.131449631 (3.0/1.0)
| | | | UG_Perf> 0.233333
| | | | | Regularity_of_the_student<=
0: 0.180589681 (3.0/1.0)
| | | | | Regularity_of_the_student> 0
| | | | | | SSC_Performance<=
0.115839: 0.156019656 (2.0/1.0)
| | | | | | SSC_Performance>
0.115839: 0.16953317 (3.0/1.0)
| | | Alcoholic_Parent> 0.171975
| | | | Regularity_of_the_student<=
0.0875
| | | | | Parents_Education<=
0.023077
| | | | | | Parents_Education<=
0.015385: 0.254299754 (3.0/2.0)
| | | | | | Parents_Education>
0.015385: 0.343980344 (3.0/2.0)
| | | | | Parents_Education> 0.023077
| | | | | | Parents_Education<=
| | | | SSC_Performance<= 0.044917:
0.033169533 (4.0/1.0)
| | | | SSC_Performance> 0.044917
| | | | | Alcoholic_Parent<= 0.079618
| | | | | |
Promptness_on_teacher_assignment<=
0.023599: 0.044226044 (3.0)
| | | | | |
Promptness_on_teacher_assignment>
0.023599: 0.025798526 (2.0/1.0)
| | | | | Regularity_of_the_student>
0.225
| | | | | | Friend_circle<= 0.102605:
0.30958231 (3.0/2.0)
| | | | | | Friend_circle> 0.102605:
0.324324324 (2.81/1.81)
UG_Perf> 1
| Medium_upto_SSC<= 0.380952
| | Behaviour_of_Student<= 0.336252:
NE (42.24/3.0)
| | Behaviour_of_Student> 0.336252
| | | UG_Perf<= 9: E (3.02/0.02)
| | | UG_Perf> 9: NE (21.12/3.0)
| Medium_upto_SSC> 0.380952
| | UG_Perf<= 6: E (3.02/0.02)
| | UG_Perf> 6
| | | Social_Problems<= 0.257009
| | | | Behaviour_of_Student<=
0.336252
| | | | | Permanent_Eisabilities<= 0.5
| | | | | |
Promptness_on_teacher_assignment<= 2:
NE (26.09/11.0)
| | | | | |
Promptness_on_teacher_assignment> 2: E
(2.0)
| | | | | Permanent_Eisabilities> 0.5:
E (2.09/0.09)
| | | | Behaviour_of_Student>
0.336252
| | | | | Social_Conciousness<=
0.477765
| | | | | |
Promptness_on_teacher_assignment<= 2
| | | | | | |
Communications_skills<= 2
| | | | | | | |
Permanent_Eisabilities<= 0.5: NE
(6.08/2.0)
| | | | | | | |
Permanent_Eisabilities> 0.5
```

Fig. 3 (continued)

```
0.030769: 0.211302211 (3.0/2.0)
| | | | | | Parents_Education>
0.030769: 0.132678133 (2.0/1.0)
| | | | Regularity_of_the_student>
0.0875
| | | | | Regularity_of_the_student<=
0.225
| | | | | | Medium_upto_SSC<=
0.126984: 0.165847666 (2.0/1.0)
| | | | | | Medium_upto_SSC>
0.126984: 0.391891892 (3.0/2.0)

| | | | | | | | | |
Regularity_of_the_student> 0.35: E
(13.08/4.08)
| | | | | | | Communications_skills>
2
| | | | | | | | Financial_status<=
0.336808: NE (14.09)
| | | | | | | | Financial_status>
0.336808: E (3.02/0.02)
| | | | | | | | | |
Regularity_of_the_student<= 0.35
| | | | | | | | | | |
Communications_skills<= 1
| | | | | | | | | | | |
Promptness_on_teacher_assignment<= 1:
E (4.0)
| | | | | | | | | | | |
Promptness_on_teacher_assignment> 1:
NE (2.0)
| | | | | | | | | | |
Communications_skills> 1: NE (2.0)
| | | | | |
Promptness_on_teacher_assignment> 2:
NE (4.0)
| | | | | Social_Conciousness>
0.477765: NE (21.12)
| | | Social_Problems> 0.257009: NE
(39.22/6.0)
```

Fig. 3 (continued)

Table 2 Student dataset used for the experimentation

Student dataset	Good students	Poor students	Total
Training	770	230	1000
Testing	120	80	200
Total	890	310	1200

Results are formulated in confusion matrix. Confusion matrix has four important terms labeled as TP, TN, FP, and FN, and those can be used in computing many evaluation measures [19, 20]. Those four terms are also called as building blocks for any classifier. These terms are summarized in the confusion matrix of Table 3.

TP is pronounced as true positive which refer to the positive tuples those are rightly classified. TN is pronounced as true negative which refers to the negative tuples those are rightly classified. FP is pronounced as false positive which refers to the negative tuples those are wrongly classified. FN is pronounced as false negative which refers to the positive tuples those are wrongly classified.

Table 3 Confusion matrix

Class	Yes	No	Total
Yes	TP	FN	P
No	FP	TN	N
Total	P^l	N^l	P + N

Table 4 Confusion matrix for testing data

Class	Yes	No	Total
Yes	116	4	120
No	5	75	80
Total	121	79	200

Accuracy of the classifier is computed using the metric accuracy which calculates the ratio of the total number of correct predictions from whole data by using Eq. 11.

$$\text{Accuracy} = \frac{\text{TP} + \text{TN}}{\text{TP} + \text{TN} + \text{FP} + \text{FN}} \tag{11}$$

Error rate is also called as misclassification rate of a classifier. It is computed as simply 1-accuracy and it is determined by using Eq. 12.

$$\text{Error rate} = \frac{\text{FP} + \text{FN}}{\text{TP} + \text{TN} + \text{FP} + \text{FN}} \tag{12}$$

After performing the experimentation on the considered dataset according to the proposed architecture, it produces 95.5% accuracy and 4.5% error rate. Resultant values for building blocks of the classifier such as TP, FP, FN, and TN are shown in Table 4.

6 Conclusion

In our research, we used data mining techniques like preprocessing, classification, and post-processing to improve the performance of student. Raw dataset is undergone through preprocessing to make ready for classification. Dataset of 1200 records are loaded, and then decision tree is employed to classify the data. The proposed method produces accuracy of 95.5% and 4.5% error rate.

References

1. M. Al-Razgan, A.S. Al-Khalifa, H.S. Al-Khalifa, Educational data mining: a systematic review of the published literature 2006–2013, in *Proceedings the 1st International Conference on Advanced Data and Information Engineering* (2013), pp. 711–719
2. M. Hofmann, R. Klinkenberg, *RapidMiner: Data Mining Use Cases and Business Analytics Applications* (Chapman and Hall/CRC Data Mining and Knowledge Discovery Series) (CRC Press, Boca Raton, 2013)
3. J.R. Quinlan, Induction of decision trees. Mach. Learn. **1**(1), 81–106 (1986)

4. H. Kaur, A literature review from 2011 to 2014 on student's academic performance prediction and analysis using decision tree algorithm. J. Glob. Res. Comput. Sci. **9**(5), 10–15 (2018)
5. N.V.K. Rao et al., A review on data mining approach used in education data mining using decision tree algorithm. J. Adv. Res. Dyn. Control Syst. 1735–1738 (2018)
6. A.B.E.D. Ahmed, I.S. Elaraby, Data mining: a prediction for student's performance using classification method. World J. Comput. Appl. Technol. **2**(2), 43–47 (2014)
7. J. Han, M. Kamber, *Data Mining Concepts and Techniques* (Morgan Kaufmann, Burlington, 2000)
8. B. Kaminski, M. Jakubczyk, P. Szufel, A framework for sensitivity analysis of decision trees. Central Eur. J. Oper. Res. **26**. https://doi.org/10.1007/s10100-017-0479-6 (2018)
9. M. Arif, A.R. Dar, Survey on fraud detection techniques using data mining. Int. J. u- e-Serv. Sci. Technol. **8**(3), 163–170 (2015)
10. Data Mining Curriculum, ACM SIGKDD. 2006-04-30. Retrieved 2014-01-27
11. D.S. Bhupal Naik, S. Deva Kumar, S.V. Ramakrishna, Parallel processing of enhanced k-means using OpenMP, in *2013 IEEE International Conference on Computational Intelligence and Computing Research* (2013), pp. 1–4
12. P. Kaur, M. Singh, G.S. Josan, Classification and prediction based data mining algorithms to predict slow learners in education sector. Procedia Comput. Sci. **57**, 500–508 (2015)
13. S. Venkatramaphanikumar et al., A novel prediction model for academic emotional progression of graduates. ARPN J. Eng. Appl. Sci. **10**(6), 45–51 (2015)
14. J. Liu, G. Sun, Q. Zhang, H. Jun, Similarity distance noise reduction of entropy based on lifting KNN classification performance. Int. J. Secur. Appl. **9**(2), 149–158 (2015)
15. N. Kalanat, M.R. Kangavari, Data mining methods for rule designing and rule triggering in active database systems. Int. J. Database Theor. Appl. **8**(1), 39–44 (2015)
16. J.-H. Seo, H.-S. Lee, J.-T. Choi, Classification technique for filtering sentiment vocabularies for the enhancement of accuracy of opinion mining. Int. J. u- e-Serv. Sci. Technol. **8**(10), 11–20 (2015)
17. W. Chao, W. Junzheng, Cloud-service decision tree classification for education platform. Cogn. Syst. Res. **52**, 234–239 (2018)
18. B.P. Battula, K.V.S.S. Rama Krishna, T. Kim, An efficient approach for knowledge discovery in decision trees using inter quartile range transform. Int. J. Control Autom. **8**(7), 325–334 (2015)
19. V. Ramakrishna Sajja, H.K. Kalluri, Brain tumor segmentation using fuzzy C-means and classification using SVM, in *Lecture notes in Networks and Systems, Proceedings of Smart Technologies in Data Science and Communication*, Visakhapatnam, LNNS, 105 (2019), pp. 197–204
20. H.K. Kalluri, M. Prasad, A. Agarwal, Palmprint identification based on wide principal lines, in *ACM International Conference Proceeding Series* (2012). https://doi.org/10.1145/2345396.2345544

Smart Farming Using IoT

D. S. Bhupal Naik, V. Ramakrishna Sajja, P. Jhansi Lakshmi, and D. Venkatesulu

Abstract With the rapid development of the world population, large area of land is utilized to develop housing and the ability of producing food is reduced. Farming has become crucial in present trend and keeps food on the tables. Farming with IoT helps in mitigating the shortage of food by demanding the existing land for stronger utilization at minimum cost. Smart farming is a notion that quickly snaps on the agricultural field. This offers an automated farming techniques, useful data collection and high-rigor crop control. Sophisticated sensor-based architecture is proposed to monitor the conditions of the farms by using sensors, and the information extracted from these sensors is stored on the Internet. This stored information is obtained and evaluated to forecast the condition of farms, according to a period of time. Based on this evaluation, the necessary improvements can be made with better farming conditions in future.

Keywords IoT · Sensors · Smart farming · Agriculture · Food · Modem · Microcontroller

D. S. Bhupal Naik (✉) · V. Ramakrishna Sajja · P. Jhansi Lakshmi · D. Venkatesulu
Department of Computer Science & Engineering, Vignan's Foundation for Science, Technology & Research, Guntur, India
e-mail: dsbhupal@gmail.com

V. Ramakrishna Sajja
e-mail: vramakrishnasajja@gmail.com

P. Jhansi Lakshmi
e-mail: laxmi.jhansi@gmail.com

D. Venkatesulu
e-mail: drvenkatesulud43@gmail.com

D. Bhattacharyya and N. Thirupathi Rao (eds.), *Machine Intelligence and Soft Computing*, Advances in Intelligent Systems and Computing 1280,
https://doi.org/10.1007/978-981-15-9516-5_34

1 Introduction

In India, farming is the only source that generates revenue for the agriculture community. Two-thirds of our population is based on farming for a livelihood. So for this reason, farming has become the backbone of Indian economy [1, 2]. Due to many problems, farmer suicides are leading to a large number across India because of the problems are faced as insufficient rainfall and irregular monsoons. These are not responsible and can be set directions to deliver by technology. At the right time, some of the problems can be solved by giving proper advices to the farmers. To harvest crops, some of the requirements like fertilizers, labors, pesticides, seeds and water are necessary. Based on the water availability, the main task of a farmer is to raise the capital. When they are unable to raise the capital in their own path, they depend on other sources like banks and some financial organizations. At this situation, the farmer depends on the crop for crediting the payments that are taken from banks and financial organizations. This may create very tough conditions if the crop fails. And sometimes, this may result in pushing the farmer into some of the critical situations and causes relentless stress [3, 4].

This type of crop failures can be avoided, and some of the preventions are to be noticed with the available resources before dumping. Only way to address these types of problems and increase the quality and quantity of crop production is by utilizing the IoT sensing technology. Farming with IoT helps in mitigating the shortage of food by demanding the existing land for stronger utilization at minimum cost. This technology makes crops to be connected through the smart farming. IoT sensors in the farms are more capable of providing information to farmers about crops and are used for monitoring the health of crops in earlier stages [5, 6].

Section 2 describes the related work carried out by various researchers, detailed explanation of IoT and smart farming is given in Sect. 3 and Sect. 4, respectively, and Sect. 5 explains the drivers and barriers raised during their integration and our perception about solutions which is in conclusion.

2 Related Work

Perera et al. [7] led a study about enabling IoT and its applications in industries by a survey. The SOA models of IoT were introduced and discussed their fundamental technologies. They also introduced some key applications by analyzing the challenges and future trends related to IoT. They mainly concentrated on the industrial applications and focused on the challenges and research opportunities which are possible for future researchers.

Gaitan et al. [8] surveyed software architectures on IoT. From the proposed architectures, the data is distributed by including a middleware layer on the Internet. These architectures meet the requirements for IoT systems by including some existing technologies. M.U. Farooq, Muhammad Waseem, Sadia Mazhar, Anjum Khairi and Talha

Kamal led study about emerging large scale of IoT technologies into networking used for monitoring the environment by embedding objects around us. They also discussed IoT vision and highlighted various technologies and few issues related to those technologies.

Shashwathi et al. [9] proposed a set of farmer-friendly technological services between the expertise of Agra-scientists and farmers. They also explain the benefits of taking the advice of agricultural experts for Chikku farm in Bangalore. This project has been spread across three years. In the first year, they have gathered the requirements and developed a model. In the second phase, some of the farmer-friendly services are deployed on the IBM cloud platform using web and mobile clients.

Kumbhar [10] led a study on IT for Sustainable Agricultural Development in India. He developed this model to help farmers to increase their food productivity, which leads to their economic growth. The model was developed to track all types of information about farmers related to crops.

3 Internet of Things

The term, IoT, was introduced by the British pioneer Kevin Ashton in the year 1998 to describe a system with large number of objects or sensors that would be connected to the Internet through communications to provide some services. In present days, IoT became more popular for characterizing the scenarios like connectivity and computing capability that extends to different objects and everyday items. The "thing" in IoT refers to any of the communicating or non-communicating object on earth [11, 12]. This object represents both physical entities and digital entities by performing some computational tasks for people. IoT supports the interaction between "things" and allows more complex structures by developing distributed applications and distributed computing. The working of IoT is done on the basis of machine to machine (M2M) communication, and this M2M is defined for communicating between two machines without the involving manpower. Even non-connected entities in IoT can turn as IoT components [13, 14]. This technology is sometimes understood as being identical with smart systems such as smart homes, smart enterprises, smart wearable, smart city and smart environment. The development of the current Internet yields information from the sensing environment and interacts with the physical world by utilizing existing Internet standards for providing services like communication, data transfer and applications (Fig. 1).

4 Smart Farming

The main goal of this smart farming is to optimize the harvesting land per unit by using modern methods to achieve the best in terms of quality, quantity and financial return. The term smart farming is also known as precision farming which uses a

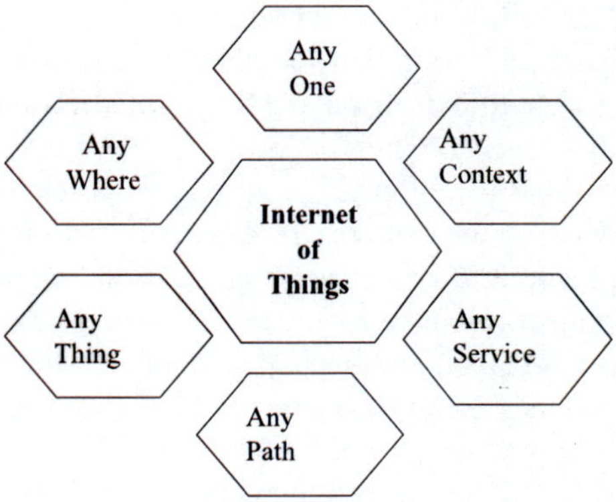

Fig. 1 Definition of IoT

wide range of technologies, including GPS services, sensors, etc. These technologies are very much required in agriculture sector which includes climate forecasting, robotics, science-based solutions, environmental controls, etc. An M2M system is necessary to analyze the data and initiates the responses to the information received [15, 16]. From this perspective of the machine to machine communication system, this technology around its vision enables some key transformations of the agriculture toward the vision of smart farming. The impact of this technology in farming delivers the connectivity between the sensors and communication systems, and these sensors spread in different areas like food traceability, land management and environmental monitoring [17, 18]. The intricacy of smart farming can reflect the agricultural ecosystem in the following ways.

Wireless Services: The wireless services include a collection of wireless sensor devices used to monitor the physical and environmental conditions [19].

Machines and Equipments: The machines are the elements used to direct the work in agriculture in improving the production and farming techniques. The most widely used machinery in agriculture work fields are tractors, combines, etc. The equipments in the field of agriculture are designed to open furrows in the earth [20, 21]. The various types of equipments are plows, fence, blade, handlebar, etc. (Fig. 2).

End Users: The end users in smart farming are the growers and managers at the back office systems. They are also considered as farmer's co-operatives which can help smaller farmers with advice and funding.

Cost: In spite of larger farms, the cost would be a major problem. The factors like soil mapping, fertilizer applications, machinery and weather data generate vast quantities of information. This information is collected by farm offices by these factors which are in the field of farming.

Stakeholders: Stakeholders role in agriculture is broad. The range of these stakeholders starts from large business fields like finance, engineering and retailers to industry which are well expertise in farming with the contribution of small suppliers.

Fig. 2 Architecture for smart farming ecosystem

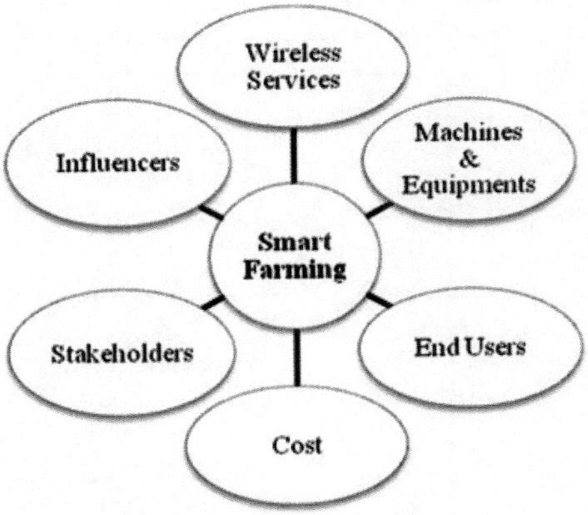

Influencers: Influencers plays an important role in word-of-mouth marketing in the field of agriculture. The benefits of influencer marketing are numerous. Partnering with an influencer can be a cost-effective form of high-quality content with built-in distribution.

The working of smart devices in farming allows farmers to apply amount of resources like water, fertilizers, etc. at the right time and right place in right time. This type of farming is a priceless tool for composing chemicals in soil. This system operates many machines and equipments very smartly and collects the data based on these equipments from the field by configuring the problems when they are detected. The devices in this field will be user-friendly and help the farmers to plan irrigation and more activities based on environmental and soil conditions. The parameters like temperature, humidity, pH value and soil moisture can be monitored by these device controllers and can be informed through some graphical representations. The devices can be configured by farmers for receiving important alerts from the field parameters in the form of SMS in his mobile through programming. In this way, most of the situations can be monitored and controlled in agriculture field by using some remotely operated devices.

The smart devices in the agriculture field consist of some key features. They are as follows:

- Different agricultural parameters like temperature, wind, humidity and moisture can be controlled and monitored by using sensors.
- Generates and schedules a plan for irrigation and fertilization.
- Data can be obtained through wireless speckles.
- Provides external memory interface for the purpose of feeding and desertification.
- Provides awareness to the farmers by alerting while he is away from the field.
- Controls the equipments in the field through sensor devices like mobile phones, tablets, computers, etc.

Figure 3 represents the architecture of the smart farming system in the agriculture. The block diagram consists of following:

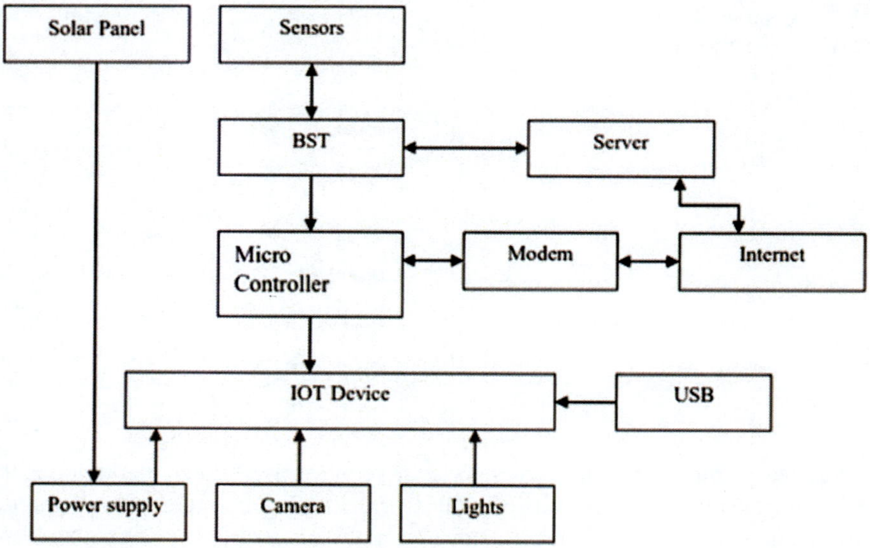

Fig. 3 Architecture of smart farming system

1. Solar panel acts as a power source and offers an opportunity to stabilize the energy cost to the farmers. These solar panels are of three types, namely:
   a. Grid solar: This reduces the final cost for the power utilization for the home and business. The energy generated from the sun is utilized by this panel and earns the credits which are used further.
   b. Hybrid solar: These are used as power bank to the systems which are used for backup purpose. The power is supplied to the grid via battery bank in the event of power outage.
   c. Off-grid solar: These types of solar panels are independently used for a remote location and are limited.
2. **Sensors**: The sensors form a network with each other. A master sensor can act as a gateway by collecting the sensor readings to the Internet using mobile connectivity. This installation requires any of the SIM cards to connect to the Internet. Node sensor is also having similar functionality as master sensor [22, 23]. Node sensor can also act as a gateway by conducting the data to the Internet, whereas master sensor pushes data directly to the Internet. Most often, spectral sensors are used to examine the growth of plants in soils with different levels of nutrients.
3. **Base Station**: In the agriculture field, base station acts as a piece of equipment with wireless communication between user and a network.
4. **Microcontroller**: The microcontroller plays a major role in field of agriculture by transferring the information about the motors in the farm. When the motor is turned ON by the controller, then the pumping functionality of the water

from the ground is activated. In the other case, it constantly monitors the digitized parameters of the various sensors and verifies those parameters with some predefined threshold values.

5. **Server**: This can act as a centralized system to store and share data among farmer and Internet.
6. **Modem**: This can be used as a wireless device for connectivity to transfer the data between source and destination. The source and destination in the farm may be office and farmer, vehicle and vehicle, etc.
7. **Internet**: This is a global computer network that provides a variety of information and communication facilities.
8. **IoT Device**: This may be any of the smart devices.
9. **USB**: It can be a flash drive that can act as portable storage device.
10. **Power**: It is one of the major sources of energy used for irrigation and controlling the motors and lights in the farming.
11. **Camera**: The role of camera in the field of agriculture plays a vital role. The key uses of these cameras are:
    - Identification of the intruders,
    - Protect the water supplies,
    - Prevent break-ins,
    - Vehicle tracking,
    - Employee monitoring.
12. **Lights**: A popular light source with eco-friendly features is LED. This LED is good for growth of plants by improving the lifetime and can have huge business opportunities by maintaining the quality. To maintain the proper growth of plants in the farms, LED screens are to be installed to provide the enough light for germination.

The applications in the area of smart farming are complicated. The structure of this smart farming is described in the Fig. 4.

The following steps describe the drivers and barriers in the adoption of smart farming into agriculture.

The drivers are:

- To increase efficiency, there should be a need to reduce waste.
- The immediate response is needed when there is a change in climate.
- Assistance is needed in a farming area from public and projects.
- Soil erosion is to be addressed before through farming.
- Many industries use an M2M-based system for monitoring and tracking purposes.
- Cost of smart devices like sensors and their connectivity must be reduced.

The barriers are:

- Investments on device installations in a farm cannot be returned once they have been directed.

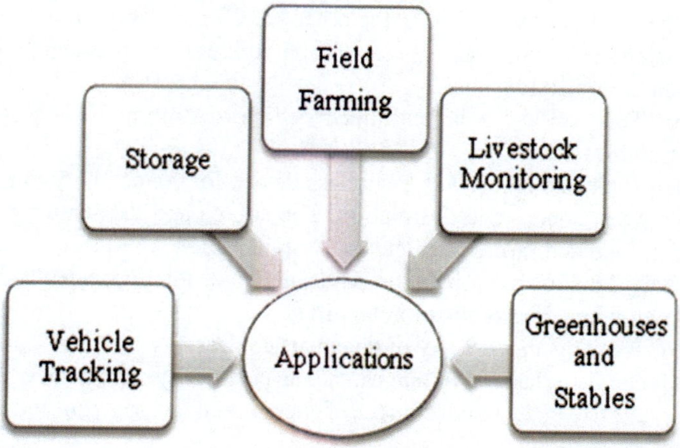

Fig. 4 Applications from smart farming

- Many issues like weather and political may create several problems by inheriting in the industry.
- Based on data collected from the owner, many queries cannot be resolved.
- Lack of wireless and broadband facilities in rural areas.
- No standard development in sensor networks.
- No maturity in specialist agricultural software.

5 Conclusion

In this paper, we discussed farming with IoT for providing many useful services to the farmers in society in future. Many of the smart applications can be developed by the IoT and vastly utilized by users. The working of smart devices in farming allows farmers to apply amount of many resources at right time and right place in right time. This type of farming is having less cost. This system operates many machines and equipments very smartly and collects the data based on these equipments from the field by configuring the problems when they are detected. Farming with IoT helps in mitigating the shortage of food by demanding the existing land for stronger utilization at minimum cost. Smart farming is a notion that quickly snaps on the agricultural field. This offers automated farming techniques, useful data collection and high-rigor crop control. The main goal of this smart farming is to optimize the harvesting land per unit by using modern methods to achieve the best in terms of quality, quantity and financial return. Finally, some of the probable solutions are examined by approaching some related research initiatives.

References

1. S.K. Dash, S. Mohapatra, P.K. Pattnaik, A survey of application of wireless sensor network using cloud computing, Int. J. Comput. Sci. Eng. Technol. **1**(4), 50–55 (2010)
2. D. Evans, *The Internet of Things How the Next Evolution of the Internet is Changing Everything*, Whitepaper, Cisco Internet Business Solutions Group (IBSG) (2011)
3. G. Kortuem, F. Kawsar, D. Fitton, V. Sundramoorthi, Smart objects and building blocks of internet of things. IEEE Internet Comput. J. **14**(1), 15–21 (2010)
4. S.B. Shen, Q.L. Fan, P. Zong, Q.Y. Mao, W. Huang, Study on the architecture and associated technologies for internet of things. J. Nanjing Univ. Posts and Telecommun. Nat. Sci. **29**(6), 1–11 (2009)
5. R. Khan, S.U. Khan, R. Zaheer, S. Khan, Future internet: the internet of things architecture, possible applications and key challenges, in *Proceedings of 10th International Conference on Frontiers of Information Technology* (Islamabad, Pakistan, 2012), pp. 17–19
6. M. Eisenhauer, P. Rosengren, P. Antolin, A development platform for integrating wireless devices and sensors into ambient intelligence systems, in *6th Annual IEEE Communications Society Conference on Sensor, Mesh and Ad Hoc Communications and Networks Workshops, SECON Workshops'09* (IEEE, 2009)
7. C. Perera, A. Zaslavsky, C. Liu, M. Compton, P. Christen, D. Georgakopoulos, Sensor search techniques for sensing as a service architecture for the internet of things. Sens. J. **14**(2), 406–420 (2014)
8. N.C. Gaitan, V.G. Gaitan, I. Ungurean, A survey on the internet of things software architecture. Int. J. Adv. Comput. Sci. Appl. **6**(12), 25–31 (2015)
9. N. Shashwathi, P. Borkotoky, K. Suhas, Smart farming: a step towards techno-savvy agriculture. Int. J. Comput. Appl. **57**(18), 51–59 (2012)
10. V. Kumbhar, IT for sustainable agriculture development in India, in *Proceedings of 3rd National Conference India-Com,* Feb 26–27 (2009)
11. B. Thirunavukarasu, Dr. T. Kalaikumaran, Dr. S. Karthik, Integration of data mining and internet of things–improved athlete performance and health care system. Int. J. Tech. Res. Appl. (2014)
12. P. Barreiro, E.C. Correa, F.J. Arranz, B. Diezma, L. Ruiz, M. Villarroel, J.I. Robla, F.J. Garcia-Hierro, Smart sensing applications in agriculture and food industry (2011)
13. S.G. Jadhav, G.N. Shinde, Web based information system for agriculture. Int. J. Inno. Technol. Creative Eng. **1**(2), 12–17 (2011)
14. R. Jensen, Information, efficiency and welfare in agricultural markets, in *Proceedings of the 27th International Association of Agricultural Economists Conference* (Beijing, China, 2009)
15. I.V. Subba Rao, Indian agriculture–past laurels and future challenges, in Indian agriculture: current status, prospects and challenges, 27th Convention of Indian Agricultural University Association, 9–11 Dec 2002
16. S. Poslad, *Ubiquitous Computing: Smart Devices, Environments and Interactions* (John Wiley & Sons Ltd., 2009), p. 75
17. V. Janamala, Parameterization of stand-alone solar PV system performance—an IoT solution. Int. J. Adv. Sci. Technol. **106**, 1–8 (2017)
18. H.K. Yang, H.J. Cha, Y.J. Song, A study of secure distributed management of sensing data in IoT environment. Int. J. Adv. Sci. Technol. **124**, 21–32 (2019)
19. S.J. Kim, S.H. Kweon, M.B. Song, Y.R. Bao, A semantic network analysis of 4th industry revolution: news frame for on drone, AI, big data and IoT news. Int. J. Adv. Sci. Technol. **111**, 115–1281 (2018)
20. A. Mehta, N. Mehta, Algae farming: a way towards prosperity. Int. J. Adv. Sci. Technol. **81**, 53–62 (2015)
21. C. Mariotti, Plate element formulation for a discrete element method. Int. J. Adv. Sci. Technol. **94**, 13–22 (2016)

22. J. Song, D. Park, Preemptive cyber response strategy and IoT forensic evidence. Int. J. Adv. Sci. Technol. **117**, 129–138 (2018)
23. S. Byun, Viability-based replication management scheme for reliable IoT data services. Int. J. Adv. Sci. Technol. **124**, 89–102 (2019)

Novel Topology for Nine-Level H-Bridge Multilevel Inverter with Optimum Switches

B. Arundhati, Madisa V. G. Varaprasad, and Vijayakumar Gali

Abstract In recent years, multilevel inverter technology expands its scope in the region of average voltage and high-power control systems. The power from renewable sources is DC in nature; however, for utilization of this power for domestic/industrial applications, it is needed to be converted into AC. The efficient conversion of power from direct current to alternating current is accomplished by multilevel inverters to reduce the harmonic content. There are several topologies of multilevel inverter, but the main short come is their intricacy and needs a greater number of power devices. Now, the proposed multilevel inverter configuration with eight switches with unidirectional nature is to obtain a voltage level of 9. The proposed topology with less no of switched improves the output waveform with reduces the harmonic content, level of electromagnetic interference (EMI). The proposed topology is modeled and validated using MATLAB/Simulink.

Keywords Modified H-bridge inverter · Multilevel inverter

1 Introduction

To attain an eminence pure sinusoidal voltage waveform at the output side of DC–AC converter with reduced number of switches, less ripple and total harmonic (THD) contents, various PMW techniques have been developed [1, 2]. Fundamental switching scheme can be used to power semiconductor switches in the inverter to create a nearly sinusoidal output. In reaction to the growing demand for medium

B. Arundhati · M. V. G. Varaprasad (✉)
Vignan's Institute of Information Technology, Visakhapatnam, India
e-mail: mvgvaraprasad@gmail.com

B. Arundhati
e-mail: b_arundhati@rediffmail.com

V. Gali
Poornima University, Jaipur, India
e-mail: vijaykumar209@gmail.com

D. Bhattacharyya and N. Thirupathi Rao (eds.), *Machine Intelligence and Soft Computing*, Advances in Intelligent Systems and Computing 1280,
https://doi.org/10.1007/978-981-15-9516-5_35

and high-power applications, multilevel inverter has been an interesting growing contemplation in variable speed wind turbine and photovoltaic systems recently. Factors like an increase in fossil fuel prices, environmental pollution and a boom in energy demand have made it necessary to consider renewable energy power supply into the utility grid. Renewable sources provide low voltage DC energy with relatively low conversion efficiency. Thus, multilevel inverter acts as a link between renewable power generation on the one side and high-power load on the other side [3] to convert low DC source into usable AC source, thereby improving its efficiency. In industrial applications, the three major types of multilevel inverters are H-bridge, flying capacitor type and diode clamped type. Cascaded H-bridge type is simpler compare to other inverter topologies because of abridged no. of switches, low harmonic distortion for higher switching frequencies, circuit complexity is reduced, low cost and size [4, 5]. The anticipated topology uses cascaded H-bridge and produces nine-level output voltage with less gate drive circuits and a smaller number of power elements in addition to less circuit layout complexity. The harmonic content can be reduced with an increase in the output level [6]. However, the number of switches is more and hence increases the complexity of the switching pattern. Also, the cost of the system will become high. To overcome these problems of conventional H-bridge multilevel inverter topology, a reduced power switches-based multilevel circuit configuration is proposed in this paper. A detailed description of switching pattern and working is explained in the following sections.

This paper is organized in the subsequent sections. Section 2 describes the operation of the proposed topology. The pulse width modulation scheme is depicted in Sect. 3. A detailed mathematical modeling of proposed topology is depicted in Sect. 4. The proposed topology is implemented in MATLAB/Simulink environment, and results are discussed in Sects. 5, and 6 explains the conclusion.

2 Novel Nine-Level Topology Operation

The configuration of the proposed nine-level multilevel inverter is depicted in Fig. 1. It is capable of generating nine voltage levels from +4 to −4 including zero level. The voltage magnitudes are indicated as V_1, V_2 and V_3.

IGBTs are used as unidirectional switches with anti-parallel diodes. The anti-parallel diode is used to block reverse current which may damage IGBT during its off state. The proposed topology is the re-design of cascaded H-topology. The switching sequence for triggering of switches is explained in the following sections. It is capable of generating nine voltage levels from +4 to −4 including zero level. The current paths in each mode for generation of nine levels have depicted following modes of operation.

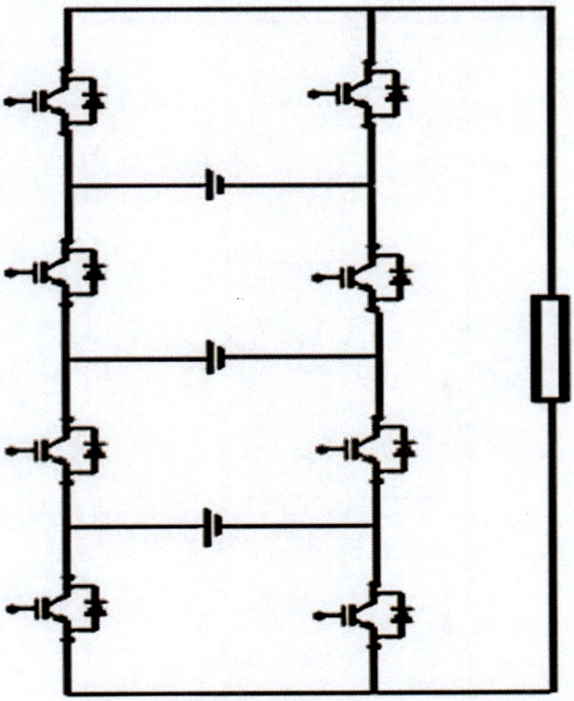

Fig. 1 Circuit diagram of nine-level multilevel inverter

Mode 1: ($V_0 = +2$ V)

In this mode, the switches Q_1, Q_6, Q_7 and Q_8 are switched ON, and the remaining switches are switched OFF. These switches enable voltage source V_1, and the resulting load voltage is given by $V_0 = +V_1 = +2$ V. The current flow is shown in the following Fig. 2.

Mode 2: ($V_0 = +3$ V)

In this mode 3 as shown in Fig. 3, the power electronic switches Q_3, Q_4, Q_5 and Q_6 are switched ON, and the remaining switches are switched OFF. These switches enable voltage source $V_{2,}$ and resulting load voltage is given by $V_0 = +V_2 = +3$ V.

Mode 3: ($V_0 = +4$ V)

The position of switch on and current flowing through switches are shown in Fig. 4. In this mode, the power electronic switches Q_1, Q_2, Q_3, Q_8 and Q_6 are switched ON, and the remaining switches are switched OFF. These switches enable voltage source V_3, and the resulting load voltage is given by $V_0 = +V_3 = +4$ V.

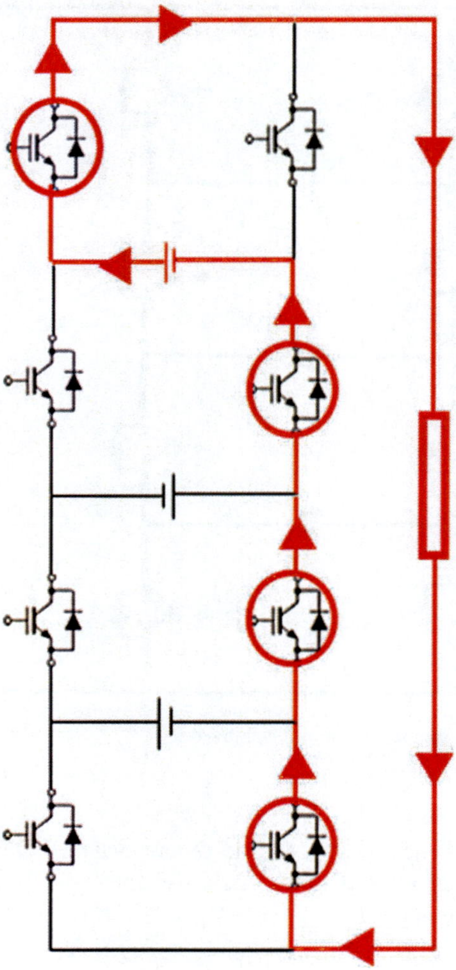

Fig. 2 Current path for mode 1

Mode 4: (V_0 = +5 V)
The mode 4 circuit status and its switch on positions are shown in Fig. 5. In this mode, the power switches Q_1, Q_3, Q_4 and Q_6 are ON, and the other switches are become OFF. These switches enable voltage sources V_1 and V_2, and the resulting load voltage is given by $V_0 = +(V_1 + V_2) = +5$ V.

Mode 5: (V_0 = 0 V)
This mode can be achieved in two combinations. Either Q_5, Q_6, Q_7 and Q_8 are ON or Q_1, Q_2, Q_3 and Q_4 are ON with remaining switches which are at OFF state. The voltage is zero (V_0=0 V) due to short circuit of load terminals. The path of the current for this mode is depicted in red color as shown in Fig. 6.

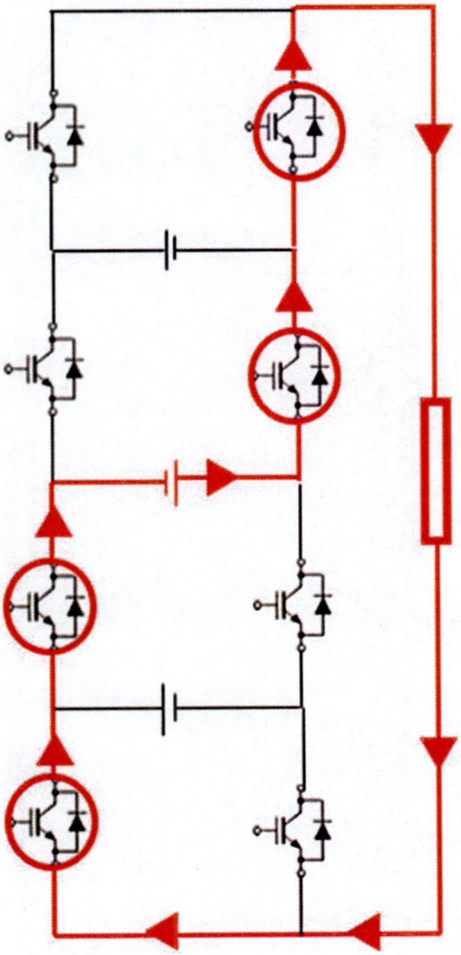

Fig. 3 Current path for mode 2

Mode 6: ($V_0 = -2$ V)

In this mode, Q_2, Q_3, Q_4 and Q_5 are ON, and the other switches are become OFF. These power electronic switches enable voltage source V_1, and the resulting load voltage is given by $V_0 = -V_1 = -2$ volts. The path of the current for this mode is depicted in red color as shown in Fig. 7.

Mode 7: ($V_0 = -3$ V)

In this mode, the switches Q_1, Q_2, Q_7 and Q_8 are switched ON, and the remaining switches are switched OFF. These switches enable voltage source V_2, and the resulting load voltage is given by $V_0 = -V_2 = -3$ V. The current paths that are active in this mode are as depicted in Fig. 8.

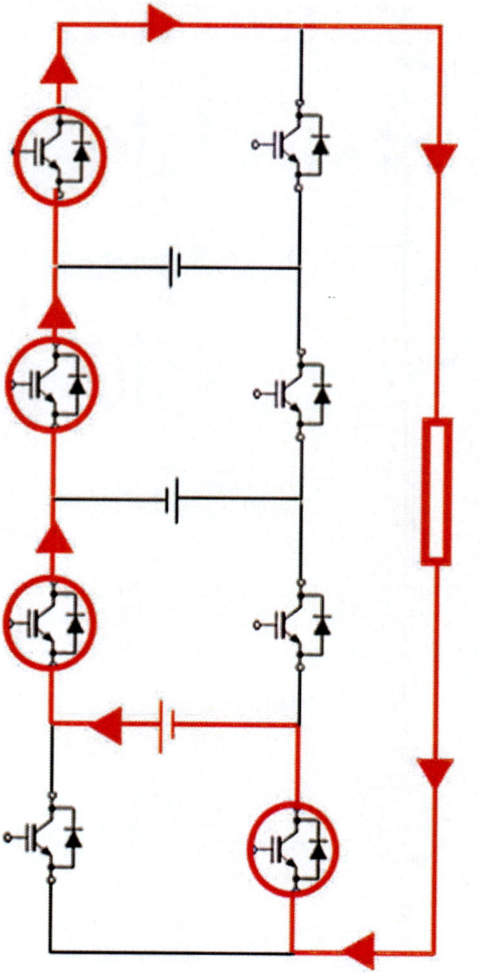

Fig. 4 Current path for mode 3

Mode 8: ($V_0 = -4$ V)

The conduction of power switches is shown in Fig. 9. In this mode, the switches Q_4, Q_5, Q_6 and Q_7 are switched ON, and the remaining switches are switched OFF. These switches enable voltage source V_3, and the resulting load voltage is given by $V_0 = -V_3 = -4$ V.

Mode 9: ($V_0 = -5$ V)

In this mode, the switches Q_2, Q_5, Q_7 and Q_8 are switched ON, and the remaining switches are switched OFF. These switches enable voltage sources V_1 and V_2, and the resulting load voltage is given by $V_0 = -(V_1 + V_2) = -5$ V. The direction of the current for this mode is depicted in red color as shown in Fig. 10 (Table 1).

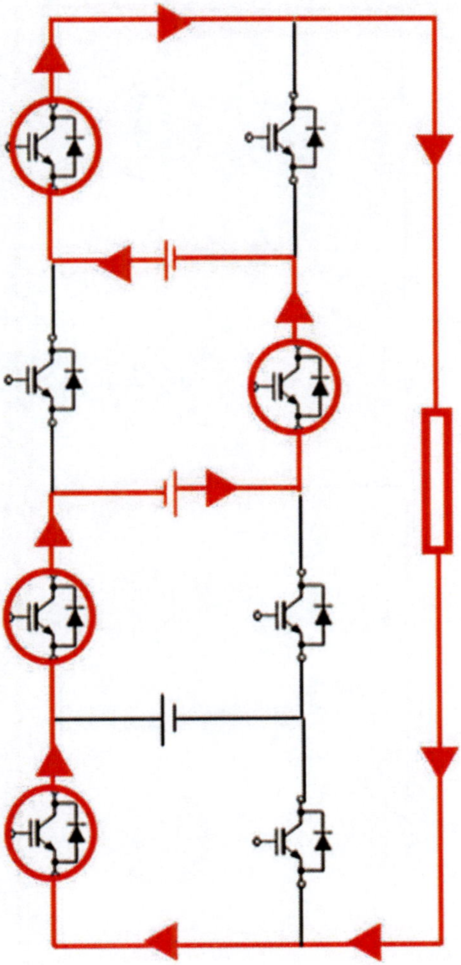

Fig. 5 Current path for mode 4

3 Pulse Width Modulation Scheme

The gate pulses for the control of each switch to obtain the desired output waveform, modulation wave (sinusoidal wave) is compared with number of carrier waves (triangular waves) equal to the number of required levels except zeroth level. Thus, the obtained waveform by comparing carrier and modulation waves is used as a reference wave for the output, and the clock pulses are generated according to it (Fig. 11).

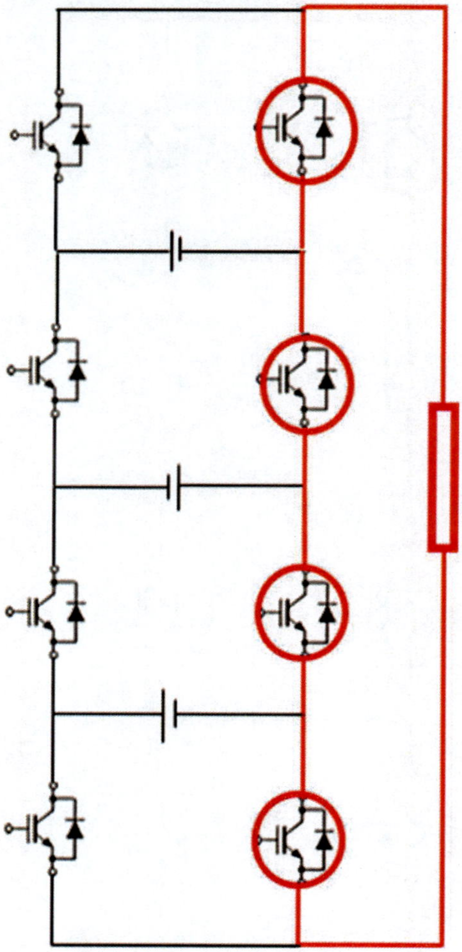

Fig. 6 Current path for mode 5

4 Mathematical Modeling

There are several levels that can be synthesized using a topology as follows:

$$L = 2^{(n+1)} - 1$$

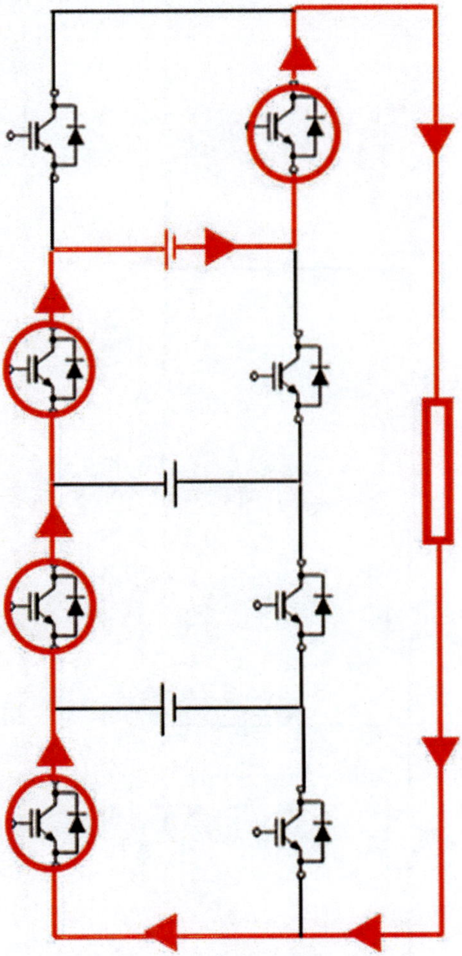

Fig. 7 Current path for mode 6

The numbers of switches used are:

$$Q = 2 * (n + 1)$$

Where

L No. of voltage levels,
n No. of voltage sources,
Q No. of switches.

Voltage Stress: The maximum voltage that each switch experiences at its OFF state, without being damaged, is called voltage stress. The voltage stress is calculated

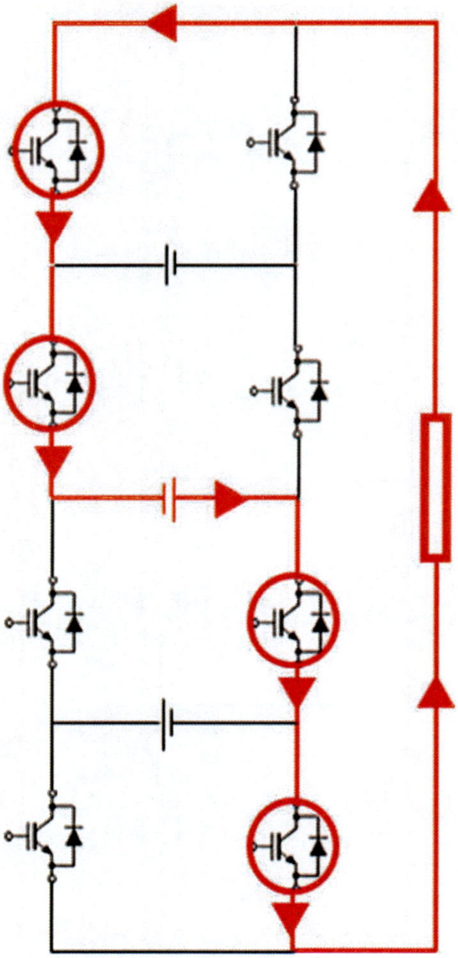

Fig. 8 Current path for mode 7

without connecting the load. The voltage stress across each switch is given in Table 2. V_1 or V_2 in Table 2, implies Q_2, Q_4 experiences maximum of V_1 or V_2 depending on the magnitude of V_1 and V_2.

Power Electronic Switch Losses: In this, IGBT switches are being used. Hence, it has conduction, and switching losses are present. These losses are being calculated as follows:

1. Conduction Loss:

$$\text{Conduction power loss } P_{\text{loss}} = V_{\text{ce}}(\text{sat}) * D * I_{\text{c}} \quad (1)$$

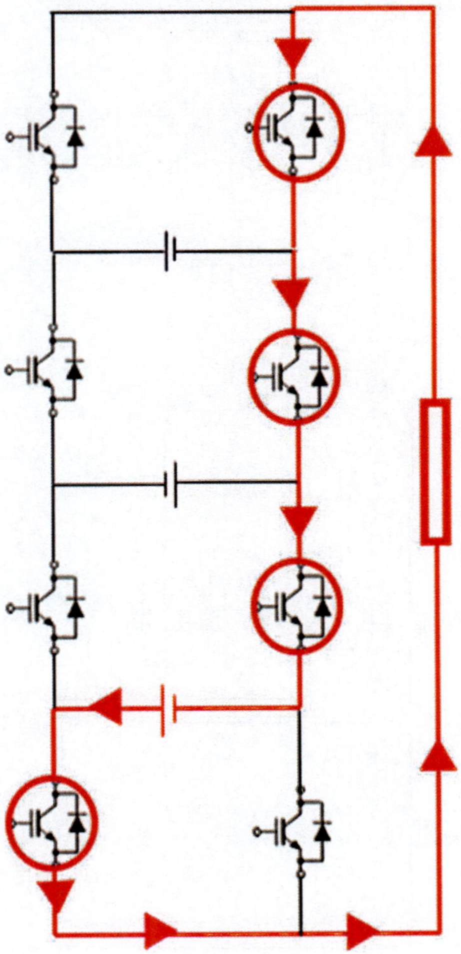

Fig. 9 Current path for mode 8

Where $V_{ce}(\text{sat})$ is the collector to emitter saturation voltage, D is the duty cycle, and I_c represents as collector current.

2. Switching Loss: The energy loss in the power electronic switch is the product of switching frequency and current and time taken in the operation. It is shown as follows:

$$w_{on} = (V_{ce} * I_c * t_{on})/6 \quad (2)$$

The power loss during turn-on time is given by

$$P_{on} = (V_{ce} * I_c * t_{on}) f_s/6 \quad (3)$$

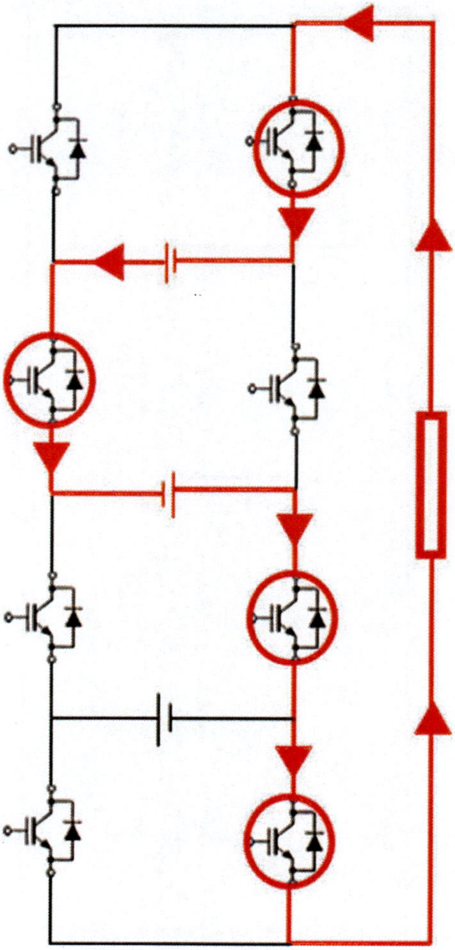

Fig. 10 Current path for mode 9

The power loss during turn-off time is given by

$$P_{\mathrm{off}} = (V_{\mathrm{ce}} * I_{\mathrm{c}} * t_{\mathrm{off}}) f_{\mathrm{s}}/6 \tag{4}$$

The total switching power loss can be calculated as:

$$P_{\mathrm{sw(total)}} = (V_{\mathrm{ce}} * I_{\mathrm{c}} * (t_{\mathrm{on}} + t_{\mathrm{off}})) f_{\mathrm{s}}/6 \tag{5}$$

Where V_{ce} is the collector to emitter voltage, I_c is the collector current,t_{on} is turn-on time,t_{off} is the turn-off time, and f_s is the switching frequency.

Table 1 Sequence of switching in nine-level topology

Q_1	Q_2	Q_3	Q_4	Q_5	Q_6	Q_7	Q_8	Output (V)
OFF	OFF	OFF	OFF	ON	ON	ON	ON	0
ON	OFF	OFF	OFF	OFF	ON	ON	ON	2
ON	ON	ON	OFF	OFF	OFF	OFF	ON	3
OFF	OFF	ON	ON	ON	ON	OFF	OFF	4
ON	OFF	ON	ON	OFF	ON	OFF	OFF	5
OFF	OFF	ON	ON	ON	ON	OFF	OFF	4
ON	ON	ON	OFF	OFF	OFF	OFF	ON	3
ON	OFF	OFF	OFF	OFF	ON	ON	ON	2
OFF	OFF	OFF	OFF	ON	ON	ON	ON	0
OFF	ON	ON	ON	ON	OFF	OFF	OFF	−2
OFF	OFF	OFF	ON	ON	ON	ON	OFF	−3
ON	ON	OFF	OFF	OFF	OFF	ON	ON	−4
OFF	ON	OFF	OFF	ON	OFF	ON	ON	−5
ON	ON	OFF	OFF	OFF	OFF	ON	ON	−4
OFF	OFF	OFF	ON	ON	ON	ON	OFF	−3

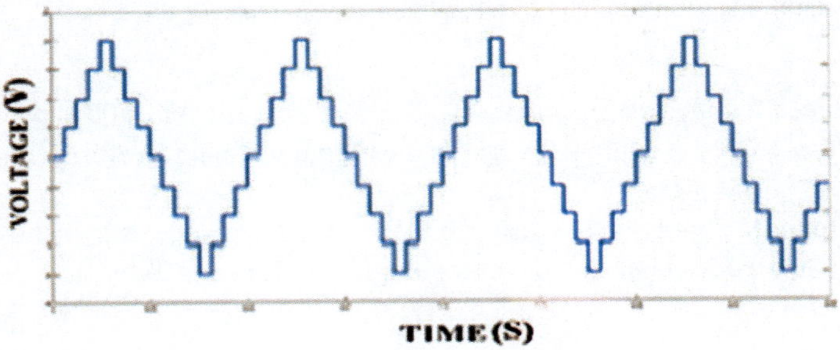

Fig. 11 Carrier sine wave modulation

5 Simulation Result and Discussion

The output voltage and current waveform of the proposed nine-level multilevel inverter are depicted in Fig. 12. The results show that the output voltage and current waveforms are very close to sinusoidal increase in the number of voltage levels and

Table 2 Detailed performance comparison of other conventional topologies

Topology/parameter	Diode clamped	Flying capacitor	Cascaded H-bridge [6]	Proposed topology
Power switches	12	12	16	8
DC sources	8	8	3	3
Capacitors	10	12	–	–
Clamping diode	24	–	–	–
THD %	22.69	20.98	15.41	13.98

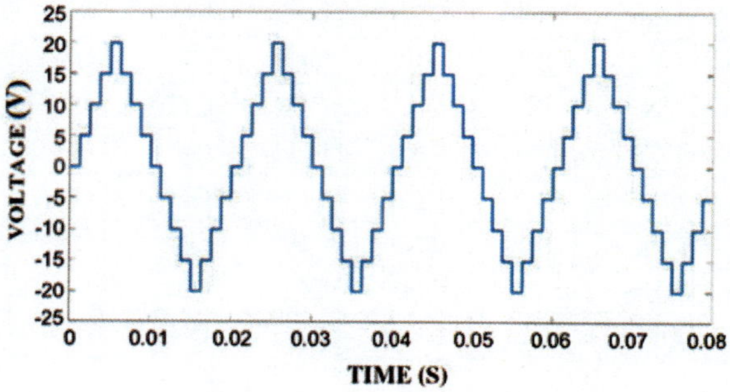

Fig. 12 Output voltage waveform of novel nine-level inverter topology

hence reduced total harmonic distortion. The FFT analysis for THD is done in order to obtain the harmonic content present in the output voltage waveform (Figs. 13 and 14).

The proposed topology has less % THD with fewer power switches and DC sources compared to other nine-level topologies as shown in Table 2.

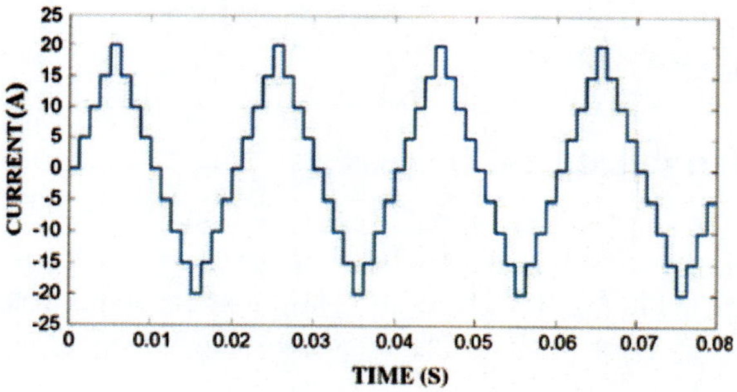

Fig. 13 Output current waveform of novel nine-level inverter topology

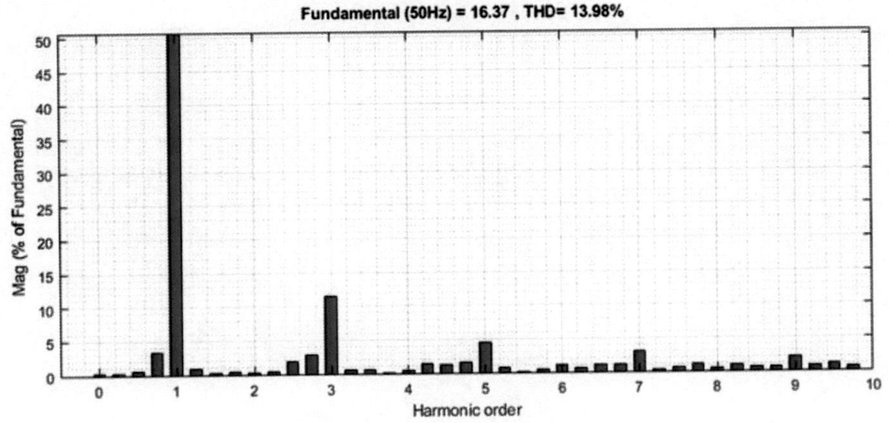

Fig. 14 THD of novel nine-level inverter topology

6 Conclusion

A novel topology for nine-level H-bridge with optimum no. of switches is proposed in this paper. This proposed topology is developed with reduced no. of switches compared with other conventional multilevel topologies. Also, this topology is developed in MATLAB and observed the voltage and current waveforms. It has been observed that with reduced no. of switches, the THD obtained is 13.98%. The THD will further be reduced by adding passive filters. Hence, the proposed topology used a very number of switches which reduce the cost of the system, complexity of control technique, etc.

References

1. K.K. Gupta, S. Jain, A novel multilevel inverter based on switched DC sources. IEEE Trans. Ind. Electron. **61**(7), 3269–3278 (2013)
2. V. Gali, N. Gupta, R.A. Gupta, Distortion free improved reference current generation algorithm for interleaved inverter based shunt APF, in *IEEE PES Asia-Pacific Power and Energy Engineering Conference (APPEEC)* (2017), pp. 1–6
3. B. Arundhati, K. Alice Mary, M. Suryakalavati, Implementation of thrust ripples reduction for a permanent magnet linear synchronous motor using adaptive feed forward controller. J. Power Electron. **14**(4), 687–694 (2014)
4. M.M. Hasan, A. Abu-Siada, S.M. Islam, S.M. Muyeen, A novel concept for three-phase cascaded multilevel inverter topologies. Energies **11**(2), 268 (2018)
5. M.V.G. Varaprasad, R.S. Ravi Sankar, K. Alice Mary, Reduction of torque and flux ripple of 3-Φ asynchronous machine using multilevel inverter. Int. J. Adv. Res. Electr. Electron. Instrum. Eng. **2**(2), 4781–4793 (2013)
6. M. Siddique, M. Saad, N.M. Shah, S. Adil, I. Atif, A. Mudasir, A new multilevel inverter topology with reduced switch count. IEEE Access **7**, 58584–58594 (2019)

Assessment of the Security Threats of an Institution's Virtual Online Resources

Daisy Endencio-Robles, Rosslin John Robles, and Maricel Balitanas-Salazar

Abstract Web application is no denying one of the most effective ways to provide essential services to clientele. However, data breach and system insecurities are increasing concern in the field of information system. The growing number of concern is the basis of the authors for this research. Several Web vulnerability scanners are used for institutions' online resources to test the security level of the online systems.

Keywords Web scanner · Web vulnerabilities · Web application · Security threats

1 Introduction

Many services of most institutions such as a university are done online through an online portal. For this project, we subject some online systems of the university to vulnerability and security tests. Hacking is usually performed by utilizing automated scripts injected into a web server. Most online resource attacks are not to simply to mess with the Web site layout or steal data, but to set up web servers to serve files of an illegal nature or utilize the server computer as an email relay spam.

For many institutions, it is until after a security breach has occurred that they consider web security practices a priority. An effective approach to Web site threats must be proactive and defensive. Data breach happens every day. In Fig. 1, Vaidya et al. shows the compiled list of 18 biggest and most significant breaches of twenty-first century [1].

R. J. Robles (✉) · M. Balitanas-Salazar
CLASE, University of San Agustin, Iloilo City, Philippines
e-mail: rjrobles@usa.edu.ph

M. Balitanas-Salazar
e-mail: msalazar@usa.edu.ph

D. Endencio-Robles
SHS Department, University of San Agustin, Iloilo City, Philippines
e-mail: dendencio@usa.edu.ph

D. Bhattacharyya and N. Thirupathi Rao (eds.), *Machine Intelligence and Soft Computing*, Advances in Intelligent Systems and Computing 1280,
https://doi.org/10.1007/978-981-15-9516-5_36

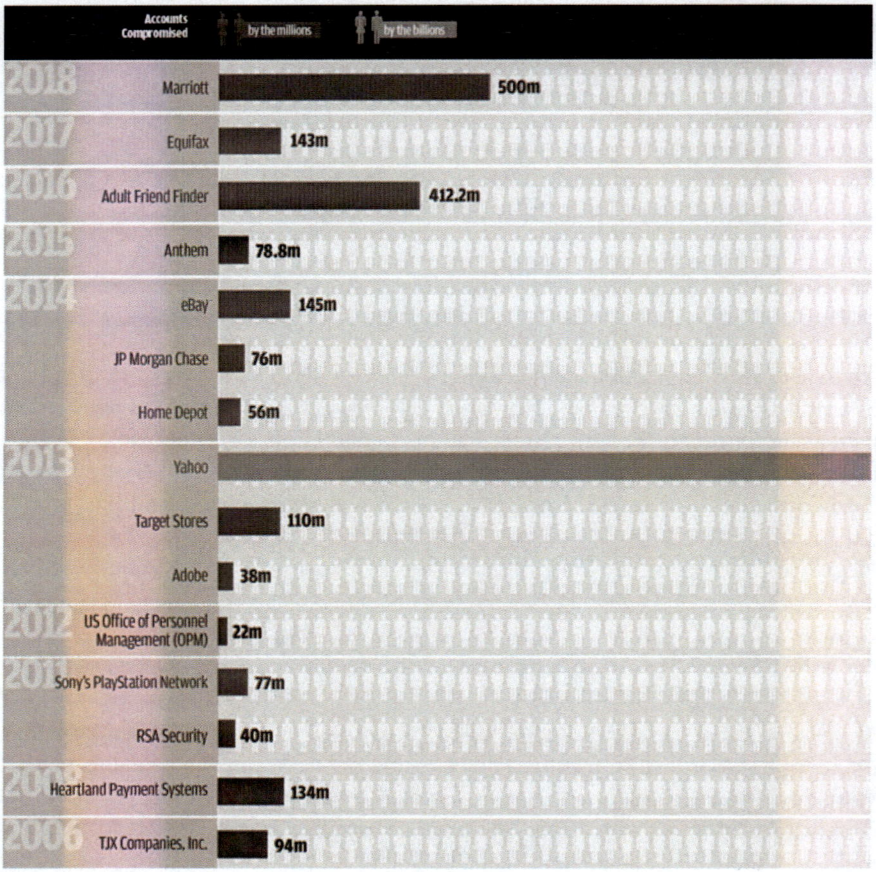

Fig. 1 Website database breaches of the twenty-first century [1]

ISO/IEC 27005, an information risk management standard, defines vulnerability as a weakness of an asset or group of assets that can be exploited by one or more security threats whereas asset is anything that has value to the organization including information resources [2].

If the institution continues to utilize vulnerable Web sites or web systems, it may lead to company losses since essential assets are insecure if not proactively and defensively protected.

Assessment of vulnerability requires deep understanding of these vulnerabilities. It becomes necessary to know the basic idea of their existence in the system, what flaws need to be address to free the system from such vulnerabilities and how the system may be resilient after an attack. Using vulnerability scanner, it may address the early detection and handling of known security problem, if a new device may be connected to the network without authorization, vulnerability scanner can help identify the rogue machine which may endanger the overall system and network

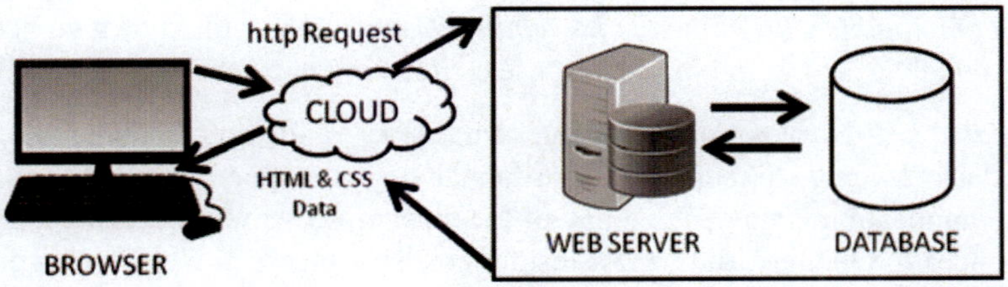

Fig. 2 Basic architecture of Web site /web application

security. Vulnerability scanner also may help to verify inventory of all devices on the network. This information is useful in security management and tracking [3].

Basic architecture of a Web site or a Web application is shown in Fig. 2. Once the user clicks or enters a URL to call the application or access the Web site, a request via communication protocol is sent to a server from the client. Web server creates a request to a backend application server (database). The result is received from the backend and returns a hypertext mark-up language result page to consumer.

Attacks may dwell right on the http request once it is connected to the Internet as presented by OWASP -Web Application Security Project. There are ten security risks identified, and the most critical security risks are associated with Web application.

The ten risks published by OWASP are [3]:

Broken authentication and session management. Misconfiguration in Web applications such as storage passwords in plain test or weak encryption of user credentials lead to form an attack [4, 1].

Injection flaws. Injection flaws allow an attacker to inject code into the vulnerable system. It then executes the injected code [5, 6].

Cross site scripting. *O*nce the injection in injection flaws is executed, the execution is made possible because the Web application is unable to properly filter input properly. This can lead to stealing of cookies, Web site defacement and session hijacking [7–9].

Insecure direct object references. Unauthentic clients are given access to restricted resources such as directories and configuration files [9, 10].

Security misconfiguration. This flaw exists if the Web applications enable certain features by default [10].

Sensitive data exposure. Using weak encryption schemes can also result in sensitive data exposure [7, 9].

Missing level access control. Application is unable to restrict resources, and this will affect the integrity of the data [11, 12].

Cross site request forgery (CSRF). *U*nauthorized request is sent from a user's browser to a Web application in which the user is currently logged on [11].

Using components with known vulnerability. Applications with known vulnerabilities are likely to be compromised because exploits might be available [11, 12].

Invalidated redirects and forwards. Due to improper validation/invalidation of user data, attacks leverage to redirect victims to malicious web pages [9, 10].

Web application is no denying one of the most effective ways to provide essential services to clientele; thus, most institutions have placed some services online. However, considering the level of sensitive information that is handled by Web application proactively and defensively an institution should make certain that data are secure and resilient to any attacks fairly by assessment of different Web services against a set of common vulnerabilities.

In this paper, authors aim to assess the virtual online resources of an institution using vulnerability scanners. The current work is a head start of a longer-term effort to develop a more effective assessment technique.

2 Materials and Methods

We subject online information systems to Web scans. The main Web site is referred to here as system1. The faculty information system is referred to as system 2, and the online library system is referred to as system 3.

The first Web scan is the Quantitative Analysis of CDPs (critical detection points). Quantitative analysis is used for assuring protection of critical, high-value assets where testing data is available. When one of these two aspects is not present—either the asset has a lower consequence of loss or the data is not available—a qualitative analysis can be used. As an example, if an asset can be temporarily lost or easily replaced, there may be limited value in looking in detail at the protection system. There may be other assets of high value, perhaps a corporate executive, where a full quantitative analysis is not performed, because it would be too disruptive to rigorously test and validate all aspects of security around an executive or time is not available to do so [13].

For quantitative evaluation of the effectiveness of security tactics, two scanning tests are performed using the Qualys test. These scanning tests launch various kinds of attacks on the application to find vulnerabilities and assess the security level of the online system.

These tests help the institution streamline their security and compliance solutions. It also builds security into their digital transformation initiatives. This tool can also check the performance vulnerability of the online cloud systems.

We used Qualys for these tests for the following reasons:

- It is trusted globally
- No hardware to buy or manage
- It is a scalable, end-to-end solution for all aspects of IT security
- Vulnerability data securely stored and processed on an n-tiered architecture of load-balanced servers

- It sensor provides continuous visibility
- Data analyzed in real time
- It can respond to threats in a real time.

The first of these tests is the Qualys OWASP scan that is normally practiced to see whether a web application works according to the security standards set by OWASP against online attackers. Qualys OWASP scan helps understand and identify vulnerabilities and support in fixing these vulnerabilities.

Scanning engine is intelligently designed to perform specific scanning tasks and avoid unnecessary vulnerability checks. Qualys scanning methodology follows the same steps as an attacker would follow (Qualys 2015). The following figure depicts the basic steps of the scanning process [14, 15] (Fig. 3).

Another test that was conducted was through the Securi Web site security check and malware scanner. Sucuri is a security platform offering business owners a suite of security tools designed to protect Web site owners from emerging online threats. Nearly 75% of all legitimate Web sites have unpatched vulnerabilities1, and websites that have been blacklisted due to a hack can experience an 80 percent reduction in traffic for more than 48 h 2. Sucuri's suite of tools provides Web site owners the solution they require to respond to hacks when they occur, while also virtually patching vulnerabilities [16].

The test scans online information system for malware and viruses and detects malicious code and infected file locations by scanning your external online information system source code. It also finds Out-of-Date software and plug-ins and identify if your online information system is running an outdated CMS or vulnerable plug-ins and extensions. Monitor online information system blacklist status to see if your online information system is blacklisted by online information system security authorities such as Google, Norton, etc. The test also detects online information system security issues and checks your online information system for security anomalies, configuration issues and security recommendations [17].

Sucuri is deeply involved in the WordPress community. The company provides an industry-leading security plug-in for the WordPress platform and actively works to advance the security awareness and management of WordPress Web sites [16].

Sucuri offerings include:

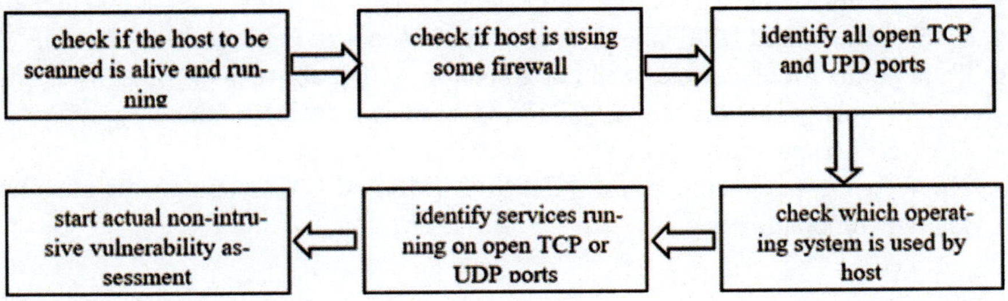

Fig. 3 Steps of the scanning process

- Virtual patching and hardening technology via Web site Application Firewall (WAF) and Intrusion Prevention System (IPS)
- Performance optimization and traffic management via a content distribution network (CDN)
- Security plug-ins for the WordPress platform
- A cloud-based backup service to ensure business continuity in the event of disasters
- Real-time mitigation of DDoS attacks
- Continuous monitoring of Indicators of Compromise (IoC)
- Professional incident response when a Web site has been compromised or blacklisted.

The last test was the Website Security Test of htbridge is a free product available online, provided and operated by htbridge. Website Security Test performs the following security and privacy checks: Non-intrusive GDPR compliance check related to Web application security, non-intrusive PCI DSS compliance check related to Web application security, analysis of CMS and its components for outdated versions and publicly-known vulnerabilities, analysis of HTTP methods that may put Web server, Web application or Web site visitors at risk [18].

Detailed analysis (syntax, validity, trustworthiness) of HTTP security headers. Analysis of altered, and thus potentially malicious, JS libraries.

Analysis of ViewState for misconfigurations and security weaknesses. Analysis of Web application cookies for security flags.

Detection of domain's presence in various blacklists. Detection of cryptojacking within JS code. Detection of WAF presence.

Scoring Methodology

- At the beginning of the test, the score is set to 100
- Points are added for good and reliable configuration of your Web site and Web server
- Points are deducted for insecure, incomplete or unreliable configuration of your Web site or Web server
- Total points for all detected CMS(s) and CMS components will not go below −50 or above +50
- Total points for all detected JS components will not go below −20 or above +20
- Total points for all HTTP methods and CSP will not go below −30 or above +30
- Total points for all cookies will not go below −10 or above +10
- No Web site may score above "C" if a vulnerable software is found
- No Web site may score above "B+" if CMS is not up to date
- No Web site may score below "C" if its CMS and CMS components have no known vulnerabilities.

3 Results and Discussions

The vulnerability assessment managed to provide information with the existing and current issues associated with Web application used by the Institution. It is important for the institution to abreast with the security issues so that respondents will learn and know the techniques to combat these security threats. With the increasing number of data breaches, a constant scanning should be done for its monitoring. Proper and adequate security measures should be in place to protect organizational Web site and client information.

Qualys SSL Labs scan discovered that sys1 works only in browsers with SNI support. Server Name Indication (SNI) allows the server to safely host multiple TLS certificates for multiple sites, all under a single IP address. It adds the host name of the server (Web site) in the TLS handshake as an extension in the CLIENT HELLO message. This way the server knows which Web site to present when using shared IPs [19] (Fig. 4).

For Sys2, the scan reveals that its server accepts RC4 cipher, but only with older protocols, and its grade is capped to B. RC4 is remarkable for its simplicity and speed in software; however, multiple vulnerabilities have been discovered making it insecure [20–22]. This server does not support forward secrecy with the reference browsers. Similar to sys1, sys2 works only in browsers with SNI support. HTTP Strict Transport Security (HSTS) with long duration is deployed on sys2's server [23].

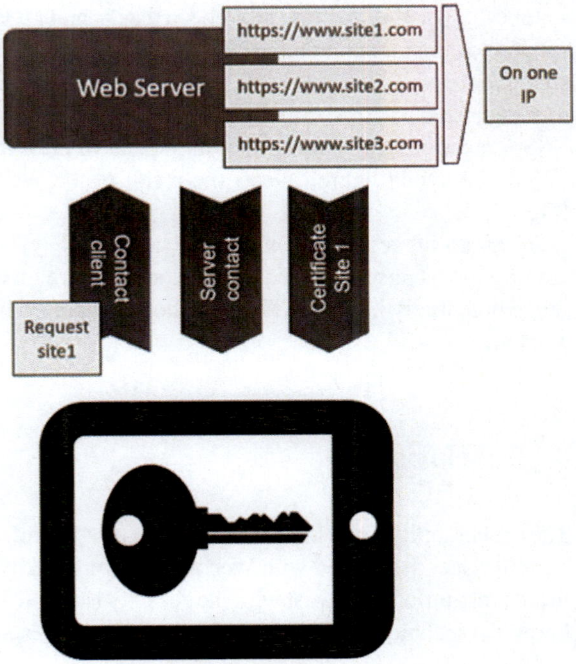

Fig. 4 TLS handshake with SNI

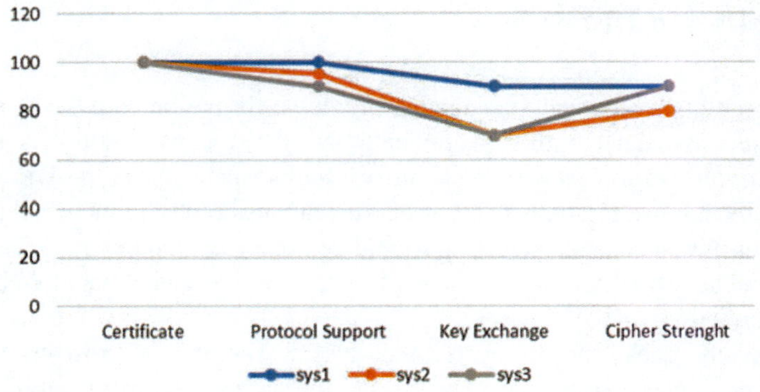

Fig. 5 Web security scan ratings

The next test that was conducted was through the Securi Web site security check and malware scanner. The scan suggests that sys1 is of Critical Security Risk. Sys1 is hosted on Weebly. Sys1 is hacked and needs immediate attention. Malicious code was detected on your site by the automated scanner on sys1. Sys1 does not appear to be blacklisted.

Unlike sys1, Securi Web site security check and malware scanner suggest that sys2 and sys3 have low security risks. Securi Web site security check and malware scanner did not detect malware on your site. Sys2 and Sys3 do not appear to be blacklisted.

No Web site application firewall detected on sys2 and sys3. It is suggested to install a cloud-based WAF to prevent Web site hacks and DDoS attacks. For protection, it is needed to consider creating an SPF record to prevent spammers from abusing your email address. If sys2 and sys3 are not used to send any emails from their domain, it is suggested to use v = spf1 –all. Sys2 and sys3 have missing security header for XSS protection and missing security header to prevent content-type sniffing.

The following figure reflects the result of the Web site security test of htbridge (Fig. 5).

The scan suggests that some HTTP headers of sys1 related to security and privacy are missing or misconfigured. Some cookies have missing secure flags or attributes. Some potentially insecure HTTP methods supported by the Web server require your attention.

4 Conclusions

The vulnerability assessment managed to provide information with the existing and current issues associated with Web application used by the institution. It is important for the institution to abreast with the security issues so that respondents will learn and know the techniques to combat these security threats. With the increasing number of data breaches, a constant scanning should be done for its monitoring. Proper and adequate security measures should be in place to protect organizational Web site and client information.

References

1. X.C.A.M. Vaidya, A survey on security and vulnerabilities of web application. Int. J. Comput. Sci. Inf. Technol. **5**(2), 1856–1860 (2014)
2. F.A. Saeed, Using WASSEC to evaluate commercial web application security scanners. Int. J. Soft Comput. Eng. **4**(1), 177–181 (2014)
3. R. Johari, P. Sharma, A survey on web application vulnerabilities (SQLIA, XSS) exploitation and security engine for SQL injection, in *International Conference on Communication Systems and Network Technologies* (2012)
4. H.A.S. Mohammad, Survey of web application and internet security threats. Int. J. Comput. Sci. Network Secur. **12**(12), 67–76 (2012)
5. Ghanacelebrities.com, Website of Registrar General's Department Hacked [Online]. Available: http://www.ghanacelebrities.com/2015/12/15/website-of-registrar-generals-department-hacked. Accessed 11 Auagust 2019
6. Y.H.A.A.J.D. Vandana, Web application vulnerabilities: a survey. Int. J. Comp. Appl. **108**(1), 25–31 (2014)
7. T. Armerding, The 18 biggest data breaches of the 21st century, 20 December 2018. [Online]. Available: https://www.csoonline.com/article/2130877/the-biggest-data-breaches-of-the-21st-century.html. Accessed 24 August 2019
8. N.A. Acquaye, Software vulnerability led to Ghana govt site hack [Online]. Available: http://www.biztechafrica.com/article/software-vulnerability-led-ghana-govt-site-hack/9583/. Accessed 11 August 2019
9. The OWASP Foundation [Online]. Available: https://www.owasp.org/index.php/Top_10_2013. Accessed 11 August 2019
10. International Organization for Standardization and International Electrotechnical Commission. ISO/IEC 27001:2005 Information technology—security techniques—information security management systems—requirement (2005)
11. K.D.A.K. Priyadharsini, A survey on security properties and web application scanner. Int. J. Comput. Sci. Mob. Comput. **3**(10), 517–527 (2014)
12. M.L. Garcia, CPP, in *Design and Evaluation of Physical Protection Systems*, 2nd edn. (2008)
13. F. Ullah, A. Johannes Raft, M. Shahin, M. Zahedi, M.A. Babar, Security support in continuous deployment pipeline, in *Proceedings of 12th International Conference on Evaluation of Novel Approaches to Software Engineering* (2017)
14. Qualys SSL Labs "SSL Server Test" URL: https://www.ssllabs.com/ssltest/index.html. Accessed: December (2018)
15. GoDaddy Inc., GoDaddy Acquires Sucuri to Advance Digital Security for Customers—Sucuri Enables Customers to Secure Websites without Being Security Experts. https://www.prnewswire.com/news-releases/godaddy-acquires-sucuri-to-advance-digital-security-for-customers-300427537.html Accessed: August 2019
16. Securi website security check & malware scanner. https://sitecheck.sucuri.net/. Accessed: December 2018
17. Symantec Internet Security Threat Report 2016. https://www.symantec.com/content/dam/symantec/docs/reports/istr-21–2016-en.pdf
18. Htbridge "Website Security Test". https://www.htbridge.com/websec/?id=FXDxffVp. Accessed: December 2018

19. R.J. Robles, S.S. Yeo, Application of role-based access control for web environment, in *Proceedings of 2008 International Symposium on Ubiquitous Multimedia Computing* (Published by IEEE, 2008), pp. 171–174
20. Andrei Popov, Prohibiting RC4 Cipher Suites. https://doi.org/10.17487/rfc7465
21. Lucian Constantin, Microsoft continues RC4 encryption phase-out plan with .NET security updates. ComputerWorld
22. D. Endencio-Robles, Analysis of regular operations application and finite automa. Int. J. Adv. Sci. Technol. **118**(1), 151–160 (2018)
23. GlobalSign Blog, What is Server Name Indication (SNI)? https://www.globalsign.com/en/blog/what-is-server-name-indication. Accessed: December 2018
24. P.S.A.A. Radaslic, A penetration test of an internet service provider. School Inf. Sci. Comput. Electr. Eng. 5–25 (2012)
25. H. Nemati, Information security and ethics: concepts, methodologies, tools, and applications: concepts, methodologies, tools, and applications, in *IGI Global* (2008)

Risk Prediction-Based Breast Cancer Diagnosis Using Personal Health Records and Machine Learning Models

Sireesha Moturi, S. N. Tirumala Rao, and Srikanth Vemuru

Abstract Breast cancer is most common in middle-aged female population. It is the fourth most dangerous cancer compared to remaining cancers. In recent years, breast cancer patients are significantly increasing, so the early diagnosis of cancer has become a necessary task in the cancer research, to facilitate subsequent clinical management of patients. The prevention of the breast cancer tumor is early detection of the tumor. Early detection of cancer can stop increase in tumor and saves lives. In the field of machine learning classification, cancer patients are classified into two types as benign or malignant. Different preprocessing techniques like filling missing values, applying correlation coefficient, synthetic minority oversampling technique (SMOTE) and tenfold cross-validations are implemented and aptly used to obtain the accuracy. The main context of this study is to identify key features from the dataset and analyze the performance evaluation of different machine learning algorithms like random forest classifier, logistic regression, support vector machine, decision tree, Gaussian Naive Bayes and *k*-nearest neighbors. Based on the results, the classification model that gives highest accuracy will be used as the best model for cancer prediction.

Keywords Breast cancer · Diagnosis · Prediction

S. Moturi (✉) · S. Vemuru
KLEF, Vaddeswaram, India
e-mail: sireeshamoturi@gmail.com

S. Vemuru
e-mail: vsrikanth@kluniversity.in

S. Moturi · S. N. Tirumala Rao
Narasaraopeta Engineering College, Narasaraopet, India
e-mail: nagatirumalarao@gmail.com

D. Bhattacharyya and N. Thirupathi Rao (eds.), *Machine Intelligence and Soft Computing*, Advances in Intelligent Systems and Computing 1280,
https://doi.org/10.1007/978-981-15-9516-5_37

1 Introduction

Breast cancer tumor is uncontrolled development of the cells and increases the tumor. This causes unusual changes or transformations, in the qualities answerable for directing development of cells and keep them healthy [1]. The qualities go about as control room of every cell which is available in every cell's core. The cells in our body supplant themselves in a routinely procedure of cell development: New cells are supplanted over as old ones kick the bucket. Yet, at some point, transformations turn on specific qualities and mood killer other cells. At the point, the changed cell has capacity to continue isolating without control or request, creating more cells and framing tumor. A tumor might be kind or dangerous tumor. Considerate tumor is not perilous to well-being and harmful tumor hazardous and considered as malignant. Breast cancer disease alludes to threatening tumor that is created from cells in the breast.

Nowadays, data is also generating more and more [2], so content extraction from the biological datasets is a challenging task in today's world [3]. Information on the web is also of different types like structured and unstructured kind of records, homogeneous, heterogeneous and mixed varieties of data and current Web sites present a larger wide variety of difficulties and complexities than traditional ones [4]. To deal with this complex data, there is a need of dominant tool like machine learning (ML) [5]. Machine learning techniques play a dominant role in disease diagnosis process [6–8].

Machine learning is one of the uses of man-made brainpower that gives personal computers (PCs), the capacity to take in naturally and improve for a fact rather than unequivocally modified. It centers around creating PC programs that can get to information and use it gain from themselves. The primary point is to permit PCs to adapt naturally without human intervention and furthermore alter activities appropriately.

2 Literature Survey

The reasons for the breast cancer are obscure and contemplates recommend that the development of cancer is influenced by estrogen. Most of the cases were observed in women over age 50. A tumor is brought about by irregular development of the tissue. Tumor may be cancerous (dangerous) or non-cancerous (non-dangerous) [9].

Kumari and Singh [10] used many approaches for detecting breast cancer and achieved accuracy of 98.6% for rough sets and backpropagation.

Sireesha et al. [11] compared different machine learning algorithms and got the highest accuracy of 93% for the New Levy Dragonfly algorithm.

Hiba Asria et al. [12] compared four machine learning algorithms SVM, NB, KNN and C4.5 and got highest accuracy of 97.3% for SVM.

3 Proposed System

3.1 Experimental Setup

This experiment was conducted on Intel® Core™ i7 processors with 64 bit Windows 10 Pro. Anaconda 5.1.0 Python distribution is used in this experiment. The dataset used is from UCI repository [13]. Wisconsin Breast Cancer Dataset (original) consists of 699 records, 2 classes (benign and malignant) and 9 integer attributes. Dataset contains 458 benign and 251 malignant instances. Figure 1 shows the process of the proposed system of breast cancer detection. Figure 2 displays the attributes, domain and description about the attributes.

Figure 3 shows the importance of each attribute in the dataset. From Fig. 3, it is observed that single epithelial cell has highest importance and bland chromatin has least importance.

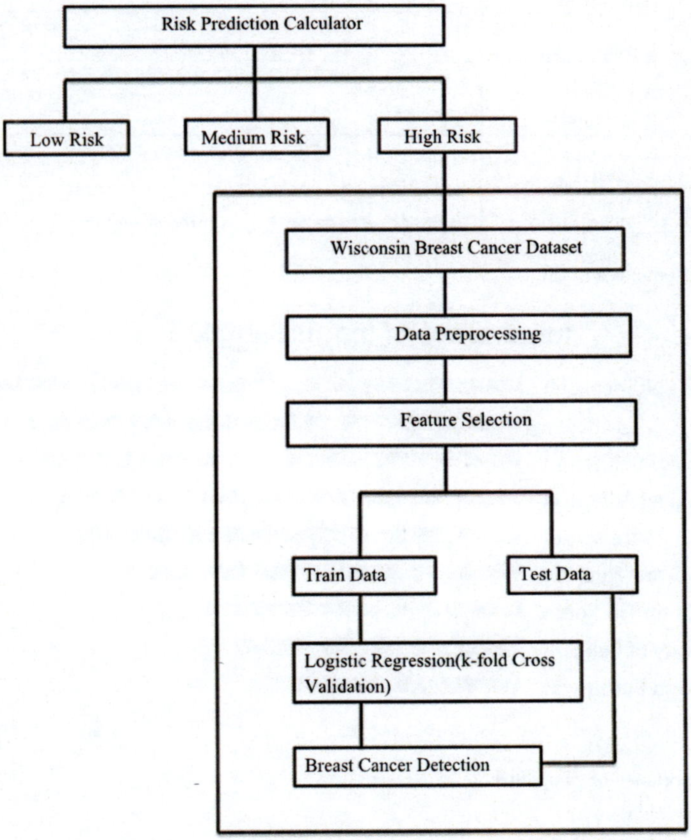

Fig. 1 Architecture of proposed system

	Attribute	Domain	Description
1	Sample Number	Id Number	Id Number
2	Clump Thickness	1 – 10	Assesses if cells are mono or multi-layered.
3	Uniformity Cell Size	1 – 10	Evaluates the consistency in size of the cells in the sample.
4	Uniformity Cell Shape	1 – 10	Estimates the equality of cell shapes and identifies marginal variances.
5	Marginal Adhesion	1 – 10	Quantifies how much cells on the outdside of the epithelial tend to stick together.
6	Single Epithclial Cell Size	1 – 10	Relates to cell uniformity, determines if epithelial cells are significantly enlarged.
7	Bare Nuclei	1 – 10	Calculates the proportion of the number of cells not surrounded by cytoplasm to those that are.
8	Bland Chromatin	1 – 10	Rates the uniform "texture" of the nucleus in a range from fine to coarse.
9	Normal Nucleoli	1 – 10	Determines whether the nucleoli are small and barely visible or larger, more visible, and more plentiful.
10	Mitoses	1 – 10	Describes the level of mitotic (cell reproduction) activity.
11	Class	2 or 4	Describes Benign or Malignant tumour

Fig. 2 Attribute description

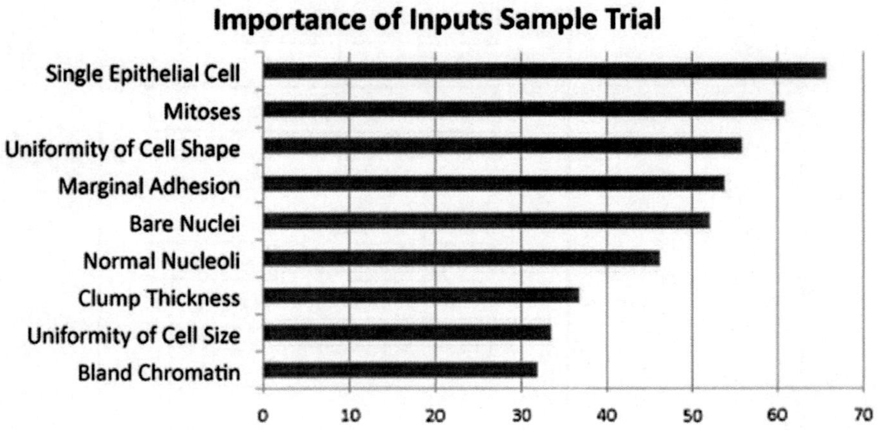

Fig. 3 Importance of each attribute in dataset

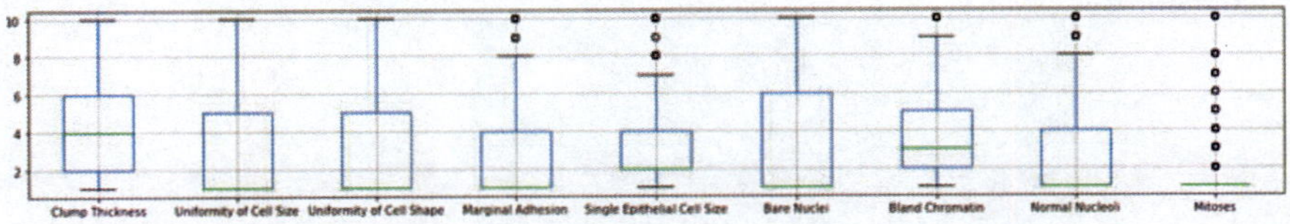

Fig. 4 Histogram of all attributes of dataset

3.2 Outlier Identification

Figure 4 is the boxplot of all attributes of dataset in which we can observe the distribution of different attributes. The above boxplot suggested that only four of the columns are not having outliers and remaining five of the columns contain abnormally high values and outliers.

3.3 Preprocessing

Preprocessing is the first step while creating the machine learning model [14]. It is the process of converting raw dataset into cleaned dataset. Raw data contains noise, missing values and duplicate values which are not suitable for machine learning model. So, preprocessing is required for cleaning the data and making it suitable for machine learning model.

Handling Missing Values

Handling missing values is important because machine learning algorithms do not support missing values in the data. Some of the techniques to handle missing values are [15]. Figure 5 shows the dataset before filling missing values. Figure 6 shows the dataset after filling missing values.

1. Replace with mean or median or mode value.
2. Back-fill or forward-fill.
3. Deleting row.

SMOTE

SMOTE is normally utilized oversampling strategies to understand imbalance problem [16]. It balances class distribution by unpredictably expanding minority samples. New minority classes are created by utilizing existing minority classes. It creates virtual preparing records by direct insertion for minority class. The virtual records are created by choosing at least one of *k*-nearest neighbors. The data is reconstructed, and classification is applied to resultant preprocessed data.

Figure 7 shows the number of instances for class-2 and class-4 before applying the SMOTE technique. Class-2 has more instances than class-4. After applying SMOTE technique, Fig. 8 shows the same number of instances for class-2 and class-3.

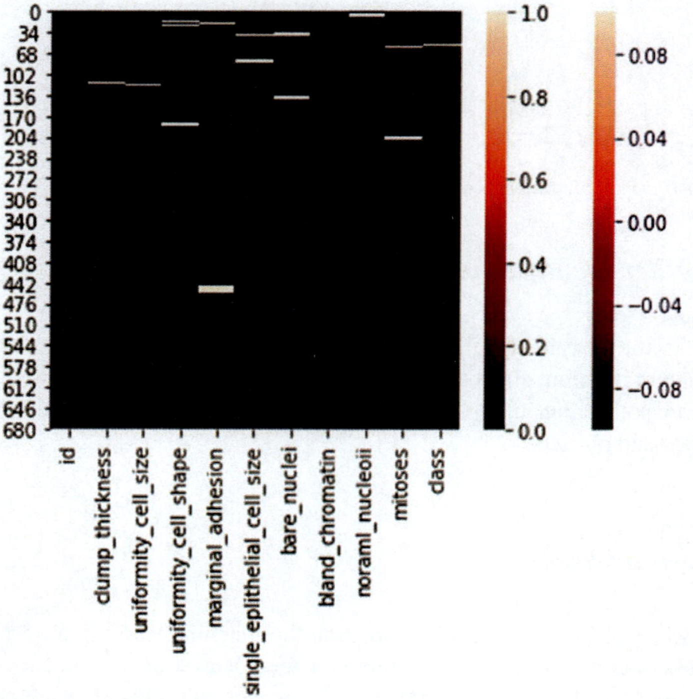

Fig. 5 Dataset before filling missing values

Feature Selection

It is the process of reducing the number of input attributes or variables when developing a model in order to reduce the computation cost of model and in some cases to increase or improve the performance or accuracy of the model.

Correlation

Correlation describes the relation between two continuous attributes [17]. It is a statistical measure to demonstrate the degree to which at least two factors change together. Figure 9 displays the correlation applied to the WBCD. After applying correlation, it is observed that two columns uniformity cell_size and uniformity_cell_shape are positively correlated, i.e., without using two attributes, one attribute in those two is sufficient. So, one of the linearly related attributes is dropped.

4 Classification Techniques

Classification is a supervised learning and can be performed on both structured and unstructured data [18]. It classifies a lot of information into classes and predicts the

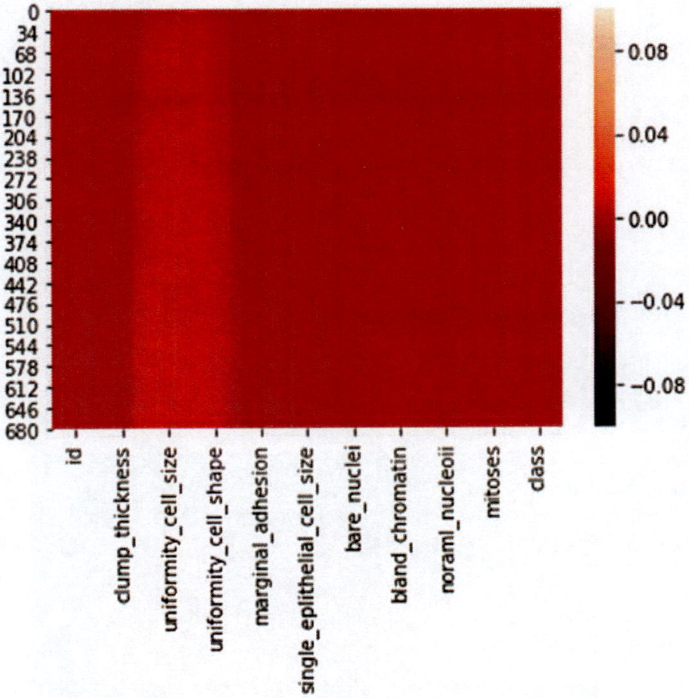

Fig. 6 Dataset after filling missing values

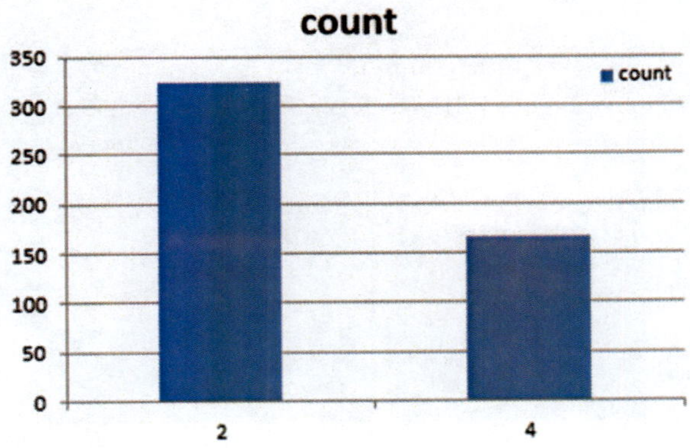

Fig. 7 Class distribution before SMOTE technique

class of given information. Classes are called as targets or marks or classifications. There are many classification algorithms available in machine learning. Compared to previous work mentioned above, our paper presents many approaches that are used for detecting breast cancer which are feature selection, *k*-fold cross-validation and six different machine learning algorithms. Three approaches give comparison between

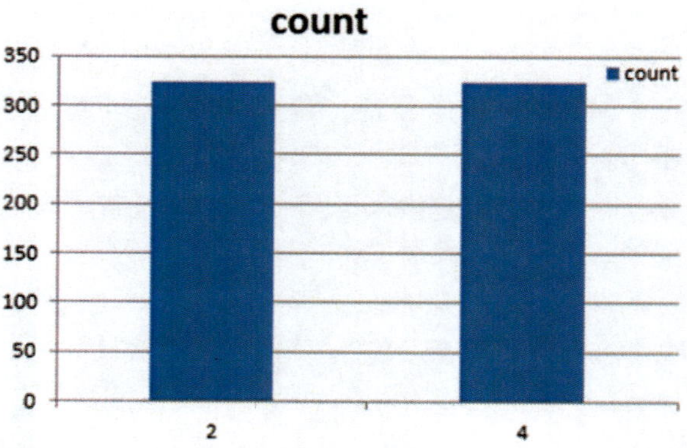

Fig. 8 Class distribution after SMOTE technique

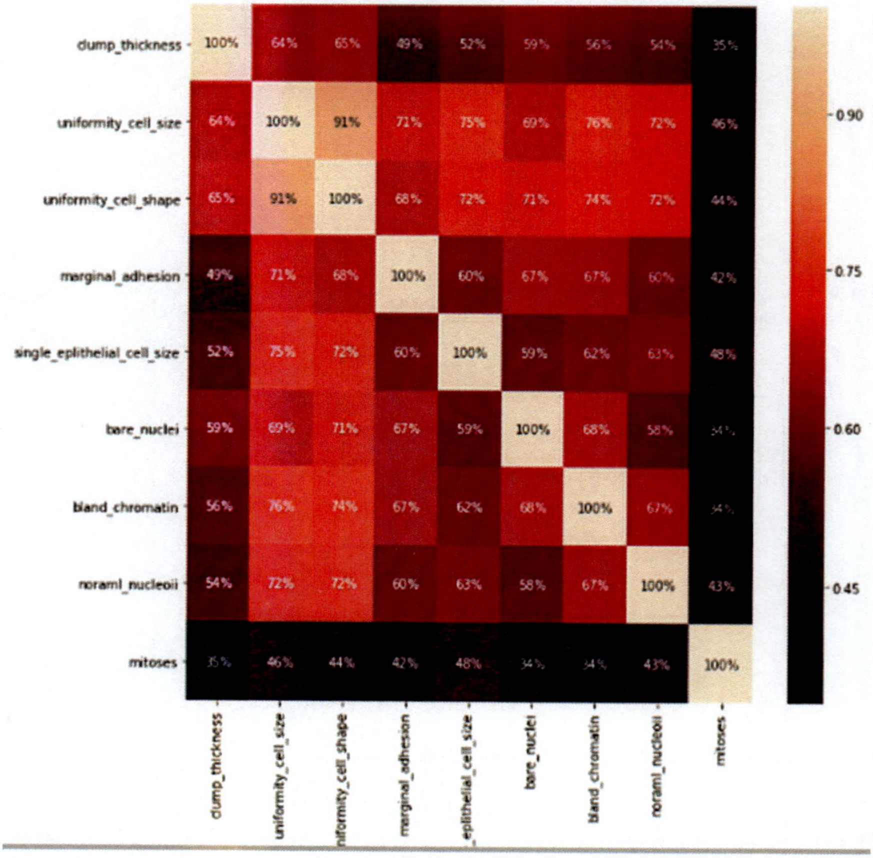

Fig. 9 Correlation for breast cancer dataset

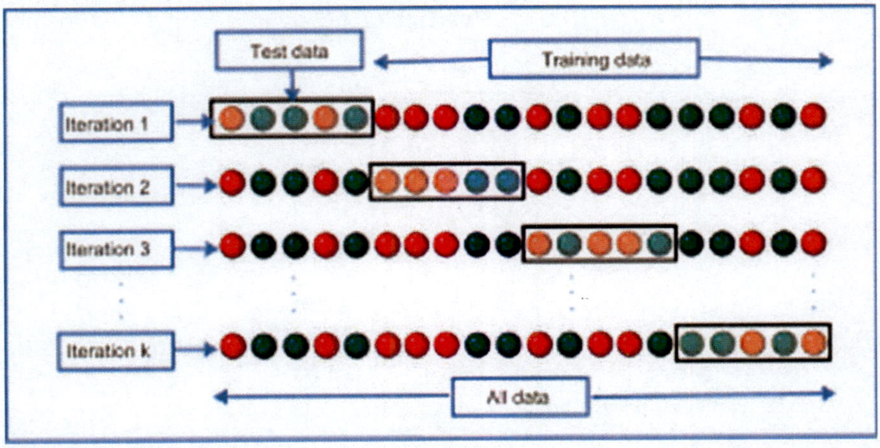

Fig. 10 *k*-fold cross-validation

six classification algorithms: logistic regression, random forest classifier, support vector machine (SVM), Gaussian Naïve Bayes, *k*-nearest neighbors and decision tree.

4.1 k-*Fold Cross-validation*

Cross-validation is a resampling process used to evaluate machine learning models [19].

Figure 10 shows how process for cross-validation is done. The process has a single parameter k, where k indicates number of iterations to be done in the process. When k value is chosen, it is utilized in place of k, such as $k = 10$ that is tenfold cross-validation. In the first iteration, $(k - 1)$ data samples are taken as training data, and kth data is taken as test data, and in the next iteration, kth data sample which is taken as test data is replaced by one of data sample in $(k - 1)$ and vice versa. The process continues until k number of iterations which covers all data samples.

4.2 *Performance Evaluation*

Performance evaluation of classification algorithm is calculated by using confusion matrix [20] as shown in Fig. 11. Confusion matrix is represented as table which describes performance based on test data for which correct values are known. Performance is calculated by considering actual and predicted class.

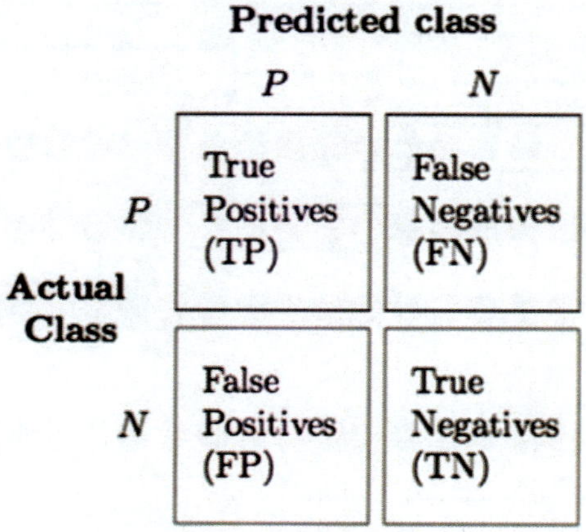

Fig. 11 Confusion matrix

5 Result Analysis

Figure 12 shows the comparison accuracy of six classifiers (random forest classifier, logistic regression, Gaussian Naïve Bayes, decision tree, SVM and KNN) before and after applying SMOTE. From Fig. 12, it is observed that random forest classifier achieves highest accuracy of 96.19%, and it is also observed that all the classifiers achieve highest performance when compared to after applying SMOTE.

From Fig. 13, it is observed that the accuracy of random forest is 95.71 when correlation is applied. The accuracy of random forest classifier before applying correlation

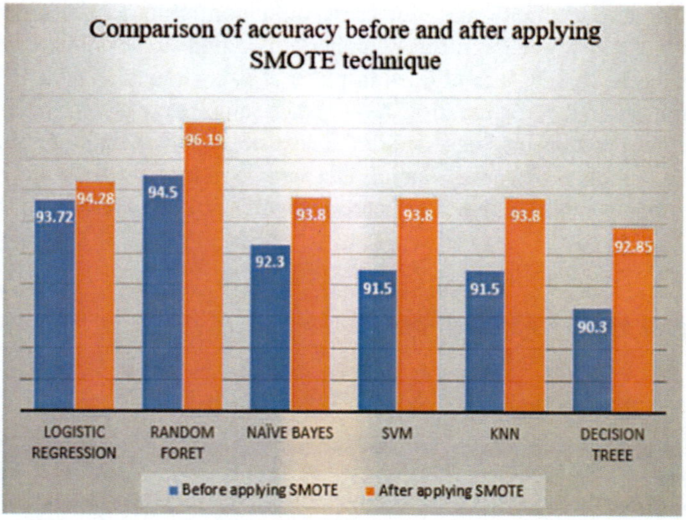

Fig. 12 Comparison of accuracy before and after applying SMOTE

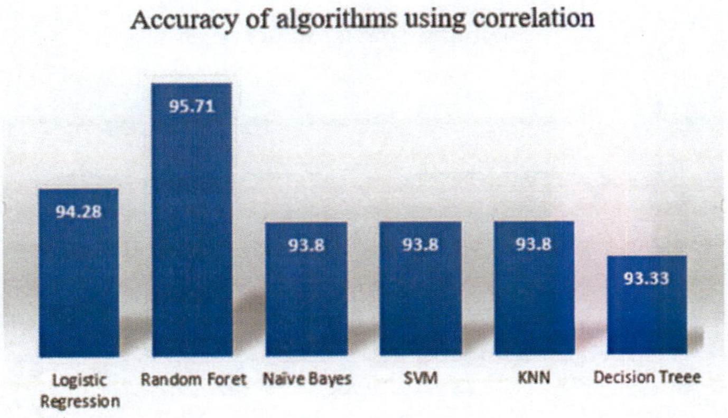

Fig. 13 Accuracy of all classifiers after applying correlation

is 94.5% as shown in Fig. 12. Similarly, all the classifier performance was improved after applying correlation.

From Fig. 14, it is observed that the accuracy of logistic regression is 100.00% when k-fold cross-validation is applied. The accuracy of random logistic regression before applying cross-validation is 93.72% as shown in Fig. 12. Similarly, all the classifier performance was improved after applying cross-validation.

Figure 15 shows the performance evaluation of different algorithms for different techniques. The highest accuracy is 100% for logistic regression when k-fold cross-validation was applied. Table 1 shows the comparison of our proposed method results with existing methods.

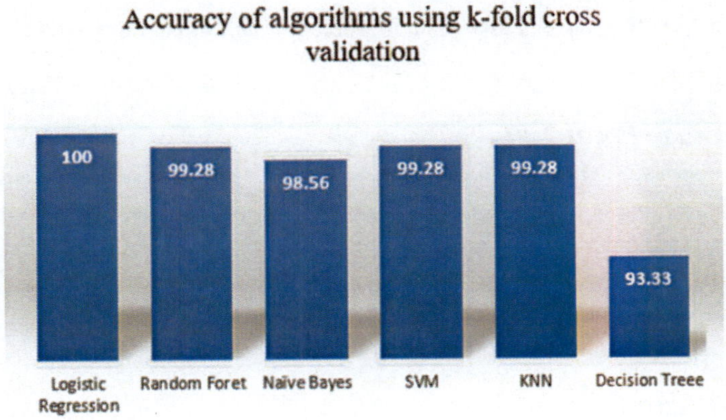

Fig. 14 Accuracy of algorithms using k-fold

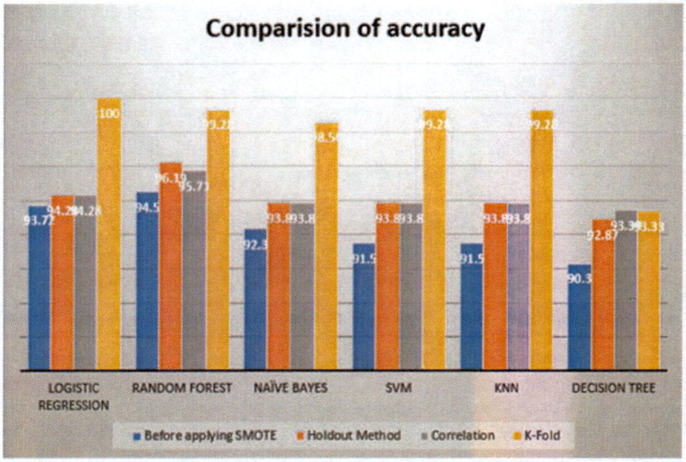

Fig. 15 Comparison of accuracy of algorithms before applying SMOTE, holdout method, correlation and *k*-fold

5.1 Risk Prediction

Figure 16 shows the risk prediction page which uses risk predictor to predict the level of risk of a person having breast cancer. The level of risk is categorized as follows:

1. High risk: If the risk score is greater than 65, then that person is having high risk of having breast cancer.
2. Moderate risk: If the risk score is between 50 and 65, then that person is having medium risk of having breast cancer.
3. Low risk: If the score is less than 50, then the patient is at low risk.

5.2 Disease Prediction

If the person is at high risk, then breast cancer detector is used to diagnose whether the person is having the disease or not with the help of clinical reports from laboratories as shown in Fig. 17, and Table 1 shows the comparison of all the results.

6 Conclusion

Accuracy is very important to diagnose the patient's disease, and it depends on the accuracy obtained when different approaches and classification algorithms are applied. In this proposed system, various preprocessing methods were applied to

Table 1 Comparison of all the results

Author	Method	Accuracy
Kumari and Singh [10]	Rough sets, backpropagation and neural network	98.60
	Feature extraction using a hybrid of *k*-mean and SVM	97.38
	LR, neural network and decision Tree	89.25
Sireesha et al. [11]	NL-DA	92
Asria et al. [12]	C4.5	95.13
	SVM	97.13
	NB	95.99
	k-NN	95.27
Our study	Logistic regression before SMOTE	93.72
	Logistic regression after SMOTE	94.28
	Logistic regression before correlation	93.72
	Logistic regression after correlation	94.28
	Logistic regression before *k*-fold	93.72
	Logistic regression after *k*-fold	100
	Random forest before SMOTE	94.50
	Random forest after SMOTE	96.19
	Random forest before correlation	94.50
	Random forest after correlation	95.71
	Random forest before *k*-fold	94.50
	Random forest after *k*-fold	99.28
	Naïve Bayes before SMOTE	92.30
	Naïve Bayes after SMOTE	93.80
	Naïve Bayes before correlation	92.30
	Naïve Bayes after correlation	93.80
	Naïve Bayes before *k*-fold	92.30
	Naïve Bayes after *k*-fold	98.56
	SVM before SMOTE	91.50
	SVM after SMOTE	93.80
	SVM before correlation	91.50
	SVM after correlation	93.80
	SVM before *k*-fold	91.50
	SVM after *k*-fold	99.28
	KNN before SMOTE	91.50
	KNN after SMOTE	93.80
	KNN before correlation	91.50
	KNN after correlation	93.80
	KNN before *k*-fold	91.50

(continued)

Table 1 (continued)

Author	Method	Accuracy
	KNN after *k*-fold	99.28

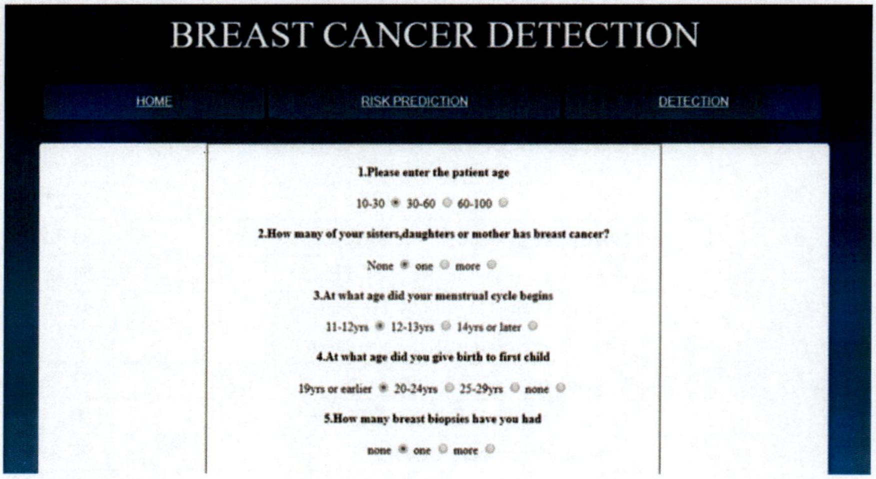

Fig. 16 Risk predictor

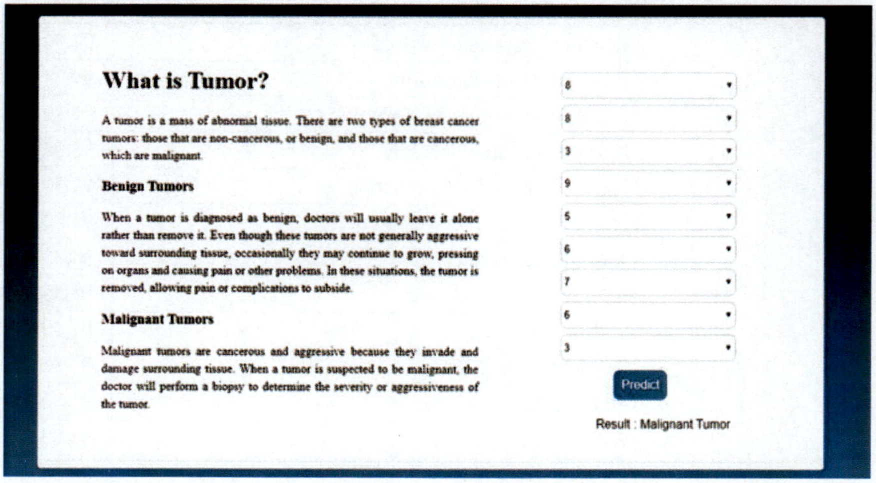

Fig. 17 Breast cancer detector

the data with different classification algorithms and observed that the highest accuracy is obtained when logistic regression is used with tenfold cross-validation that is 100.00%. This study presents more effective disease prediction which helps in developing health care by reducing cost and time. This research also ensures that the early detection of risk may improve the patient treatment to the greater extent. For this risk, calculator is used to calculate risk percentage and level of risk. If risk is very high, then it suggests to take laboratory report and those values should be used in the model to predict the stage of risk. It is also useful for the doctors to improve decision making in less time by using different machine learning techniques.

References

1. A. Jemal, T. Murray, E. Ward, A. Samuels, R.C. Tiwari, A. Ghafoor, E.J. Feuer, M.J. Thun, Cancer statistics. CA Cancer J. Clin. **55**(1), 10–30 (2005)
2. K.B. Prakash, M.A. DoraiRangaswamy, A.R. Raman, Text studies towards multi-lingual content mining for web communication, in *Proceedings of the 2nd International Conference on Trendz in Information Sciences and Computing* (2010), pp. 28–31
3. K.B. Prakash, M.A.D. Rangaswamy, Content extraction of biological datasets using soft computing techniques. J. Med. Imag. Health Inf. 932–936 (2016)
4. K.B. Prakash, Information extraction in current Indian web documents. Int. J. Eng. Technol. (UAE) (2018), pp. 68–71
5. K.B. Prakash, M.A. DoraiRangaswamy, Content extraction studies using neural network and attribute generation. Ind. J. Sci. Technol. 1–10 (2016)
6. M. Sireesha, S. Vemuru, S. N. Tirumala Rao, Coalesce based binary table: an enhanced algorithm for mining frequent patterns. Int. J. Eng. Technol. **7**(1.5), 51–55 (2018)
7. M. Sireesha, S.N. Tirumala Rao, S. Vemuru, Frequent Itemset Mining Algorithms: A Survey. J. Theoret. Appl. Inf. Technol. **96**(3), 744–755 (2018)
8. M. Sireesha, S. Vemuru, S.N. Tirumala Rao, Classification model for prediction of heart disease using correlation coefficient technique. Int. J. Adv. Trends in Comput. Sci. En. **9**(2), 2116–2123 (2020)
9. U.S. Cancer Statistics Working Group. https://www.cdc.gov/cancer/uscs/technical_notes/index.html
10. M. Kumari, V. Singh, Breast cancer prediction system, in *International Conference on Computational Intelligence and Data Science* (2018), pp. 371–376
11. M. Sireesha, S.N. Tirumala Rao, S. Vemuru, Optimized feature extraction and hybrid classification model for heart disease and breast cancer prediction. Int. J. Rec. Technol. Eng. **7**(6), 1754–1772 (2016)
12. H. Asri, H. Mousannif, H. Al Moatassime, T. Noel, Using machine learning algorithms for breast cancer risk prediction and diagnosis, in *The 6th International Symposium on Frontiers in Ambient and Mobile System* (2016), pp. 1064–1069
13. UCI Machine Learning Repository: Breast Cancer Wisconsin (Original) Data Set. https://archive.ics.uci.edu/ml/datasets/Breast+Cancer+Wisconsin+%28Original%29
14. Data Preprocessing- an overview. https://www.javatpoint.com/data-preprocessing-machine-learning
15. Handling Missing Values in machine learning. https://towardsdatascience.com/working-with-missing-data-in-machine-learning-9c0a430df4ce
16. M. Sireesha, S.N. Tirumala Rao, S. Vemuru, Predictive analysis of imbalanced cardiovascular disease using SMOTE. Int. J. Adv. Sci. Technol. **29**(5), 6301–6311 (2020)

17. In-Database Machine Learning 2: Calculate a correlation Matrix—A Data Exploration Post, Vertica. https://www.vertica.com/blog/in-database-machine-learning-2-calculate-a-correlation-matrix-a-data-exploration-post/
18. D. Lavanya, K. Usha Rani, Analysis of feature selection with classification: breast cancer datasets. Ind. J. Comput. Sci. Eng. (IJCSE) (2011)
19. Cross-validation: evaluating estimator performance. https://scikit-learn.org/stable/modules/cross_validation.html
20. J. Han, M. Kamber, *Data Mining Concepts and Techniques* (Morgan Kauffman Publishers, 2000)

Orthogonal MIMO 5G Antenna for WLAN Applications

Suneetha Pasumarthi, Srinivasa Naik Kethavathu, Pachiyannan Muthusamy, and Aruna Singam

Abstract In this paper, single radiating element is designed and simulated at resonant frequency of 5.2 GHz for 5G WLAN applications. The S_{11} parameter for single radiating element is −29.76, gain is 3.2 dB, and directivity is 6 dB. The voltage standing wave ratio is less than 2 which is achieved. The designed antenna offers good bandwidth, i.e., 200 GMHz. Further, two-element and four-element space diversity and two-element and four-element orthogonal antennas are designed, and diversity characteristics are analyzed. Compared to the two-element and four-element MIMO antenna, the size of the orthogonal antenna is reduced and achieved with good diversity characteristics.

Keywords MIMO · Directivity · Gain (DG) · ECC

S. Pasumarthi (✉) · S. N. Kethavathu
Vignan's Institute of Information Technology (A), Visakhapatnam, A.P, India
e-mail: rsss4m@gmail.com

S. N. Kethavathu
e-mail: nivas97033205@gmail.com

P. Muthusamy · A. Singam
Andhra University College of Engineering (A), Visakhapatnam, A.P, India
e-mail: pachiphd@gmail.com

A. Singam
e-mail: aruna9490564519@gmail.com

D. Bhattacharyya and N. Thirupathi Rao (eds.), *Machine Intelligence and Soft Computing*, Advances in Intelligent Systems and Computing 1280,
https://doi.org/10.1007/978-981-15-9516-5_38

1 Introduction

MIMO systems use a combination of several antennas and multiple signal paths to acquire communication channel information. The major importance of the MIMO system is increasing the channel capacity without requirement of additional power for transmission as well as for reception. But the main limitation of the MIMO system is signal interference due to the adjacent antennas signal transmission, and this is called mutual coupling interference. To overcome the limitation, generally three are used: One is by placing the antenna side by side with one-fourth wavelength distance, second one is placing the antenna elements by 90° phase shift, and third one is providing isolation. In [1], the proposed MIMO provides high isolation over the frequency range from 5.10 to 5.85 GHz by using circular polarization (truncating the patch edges) method.

Another report in [2] triple-band MIMO antenna achieved low interference by interrupting with I-shaped slot on ground plane, and in [3], a tri-band MIMO antenna is reported by avoidance of conventional method decoupling structures like DGS and EBG structures. Several antennas are designed for four-element MIMO antenna for WLAN and Wi-Max services [4–10].

2 Antenna Geometry and Design

2.1 Single Antenna Design and Analysis

A single monopole antenna is designed on FR-4 with a dielectric constant $(\varepsilon_r) = 4.4$, and thickness is 1.6 mm. The stage-by-stage design of proposed antenna is shown in Fig. 1. The antenna design parameters are $W_g = 25$ mm, $L_g = 25$ mm, $W_p = 17$ mm, $L_p = 12$ mm, $W_f = 1$ mm and $L_f = 7$ mm. The overall dimension of the single antenna is $25 * 25\ \text{mm}^2$. The simulated return loss S_{11} of all stages of proposed antenna is shown in Fig. 2a, with a return loss of −29.76 dB. The antenna covers the frequency range from 5.23 GHz to 5.43 GHz and provides a good value of gain, and VSWR at resonant frequency 5.3 GHz is 6.22 dB, 1.13, respectively, as shown in Fig. 2b–d (Fig. 3).

3 Two-Element MIMO Antenna Design and Analysis

Two element MIMO antenna is designed by taking the single element antenna. For designing MIMO Antennas there are two methods. One is space diversity technique and other technique is orthogonal. Based on these methods results are taken and compared parameters like return loss, gain, directivity, ECC and DG.

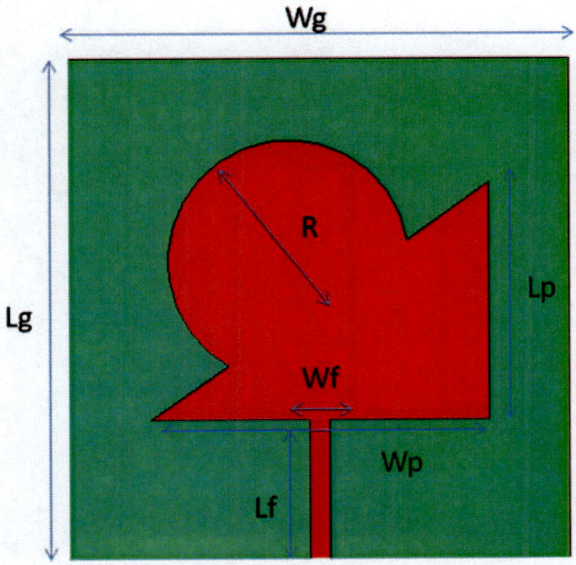

Fig. 1 Design geometry of the antenna

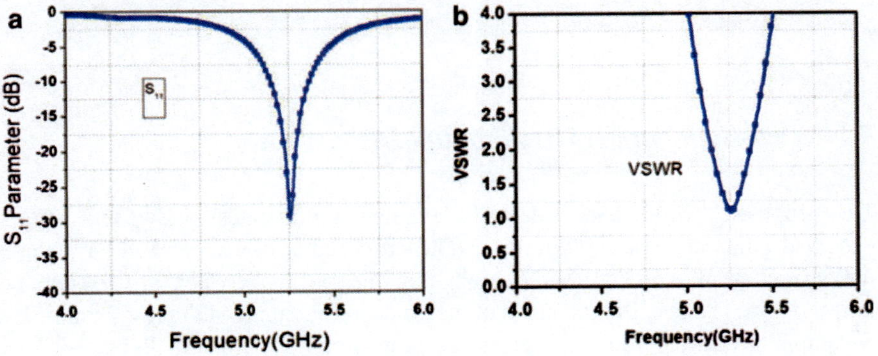

Fig. 2 **a** Simulated S-parameters of the antenna **b** VSWR for proposed work

ECC

The performance of the MIMO system is measured using envelope correlation coefficient (ECC). The value of ECC describes the correlation between the antennas. The little value of the ECC provides high separation. This could be measured by using Eq. 1 given in reference [6].

$$\rho_{eij} = \frac{\left|S_{ii}^{*}S_{ij} + S_{ji}^{*}S_{jj}\right|^{2}}{\left(1 - |S_{ii}|^{2} - S_{ij}^{2}\right)\left(1 - |S_{ji}|^{2} - S_{jj}^{2}\right)} \quad (1)$$

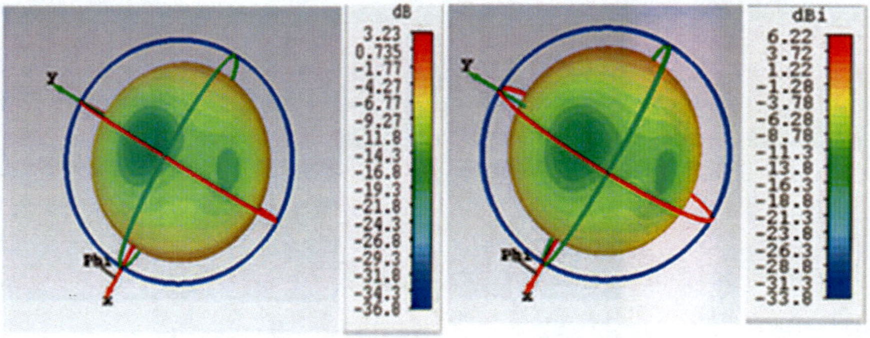

(a) (b)

Fig. 3 **a**) Simulated 3D gain plot **b**) directivity plot

Directive gain

Diversity gain (DG) measures the transmission power loss for the MIMO systems. Using Eq. 2, the diversity gain could be measured, i.e., mentioned in reference [10].

$$\mathrm{DG} = 10\sqrt{1 - \left|\rho_{eij}\right|^2} \tag{2}$$

3.1 *Two-Element MIMO Antenna Discussion*

The proposed two-element side-by-side antenna illustrates below, and all the observed parameters show Fig. 4b–e. The proposed two-element MIMO antenna dimensions are $W_g = 43$ mm and $L_g = 25$ mm. The spacing between the two elements is $D = 5$ mm. The overall dimension of the proposed antenna is 43 mm × 25 mm. The S-parameter values of the proposed antenna for S_{11} and S_{21} are −20 dB and −17 dB, respectively. The VSWR of the proposed antenna is obtained below 2 magnitudes. The ECC of the proposed antenna is closed to zero value, and DG value is greater than 9 dB.

3.2 *Two-Element Orthogonal MIMO Antenna Result Discussion*

The proposed two-element MIMO orthogonal antenna is illustrated in Fig. 5a, and all the observed parameters show Fig. 5b–e. The proposed two-element MIMO antenna dimensions are $W_g = 43$ mm and $L_g = 25$ mm. The spacing between the two

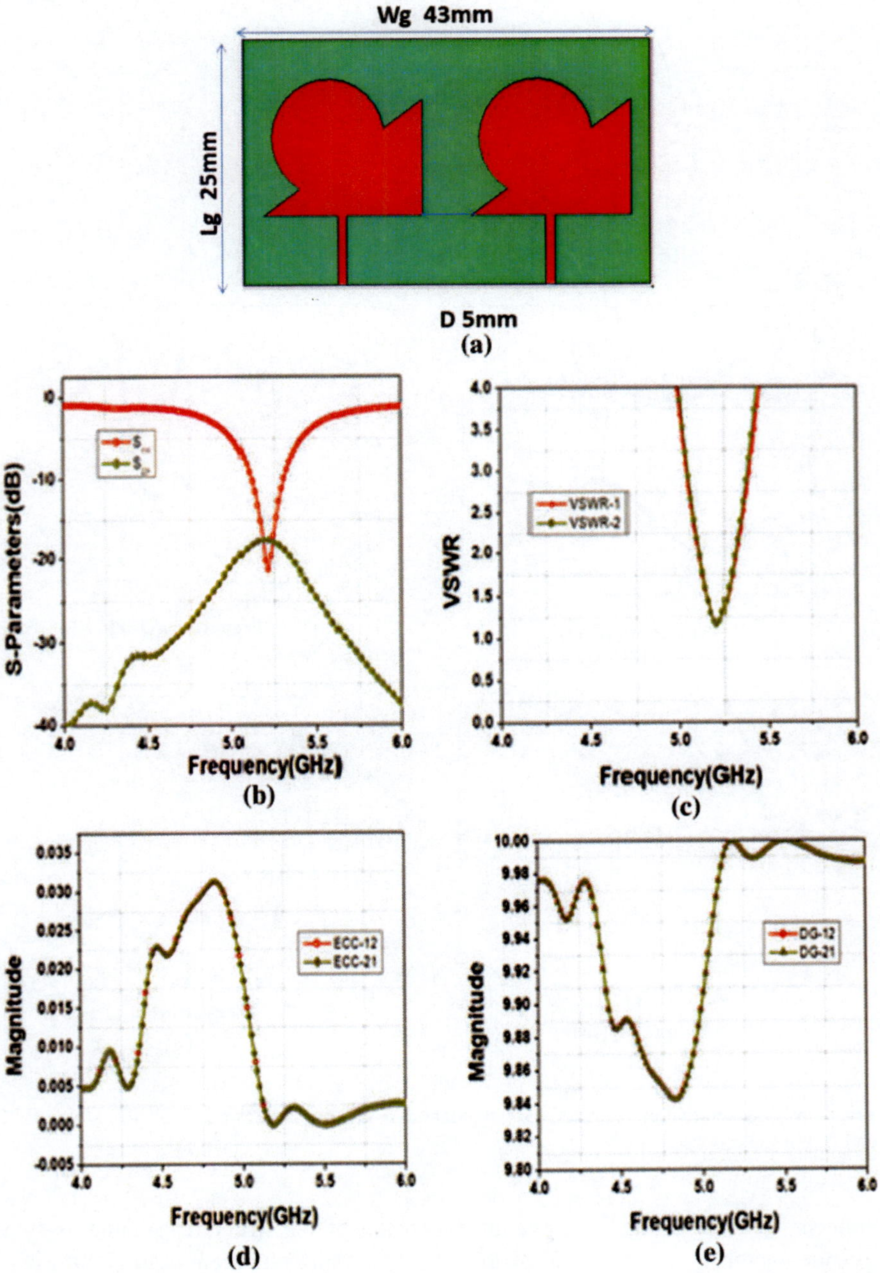

Fig. 4 **a** Two-element proposed MIMO antenna **b–c** S-parameters and VSWR of the proposed antenna **d–e** ECC and DG of the proposed antenna

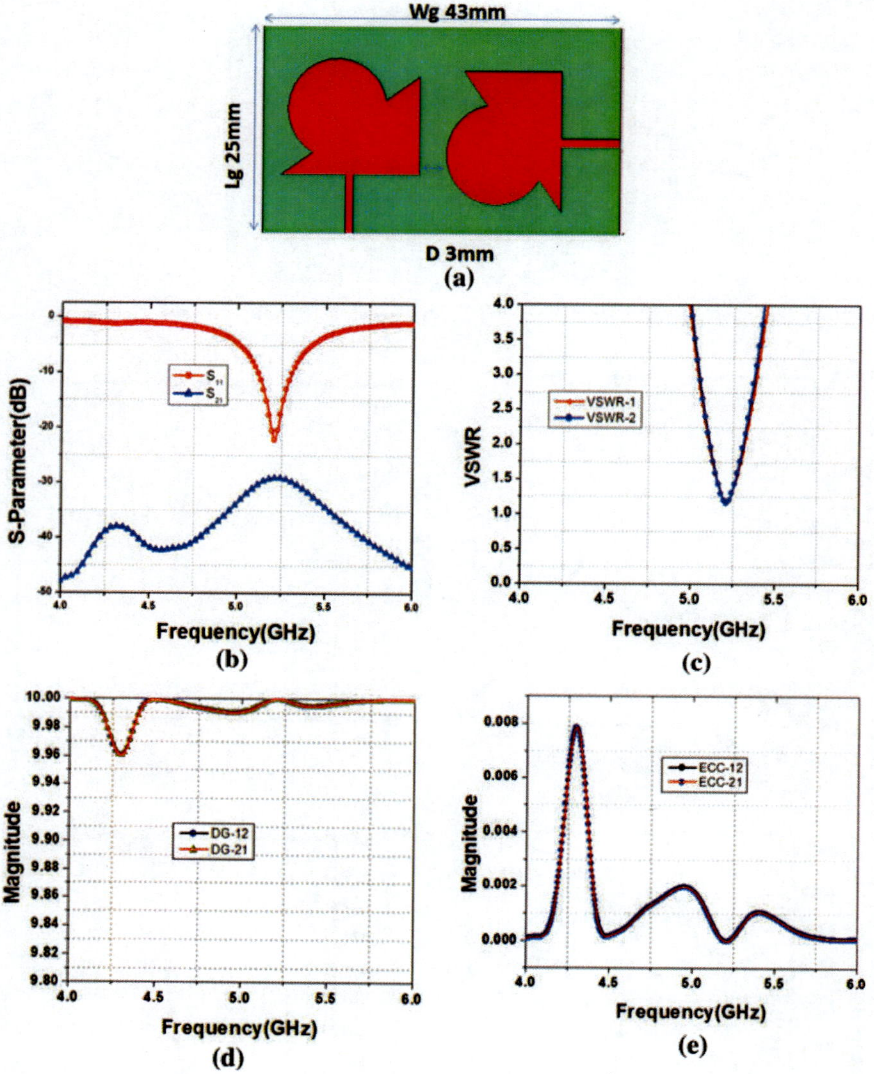

Fig. 5 **a** Proposed two-element orthogonal antenna. **b–c** S-parameters and VSWR and **d–e** ECC and DG of the proposed antenna

elements is $D = 3$ mm. The overall dimension of the structure is same as side-by-side antenna, i.e., 43 mm × 25 mm. This structure obtained higher return loss, i.e., −29 dB compared to side-by-side MIMO.

4 Space Diversity of MIMO Antenna Result Discussion

The space diversity of MIMO antenna is depicted in Fig. 6a. The distance taken between the antennas is one-tenth of the wavelength. The simulated S-parameter values and the VSWR responses of four-element space diversity antenna are shown in Fig. 6b, c. The proposed space diversity of MIMO provides envelope correlation coefficient (ECC) which is below 0.01, which is required value between the four antennas, and directive gain (DG) is 9.99 dB as shown in Fig. 6d, e.

4.1 Orthogonal MIMO Antenna Result Discussion

In this design MIMO system, all four radiating elements are placed orthogonally. The proposed orthogonal configuration is shown in Fig. 7a. The Simulated values of the designed antenna like VSWR simulated values of the designed MIMO plot and the simulated S-parameter results S_{11}, S_{21} and S_{31}, S_{41} and also between different ports provided in Fig. 7b, c and the diversity parameter values, i.e., ECC and DG provided in Fig. 7d, e, respectively. Compared to the space diversity of MIMO antenna, this proposed structure provides better diversity performance results, i.e., ECC value is 0.001 and DG > 9.8. Table 1 provided the comparison for single- and two-element and four-element MIMO antennas. Table 2 represents comparison between proposed and existing works.

5 Conclusion

In this work, the two element and four element space diversity and two element, four element orthogonal antenna was designed, started with single radiating antenna for WLAN applications. The designed antenna provides 2 GHz bandwidth. By comparing the existing work, the designed two-element and four-element orthogonal antenna is compact in size and obtained high return loss and also provides good enhanced isolation result. Overall designed results of MIMO antenna using orthogonality method satisfy all the required parameters (ECC < 0.1 and DG > 9.8) of MIMO antenna.

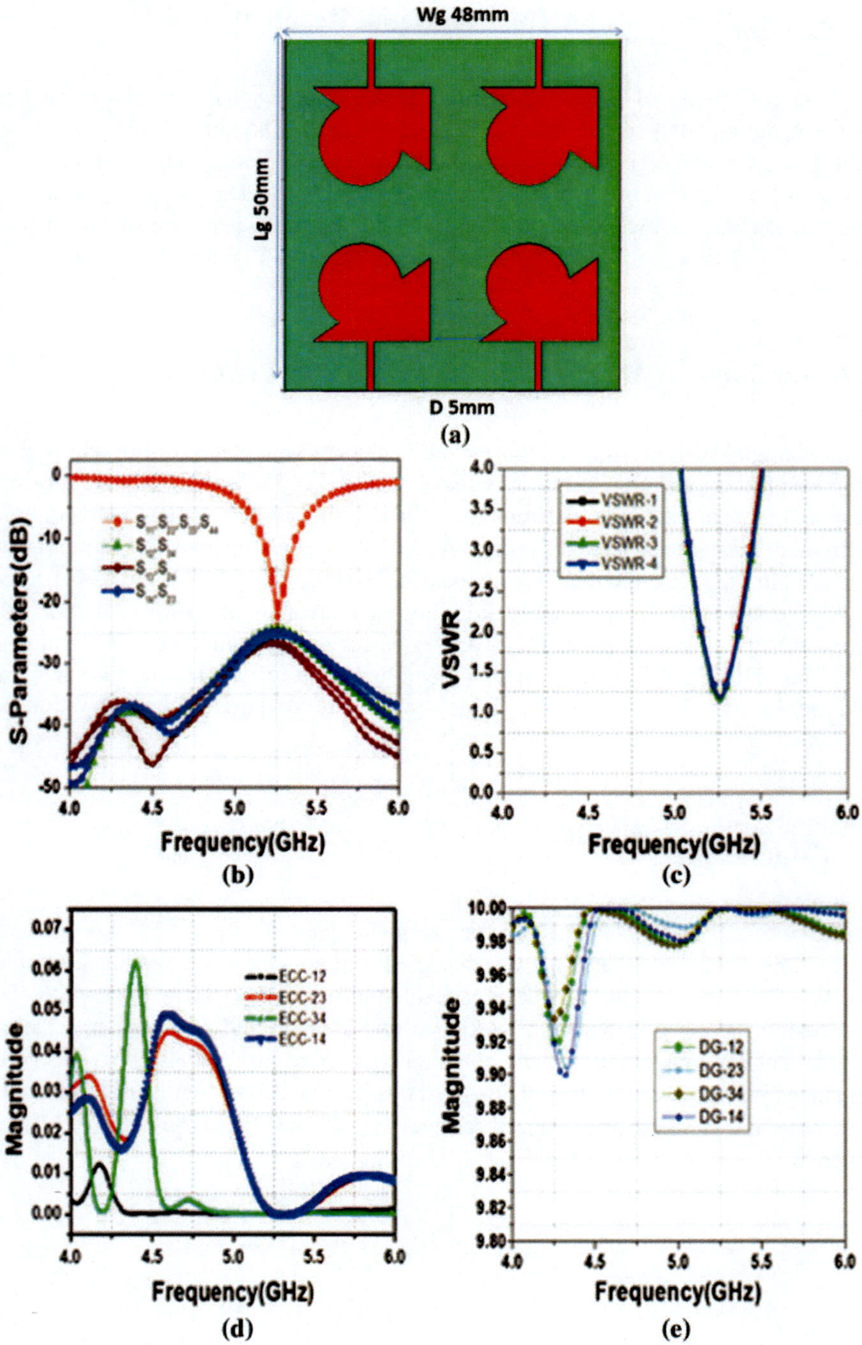

Fig. 6 **a** Space diversity of MIMO antenna structure. **b–c** S-parameters and VSWR **d–e** ECC and DG of the space diversity of MIMO antenna

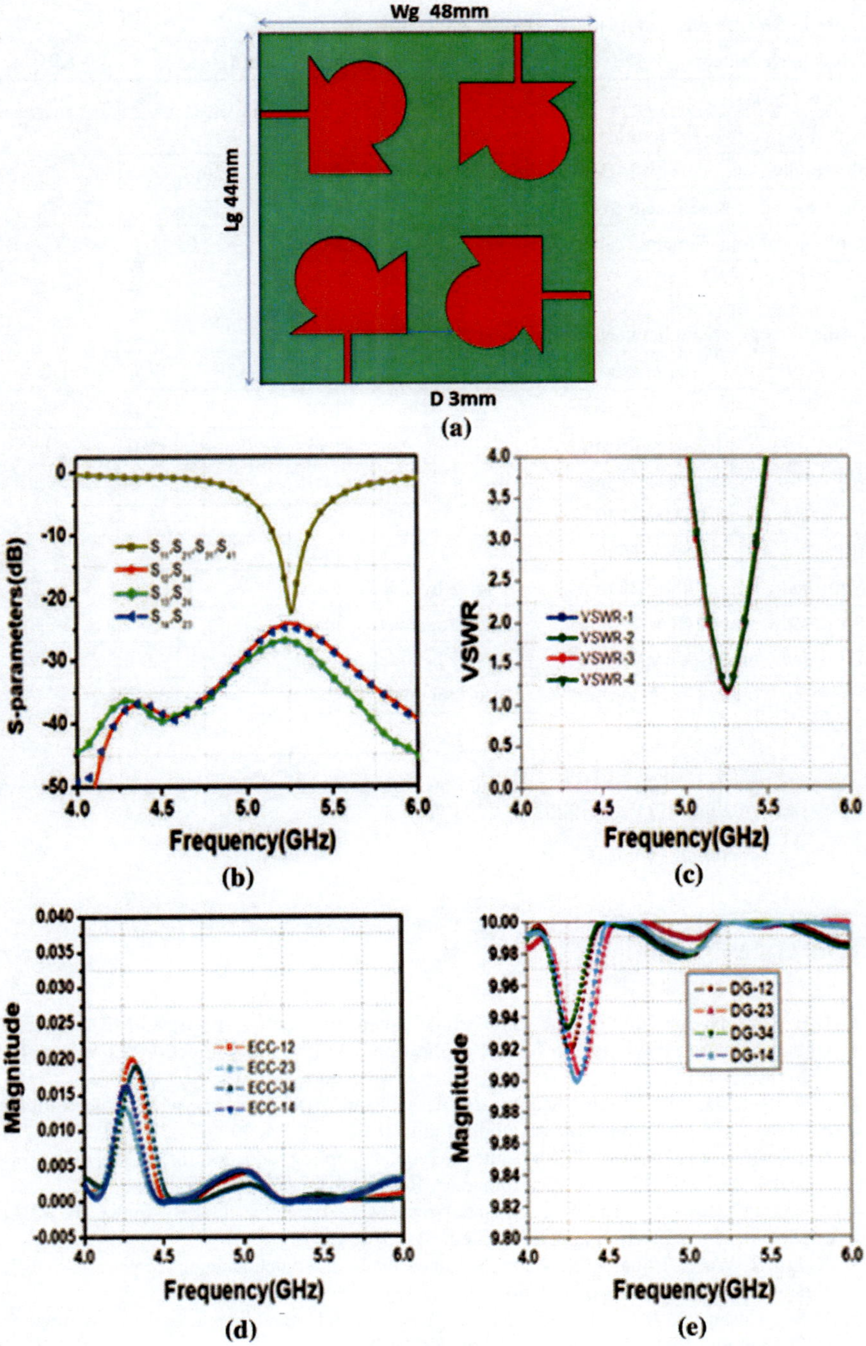

Fig. 7 **a** Orthogonal four-element MIMO antenna. **b–c** S-parameters and VSWR **d–e** ECC and DG of the orthogonal MIMO antenna

Table 1 S-parameters comparison for different configuration

Class of the antenna	S_{11} (dB)	S_{21} (dB)	S_{12} (dB)	S_{22} (dB)
Single element	−29.76	–	–	–
Side-by-side two-element MIMO	−20	−17	−18	−19
Orthogonal two-element MIMO	−22	−29	−26	−22
Side-by-side four-element MIMO	−21.51	−18.75	−34.74	−26.59
Orthogonal four-element MIMO	−22.25	−25.06	−27.01	−23.93

Table 2 Comparison between proposed works with existing work

Ref. antenna	Dimensions	No. of elements	Isolation (dB)	ECC	DG
[4]	38 × 36 mm^2	2	18	0.001	>9.8
[5]	40 × 40 mm^2	4	33	<0.02	>9.95
[6]	31 × 31 mm^2	4	<15	<0.005	>9.97
[7]	130 × 130 mm^2	4	<15	–	–
[8]	30 × 35 mm^2	4	<20	–	–
Proposed work	43 × 25 mm^2	2 (side by side)	=20	0.0002	>9.98
Proposed work	43 × 25 mm^2	2 (orthogonal)	>20	0.0001	>9.98
Proposed work	48 × 44 mm^2	4 (side by side)	<20	0.0025	>9.98
Proposed work	48 × 44 mm^2	4 (orthogonal)	>20	0.0055	>9.98

Acknowledgements This work is supported by DST Science and Engineering Research Board (SERB) with file no: EEQ/2016/000391.

References

1. H.H. Tran, N. Hussain, T.T. LE, Low-profile wideband circularly polarized MIMO antenna with polarization diversity for WLAN applications. AEU – Int. J. Electron. Commun. **108**, 172–180 (2019)
2. A.K. Gangwar, M.S. Alam, A compact size tri-band MIMO antenna with reduced mutual coupling for WLAN and Wi-MAX applications. IEEE-978-1-5090-6674-2/17 (2017)
3. N. Kumar, U.K. Kommuri, MIMO Antenna mutual coupling reduction for WLAN using spiro meander line UC-EBG. Prog Electromagnet. Res. C **80**, 65–77 (2018)
4. X. Zhang, X. Zhong, B. Li, Y. Yu A dual polarized MIMO antenna with EBG for 8 GHz WLAN application. Prog Electromagnet. Res. Lett. **51**, 15–20 (2015)
5. X. Liu, Y. Wu, Z. Zhuang, W. Wang, Y. Liu, A dual-band patch antenna for pattern diversity application. IEEE Access **6**, 51986–51993 (2018)
6. M.M. Hasan, M.R.I. Faruque, M. T. Islam, Dual band metamaterial antenna for LTE/Bluetooth/WiMAX system. Sci. Rep. **8** (2018)
7. R. Kiruthika, T. Shanmuganantham, Comparison of different shapes in microstrip patch antenna for X-band applications, in *International Conference on Emerging Technological Trends (ICETT)* (Kollam, India, 2016), pp. 1–6. https://doi.org/10.1109/icett.2016.7873722 in 2016

8. K. Kavitha, S.P. Rajan, Single band MIMO antenna for WLAN/Wi-MAX application, **04**(01), 686–1689 (2017)
9. Z.Z. Abidin, Y. Ma, R.A. Abd-Alhameed, K.N. Ramli, D. Zhou, M.S. Bin-Melha, J.M. Noras, R. Halliwell, Design of 2x2 U-shape MIMO slot antennas with EBG material for mobile handset applications, in *Progress In Electromagnetics Research Symposium Marrakesh* (2011)
10. R. Kiruthika, T. Shanmuganantham, Comparison of different in microstrip patch antenna for X-band applications, in *International Conference on Emerging Technological Trends (ICETT)*, (Kollam, India, 2016), pp. 1–6. https://doi.org/10.1109/icett.2016.7873722

Usage of KNN, Decision Tree and Random Forest Algorithms in Machine Learning and Performance Analysis with a Comparative Measure

K. Uma Pavan Kumar, Ongole Gandhi, M. Venkata Reddy, and S. V. N. Srinivasu

Abstract In The current article, the process of data science projects and the various statistical and applications of machine learning algorithms were mentioned. The usage of various parameters in the machine learning along with the suitable cases of the study was mentioned in the article. The study mainly focuses on various algorithms like KNN, Naïve Bayes, support vector machine (SVM), decision trees and random forest. The discussion mainly focused on the statistical and mathematical aspects of each algorithm, and suitability of the algorithms to certain use cases and the main drawbacks of the corresponding algorithms were mentioned in the study. The attempt we are making for the sake of summarizing the various algorithms mentioned above, and many articles and research papers were there, but we believe that our work surely give a simple and effective narration of the mentioned algorithms and will help the upcoming researchers as a base to start the research in the machine learning and data science areas. The outcome of the work is explanation of the algorithms in the extensive by covering the pros and cons of each and every aspect along with the suitable use cases. The results we are going to project are a comparison of the same dataset with all these mentioned algorithms and projecting the best performer in that use case.

K. Uma Pavan Kumar
Department of Computer Science and Engineering, Malla Reddy Institute of Technology, Hyderabad 500100, India
e-mail: dr.kethavarapu@gmail.com

O. Gandhi (✉)
Department of Computer Science and Engineering, VFSTR (Deemed to Be University), Guntur, Andhra Pradesh, India
e-mail: ongolegandhi@gmail.com

M. Venkata Reddy · S. V. N. Srinivasu
Department of Computer Science and Engineering, Narasaraopeta Engineering College, Narasaraopet, Andhra Pradesh, India
e-mail: Venkat.medikonda@gmail.com

S. V. N. Srinivasu
e-mail: drsvnsrinivasu@gmail.com

D. Bhattacharyya and N. Thirupathi Rao (eds.), *Machine Intelligence and Soft Computing*, Advances in Intelligent Systems and Computing 1280,
https://doi.org/10.1007/978-981-15-9516-5_39

Keywords Data science · Algorithms · Mathematical aspects · Machine learning · Statistical aspects

1 Introduction

The machine learning and data science words are current industry and academic buzz words, and in fact, the current and future scope of these techniques and technologies are going to generate massive revenue and notable results in various domains like banking, insurance, manufacturing, medical and retail domains [1, 2].

Machine learning and data science are having a clear distinction, as per our perspective these words are subset of the artificial intelligence (AI). AI is a process of improving commonsense reasoning and to upgrade the machines with some logic so as to take the decisions in a wise manner. Any procedure used to improve the smartness of the machines is obviously the context of AI.

Machine learning is a subfield of AI and tries to improve the learning experience of the models based on the input data items without being explicitly programmed. The outcome here is every time the algorithms learn from the data and give better performance when compared with the earlier outcomes.

Data science is another process which involves the usage of machine learning algorithms so as to complete the life cycle of the ML projects. The data science process is having various potentials like loading the input dataset, processing the data to be ready for the model implementation.

The model implementation involves the selection various categories of the algorithms like supervised and unsupervised learning methods. Once the model implementation is done, the training and testing results can be observed, and the performance tuning and hyperparameter tuning can be observed so as to tune the outcome of the model [3, 4].

Once the performance is achieved with suitable tuning of the parameters, the model can be deployed and the client can run the same in the real time scenarios.

The organization of the paper is as follows: In Sect. 2, the ML algorithms discussion with the corresponding statistical analysis was mentioned. In Sect. 3, the algorithmic performance and publication of results can be observed. In Sect. 4, conclusion and future scope were mentioned.

2 Machine Learning Perspective

The section focused on the discussion of various algorithms in the ML perspective.

In traditional programming, the data and program were given as input to the machine, and we expect the outcome, whereas in machine learning, the process is for the machine the inputs are data and output of the use case, and the expected result is the model generated by the machine which can be used a template for the future use cases [5–7].

The ML algorithms can be observed in four categories like supervised learning in which the training data includes desired outputs the examples are *k*-nearest neighbor, Naïve Bayes, decision trees, random forest and support vector machine are few to name.

The unsupervised learning in which training data does not include desired outputs the examples of this kind includes clustering; dimensionality reduction and association rule mining are few to name.

The other category includes semi-supervised algorithms like training data includes only a few desired outputs, and the final category includes reinforcement learning where the sequence of activities follows reward-based outcome.

2.1 KNN Intuitions

The KNN classifiers were defined the unlabeled examples by assigning them to most similar classes. KNN can also be known as instance-based learning or diagnosis and customer satisfaction etc [7, 8].

lazy learning. With the help of various distance measures like Euclidian, hamming, Cosine, etc., we can compute the distance and classify the instance to the appropriate class (Fig. 1).

Some of the applications where we can make use of the KNN are credit rankings, customer churn and image recognition.

The advantages of this KNN are simple and effective, fast training phase. The main problem with this algorithm is more expensive, and it has to store all the training data.

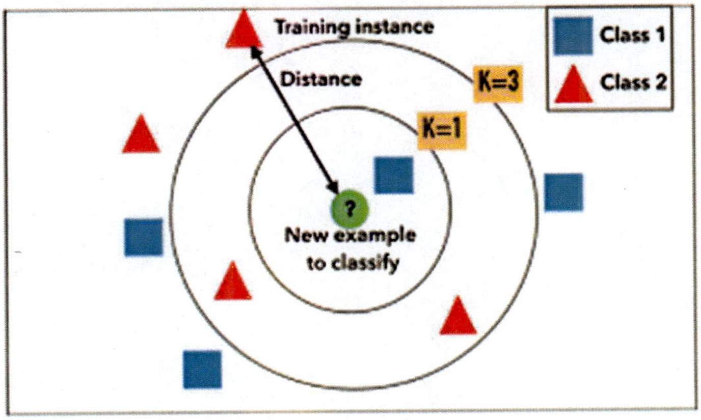

Fig. 1 KNN working model. *Source* SFL Scientific

2.2 Decision Tree Intuitions

The main purpose of this algorithm is can be suitable to classification and regression kind of the problems [9]. With the usage of recursive partitioning and divide and conquer, the process of decision tree generation and implementation of use cases is possible.

With the help of Gini index and Information Gain and chi-square, the generation of the decision tree is possible [10].

The applications involve the credit scoring models, medical condition diagnosis, customer satisfaction, etc. (Fig. 2).

The advantages of this algorithm involve there is no constraint of the data type; the variable importance is a by-product of the methodology. The problems with this

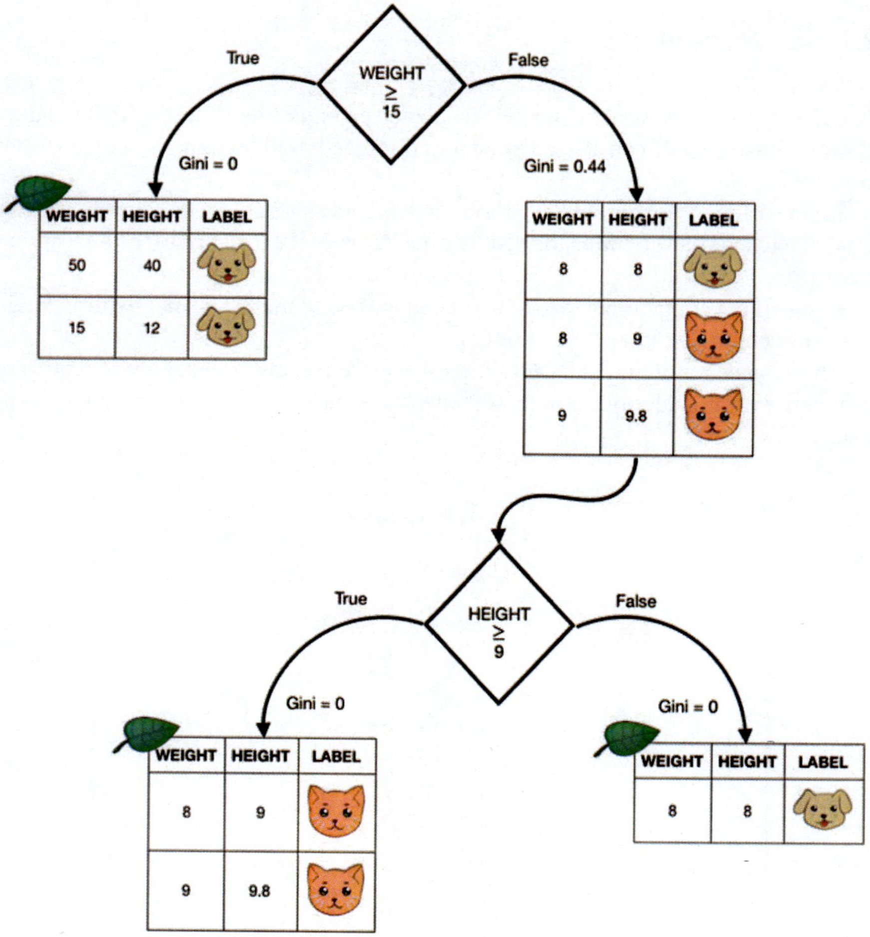

Fig. 2 Decision tree representation. *Source* Towards Data Science

approach are over-fitting like the method by hearts the data rather than learning which leads to performance degrade in the testing phase.

2.3 Random Forest Intuitions

The most helpful algorithm for the data scientist is random forest, without knowing any algorithm one can make use of this algorithm to achieve reasonable performance measures [11–13].

The algorithm is helpful for classification, regression and dimensionality reduction as well (Fig. 3).

This is ensemble learning which bootstraps various decision trees and generates a random forest with best possible performance. This algorithm gives the benefits of applicability with higher dimensions and which handles the problem of over-fitting.

The main issues with this algorithm involve computational cost, best fit for classification but not that much great for regression.

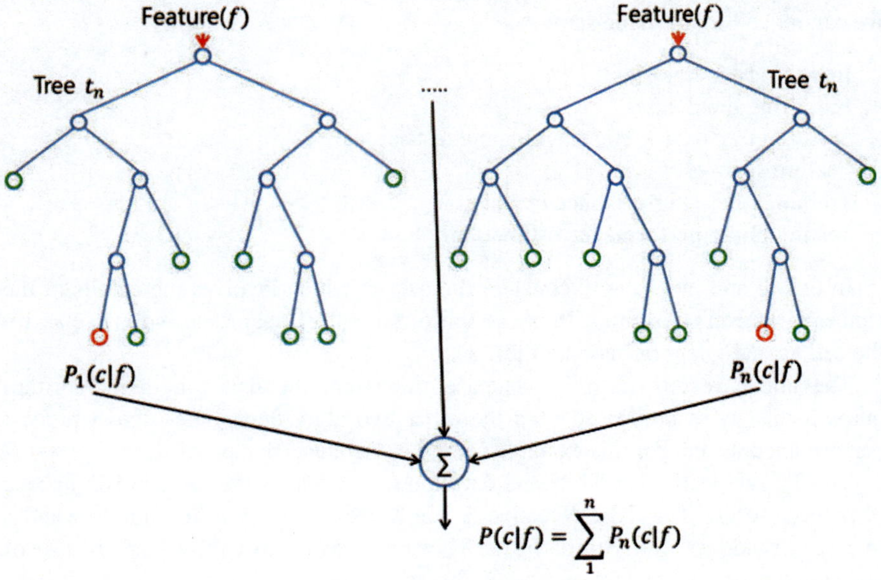

Fig. 3 Random forest working. *Source* Analytics Vidya

Table 1 Performance analysis of KNN, decision tree and random forest

Algorithm name	Accuracy	TP	TN	FP	FN
KNN	**0.947368421**	103	59	4	5
Decision tree	**0.959064327**	103	61	4	3
Random forest	**0.935672514**	102	58	5	6

3 Algorithms Performance Estimation

In Sect. 2, we have observed various algorithms and the background details of each algorithm we have seen. In this section, the performance aspects of each algorithm we are going to project.

The implementation of each algorithm involves the following steps as a template we can make use the same steps.

Importing the libraries
Importing the dataset
Splitting the dataset into training and test set
Feature scaling
Training phase performance estimation
Testing phase performance estimation.

In this article, we have to consider the dataset which involves the details of the cancer parameters like clump thinness, uniformity of cell shape, etc., so as to classify the cell as malignant or benign (Table 1).

The above results for this use case show that the decision tree algorithm outperforms when compared with the other two algorithms. The second highest performance we got for this example is 94.7 percentages in case of KNN.

The TP rate is 103 for KNN and decision tree, whereas the same is 102 in case of random forest. The False Negative is 5 in KNN, 3 in case of decision tree and 6 in case of random forest. Similarly, the TN value is 59 in case of KNN, 61 in case of decision tree and 58 in case of random forest.

4 Conclusion and Future Scope

The article deals with the concepts of machine learning and the context of the ML, with various perspectives. The work focused on the models of the algorithms like KNN, decision trees and random forest.

The work mentioned the application of the algorithms along with pros and cons of the algorithms. The article worked on the data set so as to classify the cancer in the form of malignant or benign. Each and every algorithm is compared with the accuracy and other parameters like True Positive, True Negative, False Positive and False Negative.

The future scope of the work is to work with other algorithms in the classification and as well as regression models and will try to compare the performance of the other algorithms with the parameters which give the analysis of ML algorithms.

References

1. www.Analyticsvidya.com
2. www.datacamp.com
3. Uma Pavan Kumar K (2016) Various computing models in hadoop eco system along with the perspective of analytics using R and machine learning. Int J Comput Sci Inf Secur (IJCSIS) 14:17–23
4. Sudhakar Reddy U, Uma Pavan Kumar K, Priyanka B (2019) Various recommender systems and their usage in the domains of mobile based and online applications 8(11S2)
5. Srikanth T, Nagavarapu S, Uma Pavan Kumar K, Narahari D (2019) Commonly used algorithms in data science along with internal logics and implementations through R programming 8(6S4)
6. ManikyalaRao V, Uma Pavan Kumar K, Priyanka B (2019) Statistical methods in the usage of correlation and regression of the machine learning models 8(2S11)
7. Guyon I (2013) An introduction to variable and feature selection. J Mach Learn Res 1157–1182
8. Chen CP, Zhang C-Y (2014) Data intensive applications, challenges, techniques and technologies: a survey on Big Data. Inf Sci 275:314–347
9. Mashal I, Alsaryrah O, Chung T-Y (2016) Performance evaluation of recommendation algorithms on Internet of Things services. Phys Stat Mech Its Appl 451:646–656
10. Alpaydin E (2014) Introduction to machine learning (adaptive computation and machine learning series). The MIT Press Cambridge
11. Hazelwood K (2017) Applied machine learning at Face book: a datacenter infrastructure perspective. Facebook
12. https://machinearningmastery.com/best-machinelearning-resources-for-getting-started/
13. Uma Pavan Kumar K (2018) Various issues in hadoop distributed file system, map reduce and future research directions. Int J Pure Appl Math 120(6):44414451

Bank Marketing Using Intelligent Targeting

Shaik Subhani, R. Vijaya Kumar Reddy, Subba Rao Peram, and B. Srinivasa Rao

Abstract In the banking system, bank marketing has become an essential continued existence weapon and is basically dynamic in real world. The rise of bank marketing is reused business relations, and most made banks are those who will really strong their relationships with customers. Knowledge modernization and ferocious rivalry in the midst of current banks have altered a large collection of banking facilities. Technology is neutering the relationship amid banks and its inside and outside customers. This dataset will give you the clear targets for marketing depending upon the age of the customers, salaries, duration of the call, etc. The objective of this paper is to assess the value of data in defining marketing strategies and marketing management. The technological advances in recent years offer many opportunities to marketing practitioners and researchers.

Keywords Banking system · Knowledge modernization · Marketing

S. Subhani
Department of IT, Sreenidhi Institute of Science and Technology (Autonomous), Hyderabad, India
e-mail: subhanicse@gmail.com

R. Vijaya Kumar Reddy (✉)
Department of IT, Prasad V. Potluri Siddhartha Institute of Technology, Vijayawada, A.P., India
e-mail: Vijayakumarr285@gmail.com

S. R. Peram
Department of IT, VFSTR Deemed to be University, Vadlamudi, Guntur, India
e-mail: subbarao.peram@gmail.com

B. Srinivasa Rao
Department of IT, Lakireddy Bali Reddy College of Engineering, Mylavaram, A.P., India
e-mail: buragasrinivasarao@gmail.com

D. Bhattacharyya and N. Thirupathi Rao (eds.), *Machine Intelligence and Soft Computing*, Advances in Intelligent Systems and Computing 1280,
https://doi.org/10.1007/978-981-15-9516-5_40

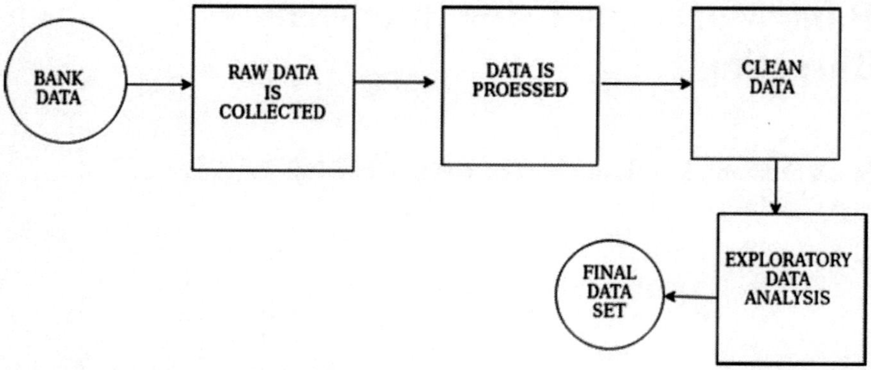

Fig. 1 Architecture of data process

1 Introduction

Banking system is actually an industry that provides varieties of banking and related facilities to its customers. The banking sector provides bank clients excess of funds and those who require a variety of financial and connected services. Those customers fit in to totally dissimilar state of economy, geographical locations and dissimilar profession and businesses. Each individual group of clients is separate from the necessities of dissimilar teams. It is essential to spot totally dissimilar solid teams and even subcluster of clients.

The position of selling within the banking system always changes. Meant for more than a few years, the first spotlight of bank promoting was advertising. Then, the major aim shifted to publicity and marketing. That was followed by target the occasion of a sales background. Though entire weather of promoting conception—client happiness, earnings integrated structure and social accountability—can stay necessary, client satisfaction should receive the best stress within years ahead. The chief considerations of most bank executives still target legal and illegal problems, consistent with most surveys (Fig. 1).

2 Background Work

Facilities offered by the banks to be found in such a way that they complete the requirements of clients. Bankers are familiar with take for granted in conditions of what banks recommend and what customer's desires. The design and deliver client required facilities, the banks should study to hunt data regarding the prevailing and prospective clients, and their apparent and latent desires on an everyday and systematic basis.

The word blueprint sensible promote services have to be bound to be correctly intended and made thus on suit a selected distinct cluster of clients. Such properly designed facilities should be listed. The customary of discharge is to be ensuring not solely from side-to-side targeted packaging, however additionally through correct client services obtainable at the bank.

However, facing abundance in data supply makes marketing analytics often similar to finding the famous needle in a haystack. The first section focuses on the challenges that come with Big Data. The author analyzes the relevance of a selection of data types: search data, user-generated content, social network data, location and sensory data, click stream data and web browsing behavior, and CRM data [1]. Section 3 uses the 4Ps of the marketing mix as a framework. For product and place, the author presents real business cases. The subsection for price will critically assess the possibility of price discrimination. Regarding promotion, the author will provide examples for how companies can use data indirectly through advertising networks to target customers [2, 3]. The overall result is presented in a table that connects data types and the 4Ps.

All bank marketing campaigns are dependent on customers' huge electronic data. The size of these data sources is impossible for a human analyst to come up with interesting information that will help in the decision-making process. Data mining models are completely helping in the performance of these campaigns. This paper introduces analysis and applications of the most important techniques in data mining; multilayer perception neural network, tree augmented Naïve Bayes known as Bayesian networks, nominal regression or logistic regression, and Ross Quinlan new decision tree model (C5.0). The objective is to examine the performance of MLPNN, TAN, LR and C5.0 techniques on a real-world data of bank deposit subscription. The purpose is increasing the campaign effectiveness by identifying the main characteristics that affect a success based on MLPNN, TAN, LR and C5.0 [4]. The experimental results demonstrate, with higher accuracies, the success of these models in predicting the best campaign contact with the clients for subscribing deposit. The performances are calculated by three statistical measures: classification accuracy, sensitivity and specificity.

3 Objectives of the Bank Sector

The objectives of the bank are to be found at different intervals. The company objectives are two types; they are short and long terms.

The short-period objectives may be

1. Growing gain of the bank year by year.
2. Spreading client base by providing novel facilities.
3. Enhancing rate of growth credit year by year.

The long-period objectives might be

1. To go up to special location in five years.
2. To turn out to be in the bank over the amount of next three years.

In general, the environmental fundamentals fall under four categories: political, economic, legal and cultural.

3.1 *Marketing ideas regarding Banking Applications*

In banking system, when we apply marketing then the marketing strategy will be mentioned to incorporate the following things [5, 6]:

- To definition target customers properly.
- Satisfy the customers at a profit for the bank.
- Design proper planning for supply markets.
- Perfect organization and administration.

3.2 *In Banking Sector: Marketing Strategy*

- Customer behavior and segmentation.
- Identification of target customers and their requirements.
- Find out the information such as the present age, sex, educational background, locality and purchase habits of the present available customers [7, 8].
- Taking a customer feedback for every few months, regarding the services provided.
- Collecting information from people who are not the actual customers of the bank.

4 Bank Dataset Preparation

Five stages of bank dataset preparation

1. In the first stage, the bank data is collected from various sources and stored in the form of a CSV file [9, 10].
2. In the second stage, data is processed.
3. In the third stage, the missing values and the improper values are detected and necessary corrections are made.
4. In the third stage, the data which is now cleaned will be processed in the form of a file and is ready to use for analysis.
5. In the fourth stage, the analyzed data is put into different blocks according to the required attributes. Now, this data is further used to make plans for strategic marketing schemes.

5 Results and Analysis

These processes done by companies create value for customers and construct strong customer relations in order to imprison cost from customers in return. Marketing campaigns are characterized by concentrate on the customer requirements and their overall approval. There are four most significant variables that should be considered when planning for a marketing campaign.

1. Section of the population: To which segment of people will this marketing campaign be useful for? This is the most important segment, which would be the only possible way to reach out to the customers.
2. Distributing the channel to reach out to the customer's place: Which segment of the population should we address? Which mode of communication should we use to get our message out? (e.g., telephones, radio, TV, social media).
3. Price: What would be the best price that we can offer to potential clients?
4. Promotional strategy: What strategy are we going to implement in order to target the potential clients? This should be made the last part of the marketing campaign analysis since there has to be an in depth analysis made available from the previous marketing campaign (Fig. 2).

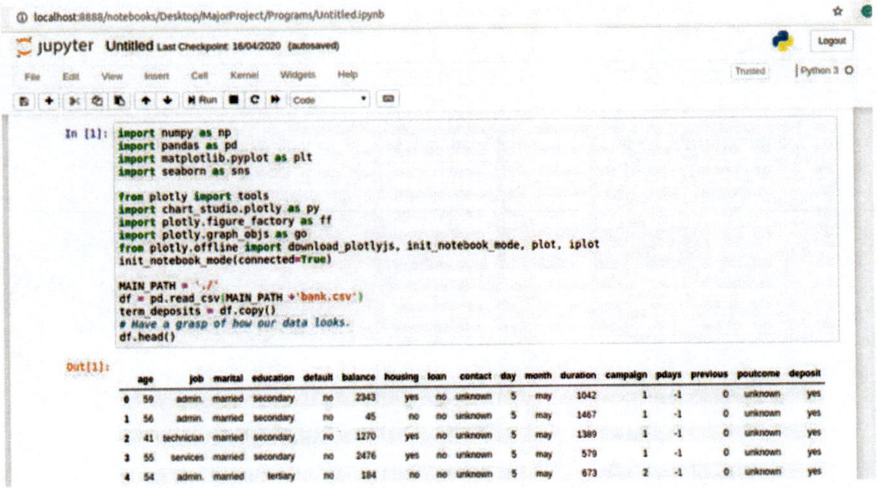

Fig. 2 Importing and checking the collected data

6 Summary on Distribution of Data Based upon Various Attributes Available

- Mean Age ≈ 42 years. (Min = 20 years and Max = 90 years old.)
- The mean balance is 1,536. The standard deviation is high. So we can understand the balance is heavily distributed across the dataset.
- Duration is obtained after the call is made to the potential client so if the target client has by no means acknowledged calls, this characteristic might not that useful (Figs. 3, 4, 5 and 6).

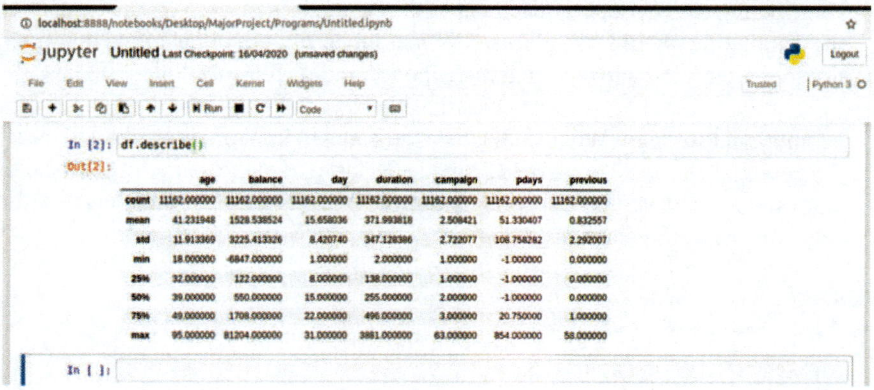

	age	balance	day	duration	campaign	pdays	previous
count	11162.000000	11162.000000	11162.000000	11162.000000	11162.000000	11162.000000	11162.000000
mean	41.231948	1528.538524	15.658036	371.993818	2.508421	51.330407	0.832557
std	11.913369	3225.413326	8.420740	347.128386	2.722077	108.758282	2.292007
min	18.000000	-6847.000000	1.000000	2.000000	1.000000	-1.000000	0.000000
25%	32.000000	122.000000	8.000000	138.000000	1.000000	-1.000000	0.000000
50%	39.000000	550.000000	15.000000	255.000000	2.000000	-1.000000	0.000000
75%	49.000000	1708.000000	22.000000	496.000000	3.000000	20.750000	1.000000
max	95.000000	81204.000000	31.000000	3881.000000	63.000000	854.000000	58.000000

Fig. 3 Checking attributes available in the loaded data

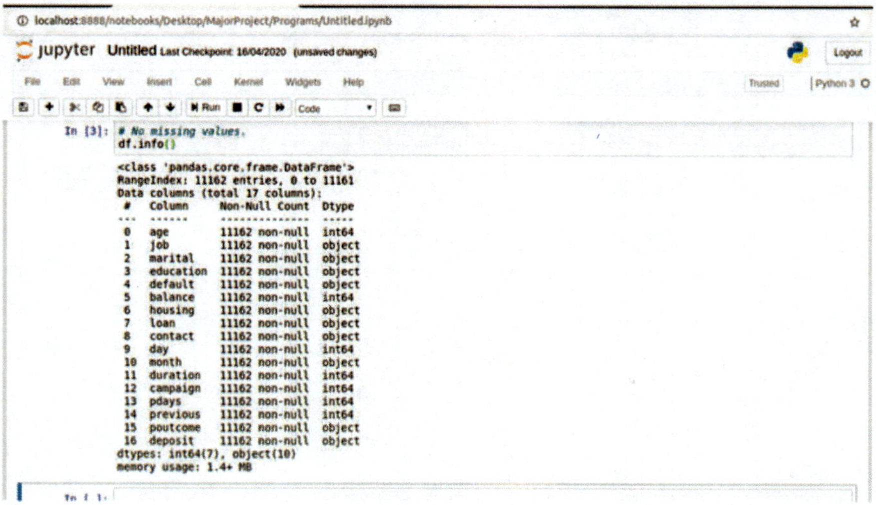


```
In [3]: # No missing values.
        df.info()

<class 'pandas.core.frame.DataFrame'>
RangeIndex: 11162 entries, 0 to 11161
Data columns (total 17 columns):
 #   Column     Non-Null Count  Dtype
---  ------     --------------  -----
 0   age        11162 non-null  int64
 1   job        11162 non-null  object
 2   marital    11162 non-null  object
 3   education  11162 non-null  object
 4   default    11162 non-null  object
 5   balance    11162 non-null  int64
 6   housing    11162 non-null  object
 7   loan       11162 non-null  object
 8   contact    11162 non-null  object
 9   day        11162 non-null  int64
 10  month      11162 non-null  object
 11  duration   11162 non-null  int64
 12  campaign   11162 non-null  int64
 13  pdays      11162 non-null  int64
 14  previous   11162 non-null  int64
 15  poutcome   11162 non-null  object
 16  deposit    11162 non-null  object
dtypes: int64(7), object(10)
memory usage: 1.4+ MB
```

Fig. 4 Checking for any missing values

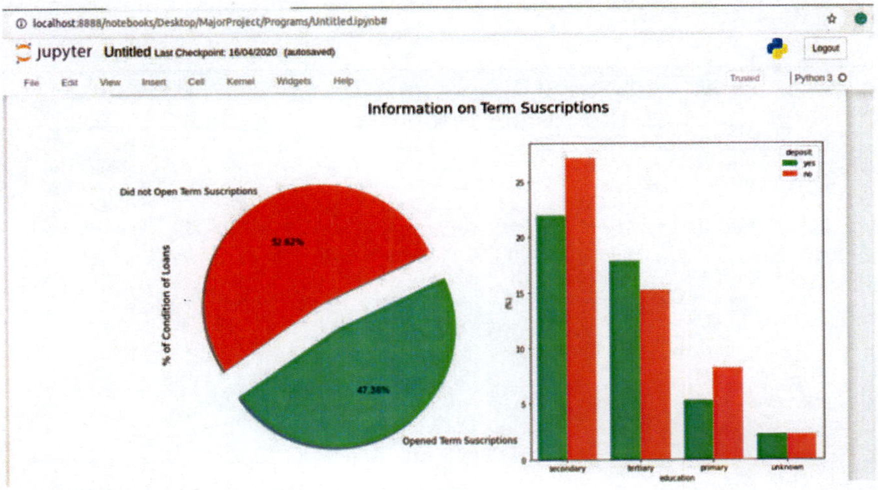

Fig. 5 Result of the information on term subscriptions

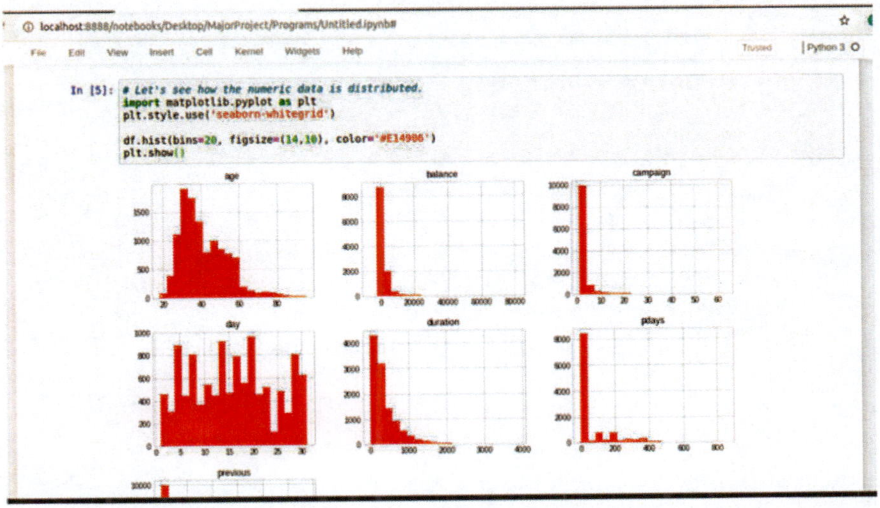

Fig. 6 Checking how attributes are distributed in the loaded data

Analysis by Profession

- **Number:** Manage the occupation that is additional prevalent in this dataset.
- **Age:** The retired age is the highest median age while student are the lowest.
- **Balance:** Manage and retirees are the highest balance in their accounts (Fig. 7).

Analysis by Marital Status

The effect of a divorce has on the balance of the individual (Fig. 8).

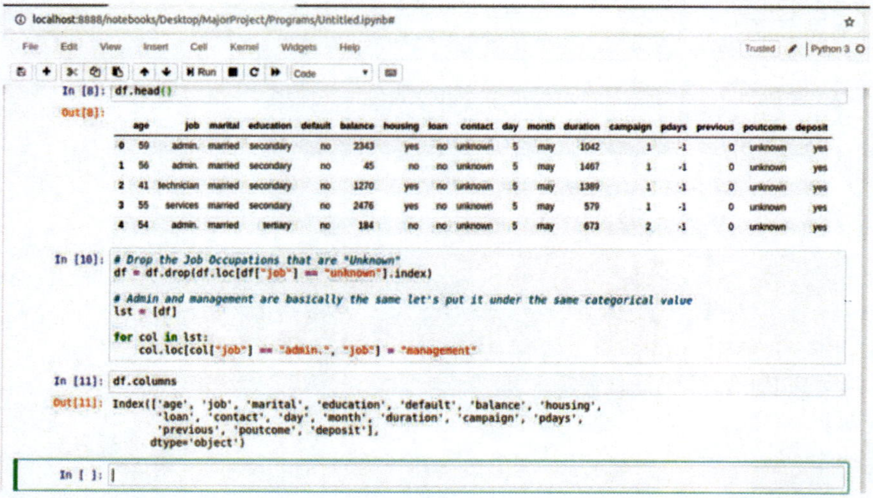

Fig. 7 Distribution of data based upon the term job occupation

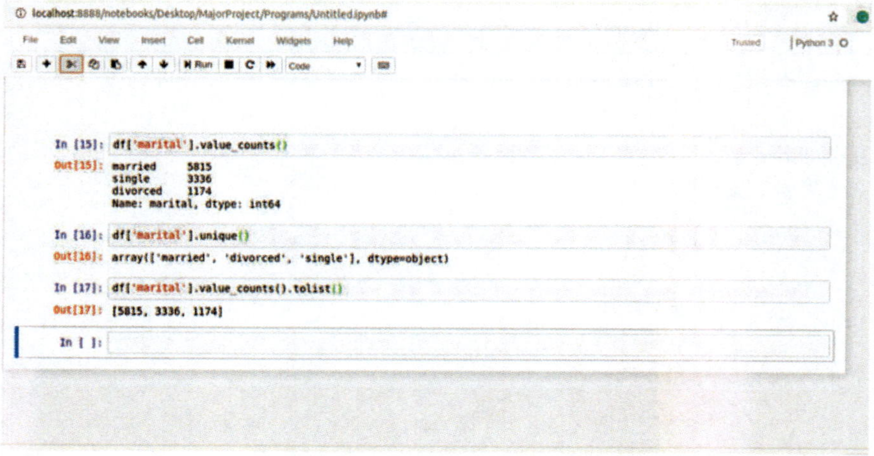

Fig. 8 Getting analyses by marital status

Analysis by Education
The level of education also has a important shock on the quantity of balance a prospect (Fig. 9).

Analysis by Loans
Whether the view has a preceding loan has a significant force on the extent of balance (Fig. 10).

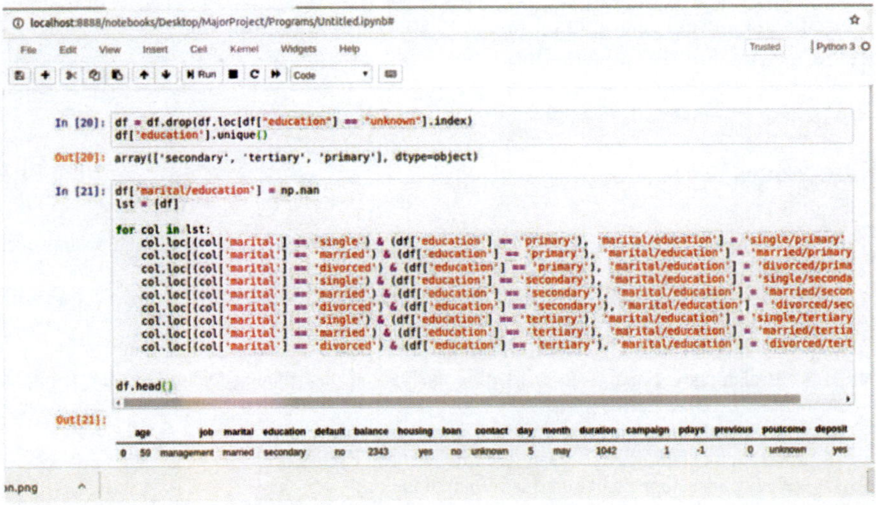

Fig. 9 Analyzing data based upon educational and marital group

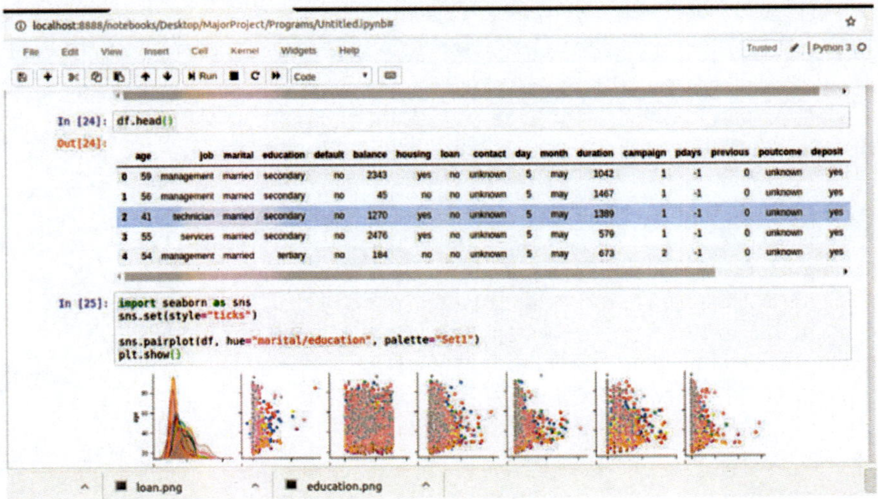

Fig. 10 Program for analysis of data by loans data and related plot graph

7 Conclusion

In banking, sector marketing is important weapon, and it is fundamentally dynamic in real world. The development of bank marketing is reused business relation and is made banks are really strong their relationships with customers. Knowledge modernization and ferocious competition among existing banks have altered a large collection of banking services. Technology is neutering the relationship amid banks and its internal and external customers. Finally, in analysis of banking data, we can generate accurate type of useful information.

References

1. https://searchcustomerexperience.techtarget.com/definition/CRM-customer-relationship-management
2. https://cloudblogs.microsoft.com/industry-blog/financial-services/2017/12/07/the-top-five-things-a-customer-needs-from-their-bank/
3. https://thefinancialbrand.com/88515/banking-product-development-design/
4. H.A. Elsalamoney, Bank direct marketing analysis of data mining techniques. Int. Comput. Appl. **85**(7), 12–22 (2014)
5. https://academia.edu/documents/in/Bank_marketing
6. https://academia.edu/upgrade?feature=search&form_navbar=true
7. https://www.investopedia.com/terms/m/marketing.asp
8. https://m.infoentrepreneurs.org/en/guides/customer-relationship-management/
9. https://www.kaggle.com/janiobachmann/bank-marketing-dataset
10. https://www.kaggle.com/

Machine Learning Application in the Hybrid Optical Wireless Networks

Deepa Naik and Tanmay De

Abstract Complex problems invariably involve big data. Machine learning (ML) deals with this big data. Machine learning does classification, regression and decision making better than human beings. Large data produced by communication networks can be utilized for deriving efficient optimized solutions. Network performance improves by usage of ML techniques for network design and management. In this article, we have summarized advanced ML techniques for network domain. Further, a survey of ML techniques in several network design problems done, and their performance is discussed.

Keywords Deep learning · Deep packet inspection (DPI) · Machine learning (ML) · Fuzzy c-mean clustering (CCM)

1 Introduction

In communication networks, each application has its own QoS requirements. The traditional algorithms are unable to deliver complex network performance parameters. Machine learning helps to build powerful algorithms and frameworks to handle the complex network problems and scenarios. Self-learning models built from ML make decision directly from the input data without any predefined rules. Traffic classification and prediction by ML help in intrusion detection and performance prediction [1], network and parameter adaption [2, 3].

D. Naik (✉) · T. De
Department of CSE, National Institute of Technology, Durgapur, India
e-mail: naiksavantdeepa@gmail.com

D. Bhattacharyya and N. Thirupathi Rao (eds.), *Machine Intelligence and Soft Computing*, Advances in Intelligent Systems and Computing 1280,
https://doi.org/10.1007/978-981-15-9516-5_41

1.1 Motivation and Contribution

Hybrid network consists of high bandwidth optical and high coverage wireless networks such as WiMax and LTE. The main objective is to study how machine learning mechanism can perform better than traditional algorithms in terms of traffic prediction, path computation, traffic flow classification and network resource management.

Traffic prediction: Accurate traffic prediction results in effective planning, operation and maintenance of networks. Traffic prediction during design phase may not be accurate as there may be under or overestimation of traffic and subsequent network loading or wastage of resources. The effective traffic prediction can optimize resource allocation based on the real-time data. Models based on ML techniques are trained to predict future traffic using historical data. Performance of these traffic prediction and resource allocation algorithms found to be better as compared to traditional algorithms.

Path computation: Incoming traffic request is assigned a proper route and allotted network resources fulfilling some quality of service (QoS) parameters without affecting the presently executed traffic requests. The traditional route computing algorithms are based on pre-computed metrics like distance, energy and delay. These are Dijkstra, Yen and Bellman–Ford algorithms. Contrary to this, the machine learning algorithms find optimized routes using real-time data about incoming requests and state of network.

Traffic flow classification: In the hybrid network, a traditional traffic classifying algorithms are unable to handle large amount of traffic flows. They may end up with under or overutilization of network resource. Machine learning algorithms classify traffic based on information extracted from data packets. Moreover, accurate flow classification results in less congestion and better QoS parameters.

Failure management: The network operators handle the network failure by rerouting the traffic over unaffected route. This involves failure detection and failure localization which enable recovery procedure and assign resources without affecting the existing running traffic requests. Continuous network activity generates huge data in network monitors and also raises certain alarms at predefined time intervals. ML helps in predicting network failures and in reducing mean time to repair (MTTR) by utilizing this derived information. Trained ML classifier algorithms recognize degraded transmissions and detect failure causes by exploiting available data on historical network failures.

This study provides information about how the network layers use machine learning tools for optical and wireless networks. Also, it discusses relevance and usage of machine learning techniques for future generation networks.

2 Previous Work

The different areas of the networking where machine learning can be used are traffic classification, network performance and congestion control, and traffic scheduling which are discussed below.

2.1 Traffic Classification

Traffic classification categorizes network traffic into several classes of traffic depending on a number of parameters (by port number or protocol, for example) [4]. Further using predetermined policies and protocols, certain QoS is guaranteed. The network classification is also helpful in detecting the intrusion [5].

2.1.1 Payload-Based Traffic Classification

Here, payload inspection techniques are used, and they are further categorized into four types.

Packet-Based No State (PBNS): In this simple payload checking techniques, parameters like port numbers are used for traffic classification. Its burden on the CPU is less. As it is per-packet-based and not application session-based, it is not always accurate.

Packet-Based per Flow State (PBFS): Flows form basis for this method. Packet stream between source destinations is called flow. All sessions are tracked using a table of five tuples (addresses and ports of source and destination, transport protocol). Multiple packets form a flow. As such, when a first packet is put in a particular class, all following packets of the flow are put in the same class.

MBFS: It is similar to PBFS but depends on messages. It needs TCP normalizer for IP fragments and TCP segments to operate on messages. As entire messages are taken into account, memory requirements increase.

Message-Based per Protocol State (MBPS): This is the method, which tracks both application and its contents. It requires status of the protocol and state of machine. Here, both CPU load and memory needed are increased. Reliability of Layer 4 port mechanism is poor as some applications use other ports.

2.1.2 Deep Packet Inspection (DPI)

DPI results in proper classification and sub-classification. For this, signature analysis is used. Each application has unique patterns called signature. A reference database of all applications is created after studying their signatures. Traffic is compared against this reference for classification purpose. Signature analysis methods depend on analysis of patterns, numerical, behavior, heuristics and protocol/states.

Pattern Analysis: Bytes/characters/string are embedded in applications in a particular pattern. Classification engines use the patterns in packet payloads. Pattern position in packets is application dependent.

Behavior and Heuristic Analysis: Traffic behavior analysis helps in understanding running applications. Heuristic analysis of the inspected packets results in classification of underlying protocols.

State or Protocol Analysis: Protocols contain a certain sequence of steps. Confirmation of GET request by a server in file transfer protocol to its client can be used for classification. For providing efficient classification, all these methods are used together. Here, all protocols have to embed with special pattern. The Packet-Based No State (PBNS) detects metrics such as port numbers. This is not accurate but results in small CPU overhead.

2.1.3 Advanced Traffic Classification Algorithms

The fundamental requirement of network management is security. Traditionally, traffic is classified using port and payload-based algorithms. Port-based techniques fail when there is a dynamic port assignment. When the traffic is encrypted, payload-based algorithms fail. Further, they suffer from security-related issues while deep inspecting the packets. As a result, these techniques are unable to solve the real-world problems. This leads to innovation in machine learning to cope up to solve the real-world problems. Most of the studies concerned with the well-known traffic classification form supervised machine learning techniques. In order to solve the zero-day traffic [6] classification, more supervised and unsupervised machine learning needs to be preferred in order to accurately classify the traffic.

[3, 7, 8] implemented deep learning techniques to control the traffic system. Traffic pattern from all routers are considered by the algorithm, and the next nodes in the routing path are predicted.

ML algorithms can accurately estimate the traffic volume. In [9] with information of network packets as input traffic is predicted using learning model.

This prediction will help in reducing the network congestion and also achieves the load balancing in the network. Similar work is found in [10] which predicts the network traffic according to the relation between the network flow and the network traffic volume.

In heterogeneous scenarios to avoid interference and to achieve higher energy, efficiency clustering is done. In [11], the authors considered a hybrid optical/wireless network. Here, wireless telecommunication traffic is reduced by utilizing high-capacity optical network. K-mean clustering algorithms were used to form a cluster of multiple access points depending on the distance from the base station.

MBFS operates on messages while Packet-Based per Flow State packets. As MBFS operates on the message, it requires the TCP normalization to handle the Internet protocol fragments and transmission control protocol segments. MBFS tracks both running of application and its contents. As such, protocol running information is necessary for classification. It consumes a huge memory and CPU time. The PBFS, MBFS and MBPS use advanced, deep packet inspection-based classification techniques in order to overcome this drawback.

Deep Packet Classification: Deep analysis of packets uses information available on Layers 3 and 4. For proper classification and sub-classification of the network traffic, the DPI needs to inspect each and every packet. DPI uses signature analysis to understand and verify different applications. A reference database is prepared after studying unique characteristics of each case. Exact applications are identified after comparing traffic against this reference. As new applications arrive and existing protocols evolve, these references are automatically updated. Methods based on analysis of patterns, behavior, heuristics, protocols and numerical research are used.

In some applications, there is a pattern of bytes/characters/string in packet payload. These patterns may vary depending on the application. Not all protocols do embed special patterns, and hence, pattern analysis is unable to classify the packets without these patterns. To overcome this drawback, numerical classification is preferred. In this method, the numerical characteristics of packets are used.

Behavior and heuristic analysis classification is based on the behavior of the application. Viruses and worms are detected using these techniques.

None of the abovementioned classification algorithms are able to classify the encrypted network traffic. In order to handle the encrypted packets, we need more intelligent and advanced classification algorithms.

2.1.4 Machine Learning in Network Resource Allocation

With the tremendous growth in Internet users, network operators are facing lots of challenges in handling the live streaming, network security issues for network applications [1]. As networks become complex network, characteristics and user demands also vary. For these different network scenarios, specific algorithms are built. The traditional algorithms are unable to handle the large networks applications in a specific time domain. Recently, machine learning algorithms played a vital role in dealing complex and large networks.

Efficient resource management, routing and congestion controls play a vital role in improving network system performance [12].

Deep reinforcement learning [13] was applied first time as an algorithm for resource scheduling. The trained model was applied to schedule the network resources to newly arrived jobs.

The machine learning algorithms can be applied to effective resource management and congestion control. The traditional dynamic bandwidth allocation (DBA) [14] algorithms used constant and linear credit bandwidth scheduling algorithms which improved the delay requirements of sensitive low traffic. In [15], traffic burst scheduling was done by Bayesian methods. But during peak hours, there were fluctuations in upstream delay. Another machine learning algorithm called k-nearest neighborhood schedules the bandwidth. However, this algorithm performs best for video traffic. In Internet traffic, there are variations in packet length. Hence, the existing predictive dynamic bandwidth assignment (DBA) algorithms suffer from over and under prediction. Former results in longer polling cycles and later leads to higher packets queuing delay. For this reason, [16] proposed ANN-based DBA algorithm. Artificial neural network was found to be the best candidate to predict the bandwidth demands for busty ON/OFF period. Compared to K-NN, it better minimizes the upstream delay and packets dropped ratio in the optical and wireless integrated networks.

Based on network state, undetermined parameters are assigned adaptable in reinforcement [17]. Louta et al. [18] WiMAX adopts the OFDMA techniques, for increasing service quality parameters of the mobile application. Jafari et al. [19] used reinforcement learning algorithms to meet the delay requirement of the sensitive traffic classes. It is found that proposed algorithm is the best solution for nrtPS and rtPS traffic classes. In [20, 21], the reinforcement learning model was applied to schedule the network resources for heterogeneous network with the objective to improve the network performance. These models are suitable for estimating the radio parameters which are associated with particular users. MIMO consists of large number of antennas. Detecting channel and estimating results in a huge search problem. This is dealt with by the SVM machine learning algorithm to estimate the channel noise condition [22]. The ML is used to predict the path loss, for wireless network based on publicly available data sets [23]. Path prediction model in Timoteo et al. [24] identifies acceptable QoS level for wireless network user.

The self-organizing network (SON) solution is incorporated with the LTE to minimize the lower packet drop ratio and dynamically handle the unpredictable traffic. In SON, ML techniques like bagging and boosted tree, SVM and linear regression are used for getting information of path loss, throughput and bandwidth allocated. Further learning-based approaches adjust these parameters to predict the performance. This increases the capacity of networks [25]. Reinforcement learning is used in heterogeneous networks for parametric tuning [20].

In the Bayesian decider [26], there is a tradeoff between delay requirements of the services and resource utilization. Here, already established connections are considered. Presently, existing connections are considered till QoS deteriorates certain level. A new light path is established to remove congestion. The author in [27] applied deep machine learning algorithms to address traffic scheduling, routing problems in software-defined routers (SDRs) with the objective to minimize delays and network overhead.

Machine learning is used in optical networks for optical signal transmission in the work presented in [28]. Machine learning shows flexibility in splitting the optical resources in the OFDM-PON architecture, based on predicted traffic requirements of heterogeneous services. Edge devices process real-time data, and using ML mitigates network congestion. They transmit valuable knowledge to base stations using its ability to analyze and make decisions [29]. The animal-human cohabitation problem was solved by applying machine learning algorithms to fiber-wireless access networks [30]. Path computation.

Route computation depends on various physical and network parameters. For machine learning, most common physical parameter used is modulation format and optical signal-to-noise ratio. Path and wavelengths are optimally selected for optical burst switching networks in [31]. This reduced probability of burst loss. In cognitive [32], traffic queries, set of modulation formats, optical noise-signal ratio, bit error rate and length of light path, set of modulation formats are considered. Similar path computation approach is found in [33, 34], whereas a ML-based monitoring was applied for software-defined network to control reconfigurable optical add/drop multiplexer (ROADM) controller.

The author in cite Song et al. [35] applied the K-means unsupervised clustering algorithm for selecting an efficient relay station for urban vehicular networks. Here, model learns device distribution, and an optimum relay station is chosen.

2.2 *Placement of Network Devices*

Base stations are placed using a K-NN ML in [36]. Unsupervised soft clustering in heterogeneous cellular networks locates the base station and relay nodes within a cell. This minimizes network latency time [37].

2.3 *Failure Handling*

ML techniques like Bayesian models are used to identify failures and their causes in gigabit passive optical networks [38] and [39]. Here, exact location and specific nature of failure can be identified [40]. In [41], ML algorithm needs the historical data as input (peak BER, threshold BER), and monitors BER of the arrived light path and detects any modification in BER. If any changes are observed, then this is considered as a failure of the network components along observed light path. In [42–44], the similar work found which identify the network failure and diagnosis the cause of failure using machine learning algorithms.

2.4 *Intrusion Detection*

Anomaly is appreciable deviation in traffic demand at a particular instance. Hierarchical clustering algorithms are proposed for identification of such anomalies Parwez et al. [45].

3 Challenges

The challenges faced by researcher during each step involved in the machine models are discussed below.

Problem Formulation: The initial step in the machine learning is the training process. The main thurst is to clearly formulate the problem. The training phase of the machine learning consumes lots of time. A problem may be of clustering, classification and decision making. This helps to decide the type of the data collected for learning the model.

Data Collection: The main challenge faced by the new research community is getting access to real-time data related to networking field. The machine learning with simulator is cheaper than working with the real-time data.

Data Analysis: In the networking field, every network-related problem has several parameters to impact the performance of the network. So, the best feature selection algorithms need to be identified that are the major contributor in effecting the network performance. For example, round trip time (RTT) and inter-arrival time of acknowledgment (ACK) are considered for determining the best possible size of the transmission control protocol (TCP) and congestion window [3]. Extracting the best features from given data requires deep knowledge which is difficult and time-consuming task.

Validation: Once the model is built, the correctness of the built model needs to be checked. The model may be under-fitting or over-fitting. When the model is over-fitting, it is needed to increase the data volume and reducing model complexity. Analyzing wrong data samples helps find the errors in the model.

The assumption in ML algorithms that the data flow follows the similar distribution may not be practically feasible in networking domain.

4 Future Directions

The cognitive networks are used neural networks for spectral white state estimation [46], hand-off decision [47] and prediction [48]. Yet, these algorithms are basic level machine learning algorithms. The genetic algorithm solves resource allocation problems in HetNets [49].

In [50], the authors used machine learning-based evolutionary algorithms for energy usage and demand prediction for smart grids. Using them, network operators can infer locations and activities of smartphone users which can help network operators to improve their efficiency with its infrastructure.

Furthermore, machine learning algorithms pose a major obstacle in realistic implementation, since many learning models assume that data follows some path and specific patterns, particularly for in-depth learning, and the researcher does not know why and how it works.

Time period over which data to be observed and patterns to be found varies depending on applications. Estimating the correct timescale for monitoring the parameters is important factor to optimize the accuracy algorithm. ML works on huge data and frequent data to produce accuracy. Traffic applications which are sensitive may be hindered by ML.

Access to real data for faulty operational conditions is difficult. The networks are typically managed via reserving the network resources while planning the network which makes failure possibility is negligible.

5 Conclusion

The machine learning algorithms can be applied in solving complex real-world problems. This article highlights the role of machine learning in solving the complex issues related to networking domain which includes the scheduling, traffic prediction and congestion control. We have discussed how ML can solve the complex problems of optical and wireless networks. The ML can be used to solve the network operator's issue while deploying the network infrastructures. In near future, trained models in base station may very well learn what type of content users may need and accordingly they may provide network resources. Trained models may adjust bandwidth of up/downlink frames according to network conditions. The future direction and challenges faced by the research community while constructing the ML models are also discussed. These can help the new researcher find the future research direction in the field of ML.

References

1. Y. Sun, X. Yin, J. Jiang, V. Sekar, F. Lin, N. Wang, T. Liu, B. Sinopoli, Cs2p: Improving video bitrate selection and adaptation with data-driven throughput prediction, in *Proceedings of the 2016 ACM SIGCOMM Conference*, pp. 272–285 (2016)
2. M. Dong, Q. Li, D. Zarchy, P.B. Godfrey, M. Schapira, fPCCg: re-architecting congestion control for consistent high performance, in *12th fUSENIXg Symposium on Networked Systems Design and Implementation (fNSDIg 15)*, pp. 395–408 (2015)
3. K. Winstein, H. Balakrishnan, Tcp ex machina: computer-generated congestion control. ACM SIGCOMM Comput. Commun. Rev. **43**(4), 123–134 (2013)
4. J. Zhang, C. Chen, Y. Xiang, W. Zhou, A.V. Vasilakos, An effective network traffic classification method with unknown flow detection. IEEE Trans. Netw. Serv. Manage. **10**(2), 133–147 (2013)
5. P.G. Bringas, Network traffic analysis for android malware detection, in *Hybrid Artificial Intelligent Systems: 14th International Conference, HAIS 2019*, Leon, Spain, September 4–6, 2019, Proceedings, vol. 11734 (Springer Nature, 2019), p. 468
6. J. Zhang, X. Chen, Y. Xiang, W. Zhou, J. Wu, Robust network traffic classification. IEEE/ACM Trans. Networking **23**(4), 1257–1270 (2014)

7. N. Kato, Z.M. Fadlullah, B. Mao, F. Tang, O. Akashi, T. Inoue, K. Mizutani, The deep learning vision for heterogeneous network traffic control: Proposal, challenges, and future perspective. IEEE Wirel. Commun. **24**(3), 146–153 (2016)
8. Z. Chen, J. Wen, Y. Geng, Predicting future traffic using hidden Markov models, in *2016 IEEE 24th International Conference on Network Protocols (ICNP)* (IEEE, 2016), pp. 1–6
9. P. Poupart, Z. Chen, P. Jaini, F. Fung, H. Susanto, Y. Geng, L. Chen, K. Chen, H. Jin, Online flow size prediction for improved network routing, in *2016 IEEE 24th International Conference on Network Protocols (ICNP)* (IEEE, 2016), pp. 1–6
10. J. Jiang, V. Sekar, H. Milner, D. Shepherd, I. Stoica, H. Zhang, fCFAg: a practical prediction system for video qoe optimization, in *13th fUSENIXg Symposium on Networked Systems Design and Implementation (fNSDIg 16)*, pp. 137–150 (2016)
11. M. Xia, Y. Owada, M. Inoue, H. Harai, Optical and wireless hybrid access networks: design and optimization. J. Opt. Commun. Netw. **4**(10), 749–759 (2012)
12. T.T. Nguyen, G. Armitage, A survey of techniques for internet traffic classification using machine learning. IEEE Commun. Surv. Tutorials **10**(4), 56–76 (2008)
13. H. Mao, M. Alizadeh, I. Menache, S. Kandula, Resource management with deep reinforcement learning, in *Proceedings of the 15th ACM Workshop on in Networking*, pp. 50–56 (2016)
14. G. Kramer, What is next for ethernet pon?, in *COIN-NGNCON 2006-The Joint International Conference on Optical Internet and Next Generation Network* (IEEE, 2006), pp. 49–54
15. M.P.I. Dias, B.S. Karunaratne, E. Wong, Bayesian estimation and prediction-based dynamic bandwidth allocation algorithm for sleep/doze-mode passive optical networks. J. Light Wave Technol. **32**(14), 2560–2568 (2014)
16. L. Ruan, S. Mondal, E. Wong, Machine learning based bandwidth prediction in tactile Heterogeneous access networks, in *IEEE INFOCOM 2018—Conference on Computer Communications Workshops*, pp. 1–2 (2018)
17. J. Jiang, S. Sun, V. Sekar, H. Zhang, Pytheas: enabling data-driven quality of experience optimization using group-based exploration-exploitation, in *14th Symposium on Networked Systems Design and Implementation*, pp. 393–406 (2017)
18. M. Louta, P. Sarigiannidis, S. Misra, P. Nicopolitidis, G. Papadimitriou, Rlam: a dynamic and efficient reinforcement learning-based adaptive mapping scheme in mobile wimax networks. Mobile Inf. Syst. **10**(2), 173–196 (2014)
19. S.M. Jafari, M. Taghipour, M. Meybodi, Bandwidth allocation in wimax networks using reinforcement learning. World Appl. Sci. J. **15**(4), 525–531 (2011)
20. G. Alnwaimi, S. Vahid, K. Moessner, Dynamic heterogeneous learning games for opportunistic access in lte-based macro/femtocell deployments. IEEE Trans. Wireless Commun. **14**(4), 2294–2308 (2014)
21. P. Sarigiannidis, A. Sarigiannidis, I. Moscholios, P. Zwierzykowski, DIANA: a machine learning mechanism for adjusting the TDD uplink-downlink configuration in XG-PON-LTE systems. Mobile Inf. Syst. **2017** (2017)
22. V.S. Feng, S.Y. Chang, Determination of wireless networks parameters through parallel hierarchical support vector machines. IEEE Trans. Parallel Distrib. Syst. **23**(3), 505–512 (2011)
23. S.Y. Han, N.B. Abu-Ghazaleh, D. Lee, Efficient and consistent path loss model for mobile network simulation. IEEE/ACM Trans. Networking **24**(3), 1774–1786 (2015)
24. R.D. Timoteo, D.C. Cunha, G.D. Cavalcanti, A proposal for path loss prediction in urban environments using support vector regression, in *Proceedings of the Advanced International Conference on Telecommunications*, pp. 1–5 (2014)
25. B. Bojovic, E. Meshkova, N. Baldo, J. Riihijarvi, M. Petrova, Machine learning-based dynamic frequency and bandwidth allocation in self-organized LTE dense small cell deployments. EURASIP J. Wireless Commun. Network. **2016**(1), p. 183 (2016)
26. V. Lopez, J.A. Hernandez, O.G. de Dios, J.F. Palacios, J. Aracil, Multilayer traffic engineering for IP over WDM networks based on Bayesian decision theory. IEEE/OSA J. Opt. Commun. Networking **2**(8), 515–529 (2010)
27. B. Mao, Z.M. Fadlullah, F. Tang, N. Kato, O. Akashi, T. Inoue, K. Mizutani, Routing or computing? The paradigm shift towards intelligent computer network packet transmission based on deep learning. IEEE Trans. Comput. **66**(11), 1946–1960 (2017)

28. J. Mata, I. de Miguel, R.J. Duran, N. Merayo, S.K. Singh, A. Jukan, M. Chamania, Artificial intelligence (ai) methods in optical networks: a comprehensive survey. Opt. Switch. Network. **28**, 43–57 (2018)
29. A. Jukan, X. Masip-Bruin, N. Amla, Smart computing and sensing technologies for animal welfare: a systematic review. ACM Comput. Surv. (CSUR) **50**(1), 1–27 (2017)
30. S.K. Singh, F. Carpio, A. Jukan, Improving animal-human cohabitation with machine learning in fiber-wireless networks. J. Sensor Actuator Netw. **7**(3), 35 (2018)
31. Y. Kiran, T. Venkatesh, C.S.R. Murthy, A reinforcement learning framework for path selection and wavelength selection in optical burst switched networks. IEEE J. Sel. Areas Commun. **25**(9), 18–26 (2007)
32. T.R. Tronco, M. Garrich, A.C. Cesar, M. de Lacerda Rocha, Cognitive algorithm using fuzzy reasoning for software-defined optical network. Photon Netw. Commun. **32**(2), 281–292 (2016)
33. A. Sgambelluri, J.L. Izquierdo-Zaragoza, A. Giorgetti, L. Gifre, L. Velasco, F. Paolucci, N. Sambo, F. Fresi, P. Castoldi, A.C. Piat, et al., Fully disaggregated roadm white box with netconf/yang control, telemetry, and machine learning-based monitoring, in *Optical Fiber Communication Conference, Optical Society of America,* Tu3D–12 (2018)
34. I. Cunha, P. Marchetta, M. Calder, Y.C. Chiu, B.V. Machado, A. Pescape, V. Giotsas, H.V. Madhyastha, E. Katz-Bassett, Sibyl: a practical internet route oracle, in *13th fUSENIXg Symposium on Networked Systems Design and Implementation (fNSDIg 16)*, pp. 325–344 (2016)
35. W. Song, F. Zeng, J. Hu, Z. Wang, X. Mao, An unsupervised-learning-based method for multi-hop wireless broadcast relay selection in urban vehicular networks, in *2017 IEEE 85th Vehicular Technology Conference (VTC Spring)* (IEEE, 2017). pp. 1–5
36. Q. Dai, L. Zhu, P. Wang, G. Li, J. Chen, A machine learning based temporary base station (bs) placement scheme in booming customers circumstance, in *International Conference on Communications and Networking in China* (Springer, 2018), pp. 562–572
37. E. Balevi, R.D. Gitlin, Unsupervised machine learning in 5g networks for low latency communications, in *2017 IEEE 36th International Performance Computing and Communications Conference (IPCCC)* (IEEE, 2017), pp. 1–2
38. S. Gosselin, J.L. Courant, S.R. Tembo, S. Vaton, Application of probabilistic modeling and machine learning to the diagnosis of FTTH GPON networks, in *2017 International Conference on Optical Network Design and Modeling (ONDM)* (IEEE, 2017), pp. 1–3
39. S.R. Tembo, S. Vaton, J.L. Courant, S. Gosselin, A tutorial on the EM algorithm for Bayesian networks: application to self-diagnosis of GPON-FTTH networks, in *2016 International Wireless Communications and Mobile Computing Conference (IWCMC)* (IEEE, 2016), pp. 369–376
40. K. Christodoulopoulos, N. Sambo, E. Varvarigos, Exploiting network Kriging for fault localization, in *Optical Fiber Communication Conference, Optical Society of America*, W1B–5 (2016)
41. A.P. Vela, M. Ruiz, F. Fresi, N. Sambo, F. Cugini, G. Meloni, L. Potı, L. Velasco, P. Castoldi, Ber degradation detection and failure identification in elastic optical networks. J. Lightwave Technol. **35**(21), 4595–4604 (2017)
42. A. Vela, B. Shariati, M. Ruiz, F. Cugini, A. Castro, H. Lu, R. Proietti, J. Comellas, P. Castoldi, S.J.B. Yoo et al., Soft failure localization during commissioning testing and lightpath operation. J. Opt. Commun. Network. **10**(1), A27–A36 (2018)
43. S. Shahkarami, F. Musumeci, F. Cugini, M. Tornatore, Machine-learning-based soft-failure detection and identification in optical networks, in *2018 Optical Fiber Communications Conference and Exposition (OFC)* (IEEE, 2018), pp. 1–3
44. D. Rafique, T. Szyrkowiec, A. Autenrieth, J.P. Elbers, Analytics-driven fault discovery and diagnosis for cognitive root cause analysis, in *Optical Fiber Communication Conference, Optical Society of America*, W4F–6 (2018)
45. M.S. Parwez, D.B. Rawat, M. Garuba, Big data analytics for user-activity analysis and user-anomaly detection in mobile wireless network. IEEE Trans. Industr. Inf. **13**(4), 2058–2065 (2017)

46. K. Tsagkaris, A. Katidiotis, P. Demestichas, Neural network-based learning schemes for cognitive radio systems. Comput. Commun. **31**(14), 3394–3404 (2008)
47. L. Giupponi, A.I. Perez-Neira, Fuzzy-based spectrum handoff in cognitive radio networks, in *2008 3rd International Conference on Cognitive Radio Oriented Wireless Networks and Communications (CrownCom)* (IEEE, 2008), pp. 1–6
48. V.K. Tumuluru, P. Wang, D. Niyato, A neural network-based spectrum prediction scheme for cognitive radio, in *2010 IEEE International Conference on Communications* (IEEE, 2010), pp. 1–5
49. N. Sharma, A. Madhukumar, Genetic algorithm aided proportional fair resource allocation in multicast ofdm systems. IEEE Trans. Broadcast. **61**(1), 16–29 (2015)
50. K. Muralitharan, R. Sakthivel, R. Vishnuvarthan, Neural network-based optimization approach for energy demand prediction in smart grid. Neurocomputing **273**, 199–208 (2018)

Author Index

D. Bhattacharyya and N. Thirupathi Rao (eds.), *Machine Intelligence and Soft Computing*, Advances in Intelligent Systems and Computing 1280,
https://doi.org/10.1007/978-981-15-9516-5